LEONID STORM RESEARCH

LEONID STORM RESEARCH

Edited by

Peter Jenniskens[1,*], Frans Rietmeijer[2], Noah Brosch[3] and Mark Fonda[1]

[1] *NASA Ames Research Center, Mail Stop 239-4, Moffet Field, CA 94035-1000, U.S.A.*
  * with the SETI Institute
[2] *Institute of Meteoritics, Department of Earth and Planetary Sciences, University of New Mexico, 200 Yale Boulevard, Albuquerque, NM 87131, U.S.A.*
[3] *The Wise Observatory and the School of Physics and Astronomy, Beverly and Raymond Sackler Faculty of Exact Sciences, Tel Aviv University, PO Box 39040, Tel Aviv 69978, Israel*

Reprinted from Earth, Moon, and Planets
Volumes 82–83, 1998
(Published in 2000)

Kluwer Academic Publishers
Dordrecht / Boston / London

A C.I.P. Catalogue record for this book is available from the Library of Congress.

ISBN 0-7923-67383

Published by Kluwer Academic Publishers,
P.O. Box 17, 3300 AA Dordrecht, The Netherlands

Sold and distributed in North, Central and South America
by Kluwer Academic Publishers,
101 Philip Drive, Norwell, MA 02061, U.S.A.

In all other countries, sold and distributed
by Kluwer Academic Publishers,
P.O. Box 322, 3300 AH Dordrecht, The Netherlands

Printed on acid-free paper

All rights reserved
© 2000 Kluwer Academic Publishers
No part of the material protected by this copyright notice may be reproduced or
utilised in any form or by any means, electronic or mechanical,
including photocopying, recording or by any information storage and
retrieval system, without written permission from the copyright owner

Printed in The Netherlands

Contents

I. INTRODUCTION

THE 1999 LEONID MULTI-INSTRUMENT AIRCRAFT CAMPAIGN – AN EARLY REVIEW
P. Jenniskens, S. J. Butow and M. Fonda   1

USAF PERSPECTIVES ON LEONID THREAT AND DATA GATHERING CAMPAIGNS
M. H. Treu, S. P. Worden, M. G. Bedard and R. K. Bartlett   27

THE ESOC "METEOR ALERT CENTER" (MALC) AND ITS APPLICATION DURING THE 1999 LEONID SHOWER
J. Zender, D. Koschny and L. Neira   39

COORDINATED OBSERVATIONS OF LEONIDS IN ISRAEL
N. Brosch   47

II. ASTROBIOLOGY

METEORS: A DELIVERY MECHANISM OF ORGANIC MATTER TO THE EARLY EARTH
P. Jenniskens, M. A. Wilson, D. Packan, C. O. Laux, C. H. Krüger, I. D. Boyd, O. P. Popova and M. Fonda   57

SEARCH FOR ORGANIC MATTER IN LEONID METEORS
R. L. Rairden, P. Jenniskens and C. O. Laux   71

OBSERVATIONS OF LEONID METEORS USING A MID-WAVE INFRARED IMAGING SPECTROGRAPH
G. S. Rossano, R. W. Russell, D. K. Lynch, T. K. Tessensohn, D. Warren and P. Jenniskens   81

COMPUTATION OF ATMOSPHERIC ENTRY FLOW ABOUT A LEONID METEOROID
I. D. Boyd   93

SCREENING OF METEOROIDS BY ABLATION VAPOR IN HIGH-VELOCITY METEOROIDS
O. P. Popova, S. N. Sidneva, V. V. Shuvalov and A. S. Strelkov   109

SEARCH FOR EXTRATERRESTRIAL ORIGIN OF ATMOSPHERIC TRACE MOLECULES – RADIO SUB-MM OBSERVATIONS DURING THE LEONIDS
D. Despois, P. Ricaud, N. Lautié, N. Schneider, T. Jacq, N. Biver, D. C. Lis, R. A. Chamberlin, T. G. Phillips, M. Miller and P. Jenniskens   129

ORGANIC MATTER IN DUST OF COMET 21P/GIACOBINI-ZINNER AND THE DRACONID METEOROIDS
N. N. Kiselev, K. Jockers and V. K. Rosenbush   141

III. COMET DUST TRAIL: METEOROID STREAM DYNAMICS

PREDICTING THE STRENGTH OF LEONID OUTBURSTS
E. J. Lyytinen and T. Van Flandern   149

GLOBAL GROUND-BASED ELECTRO-OPTICAL AND RADAR OBSERVATIONS OF THE 1999 LEONID SHOWER: FIRST RESULTS
P. Brown, M. D. Campbell, K. J. Ellis, R. L. Hawkes, J. Jones, P. Gural, D. Babcock, C. Barnbaum, R. K. Bartlett, M. Bedard, J. Bedient, M. Beech, N. Brosch, S. Clifton, M. Connors, B. Cooke, P. Goetz, J. K. Gaines, L. Gramer, J. Gray, A. R. Hildebrand, D. Jewell, A. Jones, M. Leake, A. G. LeBlanc, J. K. Looper, B. A. McIntosh, T. Montague, M. J. Morrow, I. S. Murray, S. Nikolova, J. Robichaud, R. Spondor, J. Talarico, C. Theijsmeijer, B. Tilton, M. Treu, C. Vachon, A. R. Webster, R. Weryk and S. P. Worden   167

LORENTZ SHAPED COMET DUST TRAIL CROSS SECTION FROM NEW HYBRID VISUAL AND VIDEO METEOR COUNTING TECHNIQUE – IMPLICATIONS FOR FUTURE LEONID STORM ENCOUNTERS
P. Jenniskens, C. Crawford, S. J. Butow, D. Nugent, M. Koop, D. Holman, J. Houston, K. Jobse, G. Kronk and K. Beatty   191

COMPARING METEOR NUMBER FLUXES FROM GROUND-BASED AND AIRPLANE-BASED VIDEO OBSERVATIONS
D. Koschny and J. Zender   209

LEONID STORM FLUX ANALYSIS FROM ONE LEONID MAC VIDEO AL50R
P. S. Gural and P. Jenniskens   221

THE LEONID METEORS AND SPACE SHUTTLE RISK ASSESSMENT
J. F. Pawlowski and T. T. Hebert   249

VISUAL OBSERVATIONS OF THE 1998 AND 1999 LEONIDS IN ISRAEL
A. Mikishev and A. Levina   257

OBSERVATIONS OF THE LEONID METEOROID STREAM BY A MULTISTATION FS RADIO SYSTEM
G. Cevolani, G. Pupillo, A. Hajduk and V. Porubčan   265

PRECISE TRAJECTORIES AND ORBITS OF METEOROIDS FROM THE 1999 LEONID METEOR STORM
H. Betlem, P. Jenniskens, P. Spurný, G. Docters Van Leeuwen, K. Miskotte, C. R. Ter Kuile, P. Zarubin and C. Angelos   277

PHOTOGRAPHIC LEONIDS 1998 OBSERVED AT MODRA OBSERVATORY
J. Tóth, L. Kornoš and V. Porubčan   285

TEMPORAL VARIATION IN THE ORBITAL ELEMENT DISTRIBUTION OF THE 1998 LEONID OUTBURST
M. C. De Lignie, M. Langbroek, H. Betlem and P. Spurný   295

1997 LEONID SHOWER FROM SPACE
P. Jenniskens, D. Nugent, E. Tedesco and J. Murthy   305

IV. METEOROID COMPOSITION AND ABLATION

FROM COMETS TO METEORS
J. M. Greenberg   313

COLLECTED EXTRATERRESTRIAL MATERIALS: CONSTRAINTS ON METEOR AND FIREBALL COMPOSITIONS
F. J. M. Rietmeijer and J. A. Nuth III   325

COMPARISON OF 1998 AND 1999 LEONID LIGHT CURVE MORPHOLOGY AND METEOROID STRUCTURE
I. S. Murray, M. Beech, M. J. Taylor, P. Jenniskens and R. L. Hawkes   351

FIRST RESULTS OF HIGH-DEFINITION TV SPECTROSCOPIC OBSERVATIONS OF THE 1999 LEONID METEOR SHOWER
S. Abe, H. Yano, N. Ebizuka and J.-I. Watanabe   369

JET-LIKE STRUCTURES AND WAKE IN Mg I (518 nm) IMAGES OF 1999 LEONID STORM METEORS
M. J. Taylor, L. C. Gardner, I. S. Murray and P. Jenniskens   379

GROUND-BASED LEONID IMAGING IN THE UV
E. Almoznino and J. M. Topaz   391

V. METEOR AFTERGLOW AND PERSISTENT TRAINS

TIME RESOLVED SPECTROSCOPY OF A LEONID FIREBALL AFTERGLOW
J. Borovička and P. Jenniskens   399

FeO "ORANGE ARC" EMISSION DETECTED IN OPTICAL SPECTRUM OF LEONID PERSISTENT TRAIN
P. Jenniskens, M. Lacey, B. J. Allan, D. E. Self and J. M. C. Plane   429

MID-INFRARED SPECTROSCOPY OF PERSISTENT LEONID TRAINS
R. W. Russell, G. S. Rossano, M. A. Chatelain, D. K. Lynch, T. K. Tessensohn, E. Abendroth, D. Kim and P. Jenniskens — 439

BUOYANCY OF THE "Y2K" PERSISTENT TRAIN AND THE TRAJECTORY OF THE 04:00:29 UT LEONID FIREBALL
P. Jenniskens and R. L. Rairden — 457

THE DYNAMICAL EVOLUTION OF A TUBULAR LEONID PERSISTENT TRAIN
P. Jenniskens, D. Nugent and J. M. C. Plane — 471

VI. METEOROID DEBRIS

DUST PARTICLES IN THE ATMOSPHERE DURING THE LEONID METEOR SHOWERS OF 1998 AND 1999
N. Mateshvili, I. Mateshvili, G. Mateshvili, L. Gheondjian and Z. Kapanadze — 489

RECOGNIZING LEONID METEOROIDS AMONG THE COLLECTED STRATOSPHERIC DUST
F. J. M. Rietmeijer and P. Jenniskens — 505

VII. MESOSPHERE AND LOWER THERMOSPHERE

PRELIMINARY DATA ON VARIATIONS OF OH AIRGLOW DURING THE LEONID 1999 METEOR STORM
J. Kristl, M. Esplin, T. Hudson, M. Taylor and C. L. Siefring — 525

AIRGLOW AND METEOR RATES OVER ISRAEL DURING THE 1999 LEONID STORM
N. Brosch and O. Shemmer — 535

ELF/VLF RADIATION PRODUCED BY THE 1999 LEONID METEORS
C. Price and M. Blum — 545

NOTE ON THE REACTION OF THE UPPER ATMOSPHERE POTASSIUM LAYER TO THE 1999 LEONID METEOR STORM
J. Höffner, C. Fricke-Begemann and U. Von Zahn — 555

MESOSPHERIC AND LOWER THERMOSPHERIC WINDS AT MIDDLE EUROPE AND NORTHERN SCANDINAVIA DURING THE LEONID 1999 METEOR STORM
W. Singer, P. Hoffmann, N. J. Mitchell and Ch. Jacobi   565

VIII. IMPACTS ON THE MOON

OBSERVATION AND INTERPRETATION OF METEOROID IMPACT FLASHES ON THE MOON
L. R. Bellot Rubio, J. L. Ortiz and P. V. Sada   575

Subject Index   599

Leonid MAC-99 Workshop
April 16-19, 2000
Tel Aviv University, Israel

From left to right: Joe Kristl, Margie Goss, David Asher, Liliana Formiggini, Diana Prialnik, Wayne Hocking, Daniel Fisher, Valentin Grigore, Mike Taylor, Eran Ofek, Bob Hawkes (foreground), Akiva Bar-Nun, Iain Boyd, Ohad Shemmer, Jeremy Topaz, Colin Price, Luis Bellot-Rubio, Noah Brosch, Brent Grime, David McMullan, John Zinn, Nino Mateshvili , Tracey McDaniel, Rick Rairden, Peter Jenniskens, Jacqueline Marshall, Ian Murray, Nikolai Kiselev, Kristina Smith, Marco Langbroek, Pete Worden, Jim Pawlowski, Frans Rietmeijer, Abi Har-Even, Robert Barron, Olga Popova, George Varros, Alexander Mikishev, Ronen Jacovi, Juraj Toth, Anna Levina, Lev Pustil'nik, Sergey Yeryomin, Jiri Borovicka, Ilan Manulis, Andrei Gheorghe, Pavel Spurny, Shinsuke Abe, Steve Butow, Timothy Conklin, Hajime Yano, Larry Gardner, and Mayo Greenberg.

What a wonderful night!
Thousands of Leonid meteors so nice
fly just like some glow worms.
Are we in Paradise?

                        Dimitrie Olinici

Fragments of fire
from the roar of the Lion,
alarm in the sky.

Populations of Leonids,
and populations of men
armed with feelings for a love hunt.

Be generous,
Zenith Hourly Rate,
give them all they want.

Sky fireballs, earthly observers,
Leonids and people,
a magic connection.

                    Andrei Dorian Gheorghe

She crosses flights
with rains of stars,
Lioness Temple-Tuttle...

                  Michaela Alorescu

The Great Apocalypse
of the late autumn.

                  Gelu-Claudiu Radu

Flights of lights,
whispers of stars,
the sky in the night
embracing the Earth.

Tears of light
on the sky's face,
a crying in the night
over the world.

                  Valentin Grigore

Meteor shower in a test tube.
In the Leonid Isle
every inhabitant tattoos in his palm
a small universe;
but their comet has disintegrated
into a meteor shower.

                  Diana Maria Ogescu

Arrest those fireballs, those lights,
not to burn the Earth on their flights.

                  Victor Chifelea

I gladly looked at the wave
of the blazes coming to me,
and I felt myself sinking
in the meteoric sea.

                  Tina Visarian

Man and meteor...
The sky is shaken
of their love.

                  Iulian Olaru

Astropoetry by members of the SARM -Romanian Society for Meteors and Astronomy.
Presented at the Leonid MAC Workshop, Tel Aviv, Israel, April 16-19, 2000.
English translations by Andrei Dorian Gheorghe.

THE 1999 LEONID MULTI-INSTRUMENT AIRCRAFT CAMPAIGN - AN EARLY REVIEW

PETER JENNISKENS[1], STEVEN J. BUTOW[1], AND MARK FONDA
NASA Ames Research Center, Mail Stop 239-4, Moffett Field, CA 94035-1000
[1] *with the SETI Institute*
E-mail: pjenniskens@mail.arc.nasa.gov

(Received 10 July 2000; Accepted 15 August 2000)

Abstract. Two B707-type research aircraft of the 452nd Flight Test Squadron at Edwards Air Force Base were deployed to study the Leonid meteor storm of 1999 over the Mediterranean Sea on Nov. 18. The mission was sponsored by various science programs of NASA, and offered an international team of 35 researchers observing conditions free of clouds and low altitude extinction at a prime location for viewing the storm. This 1999 Leonid Multi-Instrument Aircraft Campaign followed a similar effort in 1998, improving upon mission strategy and scope. As before, spectroscopic and imaging experiments targeted meteors and persistent trains, but also airglow, aurora, elves and sprites. The research aimed to address outstanding questions in astrobiology, planetary science, astronomy, and upper atmospheric research. In addition, USAF co-sponsored the mission to provide near real-time flux measurements for space weather awareness. First results are presented in these issues of *Earth, Moon, and Planets* in preparation for future missions that will target the exceptional Leonid returns of 2001 and 2002. An early review of the scientific achievements in the context of campaign objectives is given.

Keywords: Airborne astronomy, astrobiology, chemistry, comets, composition, elves, exobiology, instrumental techniques, Leonid MAC, Leonids 1999, lower thermosphere, meteoroids, meteor storm, meteors, mesosphere, orbital dynamics, satellite impact hazard, sprites

1. Introduction

The widely anticipated return of the Leonid shower in November of 1999 offered our best chance yet to observe a meteor storm with modern techniques. None had been seen since the Leonid storm of 1966 at the beginning of the space age. No storm was observed in November of 1998 during the first mission of the Leonid Multi-Instrument Aircraft Campaign to Okinawa, Japan, although a spectacular shower of bright fireballs unexpectedly stole the show (Jenniskens and Butow, 1999). Two months earlier, E.A. Reznikov (in a widely circulated e-mail) had predicted from model calculations the return of the October 8 Draconid shower within 10 minutes of the observed peak. That success raised hopes that the upcoming 1999 Leonid encounter might be timed as well. McNaught and Asher (1999) and Lyytinen (1999) used similar methodology and widely circulated an optimistic forecast for 1999 November 18 when the Leonid storm was to peak at about 02:08 ± 15 minutes UT. Earth was to cross the debris ejected during the return of 1899. For a peak time of 02:08 UT, the best viewing conditions would be over Europe, the Middle East, and Africa (McNaught, 1999). The storm profile was expected to have a 1/e effective duration of 0.7 hours based on past Leonid storms (Jenniskens, 1995), while a secondary peak might be visible over eastern Asia when Earth was to pass an older dust trail ejected in 1866.

The 1999 Leonid Multi-Instrument Aircraft Campaign (Leonid MAC) was a National Aeronautics and Space Administration (NASA) and United States Air Force (USAF) sponsored effort to bring an international team of researchers to a prime location for viewing the storm under guaranteed clear weather conditions. The mission unfolded as planned and became a nice example of an excellent collaboration between the two agencies. In addition, the European Space Agency (ESA), the Japanese Institute of Space and Astronautical Science (ISAS), the Israel Space Agency (ISA), and many individual institutes contributed in kind to participating researchers to make the mission an international endeavor.

Beatty (2000) gives a popular account of the mission with personal perceptions of the Leonid storm. Here, we will introduce the general approach and logistics of the mission in Section 2, with special emphasis on improvements and changes with respect to the 1998 Leonid MAC mission. In Section 3, we will present a brief overview of the 1999 Leonid shower rate and its near-real time reporting, and provide lists of detected fireballs, persistent trains, elves, and sprites. It is too early for an in-depth analysis. However, results from a first analysis of some of the data are presented elsewhere in these special issues of *Earth, Moon and Planets*. Section 4 gives a brief summary of key results in the context of the campaign research objectives and describes how results compare to other observations of the Leonid storm.

Figure 1. Participating researchers and mission crew.

2. Approach

2.1. MISSION PROFILE

A drawback of last year's approach (Jeniskens and Butow, 1999) was the use of two dissimilar aircraft, which included a propeller engine driven Electra aircraft carrying the University of Illinois airborne LIDAR. Because this LIDAR was to be deployed in Antarctica and would not be available for a second mission, we chose to team up the jet engine driven modified NKC 135-E "*Flying Infrared Signatures Technology Aircraft*" (FISTA) with a similar B707-type aircraft, the EC-18 "*Advanced Ranging Instrumentation Aircraft*" (ARIA). The USAF/452nd Flight Test Squadron operated both aircraft out of Edwards Air Force Base. The use of similar aircraft would allow stereoscopic observations while flying along a westward trajectory for a maximum number of night-time observing hours. A three night mission was called for, because of the possible return of last year's fireball shower seen half a day before the nodal passage in 1998 (Arlt and Brown, 1999; Jenniskens and Butow, 1999; Jenniskens and Betlem, 2000). Israel was chosen as our prime base, where Dr. Noah Brosch facilitated local support of Tel Aviv

University and the Israel Space Agency. The USAF/106th Rescue Wing (102nd Resque Squadron) based in New York provided a C-130 ADVON mission, which took care of advanced logistic arrangements for the group of 78 people, including 35 researchers, crew, and journalists, that took part in the 1999 Leonid MAC mission (Figure 1).

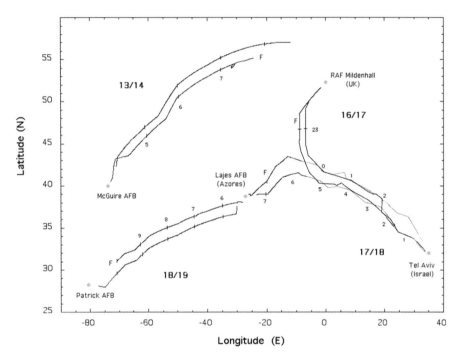

Figure 2a. Flight path of FISTA ("F") and ARIA. Times are marked in Universal Time (UT).

Following a daytime flight to McGuire AFB in New Jersey, the campaign started in the night of Nov. 13/14. The first Leonids were observed en-route to England. The FISTA aircraft briefly changed course to observe a sudden auroral display. This excercise proved a useful test of communication protocols for rapid response during the upcoming persistent train observations (Figure 2a). In the UK, Royal Air Force Station Mildenhall supported maintenance of KC135-type aircraft. At nearby RAF Lakenheath, we entertained school children with the "how's and why's" of our mission. After a rest day, the mission commenced from England to Israel's Ben Gurion International Airport in the night of Nov. 16/17. The Leonid shower showed the first signs of enhanced rates, but no sign of a fireball shower. Numerous small technical problems were solved. Immediately after departure from Ben Gurion on Nov. 17/18, Leonid rates were up and peaked to a storm several hours later. The storm was observed under excellent conditions while flying just west of Greece on our way from Israel to Lajes Air Base in the Azores (Figure 2b). Two course corrections were made by FISTA to accommodate persistent train observations. Sprites and elves were observed in an unusual lightning display over Bosnia. During the next night of Nov. 18/19, on our way to Patrick AFB in Florida, the Leonid meteor rates were almost back to pre-storm levels and the tail of the dust distribution was observed.

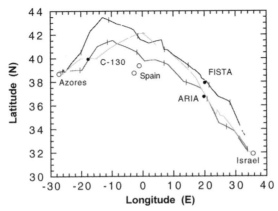

Figure 2b. Detail of Figure 2a for the night of November 18, 1999. Markers indicate 1 hour intervals and black dots mark the positions of the aircraft at the peak of the storm (02:02 UT).

2.1.1. *Ground-based research support*

At the same time, members of the Dutch Meteor Society set up ground sites for photographic multi-station photography and flux measurements at two locations in Spain (Figure 2b). Czech observers from the Ondrejov Observatory participated in this effort. In Israel, sensitive Very Low Frequency (VLF) radio sensors and Very High Frequency (VHF) radar were operated, aiming for some of the same meteors as seen from the aircraft (Brosch, 2000). ESA established a ground-based effort at measuring meteor flux at the Calar Alto observatory in Spain (Zender *et al.*, 2000), while the USAF sponsored ground-based campaign attempted to view the storm from sites in Israel, at the Canary Islands, and in the Canadian Arctic at Alert, Nunavut (Treu *et al.*, 2000).

TABLE I

Instrument	λ micron	FOV °)	Δλ/λ	Rate Hz	Alt. °)	Target	Affiliation
FISTA:							
High-res UV spectr.	0.30-0.41	37x21	250	30	30	meteor spectra	Lockheed
Intensified HD-TV	0.35-0.9	10-60	250	30	50-90	meteor (spectra)	ISAS[3]
All-sky camera	0.4-0.8	180	-.-	30	90	fireballs	Utah State U.
Filtered CCD imag.	0.4-0.9	23x18	-.-	30	30	airglow/sprites	Utah State U.
Intensified video	0.4-0.8	40,90	-.-	30	22-50	meteor flux	SETI Institute
Filtered int. video	0.4-0.9	16,10	-.-	30	80	light curves	Univ. Regina[1]
Low-res VIS spectr.	0.4-0.9	20	120	30	61	meteor spectra	SETI Institute
Low-res VIS spectr.	0.4-0.9	25	200	25	30	meteor spectra	Ondrejov Obs.[2]
High-res VIS-NIR	0.4-0.9	5	1,600	0.7	30	meteor spectra	SETI Institute
Fabry-Perrot spectr.	0.52	12x9	90,000	30	30	meteor spectra	Lockheed
Daisy NIR spectr.	0.9-2.5	20	var.	0.01	12	meteor spectra	Washington U.
Near-IR InGaAs FPA	0.9-1.67	4x4	-.-	30	20	airglow/meteor	Utah S.U./NRL
Michelson interf.	1.0-1.65	1.5	4,000	0.3	20-60	airglow/train	AFRL/SRL
Michelson interf.	1.5-3.0	1.5	2,000	0.91	20-60	airglow/train	AFRL/SRL
Mid-IR FPA imager	2.5-3.5	4x4	-.-	30	12	meteor	AFRL/SRL
MIRIS	3.0-5.5	15x5	200	16.7	40	meteor/train	Aerospace Co.
BASS	2.5-13	4	30-125	200	12	persistent train	Aerospace Co.
ARIA:							
Intensified HD-TV	0.4-0.8	37x21	-.-	30	0-60	triangulation	ISAS[3]
Filtered CCD imag.	0.35-0.8	7x8	52	30	30	trains/meteors	Utah State U.
White light CCD im.	0.4-0.9	23x18	-.-	30	30	airglow/sprites	Utah State U.
Intensified video	0.4-0.8	40	-.-	30	22-50	real-time flux	Multi-National
Intensified CCD im.	0.4-0.8	10	-.-	25	15	meteor flux	ESA/SSD[4]
Grism UV-VIS	0.3-0.8	40	200	30	0-60	meteor spectra	NOAO[3]
Slit UV-VIS spectr.	0.3-0.9	1	240	0.5	0-30	train spectra	NASA ARC
300 mm int. video	0.4-0.8	5	-.-	30	0-30	train images	SETI Institute

1 Canada; 2 Czech Republic; 3 Japan; 4 Netherlands

2.2. EXPERIMENTS

Eight experiments were performed onboard ARIA and 17 onboard FISTA (Table I). Most significant changes compared to the 1998 campaign were the replacement of the airborne LIDAR and the meteoroid cloud imaging experiment on FISTA. In their place came an international team of eight experienced meteor observers on ARIA that was assigned to gather near-real time flux measurements using video head displays and a newly designed counting tool, and transmit that information to ground locations in the USA. An intensified CCD camera provided by ESA was included to help calibrate the flux measurements. New experiments on FISTA included the DAISY Fourier transform spectrometer for near-IR spectroscopy of meteors and a near-IR InGaAs camera for near-IR imaging. Two new experiments targeted the 0.3–0.4 micron region for spectroscopy of meteors using Fabry-Perot and grism slit-less spectroscopy. An all-sky camera for fireball detection was installed on FISTA, while narrow band airglow imagers from the University of Utah flew on both aircraft. Finally, a new fiber-optic coupled spectrograph was built for meteor trains and deployed on ARIA.

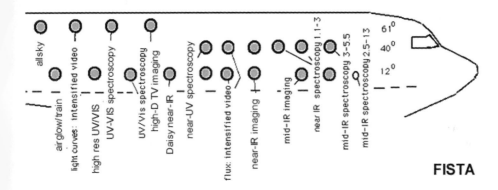

Figure 3. Relative position of instruments on-board FISTA.

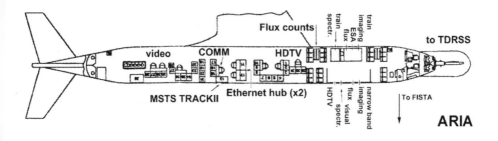

Figure 4. As Figure 3a for ARIA.

The layout of the instruments is shown in Figures 3 and 4. Specific reasons for the chosen layout on FISTA are given in Jenniskens and Butow (1999). The instruments on ARIA were positioned behind eight optical glass windows, four of which were newly installed in order to accommodate stereoscopic viewing with FISTA (Figure 4). During the Leonid storm, the ARIA aircraft was flying south of FISTA (Figure 2b). Because the FISTA windows are located on the right side of the aircraft, the overlapping area for stereoscopic work was north of, and above, FISTA's trajectory. Stereoscopic High Definition TV (HDTV) imaging was complimented with narrow band imaging of meteors in the same direction to facilitate comparison with total light output. A set of two intensified cameras was mounted at each side of the isle for flux measurements. The train spectrograph was positioned behind the only relatively large window available on ARIA, with a long focal length intensified camera co-aligned to help find and image the persistent trains seen by the spectrograph.

TABLE II

Frequency Band	Type	Purpose
VHF AM 1	Air-to-Ground	Air Traffic Control
VHF AM 2	Air-to-Air	Pilot's Discrete
UHF AM 1	Air-to-Air	PI's Discrete
UHF AM 2	Air-to-Air	Extranet Datalink (2.4 kbps)
HF 1	Air-to-Ground	Long Range C3 Back Up
HF 2	Air-to-Ground	Lajes Command Post
HF 3	Air-to-Ground	Spare
UHF SATCOM	Air-to-Space	Track II Datalink (64 kbps)
S BAND	Air-to-Space	TDRSS Video Uplink (3Mbps)
L BAND	Air-to-Air	INMARSAT SATCOM Voice / Datalink (9.6 kbps)
L BAND	Air-to-Space	IRIDIUM Satellite Telephone - ADVON
INTER 1	Interphone	Cockpit Interphone
INTER 2	Interphone	Mission Coordinator's Common
INTER 3 to 5	Interphone	Mission Scientists Discrete Channels

2.3. COMMUNICATIONS

Leonid MAC '99 was unprecedented in its use of space-based satellite assets to provide precise positioning and telemetry services in flight.

Both geo-stationary and Low-Earth Orbiting (LEO) satellite systems were utilized. Most notably, NASA's Tracking Data Relay Satellite System (TDRSS) provided a real-time video link from aircraft to NASA Ames Research Center where the compressed digital video signal was simulcast to the Internet. Voice and data communications were also transmitted via MILSTAR and INMARSAT while weather was monitored from the METEOSAT GEO European Weather Satellite System.

The communications can be characterized by three separate modes of operation: Air-to-Air, Air-to-Ground, and Air-to-Space (Table II). Communications between the two aircraft, combined with the first time use of a space-based datalink, facilitated maneuvering coordination and joint tracking of persistent trains. A unique feature of this year was the availability of MS-Track II. The Track II datalink provided continuous 3-dimensional GPS positioning of the mission aircraft. The Track II also supplied e-mail and file sharing capability between aircraft and participating ground stations. This made it possible for the HDTV team to know at all times the direction of the other aircraft during stereoscopic measurements. A VHF voice link (AM2) was used to assist alignment of the fields of view of the HDTV cameras.

A video-editing studio was set up in the back of the ARIA aircraft where the best video image of 8 different cameras was selected for live broadcast to NASA ARC. For that purpose, we used the ARIA 7-foot communication dish mounted in the nose and S-band transmission over the NASA TDRSS network. Flux measurements were transmitted over telephone lines using INMARSAT by means of a local area network on ARIA. The combined capability provided a constant source of real-time data to the Leonid Operations Control (LEOC) at NASA Marshall Space Flight Center (MSFC) and the Leonid MAC command and control center at NASA ARC. Air-to-Ground communications with the Israeli radar site served to correlate real-time surface and airborne Leonid observations during the first two hours on peak night.

A disadvantage of being inside an aircraft as opposed to being in the open air is that bright fireballs and their persistent trains are not easily noticed. To detect fireballs, an all-sky camera was connected to experimental automatic meteor recognition software. In case of a detection, a warning was transmitted over the local area network. Unfortunately, the system was not operational on peak night due to technical difficulties. Fortunately, fireballs and their persistent trains were so numerous that many were detected.

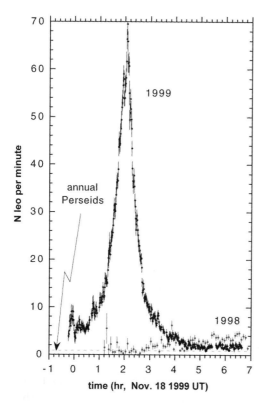

Figure 5. Rate of meteors detected during the 1998 and 1999 MAC campaigns. Dashed line shows the level of peak annual Perseid shower activity.

3. Results

The magnitude of the storm is illustrated in Figure 5, which shows the raw Leonid counts per camera on Nov. 18, 1999. Rates soared to a Zenith Hourly Rate ZHR ~ 4,000 hr^{-1} at 02:02 UT on November 18, which is twenty times higher than peak rates during the 1998 Leonid MAC mission, and 50 times higher than a typical Perseid shower (shown as a dashed line). Approaching anecdotal accounts by ground-based observers of the 1966 storm, the rates near the horizon in 1999 were phenomenal, occasionally with 7 Leonid meteors visible simultaneously in the field-of-view of intensified cameras. By all definitions the event qualified as a "storm", which is usually defined as apparent visual rates above 1 meteor per second or as ZHR > 1,000 hr^{-1}.

Near-real time flux measurements were obtained in the nights of Nov. 16 to 18, 1999. About 15,000 meteors were reported from an effective field of view of about 190,000 km^2 (4 cameras). On the peak night, meteor counts commenced at 23:30 UT shortly after takeoff, but only at 00:50 UT were we able to submit the counts in near-real time to the server at NASA ARC. The counts show the onset of the storm and trace the contour of the activity curve well. A web site presented the counts in comparison to predicted activity, with updates every 10 minutes, and later every 5 minutes (Figure 6). From NASA ARC, the counts were send to the Leonid Operation Center (LEOC) at NASA MSFC, for dissemination among NASA and USAF satellite operators.

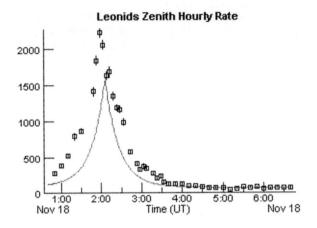

Figure 6. Leonid meteor rates reported in real time by the ARIA flux measurement team compared to the predicted activity (solid line).

The flux rates were compared on-line to predictions built in a JAVA applet called "Leonid MAC Flux Estimator", which allowed anyone on Earth to calculate the anticipated apparent Leonid rates as seen from his/her location and observing conditions. The applet was posted at the Leonid MAC website in the months leading up to the 1999 storm. The profile contained the expected storm, but also an anticipated second peak at the predicted time of passing the 1866 dust trail, with assumed peak rates of ZHR = 1,500 for the storm and ZHR = 100 for that second component. The basic ingredients were the predicted peak times (McNaught and Asher, 1999), the width from past Leonid storm profiles (Jenniskens, 1995), and peak rates estimated from past encounters. The comparison was satisfactory (Figure 6), and demonstrated the advance in recent meteor shower prediction models.

The number of fireballs and persistent trains was phenomenal. Table III gives all recorded fireballs with significant persistent trains (e.g., Figure 7). There were thirteen persistent trains lasting longer than 4 minutes. The afterglow of the 04:00:29 UT fireball, was so bright that it registered on slit-less spectrographs. The persistent train of this meteor was distorted by upper atmospheric winds into a figure "2" and was soon named the "Y2K" train. The train was well observed thanks to a flight adjustment by the FISTA operators.

Figure 7. Persistent train of the 02:13:30 UT fireball as seen from ARIA.

TABLE III

Time (UT)	Mv (magn.)		Duration (min.)	From - Until (UT)	h (°)	FISTA	h (°)	ARIA
1	(23:28:28)[a]	-.- Leo	> 6.5	23:29:54 - 35:55	-.-	no data	6	AL50R
2	00:15:09	-5 Leo	0.47	00:15:09 - 15:35	10	FL50F	-.-	no data
3	00:37:59	-7 Leo	1.8	00:37:59 - 39:45	5	FL50R	-.-	no data
4	00:50:15	-5 Leo	> 4.5	(00:52:20) - (56:50)	90	FH50F	-.-	no data
5	00:52:18	-6 Leo	> 18	(00:52:51)[b] - (11:00)			18	AR50R
				(01:00:23) - (00:35)	61	FH50R		
6	(01:52:34)	-12 Leo	> 39	(01:59:17) - (02:12:20)			12	AR50F
				(02:17:30) - (31:00)			12	AR50R
				(02:03:16) - (14:20)	17	FL50F		
7	01:57:34	-5 Leo	0.1	01:57:34	-.-	no data	14	AR50F
8	02:06:31	-5[c] Leo	0.5	01:06:31 - 07:01	-.-	no data	16	AR50R
9	(02:11:32)	-.- Leo	> 18	(02:12:20) - (30:00)	17	FL50R	-.-	no data
10	02:13:30	-9 Leo	> 13	02:13:30 - (29:30)	-.-	no data	15	AL50F
11	02:15:31	-5 Leo	> 4	02:15:31 - (19:40)	90	HDF2-2	-.-	no data
12	02:23:29	-10 Leo	> 17	02:23:29 - (34:54)			7	AR50R
				02:23:31 - (24:49)	9	FL50F		
				(02:25:07) - (26:06)	9	Daisy		
				(02:26:08) - (41:40)	9	FL50R		
13	02:24:38	-8 Leo	7.5	02:24:38 - 32:00			6	AR50R
				02:24:40 - (24:55)	9	FL50F		
				(02:25:37) - 32:00	9	Daisy		
14	02:44:32	-5 Leo	0.1	02:44:32	-.-	no data	17	AR50F
15	03:03:24	-6 Leo	0.33	03:03:24 - 03:44	6	FL50R	-.-	no data
16	(03:30:33)	-.- Leo	> 18	(03:30:52) - (48:30)	-.-	no data	3	AL300
17	03:39:18	-8 Leo	12	03:39:18 - 51:30	8	Daisy	-.-	no data
18	(03:40:12)	-.- Leo	> 4	(03:41:10) - (44:46)	-.-	no data	4	AL300
19	04:00:29	-13 Leo	> 13	04:00:29 - 13:29			13	AR50R
				(04:01:52) - (09:47)	14	FL50R		
				(04:01:15) - (01:53)	14	FH50R		

a) time from scattered light, meteor outside field of view; b) brackets indicate that the persistent train entered or left the field of view; c) flare

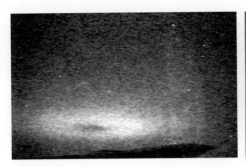 

Figure 8. Left: Elve disk (with characteristic dark central region) imaged from the FISTA aircraft at 02:10:00.79 UT on the night of the Leonid storm. Right: Bright sprite imaged from ARIA aircraft at 02:44.15.475 UT on the same night. Both events were observed over the Balkans. Images courtesy Mike Taylor and Larry Gardner, Utah State University.

TABLE IV

Time (UT)	Type	Observer / PI	Instrument
FISTA:			
01:58:11.83 [a]	elve	Gardner / Taylor	ICCD1
02:07:35.09 [a]	elve	Gardner / Taylor	ICCD1
= 02:07:34.98	elve	Smith / Jenniskens	Daisy
02:08:49.13	elve	Gardner / Taylor	ICCD1
02:10:00.79 [a]	elve	Gardner / Taylor	ICCD1
= 02:10:00.72	elve	Smith / Jenniskens	Daisy
02:15:26	sprites	Smith / Jenniskens	Daisy
02:21:28.59 [a]	elve	Gardner / Taylor	ICCD1
= 02:21:28.55	elve	Smith / Jenniskens	Daisy
02:23:10.80 [a]	elve	Gardner / Taylor	ICCD1
= 02:23:10.65	elve	Smith / Jenniskens	Daisy
02:38:22.12 [a]	elve	Gardner / Taylor	ICCD1
02:54:34.34 [a]	elve	Gardner / Taylor	ICCD1
02:57:25.59 [a]	elve	Gardner / Taylor	ICCD1
02:58:52.46 [a]	elve	Gardner / Taylor	ICCD1

a) two-station

Figure 9. The 2:21:28.57 UT Elve as seen from ARIA towards the constellation of Draco. Image courtesy: Hajime Yano (ISAS), filmed with NHK Hivision HDTV-II camera.

TABLE V

Time (UT)	Type	Obs. / PI	Instr.	Time (UT)	Type	Obs./PI	Instr.
ARIA:							
01:56:05.97	elve	Abe / Yano	HDTV	02:38:22.10	elve[a]	Abe / Yano	HDTV
01:58:11.83[a]	elve	Abe / Yano	HDTV	02:38:30.30	elve	Abe / Yano	HDTV
01:58:13	sprites	Jenniskens	AR50F	02:41:34.31	elve	Taylor	ICCD2
02:01:08.53	elve	Abe / Yano	HDTV	02:42:34.47	elve + spr.	Taylor	ICCD2
02:01:10.40	elve	Abe / Yano	HDTV	02:42:48.73	elve	Abe / Yano	HDTV
02:07:35.13[a]	elve	Abe / Yano	HDTV	02:44:15.47	elve + spr.	Taylor	ICCD2
02:10:00[a]	elve	Abe / Yano	HDTV	02:46:59.64	elve	Taylor	ICCD2
02:13:39.97	elve	Abe / Yano	HDTV	02:48:14.01	elve	Taylor	ICCD2
02:21:28.57	elve[a]	Abe / Yano	HDTV	02:49:13.47	sprite	Taylor	ICCD2
02:23:10.77	elve[a]	Abe / Yano	HDTV	02:49:34.33	elve	Taylor	ICCD2
02:26:25.00	elve	Abe / Yano	HDTV	02:50:08.53	elve	Taylor	ICCD2
02:31:21.13	sprites	Abe / Yano	HDTV	02:51:01.22	elve + spr.	Taylor	ICCD2
= 02:31:21	sprites	Jenniskens	AR50R	02:51:59.47	elve	Taylor	ICCD2
02:31:57.57	sprites	Abe / Yano	HDTV	02:52:31.24	elve + spr.	Taylor	ICCD2
= 02:31:58	sprites	Jenniskens	AR50R	02:53:34.00	elve	Abe / Yano	HDTV
02:33:48.53	sprites	Abe / Yano	HDTV	02:54:34.33	elve[a]	Taylor	ICCD2
= 02:33:48	sprites	Jenniskens	AR50R	02:55:48.54	sprite	Taylor	ICCD2
02:34:32.03	elve	Abe / Yano	HDTV	02:57:25.60	elve[a]	Taylor	ICCD2
= 02:34:32	halo	Jenniskens	AR50R	02:58:52.46	elve[a]	Taylor	ICCD2
02:36:11.57	elve	Abe / Yano	HDTV	= 02:58:53.00	elve	Abe / Yano	HDTV
02:38:10.40	elve	Abe / Yano	HDTV	03:01:52.14	elve	Taylor	ICCD2

a) two-station

The airglow was strong during much of the peak night, in contrast to the exceptionally low levels of airglow detected in 1998. The OH airglow appeared to increase in intensity in relation to the activity of the shower, but other airglow emissions remained unchanged. On the flights to and from Europe, simultaneous measurements were performed from the ground and air in collaboration with the Air Force Research Laboratory (AFRL) at Hanscom AFB (R. Huppi). The ground-based sites were located in New Hampshire (Nov. 13) and in Utah (Nov. 20).

During the meteor storm, a lightning complex over the Balkan was passed at a favorable distance to measure elves and sprites. Eleven sprites and 33 elves were recorded between 1:56 and 3:02 UT, which are listed in Table IV (FISTA) and Table V (ARIA). Elves are very brief (<1ms) duration horizontally expanding disks of light (up to ~300 km in diameter) that are induced by the absorption of electromagnetic pulse (EMP) radiation at the base of the nighttime ionosphere around 90–100 km (in vicinity of the meteor ablation level) and are due to powerful cloud-to-ground lightning discharges (Figures 8 and 9). Sprites are transient (several ms), vertically structured optical emissions that are thought to be due to large quasi-static electric field gradients over thunderstorms (Figure 8, right). They occur most often in association with large positive cloud-to-ground discharges. Sprites can extend from the lower ionosphere (~90 km) into the stratosphere (30–40 km). Both sprites and elves have identifiable ELF/VLF signatures, detected by the ELF/VLF sensors in Israel.

4. Discussion

So shortly after the observing campaign, the results of data analysis often raise more questions than answers. These special issues of *Earth, Moon and Planets* (volumes 82 and 83) contain some of the first results presented at the Leonid MAC Workshop that was held in Tel-Aviv, Israel, from April 15–19, 2000 (Rietmeijer, 2000). The first results from the 1998 mission have been published in special issues of *Meteoritics & Planetary Sciences* (Vol. 34 (6), Nov. 1999) and *Geophysical Research Letters* (Vol. 27 (13), July 2000). All tie into a comprehensive study of physical processes that, until now, have remained much ignored due to lack of data.

4.1. SCIENCE ISSUES IN ASTROBIOLOGY

In search for clues to the origin of life on Earth, a key issue is to understand all possible pathways that lead from carbon atoms to prebiotic compounds. Of traditional interest are organic molecules in the meteoroids that survive as micrometeorites. Now, attention is given to the organic matter that is deposited in the atmosphere during ablation, and the atmospheric reduced molecular compounds that are created in the path of meteors.

Last year's finding that Leonid fireballs tend to start their luminous trajectory as high as 196 km altitude (Spurny et al., 2000a) is of interest because only rather volatile materials such as organic carbon compounds are thought to evaporate at these high altitudes. Now, Spurny et al. (2000b) have discovered an unusual type of radiation above 136 km altitude, where Leonid meteors are distinctly diffuse with a V type wake.

The role of meteors in delivering organic carbon at lower altitudes is discussed in Jenniskens et al. (2000a), who searched for breakup products such as C_2 and CN. Especially CN is readily detectable even in low excitation conditions. In 1998, no CN or C_2 were detected in Leonid spectra between 700 and 870 nm. It was submitted that organic carbon compounds survive meteoric ablation in the form of large molecular compounds, in which case meteors are an efficient vehicle for delivery of organic carbon. This year, Rairden et al. (2000) provided further support for this hypothesis by reporting a strong lower limit of less than 1 CN molecule per 3 Fe atoms being produced in the meteoric plasma, based on a non-detection of the CN B–X transition at 387 nm.

Some indication that hydrogen may be lost in the process is given by this year's detection of hydrogen emission from the head of the bright 04:00:29 UT fireball at an unusual abundance of 10–20 H per Fe atom (Borovicka and Jenniskens, 2000). However, it needs to be confirmed that this hydrogen originates from organic carbons in the meteoroid rather than from water or hydrogen gas in the ambient atmosphere.

Another aspect of meteors is the production of reduced atmospheric compounds by conversion of kinetic energy. Jenniskens et al. (2000a) present the first direct measurements of the air plasma temperature in a meteor wake by comparing meteor emission spectra with air plasma models. The result, $T \sim 4{,}300$ K, is just high enough to break up CO at thermodynamic equilibrium. That can lead to an interesting crop of reduced organic compounds in a CO_2 rich early Earth atmosphere.

Following the precise temperature measurements in the wake of Leonid meteors in last year's mission, Boyd (2000) presents the first meteor models that use "Direct Simulation" Monte Carlo calculations to approach the rarefied molecular flow conditions of small bodies in air at high Mach numbers. Popova *et al.* (2000) describe the conditions of ablation. In the head of the meteor, the molecules are found to be mostly in a collision-less regime where they will cool rapidly by radiation. Indeed, Rossano *et al.* (2000) present the first mid-IR detection of meteors from measurements during the 1998 Leonid Multi-Instrument Aircraft Campaign, and find no sign of a persisting thermal emission.

Once released in the Earth's atmosphere, it is important to know the physical conditions in the meteor path and possible chemical pathways involving the interaction with the atmosphere. Borovicka and Jenniskens (2000) calculated the cooling rate of the heated gas from spectra of neutral meteoric metal atom lines in the meteor afterglow of the 04:00:29 UT fireball. Numerous non-thermal mechanisms are observed in this cooling process. Slit-less spectra obtained shortly after a 1998 Leoind fireball were published by Abe *et al.* (2000). Once the afterglow has faded, a long-lasting persistent train remains. During 1999, the first slit-less and slit spectra of persistent trains were obtained at these later evolutionary stages. Apart from sodium line emisison, a yellow continuum is observed that is interpreted as NO_2 (Borovicka and Jenniskens, 2000), suggesting an efficient production of NO, or alternatively is due to FeO emission from airglow-type chemistry (Jenniskens *et al.* 2000b). The observations appear to confirm the occurrence of airglow-type chemistry but many aspects of the dynamics and appearance of persistent trains are not yet understood (Kelley *et al.*, 2000; Jenniskens *et al.*, 2000c).

A spectacular result was the first measurement of mid-IR emission in persistent trains. Enhanced emissions of CH_4, CO_2 and H_2O were detected, which may originate from trace air compounds or materials created in the wake of the meteor (Russell *et al.*, 2000). A temperature of $T \sim 300$ K is measured minutes after the meteor, which is consistent with earlier LIDAR probes of the Doppler temperature of sodium atoms in persistent trains at the Starfire Optical Range by Chu *et al.* (2000a).

Of special interest for understanding the fate of meteoric materials is the detection of a red continuum in the 04:00:29 UT fireball spectrum that is interpreted to be caused by meteoroid debris at $T \sim 1,400$ K (Borovicka and Jenniskens, 2000). This temperature is close to the evaporation temperature of silicates at the atmospheric pressures in the upper atmosphere (Rietmeijer and Nuth, 2000). Indeed, there is evidence for continuing ablation in part of the afterglow. Given the unusual long lifetimes of neutral iron in the meteor path measured during the 1998 Leonid MAC mission (Chu *et al.*, 2000b), this emitting dust is not likely to be re-condensed meteoric vapor but rather debris from fragile cometary grains. Indeed, Rietmeijer and Jenniskens (2000) point out that certain types of spheres reported in the NASA Cosmic Dust Catalogs may be meteoroid debris of known showers. Mateshvilli *et al.* (2000) report the detection of continuum scattering of sunlight from the meteoric layer and follow what appears to be the settling of meteoric dust. Rietmeijer and Nuth (2000) describe the mineralogical and chemical diversity among incoming meteoroids.

4.2. ISSUES IN PLANETARY SCIENCE

Meteor showers are a window on the dynamical evolution of comets and provide information about the composition and morphology of relatively large cometary grains. Those grains are the nearest analog to comet surface materials and may help prepare for comet lander missions.

4.2.1. Comet dust trails and meteoroid stream dynamics

Young meteoroid streams may still carry information on their formation history and, thus, probe the mass-loss of comets. The dynamics of ejection from comets and the subsequent dynamical evolution of meteoroid orbits under planetary perturbations leads to the formation of cometary dust trails, which are ultimately responsible for the meteor storms. McNaught and Asher (1999) and Lyytinen (1999) published results of dust trail models that assumed very low ejection velocities. This approach was used before by Kondrat'eva and Reznikov (1985) to understand the Leonid showers, based on early calculations by Kazimircak-Polonskaja *et al.* (1968). However, their predictions of Leonid meteor storm occurrences, including strong returns in 2001 and 2002, were so much different from those by others that their results were not widely accepted. Now, the return of the Leonid storm at the predicted time of passing the 1899 dust ejecta and, perhaps even more so, the occurrence of a second peak at the time of passing the 1866 ejecta (Arlt *et al.*, 1999; Jenniskens *et al.*, 2000d) confirm these predictions. This establishes that a comet dust trail is composed of many "trailets" from each return of the comet. Lyytinen and Van Flandern (2000) now report new predictions for future returns of the Leonids based on refinements from a more comprehensive treatment of radiation pressure effects.

The 1999 Leonid storm peaked at 02:02 ± 02 UT and the peak ZHR was about 4,000 hr^{-1} (Arlt et al., 1999; Jenniskens et al., 2000d; Brown et al., 2000). The storm profile was smooth with no clear sign of filamentary structure when averaged over sufficient spatial scale (Jenniskens et al., 2000d). Until now, activity curves of meteor showers were either thought to be Gaussian or exponential distributions. The good statistical precision of the airborne measurements established the shape to be a Lorentzian distribution (Jenniskens et al., 2000d). It is not clear what physical process is responsible for this shape.

The cause of the second Leonid shower peak in the night of Nov. 17, 1998, appeared at odds with the current model, because the Earth passed relatively far from the various trailets. Now, Betlem et al. (2000) and De Lignie et al. (2000) find evidence that this dust may well have been a manifestation from the 1899 trailet, despite of the perturbation of that section of the 1899 trailet by the Earth in the previous return (McNaught and Asher, 1999; Lyytinen and Van Flandern, 2000).

4.2.2. Meteoroid composition and morphology

In comparing airborne Leonid light curve observations, Murray et al. (2000) report a noticeable difference in the shapes of the light curves from the1998 and 1999 missions. In both years, the light curves support the quick breakup of fragile meteoroids but this year's fragments were more abundant in larger pieces.

From narrow filter MgI imaging onboard ARIA, Taylor et al. (2000) confirm the occurrence of jet-like features in meteors, seen earlier in white light by LeBlanc et al. (2000). This points towards small meteoroid fragments being ejected at high speed.

Preliminary statistical analysis of the meteor time of incidence does not show evidence for breakup of meteoroids upon approach to Earth, which would have produced a significant increase over a Poisson distribution of short time intervals between 1/30 and 1 second (Gural and Jenniskens, 2000). Gural and Jenniskens find evidence of periodic excursions in meteor rates that may be the result of an early breakup of large grains. These excursions may be related to the fine structure in the activity profile reported independently by Singer et al. (2000), who find a yet not understood quasi-periodic behavior.

4.3. ISSUES RELATED TO THE SATELLITE IMPACT HAZARD

The '99 Leonid MAC was part of a larger US Air Force sponsored campaign to provide meteor flux data to satellite operators in near-real time (Treu et al., 2000). Leonid MAC provided the most precise flux measurements with meteor rates near the horizon up to 5.3 ± 0.4 times higher than for identical cameras pointed high in the sky at the ground-based mountain observatory of Calar Alto (Koschny and Zender, 2000). This precision enabled an early assessment of the impact hazard (Figure 6), by extrapolating the increase of flux to the predicted time of the peak, half an hour before the peak occurrence of the storm (Jenniskens et al., 2000d).

Gural and Jenniskens (2000) provide a first analysis of 17 minutes of data from one camera, calculating the distribution of meteors on the sky and providing an absolute flux measurement of 0.82 ± 0.19 particles km^{-2} s^{-1} (M_v < 6.5 magn.) at the peak of the shower. This can be compared to Arlt et al. (1999), who derived 1.4 ± 0.3, and to Brown et al. (2000), who have 1.6 ± 0.2 km^{-2} s^{-1}. In terms of impact danger, this would correspond to an impact probability of only 2.5% for the combined 600 satellites to be hit by a Leonid meteoroid of sufficient mass to cause serious or disabling damage (Yano et al., at the Leonid MAC Workshop 2000).

Smaller impacts are expected to be more frequent, but it appears that the Leonid storm was less rich in small meteoroids than implied by the magnitude distribution index of r = 3.0 derived from past Leonid storms (e.g. Jenniskens, 1995). Pawlowski and Hebert (2000) confirmed r = 3.0 ± 0.6 from Liquid Mirror Telescope data, but Arlt et al. (1999) have r = 2.2 ± 0.1 from visual observations, while Brown et al. (2000) have r = 2.44 ± 0.05 from radar observations and r = 2.40 ± 0.11 from video data. Brown et al. (2000) found this population index to be constant over the range from 1.5x10^{-6} to 2x10^{-9} kg. Bellot Rubio et al., (2000) extrapolate the size distribution to masses of 5 kg from lunar impact studies. In contrast, Gural and Jenniskens (2000) require r = 1.8 ± 0.1 for a good understanding of the observed altitude distribution of meteors as seen from the ARIA aircraft. Only Cevolani et al. (2000) find a lower magnitude population index of r = 1.5 at the peak of the shower from forward meteor scatter counts. This discrepancy means that we do not yet fully understand why aircraft observations are so effective in observing meteors near the horizon.

The satellites weathered the storm well. No anomalies were associated with the storm. From measurements by the impact detector GORID, on the geostationary Russian Express II telecom satellite, Svedhem *et al.* (2000) reported the detection of a flurry of very small meteoroids just after the Leonid storm from geocentric orbits in roughly the direction of the Leonid radiant.

In 1999, meteor storms have gone from anecdote to a space weather phenomenon. The return of the shower with two peaks at about the predicted time establishes the overall model of dust trailets (Kondrat'eva and Reznikov, 1985; McNaught and Asher, 1999; Lyytinen, 1999; Lyytinen and Van Flandern, 2000). The discovery of a Lorentz shaped dust distribution in the path of the Earth (Jenniskens *et al.*, 2000d), carries the promise of a similar, rather simple, analytical, distribution of dust perpendicular to Earth's path. If so, the November 2000 return is at the right distance from the 1866 dust trailet for determining the width and position of the trailet in this second dimension. Subsequent encounters of the 1866 trailet in 2001 and 2002 will establish a three-dimensional picture, when we can also measure the decay of dust density away from the comet position.

The discovery of numerous unusually short duration VLF emissions in the 1–20 kHz range at the time of the storm by Price and Blum (2000), has the potential to offer an automatic meteor counting system for future flux monitoring.

4.4. ISSUES IN ATMOSPHERIC SCIENCES

Meteor storms are a natural anomaly of meteoroid influx that can help trace the chemical response of the mesosphere and lower thermosphere. This year's storm provided the detection of enhanced OH airglow that closely followed the meteor storm (Kristl *et al.*, 2000). In addition, Despois *et al.* (2000) report tantalizing changes in the abundance of the upper atmosphere trace compound HCN in the night after the storm, with a promise that many more molecules can be probed in this manner by sub-mm spectroscopy. No changes were found in the sodium airglow (Brosch and Schemmer, 2000). Similarly, Höffner *et al.* (2000) show no significant enhancement of the neutral atom potassium layer at the time of the storm.

Persistent trains are natural luminous trails that trace upper atmosphere wind patterns in great detail, notably gravity waves and tides (Grime *et al.*, 2000). Jenniskens and Rairden (2000) find a vertical scale height of 8.3 km at 79–91 km altitude, in good agreement with radar wind data during the 1999 Leonid meteor storm by Singer *et al.* (2000). Further understanding of middle atmospheric chemistry will follow from the spectroscopic, spatial, and temporal analysis of persistent trains, which are now found to have airglow-type chemistry (Jenniskens et al., 2000b). Of interest, too, is the detection of warm atmospheric gasses in persistent trains by mid-IR spectroscopy, which may prove a sensitive measurement of trace gas abundances at the altitude of the meteors (Russell et al., 2000).

Also, meteors may affect the electrostatic potential between ionosphere and cloud tops. Two lightning complexes were passed during the 1999 Leonid MAC, one between 1:58 and 3:10 UT that was located over the Balkan mainland, and one between 3:25 and 04:10 UT located over the Italian peninsula. Sprites and elves were only seen to occur during the first lightning episode that coincided with the storm (Table IV and V). Further work is needed to establish a possible influence of meteor activity in the generation, or mere enhancement of visibility, of sprites and elves.

5. Implications for future missions.

With a high level of confidence we anticipate two more upcoming opportunities for viewing a Leonid meteor storm. In 2001 and 2002 Earth will cross the middle of dust trails ejected in 1866 and 1833. This chance coincidence has not occurred over the past century and promises strong shower activity. McNaught and Asher (1999) predict peak rates of ZHR = 13,000 hr^{-1} and ZHR = 25,000 hr^{-1} respectively, while Lyytinen and Van Flandern (2000) have ZHR = 6,100 and ZHR = 7,400 hr^{-1}. Those rates compare to a peak ZHR ~ 4,000 hr^{-1} during the storm of 1999. It should be noted, however, that these peak rate estimates are uncertain by an order of magnitude mainly because of a lack of information on the dispersion of dust perpendicular to the Earth's path and in the orbit of the comet. Jenniskens et al. (2000d) predict a pessimistic ZHR = 70 and ZHR = 40, from perhaps too simple geometric considerations. Clearly, further observations of this dust complex are needed to arrive at reliable flux predictions for these and other potential meteor storms.

The November 2000 Leonid encounter is the only one that can shed light on the distribution perpendicular to the Earth's path. Fortuitously, Earth will cross the 1932 and 1866 dust "trailets" exactly one day apart. Both events will be visible from the eastern USA and western Atlantic.

Peak rates can vary from ZHR = 20 to 700 hr^{-1}. Unfortunately, a last quarter Moon not far from the Leonid radiant will hamper ground-based observers. At aircraft altitudes, scattering of Moonlight is a negligible problem for purposes of meteor counting.

Acknowledgements

Our special thanks go to the team of the 452nd Flight Test Squadron at Edwards AFB, who took care of the research-aircraft operations, instrument installation and aircraft modifications, and much of mission logistics. Capt. Jeff Lampe was an outstanding Mission Commander. MSgt Greg Williams and Tsgt. Paul Poncelet were mission coordinators for FISTA and ARIA, respectively. Capts. John Hasser and Jeff Powell were the mission planners, and Pamela Cronk the contracts specialist. Much of the credit for the success of the complex communications plan is well deserved by the responsible engineers at the 452 FTS: Mr. Bob Selbrede and SSgt David Gray, as well as by Capt Dale Van Dusen at Mountain Home AFB, ID, Leslie Ambrose at NASA GSFC, and Allen Ross at NASA ARC. Bob Selbrede coordinated the video relay from FISTA to NASA Ames Research Center. Enerdyne Technologies, Inc (Tony Difede) supplied video Encoder/Decoder Equipment. The Advanced Range Telemetry Program Office provided the S-Band BPSK Transmitter (Chuck Irving). The Benefield Anechoic Facility made the S-Band Power Amplifier available (Mr. Jeff Cornell). The F16 Combined Test Force made a Spare Video Encoder available (John Reddemann). The KU-Band TDRS Equipment and initial TDRS Planning Support was provided by NASA Kennedy Space Center & Navy P3 Program Office at PT. Mugu, CA (Jerry Wiessert of QSS Group Inc.). Lt. Col. Gary Chen of the 76th SOPS at Peterson Air Force Base, CO, facilitated the Track II equipment. Operators at NASA GSFC, White Sands (New Mexico), and NASA ARC made a smooth transmission possible. Over the duration of the Leonid deployment, NASA GSFC provided the equivalent services that would be utilized during a NASA space shuttle mission. The Japanese Broadcasting Company (NHK) provided technical support and the HDTV cameras, an effort coordinated by Noriyoshi Miyazaki, Noriyuki Mizuno, and Yuichiro Ando. The following ARC video and public affairs participants played a key role. David Maurantonio (Trans Video) coordinated the video relay to Internet. Ed Schilling coordinated aircraft video operations. Dennis Murray (Web Service Group) was responsible for the flux relay to the Internet web site, using software adapted by Glenn Deardorff (Sterling Software). Kathleen Burton and Laura Lewis coordinated the outreach activities. The SETI Institute provided logistic support for the participating scientists and overall coordination. Special thanks go to project resources coordinator Debbie Kolyer, contracts and procurement manager Hal Roey, grants specialist Brenda Simmons, and senior buyer Alan Thomson. Gary Palmer of E.M. Electronics was responsible for many instrument modifications. Cindy Tatchi of The Travelsmiths assisted researchers in their travel arrangements to and from Edwards AFB. The ground-based effort in Spain was coordinated by Hans Betlem of the Dutch Meteor Society. The Israeli efforts were coordinated by Noah Brosch of Tel Aviv University, who also facilitated local arrangements for a welcome stay in Israel. Coordination at Mildenhall Air Base was in hands of Lt. Col. Jeffrey Klem, while coordination at Lajes Air Base, Azores, was in hands of MSgt Theresa Ramos and SSgt Alica A. Depuis. The ADVON mission was coordinated by the 102nd RQS, New York Air National Guard. The 1999 Leonid MAC was supported by the National Aeronautics and Space Administration Exobiology, Planetary Astronomy, and Suborbital MITM programs, the NASA's Advanced Missions and Technologies Program for Astrobiology, NASA ARC, and the USAF/XOR. Mike Meyer, Tom Morgan, Mary Mellott, and Rick Howard at NASA/HQ, Greg Schmidt at NASA ARC, and Capt. Marv Treu, Col. Pete Worden, and Lt. Gen. Carlson at USAF/HQ, facilitated logistic and financial support for the mission, while Scott Hubbard at NASA ARC facilitated public outreach financial support. Numerous contributions were provided in kind by ESA, ISAS, the Aerospace Corporation, Utah State University, AFRL, Stewart Radiance Laboratories, Ondrejov Observatory, the National Astronomical Observatory Japan, the Regina University and the SETI Institute. Our special thanks goes to Joe Kristl and this team at AFRL/SRL, who made this mission infinitely easier. Tim Conklin took care of advanced logistics arrangements. Intern David Nugent was responsible for networking on the aircraft. Charlie Hasselbach and Michelle Frank were always there to keep us sane. Finally, we thank Frans Rietmeijer for numerous helpful comments to improve this text. *Editorial handling:* Frans Rietmeijer.

References

Abe, S., Ebizuka, N., and Watanabe, J.: 2000, *Meteoritics Planet. Sci.* **35**, in press.

Arlt, R. and Brown, P.: 1999, *WGN, Journal of the IMO* **27**, 267–285.
Arlt, R., Bellot Rubio, L., Brown, P., and Gijssens, M.: 1999, *WGN, Journal of the IMO* **27**, 286–295.
Betlem, H., Jenniskens, P., Spurny, P., Docters van Leeuwen G., Miskotte, K., ter Kuile, C., Zerubin, P., and Angelos, C.: 2000, *Earth, Moon and Planets* **82–83**, 277–284.
Beatty, K.: 2000, *Sky & Telescope*, p. 42-45 (June issue).
Bellot-Rubio, L.R., Ortiz, J.L., and Sada, P.V.: 2000, *Earth, Moon and Planets* **82–83**, 575–598.
Borovicka, J. and Jenniskens, P.: 2000, *Earth, Moon and Planets* **82–83**, 399–428.
Boyd, I.D.: 2000, *Earth, Moon and Planets* **82–83**, 93–108.
Brosch, N. and Shemmer, O.: 2000, *Earth, Moon and Planets* **82–83**, 535–544.
Brosch, N.: 2000, *Earth, Moon and Planets* **82–83**, 47–56.
Brown, P., Campbell, M.D., Ellis, K.J., Hawkes, R.L., et al.: 2000, *Earth, Moon and Planets* **82–83**, 167–190.
Cevolani, G., Hajduk, A., Porubcan, V., and Pupillo, G.: 2000, *Earth, Moon and Planets* **82–83**, 265–276.
Chu, X., Liu, A.Z., Papen, G., Gardner, C.S., Kelley, M., Drummond, J., and Fugate, R.: 2000a, *Geophys. Res. Lett.* **27**, 1815–1818.
Chu, X., Pan, W., Papen, G., Gardner, C.S., Swenson G., and Jenniskens, P.: 2000b, *Geophys. Res. Lett.* **27**, 1807–1810.
DeLignie, M., Betlem, H., Jenniskens, P., and Spurny, P.: 2000, *Earth, Moon and Planets* **82–83**, 295–304.
Despois, D., Ricaud, P., Lautié, N., Schneider, N., Jacq, T., Biver, N., Lis, D., Chamberlain, R., Phillips, T., Miller, M., and Jenniskens, P.: 2000, *Earth, Moon and Planets* **82–83**, 129–140.
Grime, B.W., Kane, T.J., Liu, A., Papen, G., Gardner, C.S., Kelley, M.C., Kruschwitz, C., and Drummond J.: 2000, *Geophys. Res. Lett.*, **27**, 1819–1822.
Gural, P. and Jenniskens, P.: 2000, *Earth, Moon and Planets* **82–83**, 221–248.
Höffner, J., Fricke-Begemann, C. and von Zahn, U.: 2000, *Earth, Moon and Planets* **82–83**, 555–564.
Jenniskens, P.: 1995, *Astron. Astrophys.* **295**, 206–235.
Jenniskens, P. and Butow, S.J.: 1999, *Meteoritics Planet. Sci.* **34**, 933–943.
Jenniskens, P. and Betlem, H.: 2000, *Astrophys. J.* **531**, 1161–1167.
Jenniskens, P. and Rairden, R.: 2000, *Earth, Moon and Planets* **82–83**, 457–470.
Jenniskens, P., Laux, C.O., Packan, D.M., Wilson, M., and Fonda, M.: 2000a, *Earth, Moon and Planets* **82–83**, 57–70.
Jenniskens, P., Lacey, M., Alan, B., and Plane, J.M.C.: 2000b, *Earth, Moon and Planets* **82–83**, 429–438.
Jenniskens, P., Nugent, D., and Plane, J.M.C.: 2000c, *Earth, Moon and Planets* **82–83**, 471–488.
Jenniskens, P., Crawford, C., Butow, S.J., Nugent, D., Koop, M., Holman, D., Houston, J., Jobse, K., Kronk, G., and Beatty, K.: 2000d, *Earth, Moon and Planets* **82–83**, 191–208.
Kazimircak-Polonskaja, E.I., Beljaev, N.A., Astapovic, I.S., and Terenteva, A.K.: 1968, in Kresak L., and Millman P.M. (eds.), *Physics and Dynamics of Meteors*, IAU, pp. 449–475.

Kelley, M.C., Gardner, C., Drummond, J., Armstrong, T., Liu, A., Chu, X., Papen, G., Kruschwitz, C., Loughmiller, P., Grime, B., and Engelman, J.: 2000, *Geophys. Research Letters* **27**, 1811–1814.
Kondrat'eva, E.D. and Reznikov,E.A.: 1985, *Solar System Research* **19**, 96-101.
Koschny, D. and Zender, J.: 2000, *Earth, Moon and Planets* **82–83**, 209–220.
Kristl, J., Esplin, M., Hudson, T., Taylor, M., and Siefring, C.L: 2000, *Earth, Moon and Planets* **82–83**, 525–534.
LeBlanc, A.G., Murray, I.S., Hawkes, R.L., Worden, P., Campbell, M.D., Brown, P., Jenniskens, P., Correll, R.R., Montague, T., and Babcock, D.D.: 2000, *MNRAS* **313**, L9–L13.
Lyytinen, E.: 1999, *Meta. Res. Bull.* **8**, 33–40.
Lyytinen, E. and Van Flandern, T.: 2000, *Earth, Moon and Planets* **82–83**, 149–166.
Mateshvili, N., Mateshvili, I., Mateshvili, G., Gheondjian, L., and Kapanadze, Z.: 2000, *Earth, Moon and Planets* **82–83**, 489–504.
Murray, I.S., Beech, M., Taylor, M.J., Jenniskens, P., and Hawkes, R.L.: 2000, *Earth, Moon and Planets* **82–83**, 351–368.
McNaught, R.H.: 1999, *WGN, Journal of the IMO* **27**, 164–171.
McNaught, R.H. and Asher, D.J.: 1999, *WGN, Journal of the IMO* **27**, 85–102.
Pawlowski, J.F. and Hebert, T.T.: 2000, *Earth, Moon and Planets* **82–83**, 249–256.
Popova, O.P., Sidneva, S.N., Shuvalov, V.V., and Strelkov, A.S.: 2000, *Earth, Moon and Planets* **82–83**, 109–128.
Price, C. and Blum, M.: 2000, *Earth, Moon and Planets* **82–83**, 545–554.
Rairden, R.L., Jenniskens, P., and Laux, C.O.: 2000, *Earth, Moon and Planets* **82–83**, 71–80.
Rietmeijer, F.J.M.: 2000, *Meteoritics Planet. Sci. (editorial)* **35**, 647.
Rietmeijer, F.J.M. and Jenniskens, P.: 2000, *Earth, Moon and Planets* **82–83**, 505–524.
Rietmeijer, F.J.M. and Nuth III, J.A.: 2000, *Earth, Moon and Planets* **82–83**, 325–350.
Rossano, G.S., Russell, R.W., Lynch, D.K., Tessensohn, T.K., Warren, D., and Jenniskens, P.: 2000, *Earth, Moon and Planets* **82–83**, 81–92.
Russell, R.W., Rossano, G.S., Chatelain, M.A., Lynch, D.K., Tessensohn, T.K., Kim, D., and Jenniskens, P.: 2000, *Earth, Moon and Planets* **82–83**, 439–456.
Singer, W., Hoffmann, P., Mitschell, N.H., and Jacobi, Ch.: 2000, *Earth, Moon and Planets* **82–83**, 565–574.
Singer, W., Molau, S., Rendtel, J., Asher, D.J., Mitchell, N.J., and von Zahn, U.: 2000, *MNRAS Lett.*, in press.
Spurny, P., Betlem, H., Van 't Leven, J. and Jenniskens, P.: 2000a, *Meteoritics Planet. Sci.* **35**, 243–249.
Spurny, P., Betlem, H., Jobse, K., Koten, P., and Van 't Leven, J.: 2000b, *Meteoritics Planet. Sci* **35**, in press.
Svedhem, H., Drolshagen, G., and Grün, E.: 2000, in *IAU Coll. #181, Dust in the Solar System and Other Planetary Systems (abstracts)*, 82.
Taylor, M.J., Gardner, L.C., Murray, I.S., and Jenniskens, P.: 2000, *Earth, Moon and Planets* **82–83**, 379–390.
Treu, M.H., Worden, S.P., Bedard, M.D., and Bartlett, R.K.: 2000, *Earth, Moon and Planets* **82–83**, 27–38.
Zender, J., Koschny, D., Mokler, F., and Neira, L.: 2000, *Earth, Moon and Planets* **82–83**, 39–46.

USAF PERSPECTIVES ON LEONID THREAT AND DATA GATHERING CAMPAIGNS

MARVIN H. TREU
Space Reconnaissance Requirements Division, HQ USAF/XORR, Pentagon, Washington DC
E-mail: Marvin.Treu@pentagon.af.mil

SIMON P. WORDEN
Deputy Director of Command and Control, HQ USAF/XOC, Pentagon, Washington DC

MICHAEL G. BEDARD
Commander, 46 Weather Squadron, Eglin AFB, FL

and

RANDALL K. BARTLETT
Program Office, Assistant Secretary of the Air Force (Space), Washington DC

(Received 10 July 2000; Accepted 7 September 2000)

Abstract. The Air Force has long recognized the threat posed by the space environment to military satellite systems including the potential for disastrous effects resulting from a meteoroid impact. This concern has steadily elevated with our nation's increasing reliance on space assets for systems critical to national defense. The 1998/1999 Leonid Meteor Storm Operational Monitoring Program was initiated to address this threat. The goal of this Air Force-led, international cooperative program was to provide near real-time Leonid meteor flux measurements to satellite operators. The incorporation of these measurements with model predictions provided an approximate 2-hour lead warning of the peak storm activity, permitting satellite operators ample opportunity to exercise hazard mitigation procedures. As a result, Department of Defense (DoD) and other participating satellite operators may have helped avoid spacecraft damage. The extent of any minor damage to components impossible to detect by operators is difficult to ascertain and may not manifest itself for a period of time. Modest micrometeoroid precipitation may reduce spacecraft life expectancies as a consequence of the physical erosion or sandblasting of exterior surfaces, and damage sustained by electronic systems from concurrent high-energy plasma discharges. Later effects could take the form of premature failure of satellite sensors and other spacecraft components, leading to overall shortening of satellite mission duration. The Air Force intends to pursue further analysis of data and polling of satellite operators to fully assess the Leonid '99 event. Future U.S. Air Force involvement may include support for additional observations and analysis.

Keywords: meteor storm, satellite impact hazard, space weather, Leonids 1999, situation awareness.

1. Introduction

In response to the potentially disastrous threat posed by meteors to DoD satellite systems, the U.S. Air Force teamed with government, commercial and academic research institutions to initiate the Leonid Meteor Storm Operational Monitoring Program. Beginning as a low-level effort in 1997, it evolved into a full-scale, worldwide scientific expedition in 1998 and 1999. With space systems becoming increasingly important to U.S. national security, threats posed to our satellite constellation must be monitored and better understood. This program enabled satellite operators to monitor the storm in near real-time, providing them with an opportunity to mitigate the threat by exercising risk-reduction actions. The quality and quantity of data collected during the 1999 storm is unprecedented. Analysis of this data will provide a better understanding of meteors, improved prediction modeling of future events and determine the best strategy for future meteor storm encounters. Although the Air Force does not plan to support a large-scale deployment to observe the Leonid meteor shower in the near future, support may be provided for additional data analysis and limited additional observations.

2. Meteors - space environment threat to DoD systems

Both the U.S. military and intelligence community are becoming increasingly reliant on space systems to achieve mission success. This was shown most dramatically in the last decade during conflicts beginning with DESERT STORM and concluding with operation ALLIED FORCE. The 1999 National Security Strategy recognizes the importance of space by stating: "unimpeded access to and use of space is a vital national interest -- essential for protecting U.S. national security, promoting our prosperity, and ensuring our well-being." Space systems today enable our military leaders to dominate the battlefield by providing global communications, precise navigation, accurate meteorological data, early warning of missile launches, and near real-time signals and imagery intelligence support (Hall, 1998). Figure 1 depicts the historical use of space assets from the early 60's to the present. Space – communication, weather, intelligence, navigation, and early warning – greatly enhanced military operations in time to support the war in Iraq in the early 90's. Today's senior leaders recognize that controlling space will be a foremost objective in future military contingencies. Air Force Chief of Staff, General Michael E. Ryan recently stated, "We couldn't do the operations we're doing without space." (AF Policy Ltr, 1999).

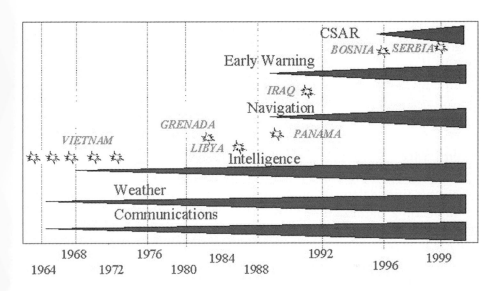

Figure 1. Integration of space with warfighting capabilities.

The trend of leveraging space for warfighting so powerfully demonstrated in DESERT STORM in the early 90's continued through the end of the century. Flexible targeting, reachback, bomb and missile impact reporting, and GPS-guided munitions are examples of how space assets were used in operation ALLIEDFORCE. Flexible targeting, the capability to redirect airborne assets to higher priority targets while in flight, is possible due to information from space-based assets being provided directly into the cockpit. Higher bandwidth and leveraging commercial satellite communication allowed unprecedented "reachback" capability for American troops. Leaving more personnel at their home bases to remotely provide information and tools needed for the combat troops is only possible through the exploitation of space. Reducing the in-theater personnel footprint is cheaper, safer, and a simpler means to conduct war. New data processing capabilities of the Defense Support Program (DSP) provided space operators with a tool to report near real-time bomb and cruise missile impacts. This allows planners to assess attack effectiveness and build future strike packages. Finally, perhaps the most dramatic example of leveraging space in "Allied Force" was the increasing use of Global Positioning System (GPS)-guided munitions. The Joint Direct Attack Munition (JDAM) provided a remarkably accurate tool for bombing a target - even in adverse weather conditions (cloud obscuration of targets) (Eberhart, 2000).

In the future, U.S. forces will rely upon space systems for global awareness of threats, swift orchestration of military operations, and precision use of smart weapons (Hall, 1998). The Air Force today is continuing its efforts to evolve toward a full spectrum, *Aerospace* Force, indicating the importance of not only using space as a medium to support military forces, but a medium through which to employ and maintain power.

With this increasing reliance on space systems for successful accomplishment of military missions, threats to DoD and other spacecraft utilized for defense purposes must be taken seriously. One example of a natural threat is from potential meteoroid impacts on spacecraft. Unfortunately, meteor storms are one of the most poorly understood phenomena in our Solar System (Beech *et al.*, 1995). The damage potential ranges from surface sandblasting and possible mechanical damage to complete loss of the spacecraft. The most probable risk, however is the result of impact-induced electrostatic discharge. This occurs when a meteoroid impacts the satellite surface, generating a charged plasma capable of producing a current "spike." The damage caused by this discharge can sometimes prove to be fatal to a satellite. This type of event is believed to have caused the end-of-life anomaly on Olympus, the European Space Agency's experimental communications satellite on the night of 11-12 August 1993 (Caswell *et al.*, 1998). On this night of the predicted peak of the Perseid meteor shower, Olympus lost its earth pointing ability and entered into a spin. Efforts to recover control of the spacecraft depleted its fuel. With insufficient propellant available to recover and continue operations, the spacecraft was removed from its orbit - ending the mission. Although a Perseid impact cannot be conclusively shown, one of several meteor effects (structural damage, momentum transfer, creation of a plasma cloud, and the triggering of discharges of previously charged surfaces) is the most probable explanation for the demise of Olympus (Caswell *et al.*, 1998).

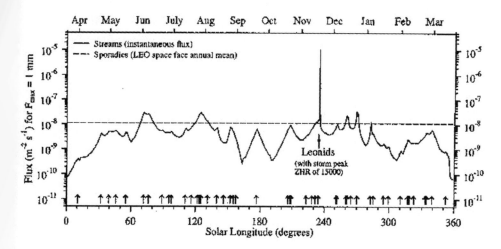

Figure 2. Penetrating fluence of the natural meteoroid complex and a Leonid storm with peak ZHR of 15,000. Reprinted from: Planetary Space Science, Vol. 47, 1999, McBride *et al.*, "Meteoroid impact on spacecraft: sporadics,...", p. 1011, (c) 1999, with permission from Elsevier Science.

The Leonid meteor shower is of particular concern to the DoD due to the potential for considerable risk to the satellite population during 'storm' conditions (Beech and Brown, 1993). The cause of the concern stems from hypervelocity impact and the anticipated flux. The relative geocentric encounter velocity of a Leonid at the top of the atmosphere is ~71 km/s (McBride and McDonnell, 1999). With meteor penetration potential varying with the square of the velocity (V^2) and current production proportional to V^4, a Leonid meteor impact clearly poses a high threat to spacecraft. The threat from a low number of high velocity meteors would not warrant significant concern to satellite operators. The flux from the Leonid shower however easily distinguishes this threat from other meteor or sporadic events. The instantaneous storm flux required to penetrate a surface can exceed the background by several orders of magnitude (Figure 2). A good way to translate the pertinent variables of velocity, size distribution and flux threat to officials in the Air Force and U.S. Government is to describe the exposure risk to the satellites. The penetrating fluence encountered during a two-hour peak of a Leonid storm is equivalent to a year of exposure risk from background meteors.

3. Leonid storm operational monitoring program

In response to the increasing reliance of the DoD (as well as civil and commercial organizations) on space combined with limited understanding and data available on meteor storm threat to spacecraft, the Leonid Meteor Storm Operational Monitoring Program was developed. The dual purpose of this Air Force/NASA co-sponsored program was to: (1) provide near real-time Leonid meteor flux measurements to satellite operators, and (2) use the data to better understand and model these events. While the Air Force interest is more operationally focused toward satellite impact hazard, NASA has additional scientific interests relevant to astrobiology and the origin of the solar system. The data gathering campaign was subsequently designed to provide near real-time data to satellite operators and collect data for research purposes. Ground-based radar and electro-optical instrumentation were deployed to support the near real-time flux measurements, while two USAF aircraft collected both real-time data and scientific mission data. Collecting data was the first step in the program. Processing the data into a useful form and disseminating it to the satellite operators requires as much consideration and planning. The users themselves were responsible for accomplishing the final step - how to use the data to protect the spacecraft.

There are a number of options available to protect or mitigate the effects of meteor impacts on a spacecraft. The most effective is spacecraft and sensor hardening, however this can be a costly solution and is obviously not a possible option for satellites already in orbit. Other mitigation procedures were employed by the Air Force and documented in operations plans (OPlans). First, the meteor flux data provided satellite operators with situational awareness. Just knowing the threat and being alert to the possibility of meteor impact or effects proved invaluable to operators. If an anomaly were to occur during peak meteor activity, critical time could be saved in determining the specific cause and most appropriate corrective action. Proactive measures include turning off critical or sensitive sensors and avoiding command and control contacts during the peak of the storm event. Doing this reduces the opportunity for bit-flips, single-event upsets, or other disruption in a command sequence to place the satellite at increased risk of mission downtime. Reorientation of larger spacecraft is an option considered to minimize the surface profile exposed to the meteor stream. Some systems propose orienting their solar panels parallel to the stream to reduce the probability of impact. A final action pursued in the Oplans is to either bring in or place "on-call" the most experienced satellite operators during the storm peak. The most experienced personnel are best qualified to quickly diagnose and react to anomalous satellite behavior induced by meteor impact and associated effects.

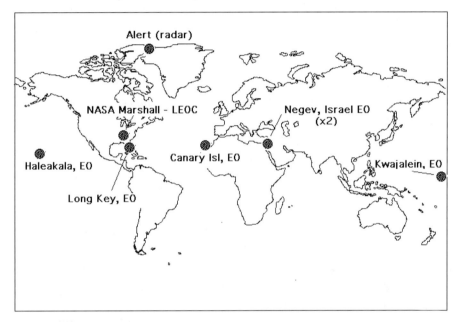

Figure3: Map showing the location of the LEOC (Leonid Environment Operations Control center) at NASA Marshall Space Flight Center in Huntsville, AB, the six electro-optical (EO) sites around the world, and the location of the dual-frequency radar in Alert, Canada. From: Brown, *et al.* (2000).

4. Campaign summary

The Leonid Meteor Storm Operational Monitoring Program consisted of three important components: the ground-based campaign, the airborne campaign, and the data processing and dissemination segment.

Dr. Peter Brown of the University of Western Ontario developed the Concept of Operations (CONOPS) and led the ground-based observation and data collection campaign. The CONOPS was perfected over the past three years and worked extremely well in 1999 despite additional observing locations and the challenge of adverse weather at multiple sites. Both electro-optical (low-light television) and radar equipment were used to collect data (Brown *et al.*, 2000). Human observer and Science Applications International Corporation (SAIC) real-time detection algorithms were used to provide meteor counts. The image intensified charge-coupled device (CCD) detectors were deployed to the Negev Desert, Israel (primary site); Las Palmas, Canary Islands, Spain; Key West, FL; Haleakala, HI; and Kwajalein Atoll, Marshall Islands (Figure 3). The locations were selected based on climatology (to offer the best chance for cloud-free, unobstructed observing), and maximum viewing time of the event – 22 hours. For full 24-hour, all-weather coverage, a mobile, multi-high frequency (HF) radar was placed at Alert, Nunavut, Canada. These ground sensors provided proven, extensive coverage for the storm, while NASA airborne sensors offered a means to mitigate the risk of weather obstruction in addition to providing complementary science collection value.

Dr. Peter Jenniskens and Capt Steve Butow, both with the Search for Extraterrestrial Intelligence (SETI) Institute, led the 1999 NASA Leonid Multi-Instrument Aircraft Campaign (Leonid MAC), with a prior mission in 1998. Their 35+ member airborne team performed high altitude stereoscopic observations of the Leonid meteor components using low light level video, UV/visible spectrometers, High Definition Television (HDTV) imagers, and near and mid-infrared spectrometers and scanning imagers (Jenniskens *et al.*, 2000). The purpose was to determine meteor flux, orbital elements, composition, structure, ablation properties and disposition of meteoric debris for studies related to the origins of life and space weather hazards. HQ USAF/XO co-sponsored the airborne campaign specifically to collect flux data from above obscuring clouds. The high altitude also provided the most precise counts, showing the early rise of the storm profile. The Leonid MAC flux data were transmitted in near real-time to the operations center through space-based telemetry systems including INMARSAT and the Multi-Source Tactical System (MSTS) TRACK II. The MSTS TRACK II provides en route aircraft tracking as well as burst transmission of e-mail via secure UHF SATCOM between aircraft and ground-based mission command and control at NASA Ames Research Center (ARC) in California.

Dr. Bill Cooke (Computer Sciences Corporation, Marshall Space Flight Center/Space Environment Group) and Dr. Rob Suggs (Marshall Space Flight Center/Space Environment Group) established and led the Leonid Environment Operations Center (LEOC). The LEOC collected the observed data from the worldwide network of ground optical and radar observers and the Leonid MAC airborne observers, performed real-time data analysis and transmitted the results to satellite operators via e-mail. This information was also posted on a secure internet site (Figure 4). Approximately 100 authorized web users and 200 e-mail customers received the LEOC data. The incorporation of these measurements with model predictions provided ample lead warning during the onset of the storm peak. The Marshall team developed the CONOPS for the data assimilation and analysis task responsible for the Leonid flux data getting to the user in near real-time.

Providing the resources for a program of this magnitude proved challenging. To accomplish all the goals of the mission, over 25 international government, educational and private organizations cooperated by providing funding, personnel, or other in-kind support. The ground-based (including the LEOC at MSFC) and airborne campaigns cost nearly $2M (US). This purchased near real-time reporting, deployment (equipment, personnel, and USAF aircraft), analysis, optimal equipment re-configuration, modeling and management infrastructure. The USAF provided 45% of the funding required to support the operational aspect of the campaign. NASA (HQ, MSFC, and ARC) provided 49% of the funds supporting both the airborne effort as well as operation of the LEOC. Required personnel came from USAF, NASA, government contractors, university researchers and even student volunteers. The in-kind support from Canadian Forces and Israel was essential to the mission.

The Leonid Meteor Storm Operational Monitoring Program was a success. The 1999 Leonid meteor shower reached modest storm strength (ZHR ~3,700) and lasted approximately 90 minutes. Predictions of the peak were remarkably accurate. McNaught and Asher (1999) predicted the actual peak time of 02:08 ± 15 min. UTC (Universal Time, Coordinated) on November 17, while Brown's (1999) forecast was nearly as accurate at 02:20 UTC. All customers received the data they required in a timely fashion and reported no satellite impacts or anomalies. An intensive post-event data reduction effort is underway on a very unique data set. The data collected during the 1999 Leonid meteor storm marks the first time in history radar and electro-optical real-time meteor observations were combined in a global manner to produce a snapshot of the meteoroid environment in near-earth space. In addition, this is one of the largest data sets ever collected on a meteor shower from a single campaign. More data now exists from the 1999 Leonids than from all the data from all previous data collection efforts.

Lessons learned in the aftermath of the program identified a few areas for improvement. The amount of coordination between the airborne and ground components of the campaign prior to the shower was underestimated. Telemetry testing between the LEOC at NASA MSFC and NASA Ames required advance testing but was not possible due to aircraft and avionics availability. More time and work was also required to determine the proper means of synthesizing video, radar, and visual near real-time meteor observations. A software problem when the ZHR exceeded 150 and a power failure on one of the radar frequencies were unanticipated events that prevented the reporting of absolute flux values. This resulted in an underestimated flux by a factor of 2-3. Finally, it was determined that while automated detection instrumentation is useful, the most reliable and best sensor is the trained meteor observer.

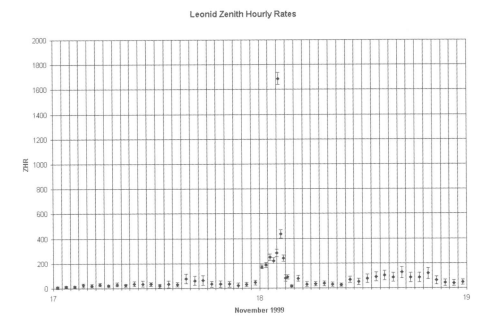

Figure 4. Near real-time data output from the Leonid Environment Operations Center.

5. Future plans

The Air Force is not planning to participate in a Leonid shower operational monitoring and data gathering campaign in November 2000. Because of the low peak rates expected, resources will be better spent investigating the spatial density and other characteristics of the 1866 Leonid stream. This stream is projected to be the primary contributor to the Leonid event in 2000. This work will also help refine the Leonid meteor forecast development for years 2001-2002. The science campaign should de-emphasize ground video and visual observations in 2000 because of moon phase (last quarter, high in the sky and close to the radiant of the shower).

6. Conclusions

The 1999 Leonid Operational Monitoring Program was a complete success. A consortium of over 25 multi-national government organizations teamed with academia and industry to deploy scientists, researchers, and volunteers around the world. This team observed from the ground and in the air the largest meteor storm in nearly 35 years. The data collected was quickly analyzed and distributed to satellite operators for their situational awareness and protection of critical spacecraft. In addition to developing preventative measures to minimize the risk to satellites for this storm, tremendous benefit was gained by documenting procedures and OPlans to address a naturally occurring threat in space. The experienced gained from this campaign will prove to be useful in the future.

A secondary goal of the program was to assemble a comprehensive data set from ground and airborne observations to improve our understanding and modeling of future meteor events. The unique, most comprehensive data set collected to date is anticipated to meet the expectations of researchers. Before anticipating a follow-on meteor observing campaign of this magnitude, the Leonid's 1999 data set must be completely reduced. The focus must now be on reducing this existing data, not planning the next operational program. If more data is required to better understand the threat of these meteor events to the operation of artificial satellites, a clear and convincing argument must be provided to support the remaining gap in understanding. Until this time, the successful partnerships forged in this campaign must continue. Future Air Force participation could include possible low-level support for data reduction, including some potential low-level observing support from a proper USAF source, such as the Air Force Research Laboratory.

Acknowledgements

Editorial handling: Mark Fonda.

References

Beech, M. and Brown, P.: 1993, *MNRAS* **262**, L35–L36.
Beech, M., Brown, P., and Jones, J.: 1995. *Q. J. R. astr. Soc.* **36**, 127–152.
Brown, P.: 1999, Ph.D. Thesis University of Western Ontario, University of Western Ontario, Canada.
Brown, P., Campbell, M.D., Ellis, K.J., Hawkes, R.L., Jones, J., Gural, P., Babcock, D., Barnbaum, C., Bartlett, R.K., Bedard, M., Bedient, J., Beech, M., Brosch, N., Clifton, S., Connors, M., Cooke, B., Goetz, P., Gaines, J.K., Gramer, L., Gray, J., Hildebrand, A.R., Jewell, D., Jones, A., Leake, M., LeBlanc, A.G., Looper, J.K., McIntosh, B.A., Montague, T., Morrwo, M.J., Murray, I.S., Nikolova, S., Robichaud, J., Spondor, R., Talarico, J., Theijsmeijer, C., Tilton, B., Treu, M., Vachon, C., Webster, A.R., Weryk, R. and Worden S.P.: 2000, *Earth, Moon and Planets* **82–83**, 167–190.
Caswell, R. D., McBride, N., and Taylor, A.: 1998, in Lynch *et al.* (eds)., *Leonid Meteoroid Storm and Satellite Threat Conference Proceedings (no pagenumbers).*
Eberhart, R. E.: 2000, *Statement by Commander-In-Chief, North American Aerospace Defense Command and United States Space Command Before the United States Senate Armed Services Committee, Strategic Subcommittee* 8 March 2000.
Hall, K. R.: 1998, Statement by Assistant Secretary of the Air Force (Space) and Director, National Reconnaissance Office, Keith R. Hall, on U.S. Spacepower in the 21st Century 29 September 1998.
Jenniskens, P., Butow, S.J., and Fonda, M.: (2000) *Earth, Moon, and Planets* **82–83**, 1–26.
McBride, N. and McDonnell, J.A.M.: 1998, *Planet. Space Sci* **47**, 1005–1013.
McNaught, R.H. and Asher, D.J.: 1999, *WGN, Journal of the IMO* **27**, 85–102.

THE ESOC "METEOR ALERT CENTER" (MALC) AND ITS APPLICATION DURING THE 1999 LEONID SHOWER

JOE ZENDER AND DETLEF KOSCHNY

Space Science Department, ESA, 2200 AG Noordwijk, The Netherlands
E-mail: jzender@estec.esa.nl

LUIS NEIRA

Universidad de Vigo, E.T.S.I Telecommunicacion, 36200 Vigo, Spain

(Received 22 June 2000; Accepted 7 August 2000)

Abstract. Spacecraft operators were concerned that the high number density of meteoroid particles during the anticipated 1999 Leonid meteor storm might result in damage to spacecraft. Switching off a spacecraft is expensive and operators try to avoid it, which created a need for real time monitoring systems. At the Space Science Department of ESA, we designed a near-real time meteor monitoring system, displaying observational information at the European Space Operations Center (ESOC) in Darmstadt, Germany. The system consisted of software tools that connect automatic video systems in the field with a central data node. Here, we describe the design, implementation, setup and results of the system. Wireless communication was implemented by means of the Global System for Mobile (GSM) communication. Unfortunately, during the operational phase this communication system failed. Conventional transmission by telephone was used instead. Results of the near real time reporting are presented and discussed.

Keywords: Leonids 1999, meteors, spacecraft hazards

1. Introduction

After the reduction of observational data from the November 1998 Leonid meteor shower, models were improved and predictions for a meteor storm on November 18, 1999, were made by several scientists (McNaught and Asher, 1999, Cook, 1999). The predicted peak zenith hourly rate (ZHR) was of order 1,200-1,600. At the Space Science Department of ESTEC (European Space Research and Technology Centre) in Noordwijk, The Netherlands, preparations were started in early 1999 for a Leonid meteor storm observation campaign to explore the storm's scientific potential. The selected observation sites were in southern Spain for best chances of clear weather.

The Space Science Department also participated in the 1999 Leonid Multi-Instrument Aircraft Campaign (Leonid MAC) (Jenniskens *et al.*, 2000). On the ARIA aircraft, ESA representative Michael Schmidthuber operated a similar intensified CCD camera as operated from the ground in Spain for absolute calibration of meteor fluxes. Results are reported in Koschny and Zender (2000).

In parallel, spacecraft operators considered measures to mitigate possible spacecraft hazards, which created a need for near real time reporting of meteor flux (Jenniskens, 1999; Cooke, 1999). The spacecraft's most vulnerable parts (e.g. the solar cells) can be turned parallel to the flight direction of the meteoroids and high power voltages can be shut off. The observations were also to confirm the model predictions. If the predictions were precise enough, it would be able to design alternative spacecraft orbits to mitigate future meteor storms (McNaught, 1999; McNaught and Asher, 1999) or incorporate the shower's presence in attitude control during flight scheduling.

ESOC decided to monitor the Leonid activity in near real time, and change spacecraft attitudes or switch of power supplies only if considered necessary (Jeanes, 2000). Several monitoring sources were used for near real time reporting. One was the activity reported from the airborne Leonid MAC, data of which were send to ESOC (Jenniskens *et al.*, 2000). Leonid MAC was part of a larger effort of ground-based radar and electro-optical observations with a data node at NASA/Marshall, which was also supported by ESOC. Because of disappointing results during the 1998 encounter and unproven technology in the new airborne effort, ESOC and the Space Science Department of ESA developed its own "Meteor Alert Center" as well. The MALC software system would allow near real-time transmission of reduced observational data from our locations in southern Spain to ESOC.

2. Requirements and Design

2.1. The Meteor Detection Sites

Observing locations were chosen at the high altitude mountain sites of the Observatorio de Sierra Nevada (OSN) at longitude 3.55083° West, latitude 37.0583° North, and altitude 2896m) and the Centro Astronomica di Hispannia en Alemania (CAHA) at longitude 2.54625° West, latitude 37.2236° North, altitude 2168m), both located in southern Spain.

The instrumentation deployed during the ESA-SSD's Leonid campaign ranged from a still-image CCD camera to several video-intensified cameras with a range of optics (see Table I). In table I, ICC stands for intensified CCD camera, LCC stands for low cost intensified video camera. The SUMO1 instrument is also a low cost intensified video

Table I.

Instrument	Optics	FOV (deg^2)	Limit. magn.	Science Target	Site
ICC1	50mm f/0.75	12x9	8.5	ZHR, flux	ARIA
ICC2	50mm f/0.75	12x9	8.5	ZHR, orbits	CAHA
ICC3	80mm f/1.00	7.5x5.6	8.8	ZHR, flux, orbits	OSN
ICC4	80mm f/1.00	7.5x5.6	8.8	ZHR, flux, orbits	OSN
ICC5	50mm f/0.75	12x9	8.5	flux, orbit	OSN
LCC1	28mm f/2.0	30 ø	6.0	persistent trains	OSN
1.5m Ritchey-Chretien Tel.				spectr., pers. trails	OSN
LCC2	28mm f/2.0	30 ø	6.0	3D observ., pers. trains	CAHA
LCC3	28mm f/2.0	30 ø	6.0	PR camera	CAHA
Photometrics CCD camera	2kx2k,12bit	30 ø		PR camera flux	CAHA
SUMO1		40x40		spectr. of pers. trains	OSN
Fish-eye cam.		180 ø		flux, radiant det.	CAHA

camera, but equipped with a grating. Cameras LCC1, operated at CAHA, and LCC2, operated at OSN, could be pointed by a visual observer to persistent trains of bright Leonids. Unfortunately, no persistant trains were observed. Cameras ICC2, operated at CAHA, and ICC5, operated at OSN, were unguided and pointed to the same area in the atmosphere for stereoscopic observations and the determination of orbits. All camera data were recorded on VHS tapes.

Although we selected two fixed observation sites, our equipment was portable and could be operated using batteries as well as a standard power supply. Under bad weather conditions this would have allowed us to move to alternative observation sites.

2.2. Communication with Meteor Alert Center

At the observation location, the video signal of an intensified video camera was fed into a frame grabber running on a Personal Computer (PC/DOS). The automatic meteor detection software *MetRec* was used to detect meteors in real time (Molau, et al., 1996). The brightness was estimated and the association with one of the meteor showers done automatically.

The result was then transmitted over a serial cable to another PC (PC/WIN98), responsible for buffering and transmitting the data to the system at MALC in Darmstadt. The camera operator could set and change key parameters such as the limiting magnitude, the cloud factor and the elevation angle of the camera via a small graphical user interface and could see the status of the automatic meteor detection software as well as the communication progress.

The communication between the monitoring center and the observation sites was based on wireless communication infrastructure, aimed at being mobile and flexible. The extreme observing conditions at the high altitude sites in November dictated some flexibility for short-duration interuptions from equipment failure or operations. For data transmission from the observing locations to Darmstadt, a Siemens M20 data engine was used. This is a small portable cellular device for digital data transmission.

2.3. MALC Design

The Meteor Alert Center (MALC) was a software tool installed at ESOC, Darmstadt. MALC's task was to collect the counts and plot the result in terms of Zenith Hourly Rate (ZHR) as well as flux in near real time. The plotting intervals for these parameters was adjustable, in order to allow higher temporal resolution when the meteor rates would increase.

MALC responded to a number of Meteor Detection Sites (MDS) that were reporting to it. The number of observation sites equipped with a MDS component is configurable in the MALC software. MALC was executed on a Personal Computer (PC,WIN98). MALC calls in automatically to the data engines of the MDS component, one by one. To avoid a translation into another communication protocol, we used the same data engine for the MALC component as for the MDS component.

At the MALC, each MDS was defined through a graphical user interface (GUI). The MALC-operator was to set the observational and equipmental parameters for each MDS and select the plot specifications. Those parameters are the identification of the observation site (MDS), which can be defined by a phone number, a location from a pull-down menu, or by geographic coordinates. Other parameters included the camera name from a pull-down menu, the number of square degrees of the field of view, the communication interval, and the meteor showers of interest. The communication interval specifies the interval after which the MALC system should automatically call the MDS site and collect the newly acquired data. The pull-down list for the meteor showers of interest included entries for Leonids, but also for sporadics and for all meteors together.

The plot specifications allowed a choice for the information display being either a text window or a ZHR plot. The option to plot the meteor flux was not yet in operation at the time of the shower. If a text window is selected, then the raw output of the *MetRec* software is shown there. If a ZHR plot window is selected, then the effective ZHR is displayed for a selected time range. The following formula was used to compute the effective ZHR ($eZHR$) (Ceplecha et al., 1998):

$$eZHR = \frac{n_{met}}{T_{eff}} F \, r^{6.5-lm} \frac{1}{sinz} \qquad (1)$$

Here, n_{met} represents the number of shower meteors; T_{eff} the interval time; F the correction factor for cloud cover; r the population index of meteor stream and lm is the limiting magnitude. Each data point in the effective ZHR plot is calculated from the detected meteors, n_{met}, within a time interval T_{eff}. The error bar for each data point is calculated by $\frac{n_{met}}{\sqrt{n_{met}}}$ within each interval. We set the population index to a constant and assumed r = 2.1 for the Leonid meteor shower.

The effective ZHR does not match the visual Zenith Hourly Rate due to software and camera effects, e.g. different field of view than human observer, spectral response of intensifier. At the time of the 1999 shower, no effort had yet been made to calibrate the response due to lack of time.

We implemented both MALC and MDS software components in JAVA and tested the software on SPARC/Solaris and Intel/Windows98. In general, the software development in JAVA went smoothly, but unimplemented details in the libraries suprised us often during the software development.

3. MALC Results during the Leonids

A typical result of the process is shown in Fig. 1. The plot represents the data received from camera ICC2, operated at CAHA in the night of Nov. 17/18, 1999. Such plots were updated, after each successful communication between MALC and one of the Meteor Detection Sites. Plots with several different user-defined temporal resolutions are easily generated.

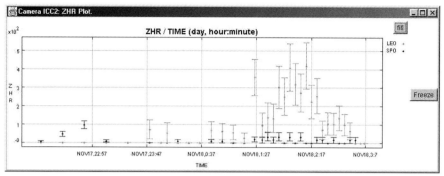

Figure 1. A typical MALC effective ZHR plot.

Figure 1 is based on the original *MetRec* data from the observing night. It contains data points for the sporadic meteor background and the detected Leonids meteors. The ZHR of the sporadic meteors stayed nearly constant over the whole night, whereas the Leonid ZHR rose continuously from 1:30 UT to reach their peak at the interval just after 2:00 UT.

4. Discussion of performance

The detection software *MetRec* was extensively tested in our lab with intensified video camera data from earlier campaigns. If we define the detection quality as:

$$quality = \frac{n_c^0 - n_c^1}{n_v} \qquad (2)$$

then we obtained quality values higher than 0.7. Here, n_c^0 is the number of correctly identified meteors; n_c^1 is the number of incorrectly identified meteor events; n_v is the number of visually counted meteors after tape inspection. A visual inspection on the tapes from November 1999 showed, however, a quality value of 0.5. Owing to the number of faint meteors, only half of the visually found meteors were detected by *MetRec*. In addition, there were false detections of meteors at a rate of about 5 per hour in the tests as well as on the tapes from November 1999. We find the quality values adequate for monitoring purposes, as we can determine the ZHR to within at least a factor of 2.

The output of the intensified cameras ICC2 at CAHA and ICC5 at OSN (see Table 1) were used as input for the MALC monitoring system during the near real time reporting. Only minor problems were encountered with the camera equipment and the computer setup and these could be fixed quickly.

The communication between the MDS and MALC worked satisfactorily when testing was done between Noordwijk, the Netherlands, and Darmstadt, Germany. However, the communication links between Germany and Spain were a complete failure. The data engines had GSM coverage at both locations in southern Spain, but the data connection could only be established once over a period of several days. Surprisingly, we could use the usual (voice) portable GSM phones from both locations. Therefore, we believe that a protocol switch took place between one of the national GSM providers involved and that this disrupted all data links. In the end we had to fall back on person-to-person telephone conversation to report the actual rates at the location sites.

We will continue to develop the MALC/MDS software components and system and hope to demonstrate its value in the coming years.

Acknowledgements

Members of the Dutch 'Werkgroep Meteoren' and the International Meteor Organization joined the ESA/SSD team with their own instrumentation, great spirit and humour. Special thanks goes to Sirko Molau for support of the software *MetRec*. We received generous support from the division of Cellular Phones, Siemens, Munich, who provided the complete communication equipment. DEP Rhoden, the Netherlands, provided support with the intensifier equipment. Financial support was obtained from ESOC, Darmstadt. The directors of the Spanish astronomical observatories made deployment at OSN and CAHA possible. We thank Peter Jenniskens for many improvements of the manuscript. *Editorial handling:* M. Fonda.

References

Ceplecha, Z., Borovicka, J., Elford, W.G., ReVelle, D.O., Hawkes, R.L., Porubcan, V., and Simek, M.: 1998, *Space Science Reviews* **84**, 327–471.
Cooke, W. J. and Suggs, R. M.: 1999, in D.K. Lynch, *et al.* (eds.), *Proceedings of the 1999 Leonid Meteoroid Storm and Satellite Threat Conference*, May 11-13, 1999, AIAA Manhattan Beach, CA (no pagenumbers).
Jeanes, A.: 2000, *Leonid 1999 Airborne tracking campaign*, Final Report of ESA/ESOC Contract No. 13807/99/D/CS
Jenniskens, P.: 1999, *Adv. Space Res.* **23**, 137–147.
Jenniskens, P., Butow, S.J., and Fonda, M.: 2000, *Earth, Moon and Planets* **82–83**, 1–26.
Koschny, D. and Zender J.: 2000, *Earth, Moon and Planets* **82–83**, 209–220.

McNaught, R.: 1999, *Space Industry News*, Issue 84, September 1999, p. 10.
McNaught, R. H. and Asher, D. J.: 1999, *WGN, Journal of the IMO* **27**, 85–102.
Molau, S. and Nitschke, M.: 1996, *WGN, Journal of the IMO* **22**, 119–123.

COORDINATED OBSERVATIONS OF LEONIDS IN ISRAEL

NOAH BROSCH

The Wise Observatory and the School of Physics and Astronomy,
Beverly and Raymond Sackler Faculty of Exact Sciences,
Tel Aviv University, Tel Aviv 69978, Israel
E-mail: noah@wise.tau.ac.il

(Received 22 June 2000; Accepted 18 August 2000)

Abstract. Ground-based meteor observations from Israel were coordinated with the 1999 Leonid Multi-Instrument Aircraft Campaign. For the first time, a large concentration of observational means for meteor astronomy was gathered in a small country. Many observations were specifically keyed to the peak of the Leonid activity, when the predictions indicated a possibly strong storm component that would be ideally observable from the Middle East. This unique opportunity allowed us to forge collaborations with the amateur astronomers community and with high-technology industries, who contributed unique observational means. The elaborate preparations led to an unprecedented success, yielding unique observational results. Some of these are included in these proceedings, but their interpretation and incorporation in a coherent picture of the Leonid meteors is still to come.

Keywords: Ground-based observations, Leonids 1999, meteor, meteor shower, meteor storm, Multi-Instrument Aircraft Campaign

1. Introduction

The recovery of comet P55/Tempel-Tuttle in 1997 (Hainaut et al., 1998) raised the hope that a Leonid storm may take place in 1998 or 1999. The passage of the Earth near the node of the comet's orbit around November 17, 1998, at about 19 hours UT, was expected to be a good time to view the most intense meteor activity. Brown (1999) predicted that a strong meteor storm, as seen in 1966, would be unlikely but that ZHRs could reach a few 1,000 in 1998, with a peak over eastern Asia. Rather, the most spectacular activity came from a broad shower of bright fireballs that had its peak over Europe (Arlt et al., 1998; Jenniskens, 1999). For the next return, McNaught and Asher (1999) expected ZHRs of order 1,500 in 1999 as a result of mass ejection that took place in 1899. They put the peak of the 1999 event at 02:08UT±15 minutes, in the early hours of November 18, a favorable location for Israel.

In Nov. 1998, amateur groups from the Israel Astronomical Association deployed in a number of locations, mainly in the Negev desert, near the town of Mizpeh Ramon where the Wise Observatory of the Tel Aviv University (TAU) is located. The visual observers in Israel saw the Leonid outburst rich in bright fireballs nearly one day before the predicted maximum, when hundreds of meteors appeared during a short period (Mikishev and Levina, 2000).

A first for meteor research in the 1998 Israeli campaign was the use of a large synthetic-aperture radar manufactured by ELTA (a division of the Israel Aircraft Industries) to acquire and track the meteors. The radar observations were done in the night of 17–18 November 1998, from about midnight to 01:30 LT. All-sky hourly rates of ~7,500-16,000 were observed, with mostly sporadic meteors. An interesting aspect of these observations, reported at the August 1998 URSI meeting in Toronto (Rosenkrantz, 1999), was the detection of a few persistent radar trails lasting a number of minutes. They are reminiscent of the persistent trains reported by visual observers.

The 1999 return of the shower was predicted to peak over Europe, North Africa, and the Middle East, with best observing conditions from the longitude of Israel. Israel offered one of the most accessible airports on the eastern Mediterranean. In coordination with Tel Aviv University and the Israeli Space Agnecy, the Leonid Multi-Instrument Aircraft Campaign was invited to stage out of Ben Gurion airport. This would make it possible to coordinate ground-based radar, radio and optical observations with the airborne measurements.

Here, I report briefly on the scope of the observing campaign and provide a first review of observational results and accomplishments. Results were presented at the Leonid MAC Workshop 2000, which was hosted at TAU from April 16–19, 2000.

2. The campaign

2.1. Coordination with Airborne observations

A short window for coordinated observations existed at the beginning of the flight, shortly after takeoff. From about 23:20 until 00:45 UT, the aircraft would circle near the coast of Israel in order not to be too far west at the time of the storm. An elongated pattern towards west-north-western direction was chosen to maintain the best relative orientation of the two aircraft for stereoscopic measurements (Figure 1). The relative position with respect to the radar would enable specular reflections in a narrow track above this pattern for meteors oriented in the same way as the Leonids.

Also, upon arrival in the morning of November 17, there was a brief opportunity for combined observations (Figure 2). However, because of an approach in dawn, no further flight changes were made to increase the period of coordinated observations.

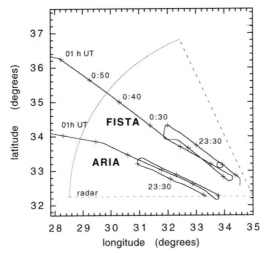

Figure 1. Flight track of the ARIA and FISTA aircraft on departure from Tel Aviv during the night of Nov. 17/18, with coverage of ground-based radar system. Time markers on the tracks are in units of 10 minutes.

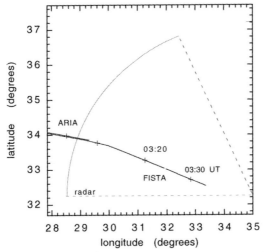

Figure 2. As Figure 1. Flight track of the ARIA and FISTA aircrafts during approach of Tel Aviv on Nov. 16/17.

2.2. WISE OBSERVATORY

At the same time, ground-based observations were set up at several sites in Israel. The Wise Observatory (WO) acted as the center of coordination. The Wise Observatory is equipped with a 1.0-meter telescope and with a CCD camera. The CCD installed on the WO telescope is a good Tektronix product, with 1024^2 pixels. This detector can be used for either imaging or spectroscopy. In consultation with Dr. Peter

Figure 3. Location of sites mentioned in the text. Also shown is the schematic coverage of the three radar systems used for Leonid observations.

Jenniskens, we decided to use the FOSC (Faint Object Spectrometer Camera) to image the airglow spectrum during the night, and to try to obtain high-resolution spectra of persistent trains. The FOSC is a high-throughput transmission-type spectrometer.

Another instrument deployed for the Leonid Storm research was a two-star photometer (PHTM) attached to a laptop computer, to allow data collection to proceed without tying up a normal observatory computer. The photometer could be operated without imaging optics, looking straight up to the zenith. A bright fireball would register in the PHTM as a brief enhancement of the sky brightness due to light scattering by the atmosphere, and thus would be recognizable even if it would not pass exactly overhead. GPS timing of such events could be achieved by setting the controlling computer to the master GPS receiver at the WO. Fainter meteors would register only if they passed through the $\sim 8°$ diameter field.

With the help of Dr. Moshe Oron, Chief Scientist of El-Op (Electro-Optical Industries, Ltd.), the campaign was promised two FLIR (Forward - Looking InfraRed) staring cameras for the Leonid shower. The El-Op cameras would allow us to attempt meteor observations in the 3-5 and 8-12μm bands. In addition, Mr. Jeremy Topaz from El-Op was able to borrow a solar-blind UV camera manufactured by DEP (Delft Electronische Producten) and video recorder from the Ophil company. This would open up for observation a spectral band from 310 to $\sim$350 nm, mainly UV-A with a small contribution from UV-B type of ultraviolet radiation, in which the meteors could be observed.

Optical high-data-rate imaging at the Wise Observatory would be obtained with a Xybion camera belonging to a high-tech industry in Israel and loaned to Dr. Adam Devir for the event.

Wise Observatory was also the main observing site of the USAF ground-based observing campaign in Israel, coordinated by Peter Brown and Bob Hawkes. On the Observatory terrace, four cameras belonging to the Canadian/USAF team operated to obtain intensified-video coverage of large parts of the sky. One of these was equipped with a diffraction grating for meteor spectroscopy. The meteor rates, counted visually by an observer on the terrace of the Wise Observatory, would be phoned-in to the Leonid center at the Marshall Space Center (MSC).

The pointing of the optical instruments was coordinated to obtain simultaneous measurements of the same meteors. Notably, the two mid-IR cameras would be imaging in the same field of view as one of the Canadian cameras, and in the same direction as the UV camera.

Finally, an Extremely Low Frequency (ELF) receiver was operated from the Wise Observatory site and a Very Low Frequency (VLF) receiver was operated in Sde Boker, a kibbutz approximately 20 km north of Mitzpe Ramon, by Colin Price and colleagues of Tel Aviv University. This part of the program used existing experimental equipment belonging to TAU's Department of Geophysics & Planetary Sciences, normally used to study world-wide lightning activity through VLF and ELF radio emissions. The global capacitor made by the ionosphere and the ground resonates whenever a lightning strike happens. These resonating electromagnetic waves, with frequencies of a few Hz to a few tens of kHz, propagate around the globe. Radio emission from meteors could be connected with the electrophonic sounds detected by Vinkovic (1999). The ELF and VLF detectors could be used to measure the strength of the electric fields during the meteor storm. This output could then be correlated with the visual and radar detections and, hopefully, help identify some bright Leonids with the production of electromagnetic waves, by correlation with the fireballs detected from the Leonid MAC aircraft, which covered a nearby large surface area.

2.3. SEA OF GALLILEE AND HERMON MOUNTAIN

In northern Israel, a team from the Emek HaYarden Regional College, headed by Dr. Lev Pustilnik, planned to deploy two PAL video cameras, one to operate from the shore of the Sea of Gallilee and the other from the Hermon mountain. This would allowed a two-station capability for stereoscopic imaging, from which space trajectories and light curves for the brighter meteors could be derived. This part of the operation had no assured radar coverage. We would attempt to adapt a weather radar operating from the TAU campus to cover this part of the country.

2.4. NEGEV DESERT

In southern Israel, teams of the Israel Astronomical Association planned to deploy in the Negev desert. Some would be located on the premises of the WO, while others would deploy further south, in a location with reasonable accessibility and fainter sky background illumination. They planned to perform mainly visual observations, such as plotting a panoramic view of the meteor shower at low activity levels, and switching to meteor counts when the storm component would appear. These teams would be coordinated by Mr. Ilan Manulis and Ms. Anna Levine.

The Revivim Observatory $\sim$45-km north of Mizpe Ramon hosted the remote site of the USAF ground-based campaign, for stereoscopic observations with Wise Observatory. This team would also phone-in their meteor counts to MSC.

Halfway between these sites, at the Desert Research Institute in Sde Boker, a group of amateur astronomers would deploy for their observations. The same location hosted the VLF facility.

2.5. TEL AVIV UNIVERSITY

One expectation from this shower was that, if the meteor activity would be high enough, one could expect a significant enhancement of the ionospheric electron density. This could be detectable, in principle, through the measurement of signal delays from the GPS satellite system. Israel is operating a nation-wide network of high-precision GPS receivers for geophysical purposes (Dr. Shimon Wdowinski, PI) and the monitoring of GPS satellite signal delays would be part of the Leonid effort. This experiment was to be performed from Tel Aviv University.

3. Results

A Leonid homepage in Hebrew was created by the Center for Educational Technology (CET), so that school children and the general public could read the basic facts about the expected event. A downloadable report form for visual observations by the general public, patterned after the similar NASA report form, was prepared and installed on the website.

During the week preceeding the Leonid shower a number of articles were published in the general press. Radio and TV aired interviews. A press conference was called on November 17 at 15:00 UT. The conference was attended by TAU officials and by NASA and USAF personnel involved in the MAC-99 campaign. The press conference was the culmination of an intensive and coordinated outreach effort. It is difficult to see how anyone in Israel could remain unaware of the Leonid scientific activity, as well as of the practical aspects (such as dangers to space-borne platforms from meteors), given this intensive exposure.

Ohad Shemmer was the WO observer during the first half of the night. He planned to monitor spectroscopically a few active galactic nuclei (AGNs) with the 1.0-meter telescope, then carry on during the second half of the night with airglow spectroscopic monitoring. These results, as well as those from the PHTM, are presented in Brosch and Shemmer (2000).

As the time drew closer to midnight, the Canadian/USAF team's computers began screening out meteors belonging to the Leonid shower. Results of their counts are published in Brown *et al.* (2000). It seemed that the meteors tended to be on the faint side, with no fireballs showing.

In this issue of *Earth, Moon and Planets*, Mikishev and Levina (2000) report on some of the visual observations. The rate of meteors picked up close to the predicted time of maximum by McNaught and Asher (1999). We started seeing a number of meteors every second, most belonging to the Leonid shower. The meteors were coming so fast that it was no problem spotting the location of the radiant by tracing them back in our minds. Some meteors showed fascinatingly reddish trails. The number of visible fireballs was relatively low.

Sometime after 02:30 UT (04:30 local time) it was clear that the peak Leonid activity was past. The meteor rates were dropping very fast; we now were seeing only one meteor every ten seconds or thereabouts. In a very short time, the show was over. At about 03:15 UT dawn began.

During the night, Dr. Mark Rosenkrantz reported that the radar systems were operating fine. While tracking the air space above the Leonid MAC aircraft, one system was seeing thousands of radar returns.

At the URSI (Union of Radio Sciences) yearly meeting in Tel Aviv, on February 15 2000, a few preliminary results were presented. They showed that the distribution of the meteor radar returns with height was double-peaked: one peak corresponded to the "normal" meteors, at altitudes between 90 and 130-km. The other peak, with a similar amplitude but wider distribution, corresponded to a mean altitude of 250-km. However, the measured meteor velocities of this higher peak, derived from line-of-sight Doppler velocities combined with the motion perpendicular to the line of sight derived from the tracking data, were never higher than 12 km/s, and typically 4–5 km/s, suggesting that some instrumental problem might be responsible for this. Numerous direct meteor detections still wait analysis and correlation with optical observed meteors from the aircraft.

A few VLF results were reported by Dr. Colin Price at the same URSI meeting (Price and Blum, 2000). Preliminary results show that large numbers of short VLF emissions were detected, at radio frequencies which were significantly different from those of regular lightning. The brief VLF signals became much more frequent close to the Leonid peak activity. It is clear that the VLF measurements will open new ways of studying the interaction of meteors with the atmosphere.

Optical observations were successful too. The two-site video campaign by Lev Pustilnik *et al.*, succeeded in detecting a number of meteors from both locations, despite the lack of an intensifier stage on their cameras. These results will provide accurate stereoscopic measurements from a second location over Israel.

Jeremy Topaz recorded several meteors in the near UV between 300 and 350 nm (Almoznino and Topaz, 2000). Unfortunately, none of the meteors was recorded also by optical techniques. No mid-IR detections were obtained.

4. Discussion

Attempts to link up with neighboring countries, through the national meteor observer associations or the local scientific institutions, went unanswered despite the intervention of the UN Office for Outer Space Affairs and of ESA. We hope that better relations in the future will enhance contacts between neighboring astronomical communities, where only tens to hundreds of km separate observing teams, to further coordinated and stereoscopic observations of meteors.

Although publicity was intense in Israel and many persons probably took the trouble to wake up in the early hours of the morning of November 18 to look at meteors, no forms reporting observations were received. Much work needs to be done to develop the coordination of visual meteor observations in Israel and involve more people in the research.

The splendid shower and the fruitful scientific output of the effort will help a lot in creating a long-term interest in meteor observations among Israeli researchers. The coordination of Leonid MAC and ground-based campaigns provided strong incentive for future work. The large amount of data obtained by non-academic participants, who are not yet funded for their part in this project, will need to be addressed. In order to prevent this from happening in future campaigns, it is advisable to pair-up industry and academy teams very early in a project.

Lessons learned from this coordination exercise should be applied in future campaigns, both for the 2001 and 2002 Leonids and for observing other high-visibility astronomical phenomena.

Acknowledgements

I am grateful for help with the Leonid campaign in Israel from the Israel Space Agency. Local financial support for the Leonid MAC Workshop was provided by the following bodies, which are thanked for their generosity: Tel Aviv University, the Israel Space Agency, and USAF-EOARD. Kibbutz Revivim and the operators of the Revivim Observatory are thanked for their help to the Canadian/USAF Leonids team deployed there. Prof. Dan Maoz, Director of the Wise Observatory, is thanked for allowing the Leonid campaign to be conducted from the Wise Observatory. The issue of public outreach was addressed with the help of Dr. Meir Meidav from the School of Education of TAU, and with significant assistance from the Center for Educational Technology (CET), with the help of Dr. Yoav Yair. Last but not least, it is a pleasure to acknowledge the organizational help of Ms. Margie Goss. *Editorial handling:* Mark Fonda.

References

Almoznino, E. and Topaz, J.M.: 2000, *Earth, Moon and Planets* **82–83**, 391–398.
Arlt, R.: 1998, *WGN, Journal of the IMO* **26**, 239–248.
Brown, P.: 1999, *Icarus* **138**, 287–308
Brown, P., Campbell, M.D., Ellis, K.J., Hawkes, R.L., et al.: 2000, *Earth, Moon and Planets* **82–83**, 149–166.
Brosch, N. and Shemmer, O.: 2000, *Earth, Moon and Planets* **82–83**, 539–548.

Hainaut, O.R., Meech, K.J., Boehnhardt, H., and West, R.M.: 1998, *Astron. Astrophys.* **333**, 746–752.
Hunten, D.M., Cremonese, G., Sprague, A.L., Hill, R.E., Veranui, S., and Kozlowski, R.W.H.: 1998, *Icarus* **136**, 298–303.
Jenniskens, P.: 1999, Meteoritics Planet. Sci. **34**, 959–968.
Jenniskens, P. and Butow, S.J.: 1999, Meteoritics Planet. Sci. **34**, 933–943.
McNaught, R.H. and Asher, D.J.: 1999, *WGN, Journal of the IMO* **27**, 85–102.
Mikishev A. and Levina, A.: 2000, *Earth, Moon and Planets* **82–83**, 257–264.
Price, C. and Blum, M.: 2000, *Earth, Moon and Planets* **82–83**, 549–559.
Rosenkrantz, M. 1999, lecture at the URSI meeting in Toronto.

METEORS: A DELIVERY MECHANISM OF ORGANIC MATTER TO THE EARLY EARTH

PETER JENNISKENS AND MIKE A. WILSON
NASA/Ames Research Center, Mail Stop 239–4, Moffett Field, CA 94035
E-mail: pjenniskens@mail.arc.nasa.gov

DENNIS PACKAN, CHRISTOPHE O. LAUX
AND CHARLES H. KRÜGER
High Temperature Gasdynamics Laboratory, Building 520, Mechanical Engineering Department, Stanford University, Stanford, CA 94305–3032

IAIN D. BOYD
Department of Aerospace Engineering, University of Michigan, 3012 Francois-Xavier Bagnoud Building, Ann Arbor, MI 48109–2140.

OLGA P. POPOVA
Institute for Dynamics of Geospheres RAS, Leninsky prospekt 38, bld.6, Moscow 117979, Russia

and

MARK FONDA
NASA/Ames Research Center, Mail Stop 239–4, Moffett Field, CA 94035

(Received 29 May 2000; Accepted 31 July 2000)

Abstract. All potential exogenous pre-biotic matter arrived to Earth by ways of our atmosphere, where much material was ablated during a luminous phase called "meteors" in rarefied flows of high (up to 270) Mach number. The recent Leonid showers offered a first glimpse into the elusive physical conditions of the ablation process and atmospheric chemistry associated with high-speed meteors. Molecular emissions were detected that trace a meteor's brilliant light to a 4,300 K warm wake rather than to the meteor's head. A new theoretical approach using the direct simulation by Monte Carlo technique identified the source-region and demonstrated that the ablation process is critical in the heating of the meteor's wake. In the head of the meteor, organic carbon appears to survive flash heating and rapid cooling. The temperatures in the wake of the meteor are just right for dissociation of CO and the formation of more complex organic compounds. The resulting materials could account for the bulk of pre-biotic organic carbon on the early Earth at the time of the origin of life.

Keywords: Ablation, astrobiology, exobiology, meteors, meteoroids, origin of life

1. Introduction

Accretion of extraterrestrial matter has long been of interest as a source of pre-biotic organic carbon for the origin of life on Earth (Oró, 1961; Sagan, 1974; Lewis *et al.*, 1979; Anders, 1989; Pepin, 1991; Huebner and Boice, 1992; Delsemme, 1992; Chyba and Sagan, 1992, 1998; Oberbeck and Aggarwal, 1993; Chang 1993, Whittet, 1997; Oró and Lazcano, 1998). Most organic carbon is thought to be accreted by impacts of comets and primitive asteroids (Oró, 1961; Chyba *et al.*, 1990). However, the high-speed impacts are expected to be so energetic that the ensuing fireball destroys virtually all molecular species in the impacting object and subsequent synthesis upon cooling in a CO_2 rich atmosphere is not efficient (McKay and Borucki, 1997; Chyba and Sagan, 1998). Moreover, such impacts are infrequent and have strong perturbing effects on the pre-biotic environment at the time of delivery (Maher and Stevenson, 1988; McKinnon, 1989; Chyba, 1993).

Accretion of meteoroids is a more gentle and continuous mechanism for delivery of organic carbon species, especially in the case of the Interplanetary Dust Particles (IDP) that are collected mostly intact in the Earth's atmosphere. Among this collected debris of small solar system bodies there are chondritic IDPs with organic carbon abundances of about 10 mass percent on average (Anders, 1989; Gibson, 1992). The organic carbon contains complex aromatic molecules up to 500 a.m.u. (Clemett *et al.* 1993). However, all IDPs represent a small fraction (< 8 %) of the incoming mass, while low encounter velocities favor asteroidal particles that are relatively poor in organic carbon compared to cometary matter (Bradley *et al.*, 1988). Comet Halley's dust has been measured to contain up to 50 mass percent refractory organics (Krueger and Kissel, 1987; Greenberg, 2000).

Here, we consider the possibility that meteors could have delivered organic carbon to the early Earth. Meteors are the luminous phase that represents the process of ablation and fragmentation of the meteoroids' interaction with the atmosphere. Of particular interest are those meteoroids that are too small to cause destructive high-pressure and high-temperature shock-induced chemistry (Menees and Park, 1976, Park and Menees 1978).

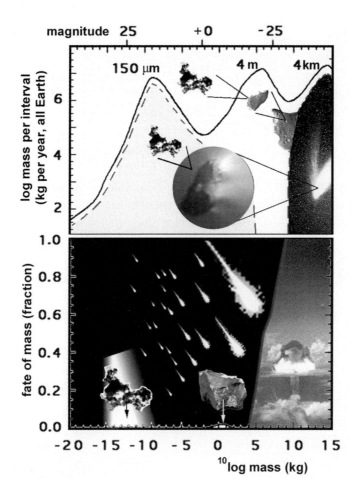

Figure 1. Top panel: the current annual (log) mass influx per unit (log) mass interval of all types of incoming matter in Earth's atmosphere, compiled from Love and Brownlee (1993) and Ceplecha (1992, 1994) Bottom panel: the fate of this matter upon accretion, as a fraction of the total. Fractions are derived from estimated meteorite yields (Oberst *et al.*, 1998; Halliday *et al.*, 1984: Bland *et al.*, 1996), impact limit (Ceplecha, 1992, 1994), and entry heating (Flynn, 1989; Love and Brownlee, 1991), the latter for the new present-day velocity distribution (Taylor and Elford, 1998).

The fraction of mass of meteoroids that are too large and too fast to survive ablation, but too small to cause shock-inducted modification of the incoming meteoroid, is schematically shown in Figure 1. The figure shows the present annual mass influx per mass interval. The various sources of matter are identified. Impacts become progressively infrequent for more massive objects. The present-day distribution is thought to be representative of the dust influx on the early Earth, except that the relative contributions from asteroids and comets may have differed (Chyba and Sagan, 1998). The mass distribution has peaks at 150 microns (meteoroids), 4 meters (asteroid fragments and comet boulders), and 4 km (comets and asteroids). Meteoroids survive as IDPs or ablate during interaction with the atmosphere while in the molecular flow regime, and are recovered as micrometeorites that are mostly CM-type materials and lesser amounts of CI-type materials (Kurat *et al.*, 1994; Engrand and Maurette, 1998). The shaded area reflects organic-rich cometary dust. The meter-sized meteoroids lose most of their mass in the continuum flow regime, where they develop shocks because of their subsonic entry speeds. Some asteroidal fragments will survive as meteorites. Impacting comets and asteroids will catastrophically fragment in an airburst or explode upon impact (Chyba *et al.*, 1990).

Based on this mass influx distribution, and the fate of accreting matter, we submit that meteoroids and small comet boulders can account for the bulk of the organic carbon on the early Earth – that is, if the survival of exogenous organic carbon, or the creation of reduced molecules by atmospheric chemistry, is efficient. Little is known about whether organic carbons can survive the meteor phase. Conditions of free molecular flow (in case the mean free path in air is larger than the typical dimensions of the object) and high Mach number conspire against theoretical and laboratory studies of meteoric plasmas. Factual information relies on remote sensing, to which few modern techniques have been applied.

Here, we report measurements of air plasma temperatures in the wake of meteors, which are compared to models of air plasma emission. The data are interpreted by means of direct simulation by Monte Carlo modeling of meteoroids in the free molecular flow regime. We conclude that meteors offer interesting pathways for the survival of organic carbon that warrant further study.

2. Spectroscopy of Meteoric Plasmas

We deployed from an airborne platform a new un-intensified slit-less CCD spectrograph for near-infrared and visible wavelengths at the time of the intense 1998 Leonid shower during the 1998 Leonid Multi-

Instrument Aircraft Campaign (Jenniskens and Butow, 1999). The spectrograph consisted of a 600 l/mm grating, a Nikon f2.8/300 mm lens, and a Pixelvision two-stage thermoelectrically cooled 1024 x 1024 pixel CCD camera. The same spectrograph was used again to probe Perseid meteor spectra from a ground site in August of 1999. Our best spectra probe 1 cm-sized meteoroids with entry velocities of 61 km/s (Perseids) and 72 km/s (Leonids) at altitudes 90–100 km. The spectra cover part of the wavelength range 580–900 nm at a relatively high 0.5 nm resolution (Figures 2 and 3).

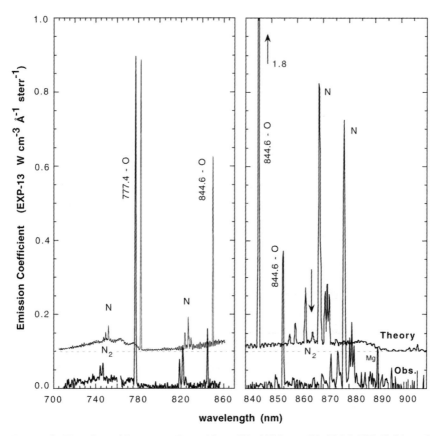

Figure 2. Two Leonid spectra from Nov. 17, 1998, at 17:47:06 UT (left) and 18:08:47 UT (right). The spectra are compared with NEQAIR2 model calculations, which is slightly displaced to facilitate comparison. Note the different line intensities of the OI line emission at 844.6 nm and the NI line emission at 865.6 nm (arrow). The line at 880.7 nm is of MgI.

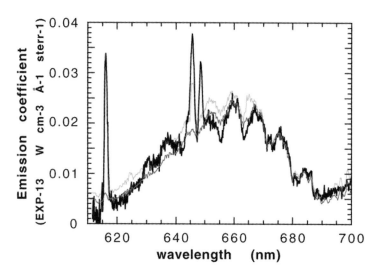

Figure 3. The Δv = 3 first positive band of N_2 as resolved in the spectrum of a magnitude −1 Perseid meteor from 00:25:26 UT, Aug. 13, 1999 (dark line). Emission lines at 615.7, 626.2, 645.4, and 665.4 nm are from atomic oxygen. The line at 648.9 nm may also be of OI. The two gray lines show the NEQAIR simulations at 4,300 K (dark, down) and 4,670 K (light, up), respectively. The simulation considers pure air (79% N_2 and 21% O_2) at the standard atmosphere pressure of $P = 10^{-6}$ atm at the altitude of the meteor (95 km). The spectrum was convolved with a triangular slit function of FWHM = 0.5 nm, which matches the line width measured on the observed spectra.

The dominant emission is from atmospheric lines of O and N and the first positive bands of N_2, in contrast to prior studies at shorter wavelengths where the emission lines of ablated meteoric metal atoms dominate. The observed lines and bands are well matched by the NEQAIR2 radiation model of heated air in thermodynamic equilibrium (Park, 1985; Laux, 1993). The match implies that the bulk of emission is from gas in near thermal equilibrium despite the high Mach number flow. The observed ratio of atomic and molecular nitrogen in the Leonid spectra (Figure 2) is a sensitive measure of temperature and implies a chemical equilibrium temperature of $T_c = 4,340 \pm 100$ K. The N_2 band contour (and NI lines) of the Perseids are well matched by a simulation at $T_v = 4,300 \pm 40$ K (Figure 3). All values are similar to temperatures estimated from meteoric metal atom emission lines at $T_e = 4,500 \pm 500$ K (Borovicka *et al.*, 1999).

The data are sufficiently precise to recognize numerous signs of non-equilibrium behavior. There is excess emission at high v levels in the N_2 molecular band as a result of recombination processes (Figure 3). The OI line intensities are not always well matched (Figure 2). Notably, the OI line at 844.6 nm is a factor of 3 fainter than that calculated in all Leonid spectra and different from laboratory LTE air plasmas (Park et al., 1997, 1998), while OI lines between 600 and 700 nm are stronger (Figure 3). Also, the NEQAIR model with initial 0.03% atmospheric CO_2 predicts CN emission comparable to the first positive N_2 bands, but no CN (or isoelectric N_2^+) is observed.

Another sign of non-equilibrium is the intensity of the OI 777.4 nm line emission. The measured intensity of the line, relative to the intensity of recorded background field stars, can be compared to the volume emission coefficient of the plasma in the model (vertical scales in Figure 2). From this, we derive a volume for the emitting gas of 1×10^{13} cm^3, assuming the gas is at 4,300 K. Initial-train-radius theory predicts a volume of only about 3×10^7 cm^3 for the head of the meteor (Jones, 1995).

3. Theoretical Model of rarefied Flow

The source region of the T ~ 4,300 K emission was identified using the direct simulation Monte Carlo (DSMC) technique, which was applied to the two-dimensional flow about a 1 cm-sized Leonid meteor (density 1 g/cm^3). For many years DSMC has been developed and applied to a variety of rarefied flows (Dietrich and Boyd, 1996), but this is the first attempt to apply the technique to a computation of this type. Two cases were considered: one with no ablation, and another one with a simple Bronsten ablation model (Bronsten, 1983), in which the ablated material is assumed to be magnesium. In the case without ablation (Figure 4a), we find a rapid decline of temperatures outward, where multiple collisions quickly stop the accelerated air molecules. This is the process described by the initial meteor-train-radius theory, which in light of this model might still apply well to the near spherical source of radar head echoes (Jones, 1995).

Including meteor ablation result in dramatic changes when applying this model (Figure 4b). We find that ablation increases the flow field temperature around the meteor over an extended area in a wake behind the meteor, with elevated values around 5,000 K.

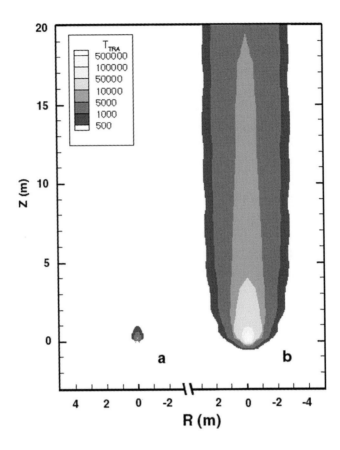

Figure 4. Translational temperature field from a rarefied flow model of a −1 magnitude Leonid meteor at 95 km altitude shown without ablation (a), and with ablation of Mg atoms (b).

The wake is caused by air molecules penetrating the skin of a dense plasma of ablated material in front of the meteoroid that collides with the meteoric plasma several times in a process of thermalisation. The collisions drive the meteoric ablation products and heated air past the meteoroid, where they expand into the meteor wake. The rotational temperatures are typically less than translational temperatures in the meteor's head but they equilibrate in the wake. We conclude that most of the observed meteor emission originates from this nearly equilibrated gas.

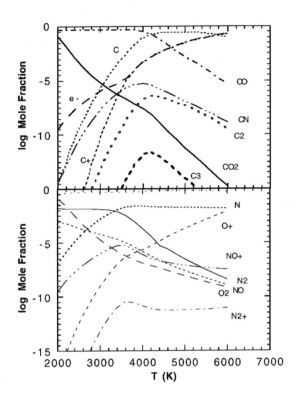

Figure 5. Molecular abundances for equilibrium air plasma at 95 km altitude (P = 10^{-6} atm) in a range of Local Thermodynamic Equilibrium temperatures and for an assumed Mars-like early-Earth atmosphere of particle number composition $O_2/CO_2/N_2/Ar/CO = 0.13/95.32/2.7/1.6/0.08\%$.

4. The Delivery of Organic Matter to the early Earth.

Interestingly, the study of fast shower meteors can help clarify a role of meteors in creating pre-biotic conditions on Earth, which involved a wide range of meteoroid masses and entry velocities. This is because the wake temperatures of all meteors are in the same narrow range of 3,900 ± 900 K as derived from the well studied meteoric metal atom emission lines (Borovicka and Bocek, 1995; Borovicka and Betlem; 1997; Harvey, 1973). There is no obvious trend with meteor magnitude (mass) or entry velocity. The observed Leonid spectra, too, do not change significantly with altitude or meteor brightness over the observed range.

We submit that the mixing time-scale of air and ablation plasma does not vary much with meteor mass and velocity. Meteors represent two sources of pre-biotic carbon: a) Direct influx of organic carbon and metallic compounds in a rarefied high Mach number flow, and b) kinetic energy induced atmospheric chemistry involving CO_2 and N_2, which is also of interest to the issue of nitrogen fixation (Chyba and Sagan, 1998).

Our observations show that relevant atmospheric chemistry can occur in two regimes: (1) the extended wake of the meteors at temperatures of about 4,300 K and (2) at the interface layer between impinging air and ablation products at temperatures of about 10,000 K. The Mg^+ emission line in bright meteors was traced to a component that increases in relative intensity at $T\sim10,000$ K with meteor brightness and entry velocity (Borovicka, 1994). This component is due to meteoric vapor that builds up in front of the ablating meteoroid, rather than to the formation of a shock front as is commonly believed. In both cases, the not fully equilibrated chemical reactions occur at higher temperatures and longer time scales than in the 1-D equilibrium models by Menees and Park (1976) and Park and Menees (1978).

Unfortunately, no models are yet capable of reliably handling the types of non-equilibrium chemistry implied by our observations. An important find, however, is that the observed excitation temperatures are close to the dissociation equilibrium of CO. As a result, our equilibrium chemistry simulation of the meteor plasma in a CO_2 rich atmosphere (Figure 5) results in relatively high yields of potential pre-biotic molecules. This air composition may reflect that of the early Earth (Chyba and Sagan, 1998) and is certainly the least favorable case for reaction chemistry. Right at about 4200 K is where the production of linear carbon chains such as C_2 and C_3 peaks. Under these conditions, small amounts of aromatic hydrocarbons are expected to be formed upon cooling, as well as compounds rich in C=O and C-N groups. Such compounds could offer numerous chemical pathways to yield other reduced molecules of potential significance to the origin of life on the early Earth.

Finally, assuming organic compounds are common to Leonid meteoroids, the lack of observed C, C_2 and CN emission from the combustion of organic matter in the ablating meteoroids implies that organic compounds survive as large molecular fragments. It is possible that they are lost early in the meteor trajectory at low temperature $T < 500$ K when no optical emission occurs. However, meteors do not typically show differential ablation (Borovicka *et al.* 1999). This suggests that all mineral compounds are lost simultaneously irrespective of volatility in a process rather like sputtering by impinging meteoric ablation products instead of complete evaporation of the grain. Hence, most organic carbon, too, is expected to be lost at 80–110 km altitude under the physical conditions described in this paper.

Perhaps the survival of organic carbon is an analog with the common technique of laser induced desorption of large molecules, whereby they are only momentarily heated in a chemically reducing environment and subsequently quickly cooled by collisions with gas molecules and radiative cooling. For example, laser pulsed heating of aniline on a sapphire surface at 10^8 K/s leads to a peak surface temperature of 1,000 K, but with aniline remaining intact at 360 K internal vibrational temperature (Maechling et al., 1996).

5. Relevance of Delivery through Meteors

At the time of the origin of life on Earth 4 Gyr ago, the mass influx of meteoroids was about 200 times the present day mass influx of 4×10^7 kg/yr (Love and Brownlee, 1993). The factor of 200 follows from the lunar impact record and a linear scaling between meteoroids and parent bodies (Chyba and Sagan, 1998). Approximately half of this influx is thought to be carbon-poor asteroidal matter with at best 4 mass % organic carbon in rare CI chondrites and the other half is carbon-rich (25–50 %) cometary matter (Krueger and Kissel, 1987; Delsemme, 1991). Hence, meteors contributed at least 1×10^9 kg/yr of organic carbon to the early Earth if all organic carbon survived the ablation process.

This outweighs the yield from all other exogenous and terrestrial sources of organic carbon on the early Earth as estimated in the recent review by Chyba and Sagan (1998). The main exogenous source, interplanetary dust particles, was estimated to yield at best only 2×10^8 kg/yr, because less than 8 mass percent of small <1 gram particles survive atmospheric entry heating (Figure 1). Meteorites are a negligible source because only 0.04 mass percent of matter >1 g arrives to Earth in the form of meteorites (Oberst et al., 1998; Halliday et al., 1984; Bland et al., 1996). For even larger bodies, only negligible organic carbon survives the shock chemistry (Chyba and Sagan, 1998). Hence, meteors could account for at least 80% of all exogenous organic carbon.

A relatively high production yield of C_2 and CN from atmospheric chemistry is derived from Figure 4. For a CO_2 rich atmosphere, we calculate a yield of carbon of 1×10^{-9} kg/J. For a total kinetic energy of 1×10^{18} J/yr from meteors 4 Gyr ago, the yield is 1×10^9 kg/yr. This is only an upper limit because the chemical reactions are not expected to reach full thermodynamic equilibrium, which is supported by the absence of CN or C_2 emission in the observed spectra.

On the other hand, alternative scenarios imply low yields as well. From organic residues at the K/T boundary, the yield of organic carbon in giant impacts 4 Gyr ago in a mildly reducing atmosphere has been optimistically estimated at only 2×10^8 kg/yr (Chyba and Sagan, 1998).

Terrestrial sources are not much more efficient in this regard. Electric discharges contribute a mere 3×10^7 kg/yr and UV photolysis by sunlight in a mildly reducing atmosphere can account for 3×10^8 kg/yr (Chyba and Sagan, 1998). Furthermore, the analysis of effluents from bio-organically uncontaminated hot or cold springs has failed to show any organic molecules besides methane (Mojzsis *et al.*, 1999).

That leaves meteors as a potential source for more than 2/3 of the pre-biotic organic carbon on the early Earth.

Of particular interest for future studies of meteors is the possible detection of ablated organic carbon by its 3.4-micron feature, as molecules in stratospheric aerosols, or indirectly from its interaction with the airglow chemistry. Additional Leonid storms are anticipated in November 2001 and 2002 (McNaught and Asher, 1999), when meteors may fill the sky again as frequently as they did at the time of the origin of life.

Acknowledgements

We are grateful for the constructive reviews by Jiri Borovicka, Kevin Zahnle, and an anonymous referee to improve the presentation of this paper. The spectrograph was developed with support of the NASA Ames Research Center Director's Discretionary Fund. The 1998 Leonid Multi-Instrument Aircraft Campaign was NASA's first Astrobiology Mission. The campaign was made possible by grants from NASA's Exobiology program and the Advanced Missions and Technology Program for Astrobiology, as well as by NASA's Planetary Astronomy program, NASA Ames Research Center, and the U.S. Air Force. *Editorial handling*: Frans Rietmeijer.

References

Anders, E.: 1989, *Nature* **342**, 255–257.
Bland, P.A., Smith, T.B. and Jull, A.J.T., Berry, F.J., Bevan, A.W.R., Cloudt, S., Pillinger, C.T.: 1996, *MNRAS* **283**, 551–565.
Borovicka, J.: 1994, *Planet. Space Sci.* **42**, 145–150.
Borovicka, J. and Betlem, H.: 1997, *Planet. Space Sci.* **45**, 563–575.
Borovicka, J. and Bocek, J.: 1995, *Earth Moon, and Planets* **71**, 237–244.
Borovicka, J., Stork, R., and Bocek, J.: 1999, *Meteoritics Planet. Sci.* **34**, 987–994.
Bradley, J.P., Sandford, S., and Walker, R.M., 1988, in J.F. Kerridge and M.S. Matthews (eds.), *Meteorites and the Early Solar System*, pp. 861-898, University of Arizona Press, Tucson, AZ.
Bronshten, V.A.: 1983, *Physics of Meteor Phenomena*, D. Reidel, Dordrecht, pp. 356.
Ceplecha, Z.: 1992, *Astron. Astrophys.* **263**, 361–366.
Ceplecha, Z.: 1994, *Astron. Astrophys.* **286**, 967–970.
Chang, S.: 1993, in J.M. Greenberg et al. (eds.), *The Chemistry of Life's Origins*, p. 259–299.
Chyba, C.F.: 1993, *Geochim. Cosmochim. Acta* **57**, 3351–3358.
Chyba, C.F. and Sagan, C.: 1992, *Nature* **335**, 125–132.
Chyba, C.F. and Sagan, C.: 1998, in P.J. Thomas, C.F. Chyba and C.P. McKay (eds.), *Comets and the Origin and Evolution of Life*, Springer Verlag, p. 147–173.
Chyba, C.F., Thomas, P.J., Brookshaw, L. and Sagan, C.: 1990, *Science* **249**, 366–373.
Delsemme, A.H.: 1991, in R.L. Newburn, Jr. et al. (eds.), *Comets in the Post Halley Era*, Kluwer, the Netherlands, pp. 377–428.
Delsemme, A.H.: 1992, *Origins of Life* **21**, 279–298.
Dietrich, S. and Boyd, I.D.: 1996, *J. Comp. Phys.* **126**, 328–342.
Engrand, C., and Maurette, M.: 1998, *Meteoritics Planet. Sci.* **33**, 565–580.
Flynn, G.J.: 1989, *Icarus* **77**, 287–310.
Gibson, E.K.: 1992, *J. Geophys. Res.* **97**, 3865–3875.
Greenberg, J.M.: 2000, *Earth, Moon and Planets* **82-83**, 313–324.
Halliday, I., Blackwell, A.T. and Griffin, A.A.: 1984, *Science* **223**, 1405–1407.
Harvey, G.A.: 1973, in C.L. Hemenway, P.M. Millman and A.F. Cook (eds.), *Evolutionary and Physical Properties of Meteoroids*, NASA SP 319, NASA, Washington D.C., pp. 103–129.
Huebner, W.F., and Boice, D.C.: 1992, *Origins of Life* **21**, 299–316.
Jenniskens, P., and Butow, S. J.: 1999, *Meteoritics Planet. Sci.* **34**, 933–943.
Jones, W.: 1995, *MNRAS* **275**, 812–818.
Krueger, F.K. and Kissel, J.: 1987, *Naturwissenschaften* **74**, 312–316.
Kurat, G., Koeberl, C., Presper, T., Brandstätter, F., Maurette, M., 1994, *Geochim. Cosmochim. Acta* **58**, 3879–3904.
Laux, C.O.: 1993, '*Optical Diagnostics and Radiative Emission of High Temperature Air Plasmas*', Ph.D. Dissertation, HTGL Report T288, Stanford University, CA.
Lewis J., Barshay S.S. and Noyes B.: 1979, *Icarus* **37**, 190–206.
Love, S.G. and Brownlee, D.E.: 1991, *Icarus* **89**, 26–43.
Love, S.G. and Brownlee, D.E.: 1993, *Science* **262**, 550–553.

Maechling, C.R., Clemett, S.J., Engelke, F. and Zare, R.N.: 1996, *J. Chem. Phys.* **104**, 8768–8777.
Maher, K.A. and Stevenson, D.J.: 1988, *Nature* **331**, 612–614.
McKay, C.P. and Borucki, W.J.: 1997, *Science* **276**, 390–392.
McKinnon, W.G.: 1989, *Nature* **338**, 465–466.
McNaught, R.H. and Asher, D.J.: 1999, *WGN, Journal of the IMO* **27**, 85–102.
Menees, G.P. and Park, C.: 1976, *Atmospheric Environment* **10**, 535–545.
Mojzsis, S.J., Krishnamurthy, R. and Arrhenius, G.: 1999, in R.F. Gesteland, T.R. Cech, J.F. Atkins (eds.), *The RNA World*, Second edition. Cold Spring Harbor Laboratory Press, cold Spring Harbor, NY, 709 pp., p. 1–47.
Oberst, J., Molau, S., Heinlein, D., Gritzner, C., Schindler, M., Spurny, P., Ceplecha, Z., Rendtel, J. and Betlem, H.: 1998, *Meteoritics Planet. Sci.* **33**, 49–56.
Oberbeck, V.R. and Aggarwal, H.: 1993, *Origins of Life & Evol. Biosphere* **21**, 317–338.
Oró, J.: 1961, *Nature* **190**, 389-390.
Oró, J., and Lazcano, A.: 1998, in P. J. Thomas, C. F. Chyba and C. P. McKay (eds.), *Comets and the Origin and Evolution of Life*, Springer Verlag, pp. 3–27.
Park, C.: *1985, Nonequlibrium Air Radiation (NEQAIR) Program: User's Manual*, NASA TM 86707, July 1985.
Park, C. and Menees, G.P.: 1978, *J. Geophys. Res.* **83**, 4029–4035.
Park, C.S., Newfield, M.E., Fletcher, D.G., Gökçen, T. and Sharma, S.P.: 1997, *AIAA Paper* **97–0990**.
Park, C.S., Newfield, M.E., Fletcher, D.G., Gökçen, T. and Sharma, S.P.: 1998, *J. of Thermophysics and Heat Transfer* **12**, 190–197.
Pepin, R.O.: 1991, *Icarus* **92**, 2–79.
Sagan, C.: 1974, *Origins of Life* **5**, 497–505.
Taylor, A.D., and Elford, W.G.: 1998, *Earth Planets Space* **50**, 569–575.
Whittet, D.C.: 1997, *Orig. Life & Evol. Biosphere* **27**, 249–262.

SEARCH FOR ORGANIC MATTER IN LEONID METEOROIDS

RICHARD L. RAIRDEN

Lockheed Martin Space Sciences Laboratory, Dept L9-42, Bldg 255, 3251 Hanover Street, Palo Alto, California 94304
E-mail: rairden@spasci.com

PETER JENNISKENS

SETI Institute, NASA ARC, Mail Stop 239-4, Moffett Field, CA 94035
E-mail: pjenniskens@mail.arc.nasa.gov

and

CHRISTOPHE O. LAUX

High Temperature Gasdynamics Laboratory, Building 520, Mechanical Engineering Department, Stanford University, Stanford, California 94305-3032
E-mail: laux@saha.stanford.edu

(Received 13 July 2000; Accepted 29 August 2000)

Abstract. Near-ultraviolet 300–410 nm spectra of Leonid meteoroids were obtained in an effort to measure the strong B $\to$ X emission band of the radical CN in Leonid meteor spectra at 387 nm. CN is an expected product of ablation of nitrogen containing organic carbon in the meteoroids as well as a possible product of the aerothermochemistry induced by the kinetic energy of the meteor. A slit-less spectrograph with objective grating was deployed on FISTA during the 1999 Leonid Multi-Instrument Aircraft Campaign. Fifteen first-order UV spectra were captured near the 02:00 UT meteor storm peak on November 18. It is found that neutral iron lines dominate the spectrum, with no clear sign of the CN band. The meteor plasma contains less than one CN molecule per 3 Fe atoms at the observed altitude of about 100 km.

Key Words: Astrobiology, CN, exobiology, Leonids 1999, meteors, meteoroids, origin of life, spectroscopy, ultraviolet

1. Introduction

Refractory organic carbon is thought to be an abundant compound of cometary meteoroids, making up some 23 % of the comet mass fraction and some 66% of meteoroid mass once the volatile compounds have evaporated (Greenberg, 2000). This organic carbon has its origins in accretion processes in the interstellar medium (Jenniskens et al., 1993). The GIOTTO and VEGA spacecraft probes measured the elemental composition of dust grains in the coma of comet 1P/Halley and found many to be rich in the elements C, H, O and N (Kissel and Krueger, 1987; Jessberger and Kissel, 1991). This organic carbon was mixed intimately with the silicate component and has high molecular mass. Most of it is expected to survive exposure to the vacuum of space and the gentle warming when grains come as close to the Sun as Earth's orbit (T_{max} ~ 300 K). When these meteoroids encounter the Earth's atmosphere, the organic carbon is ablated either 1) early on at a relatively cool temperature (400-750 K) in the classical case where the grains are heated throughout and the evaporation is the key ablation process, or possibly 2) together with the silicate grains when the ablation process is like sputtering and thermal conductivity is low (Jenniskens et al., 2000a).

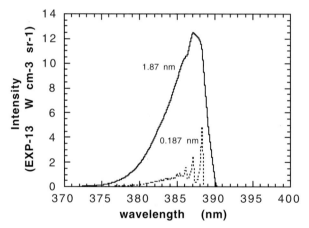

Figure 1. CN B → X emission band in a theoretical spectrum of a 4,300 K air plasma in LTE at 95 km altitude for a total concentration of CN equal to 1.0×10^5 molecules of CN per cubic centimeter. The result is shown for two instrumental resolutions with a Gaussian profile, at FWHM = 1.87 nm, as for our data, and at smaller FWHM = 0.187 nm to bring out the vibrational band structure. The CN 0-0 band head peaks at 388.3 nm, while the 1-1 band is at 387.1 nm.

When organic carbon, containing some fraction of nitrogen and oxygen, is heated to high temperatures, or sputtered by fast ions or energetic photons, small fragments are lost such as H, O, H_2, O_2, OH, H_2O, CO, CO_2 and CH_4 in a process called carbonization (Jenniskens et al., 1993; Fristrom, 1995, p. 318). Further heating (up to 700 K) leads to further loss of H and H_2 in a process of polymerization, with growth of aromatic ring structures, and some graphitization by stacking of those rings (e.g. Koidl *et al.*, 1990). Ultimately, temperature increase leads to decomposition by loss of H_2, CN, CH, C_2, and C_2H. Hydrocarbon radicals higher than C_2 are so excited by their formation that they rapidly fission into lower products, leaving only the thermally stable single and double carbon radicals, unless prevented by rapid radiative cooling. Carbon atoms are present too, but at lower abundance because of thermodynamic considerations.

Of these, the CN radical is the most easily detected because of a strong B → X transition of low energy potential. This electronic transition has a band head at 388.3 nm in the near ultraviolet. Figure 1 shows simulated emissions for a typical meteor wake temperature of 4,300 K (Jenniskens *et al.*, 2000a). The use of two instrumental resolutions in Figure 1 shows how the shape of the band is affected by a lower spectral resolution.

The CN radical is also a product of aerothermochemistry in an N_2 / CO_2 atmosphere, so CN may also be formed from the deposition of kinetic energy of the meteor in the air at altitude.

Unfortunately, the CN band is found in a part of the meteor spectrum that is rich in atomic iron lines from ablated meteoric metals. This demands high-resolution spectroscopy in order to differentiate between iron lines and the CN band. Here, we report on one experiment to study the possible presence of CN in the spectra of Leonid meteors. These results were obtained during the 1999 Leonid meteor storm, from the perspective of FISTA during the Leonid Multi-Instrument Aircraft Campaign (Jenniskens *et al.*, 2000b).

2. Method

The camera consisted of the following parts: An ATE Noctron Image Intensifier with a UV-Nikkor 105 mm f/4.5 objective lens, a 600 lines/mm diffraction grating mounted in front of the objective, and a Nikon 230–410 nm passband filter mounted in front of the grating. The intensifier was optically coupled to a Cidtek CID video camera model 3710-D, and data were stored on Hi-8 video tape cassettes. The camera had a field of view of 8 x 11 degrees and gave a better than 2 nm resolution in first order over the wavelength range of 290 to 410 nm. The objective bandpass filter prevented visible wavelengths from overlapping higher order UV spectra.

The wavelength scale was calibrated in the laboratory. A plot of the relative spectral sensitivity is shown in Figure 2. Sensitivity at the short end of the spectrum is limited by the BK-7 window on the aircraft. For analysis the video images are digitized as 640 x 480 pixel arrays at 8-bit resolution. The first order spectral dispersion at 370 nm was measured to be 0.468 nm per pixel. The instrumental broadening was measured from the camera response to stars and from laboratory measurements of the response to calibration lamps. The FWHM of the instrumental convolution function equals 4 pixels, or 1.87 nm.

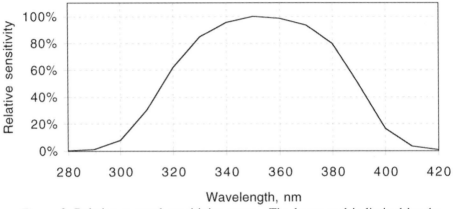

Figure 2. Relative spectral sensitivity curve. The lower end is limited by the BK-7 window, and the upper end by the visible wavelength blocking filter.

3. Results

A total of 15 (partial) spectra were recorded during the peak of the storm. Table I gives their magnitude and time of peak brightness as derived from the zero order imaging by camera FL50F (Jenniskens *et al.*, 2000c), as well as some remarks on the quality of the spectrum. We used co-aligned wide-angle video cameras to estimate the apparent visual magnitude in zero order. Particularly nice spectra were obtained at 02:09:26, 02:36:40, and 03:09:58 UT. Figure 3 shows composite images of several video frames for two of these events. All spectra are similar in appearance. Two strong emission features are visible at the longer wavelengths, while a forest of lines extends to shorter wavelengths. All features are broader than the instrumental response curve and appear to be made up of multiple emission lines. One spectrum (01:54:24 UT) was recorded while the zero order meteor image passed the field of view during part of its trajectory. This enabled us to set the zero point of the wavelength scale. All bright meteors were imaged also with coaligned intensified cameras (Table I).

SEARCH FOR ORGANIC MATTER IN METEOROIDS

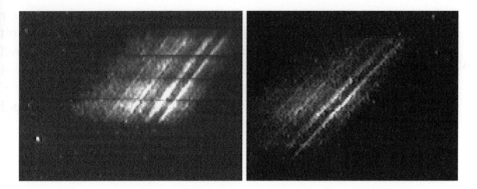

Figure 3. Cropped composite of four video fields for the 02:36:40 UT (left) and the complete 02:09:26 UT (right) meteors. The bright star is beta Cephei. Meteors entered the frame from the upper right. Spectral dispersion is horizontal, with shorter wavelengths toward the left.

TABLE I

Time (UT)	Mv*	Remarks	Time (UT)	Mv	Remarks
01:49:29	0	high altitude only	02:09:26	−2	first order
01:51:57	+1	first order	02:21:13	+1	first order
01:54:24	−1	zero +part first	02:28:04	+1	first, faint
01:55:51	0	part first, long λ	02:34:12	-.-	2nd order, faint
01:57:33	−1	2nd order, faint	02:36:40	−2	first order
02:00:24	0	first order	02:47:57	−1	high altitude only
02:00:35	0	part first, long λ	03:09:58	0	part first, long λ
02:08:28	0	first order			

* Assumed meteor height = 100 km for normalized distance = 100 km (magn.).

The bright features are at 357, 373 and 382 nm and no single emission line is responsible for the features. Indeed, they are broader than the instrumental resolution. There are small changes in the profile of the spectra at different positions along the trajectories that help discriminate between the various contributions in the spectrum. Figure 4 shows two traces of the 02:36:40 meteor. The main features remain at about constant relative intensity. However, features near 345 and 355 nm are relatively stronger in the latter spectrum, while features at 332, and possible the Ca II doublet at 392 and 395 nm are weaker.

The expected Leonid spectrum over the range 350–410 nm was calculated by Jiri Borovicka and is given in Jenniskens *et al.* (1998).

Only transitions with upper states < 5 eV are considered, and only transitions that originate from the wake of the meteor. These emission lines (Fe I, Mg I, Mn I, Cr I, etc.) are characterized by an electronic excitation temperature of T ~ 4,500 K. The T ~ 10,000 K component of the meteor head plasma (Ca II, Fe II, Si II) is not evident in the spectra. We convolved each line with the instrument point spread function and derived a synthetic spectrum over this wavelength range (Figure 5).

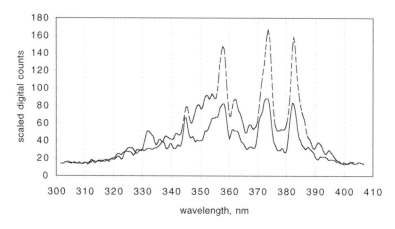

Figure 4. Spectral profiles of 02:36:40 UT meteor at two different altitudes. The brighter profile is at lower altitude.

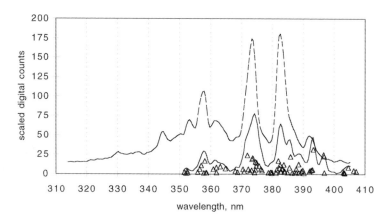

Figure 5. The observed Leonid spectrum (mean of 02:36:40 and 02:09:26 UT – upper curve) compares well with a synthetic spectrum (lower curve) produced by convolution of the instrument spectral resolution with the expected atomic lines (mostly Fe). Triangle symbols indicate individual line strengths in the model. Calcium lines appear weak in the observed spectra.

The strong features are readily identified as due to clusters of numerous Fe emission lines. There are some disparities. The 382 nm cluster is stronger in the observations than the 373 cluster. Also, there is strong diffuse emission observed between 330 and 370 nm that is not matched by the synthetic spectrum. Figure 6 shows the difference between synthetic and a mean of all observed meteor spectra. The meteor head plasma Ca II lines in the +0 magn. spectrum published in Jenniskens *et al.* (1998) are too strong in comparison to the observations, suggesting that there is less contribution from the meteor head than expected.

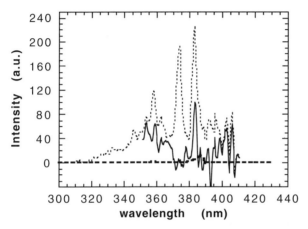

Figure 6. Difference between synthetic and observed spectrum. The dotted line is observed spectrum, while the solid line is observed minus synthetic. The dark dashed line shows the highest possible CN contribution.

The broad emission at short wavelength is the most prominent feature. It may have a band reversion at about 370 nm and stretches to 320 nm. The origin of this emission remains unknown. Excess emission at 382 nm is also clearly present, overlapping one of the Fe bands, the source of which is unknown also. The 382 nm excess is not caused by CN emission. Figure 1 shows that the expected CN emission band at our instrumental resolution should peak at about 389 nm.

4. Discussion

4.1. Abundance of CN in the Meteoric Plasma

Fortunately, the CN band head coincides with a minimum of Fe line density (Figure 5). The CN emission would have been detected if the integrated emission was stronger than 9% of the integrated Fe I line intensity between 370 and 375 nm (a 9% contribution would be the dark dashed line in Figure 6). This part of the spectrum is thought to be exclusively due to Fe I line emission and will be used for calculating the upper limit for the CN production per Fe atom in the meteor wake.

We calculated the Fe emission intensities using all neutral iron lines in the NIST Atomic Spectra Database Version 2.0 (NIST Standard Reference Database #78). The CN band emission was calculated using the NEQAIR2 model of heated air in thermodynamic equilibrium (Park, 1985; Laux, 1993). With ICN the total integrated emission intensity of CN in the range 375–393 nm, and IFe the total integrated emission intensity of Fe in the range 370–376 nm, we find:

$$IFe/nFe = 2.16 \times 10^{-16} \text{ W/sr}$$
$$ICN/nCN = 5.58 \times 10^{-17} \text{ W/sr}$$

where nFe is the total concentration of Fe atoms in cm^{-3} and nCN is the total concentration of CN molecules per cm^{-3}. The above values are for T = 4,500 K. For T = 4,300 K, the numerical factors are to be replaced by 1.42×10^{-16} for Fe and 3.93×10^{-17} for CN. These are concentrations per total Fe atoms and total CN molecules (that is, not per molecule of CN in the excited electronic B state). The calculations assume that the emitting levels are in Boltzmann equilibrium with the ground state of Fe and CN at the given temperature.

From this, we have calculated the abundance of CN atoms per Fe atom in the meteor plasma for given ratio of measured features:

$$[CN]/[Fe] = 3.88*(ICN/IFe) \text{ at } 4,500 \text{ K}$$
$$[CN]/[Fe] = 3.60*(ICN/IFe) \text{ at } 4,300 \text{ K}$$

where [CN] and [Fe] denote the concentrations or total number of atoms of CN and Fe. Hence, if the integrated emission intensity of CN is less than 9% of the integrated emission of Fe, then our upper limits are [CN]/[Fe] < 0.35 at 4,500 K and [CN]/[Fe] < 0.32 at 4,300 K. Thus, less than 1 CN molecule is present per three Fe atoms in the meteor plasma.

4.2. EXPECTED ABUNDANCE FROM METEOROID ABLATION

Comet Halley's dust has a nitrogen abundance of [N]/[Fe] = 0.79 ± 0.02 (Delsemme, 1991). All of that nitrogen is part of the organic carbon. We expect that all of the nitrogen is released as CN. In that case, our upper limit is strong enough to say something significant about the organic matter in meteoroids.

This is with the caveat that some of the Fe may not be released as atomic iron atoms, but stay in a solid form in the form of meteoric debris. That would imply even lower CN abundances. Less likely is that some organic carbon may have been lost at high altitudes, where the excitation temperatures are low enough to cause relatively faint luminosity. Possibly, a fraction of nitrogen is also released as NH and N_2. Clearly, other techniques at detecting organic matter in general, or nitrogen in particular, have to be explored to address this possibility.

4.3. EXPECTED ABUNDANCE FROM LTE AEROTHERMOCHEMISTRY

The −2 magnitude Leonid meteors studied here have a mass of about 0.44 gram, of which 0.033 gram is iron (Delsemme, 1991). That iron is ablated and distributed over a path length of about 17 km (Betlem et al., 2000), in a cylindrical al volume of at least 2 meter radius (Jenniskens et al., 2000a). That gives a mean density of [Fe] = 2 x 10^9 atoms / cm^3. The calculated CN abundance for a 4,300 K air plasma in LTE at 95 km altitude for a total concentration of CN equals 1.0x10^5 molecules of CN per cubic centimeter. Hence, the expected [CN]/[Fe] is about 5 x 10^{-5}.

At first sight, this implies that our observations do not put a strong upper limit on the efficiency of aerothermochemistry in the meteor plasma. However, the effective volume of air that may be processed directly or indirectly by the kinetic energy of the meteor may well be much larger than suggested above. In fact, Jenniskens et al. (2000a) found an effective volume 10^6 times larger from the intensity of OI line emission. This would increase the [CN]/[Fe] to 50 atoms / cm^3, significantly above our detection limit. It is crucial, therefore, to measure the effective volume of air processed by the meteors in a direct manner in order to measure the efficiency of the aerothermochemistry process.

5. Conclusions

CN atoms are detected at less than 1 per 3 Fe atoms in the plasma of 1899 ejecta Leonid meteors at altitudes of about 100 km. This implies that either the Leonid meteoroids contain less than half the organic carbon that is present in comet Halley's dust, or a significant fraction of the nitrogen is not released in the form of CN radicals.

Acknowledgements

The camera, UV intensifier and lens were supplied by S. B. Mende of the Space Science Laboratory at University of California, Berkeley. Preparations and analysis were supported in part by Lockheed Martin internal research funds. The 1999 Leonid MAC was sponsored by the NASA Exobiology, Planetary Astronomy, and Suborbital MITM programs, by the Advanced Missions and Technologies for Astrobiology Program, and by NASA Ames Research Center. Aircraft flight time was supported by the USAF/XOR. *Editorial handling:* Mark Fonda.

References

Betlem, H., Jenniskens, P., Spurny, P., Docters van Leeuwen, G., Miskotte, K., Ter Kuile, C.R., Zerubin, P., Angelos, C.: *Earth, Moon and Planets* **82–83**, 277–284.
Borovicka, J., Stork, R., and Bocek, J.: 1999, *Meteoritics Planet. Sci.* **34**, 987–994.
Delsemme, A.H.: 1991, in R.L. Newburn, Jr., M. Neugebauer, J. Rahe (eds), *Comet in the Post-Halley Era,* **1**, 377–428.
Fristrom, R.M.: 1995, *Flame Structure and Processes.* NY: Oxford University Press. 510 pp.
Greenberg, J.M.: 2000, Earth, *Moon and* Planets **82–83**, 313–324.
Jenniskens, P. and Butow, S.J.: 1999, *Meteoritics Planet. Sci.,* **34**, 933–943.
Jenniskens, P., Baratta, G.A., Kouchi, A., de Groot, M.S., Greenberg, J.M., and Strazzulla, G.: 1993, *Astron. Astrophys.,* **273**, 583–600.
Jenniskens, P., de Lignie, M., Betlem, H., Borovicka, J., Laux, C.O., Packan, D., and Krüger, C.H.: 1998, *Earth, Moon and Planets,* **80**, 311–341.
Jenniskens, P., Packan, D., Laux, C., Wilson, M., Boyd, I.D., Kruger, C.H., Popova, O., and Fonda, M.: 2000a, *Earth, Moon and Planets* **82–83**, 57–70.
Jenniskens, P., Butow, S.J., and Fonda, M.: 2000b, Earth, Moon and Planets **82–83**, 1–26.
Jenniskens, P., Crawford, C., Butow, S.J., Nugent, D., Koop, M., Holman, D., Houston, J., Jobse, K., Kronk, G., and Beatty, K.: 2000c, *Earth, Moon and Planets* **82–83**, 191–208.
Jessberger, E.K. and Kissel, J.: 1991, in R. Newburn, M. Neugebauer, J. Rahe (eds.), *Comets in the Post-Halley Era,* Kluwer, Dordrecht, Vol. 2, 1075–1092.
Kissel, J., Krueger, F.R.: 1987, *Nature* **326**, 755–760.
Koidl, P., Wild, Ch., Dischler, B., Wagner, J., and Ramsteiner, M.: 1990, *Materials Science Forum,* **52–53**, pp. 41–70.
Laux, C.O.: 1993, *'Optical Diagnostics and Radiative Emission of High Temperature Air Plasmas',* Ph.D. Dissertation, HTGL Report T288, Stanford University, CA.
Park, C.: *1985, Nonequilibrium Air Radiation (NEQAIR) Program: User's Manual,* NASA TM 86707, July 1985.

OBSERVATIONS OF LEONID METEORS USING A MID-WAVE INFRARED IMAGING SPECTROGRAPH

GEORGE S. ROSSANO, RAY W. RUSSELL, DAVID K. LYNCH,
TED K. TESSENSOHN, AND DAVID WARREN

The Aerospace Corporation, M2/266, 2350 E. El Segundo Blvd., El Segundo, CA 90245
E-mail: george.s.rossano@aero.org

and

PETER JENNISKENS

SETI Institute, NASA Ames Research Center, Mail Stop 239-4, Moffett Field, CA 94035
E-mail: pjenniskens@mail.arc.nasa.gov

(Received 7 July 2000; Accepted 20 July 2000)

Abstract. We report broadband 3-5.5 µm detections of two Leonid meteors observed during the 1998 Leonid Multi-Instrument Aircraft Campaign. Each meteor was detected at only one position along their trajectory just prior to the point of maximum light emission. We describe the particular aspects of the Aerospace Corp. Mid-wave Infra-Red Imaging Spectrograph (MIRIS) developed for the observation of short duration transient events that impact its ability to detect Leonid meteors. This instrument had its first deployment during the 1998 Leonid MAC. We infer from our observations that the mid-wave IR light curves of two Leonid meteors differed from the visible light curve. At the points of detection, the infrared emission in the MIRIS passband was 25 ± 4 times that at optical wavelengths for both meteors. In addition, we find an upper limit of 800 K for the solid body temperature of the brighter meteor we observed, at the point in the trajectory where we made our mid-wave IR detection.

Keywords: Ablation, Leonids 1998, meteors, meteoroids, mid-IR emission,

1. Introduction

The Aerospace Corporation Mid-wave Infra-Red Imaging Spectrograph (MIRIS) was initially developed as a proof of concept instrument to test the feasibility of producing a mid-wave infrared spectrograph using a binary-optic grism (a transmission grating produced on a shallow angle prism) fabricated using semiconductor etching technology as the dispersing element (Warren *et al.*, 1994; 1998). Having successfully demonstrated the practicality of using such a grism as a dispersing element, MIRIS was developed into an operational instrument which was optimized for the study of transient events with a time resolution as short as 1 msec, making MIRIS uniquely suited for mid-wave IR observations of meteors. The first fully operational use of MIRIS was to observe the Leonid meteor storm as part of the 1998 Leonid Muli-instrument Airborne Campaign (Jenniskens and Butow, 1999). We describe here the observations made during the intense 1998 Leonid shower in which two broadband detections of Leonid meteors were obtained. As far as we are aware, these detections are the first published mid-wave IR observations of meteors.

Figure 1. MIRIS installed on FISTA during the Leonid meteor observations. Intensified visible video camera (A) mounted on top provides visible reference images of the field of view. Rectangular vacuum case (B) contains cooled reimaging and dispersing optics. F/15 Objective lens (C) is mounted outside the vacuum case. InSb infrared camera head (D) is mounted on the left side of the case. 228 mm f/15 objective lens at front of case views the sky through an observing port in the aircraft skin.

2. Instrumentation

2.1. INSTRUMENT LAYOUT

MIRIS (Figure 1) consists of three modules: uncooled refractive collecting optics; cooled reimaging optics; and an InSb detector array system (camera head). The camera head and reimaging optics can be cooled independently of each other and the system can be operated with warm reimaging optics for alignment and testing.

The objective lens is made up of two Si elements and one Ge element with an effective focal length of 228 mm and a speed of f/15. All lens elements in the system are anti reflection coated for the 3.0–5.5 μm band pass of the instrument.

The reimaging optics and grism are located in a rectangular LN_2 dewar and are cooled to an operating temperature of 150 K. This temperature was chosen to reduce the background from the reimaging optics to a manageable level without fully stressing the grism to the extent it would have been by cooling it to 77 K.

The camera lens focal plane lies 1 cm inside the reimaging optics dewar and is illuminated through a 3 in diameter antireflection coated ZnSe window. The reimaging optics reduce the image by a factor of 6 to produce an effective system focal length of 38 mm at the detector focal plane. The objective lens can be removed and other collecting optics (e.g., an astronomical telescope) can be used instead.

The camera head is a modified Amber Model 4256 InSb infrared camera. The final element of the reimaging optics and the system's 3.0–5.5 μm bandpass filter are located within the Amber LN_2 dewar and are cooled to 77 K, as is the detector array. The detector array in the camera head is a 256 by 256 element array with 38 μm pixels.

The optics are configured so both the zero and first orders are imaged onto the detector array. The images are optimized for first order so that the spectrum of a point source will illuminate a single row of the detector array in first order. Residual aberrations in the zero order image produce a point spread function that illuminates four (2 by 2) pixels. The full unvignetted field of view overfills the detector array in both zero and first order. The optics are configured so that the zero order is mainly imaged onto the right hand side of the array and the first order spectrum is mainly dispersed across the array on the left hand side of the array.

In this slitless mode of operation some sources seen in zero order will be dispersed entirely off the left hand side of the array, and some sources in zero order imaged entirely off the right hand side of the array will be dispersed onto the middle of the array. In other words, while for a large portion of the detector the zero order images of sources and their first order spectra are imaged onto the detector array, for some positions only the zero order image falls on the detector while for other positions only the first order spectrum falls on the detector. In slitless mode the unvignetted field of view is a 12 degree diameter circle. This mode of operation was used for the 1998 Leonid meteor observations to maximize the chances of a meteor being observed. The slitless mode of operation is intended for the observation of point sources and line sources. Extended sources can also be observed, but with degraded spectral resolution depending on the width of the source in the dispersion direction. Both the spectral and spatial resolution is limited by the pixel size. For 38 μm pixels, the spectral resolution is 0.025 μm and the spatial resolution is 1 mrad. The measured single pixel system sensitivity using the model 4256 camera head in zero order is 1.1×10^{-14} W for a signal-to-noise of 1 in 1 second.

Observations during the 1998 Leonid meteor storm were made aboard the Flying Infrared Signatures Technology Aircraft (FISTA), a modified KC-135E aircraft, as part of the 1998 Leonid MAC. MIRIS was located on a steerable tray affixed to one of the aircraft's viewing windows (Jenniskens and Butow, 2000). The window material through which the observations were made consisted of a 1 cm thick zinc sulfide plate. The instrument was mounted in such a way that the majority of the meteor trails were oriented roughly perpendicular to the dispersion direction.

2.2. DATA ACQUISITION

Data from the camera controller electronics were acquired using a commercially available frame grabber. The data acquisition software was developed at the Aerospace Corporation. The maximum operating speed of the model 4256 camera is 60 Hz resulting in a data rate of 7.5 MB s^{-1}.

Four data acquisition modes are available: snapshot, continuous, burst, and transient. In the snapshot mode a single frame or single coadded group of frames is recorded under user control. In continuous mode a sequence of individual frames or a sequence of coadded frames is recorded under user control. For very high speed data capture the burst mode allows the capture of 64 MB of data at rates of up to 48 MB per second. User triggered transient events can be captured using the transient mode. A cyclic buffer is used allowing pre- and post-event data buffering. When a transient event occurs the user triggers the capture of a sequence of data frames. The user can select the amount of data (number of frames) before and after the trigger that is recorded.

For the Leonid observations we operated in the continuous mode with the data saved to disk in 3 minute blocks. The detector array was operated at a frame rate of 60 Hz and an integration time of 10 msec or less – depending on the background signal level which varied somewhat during the duration of the flight. Data recording was limited at that time to the use of 2 GB JAZ media and the captured data rate was limited by the speed of the JAZ drive. Sequences of individual frames were recorded at a rate of 16.7 Hz; slightly better than every fourth frame from the detector array.

2.3. CALIBRATION

For radiometric calibration we use two different types of blackbody sources - warm plates and hot cavities. We simulate a point source using a blackbody cavity with pinholes of varying sizes. In addition to providing a radiometric calibration, this source is used for focus and alignment checks, and in one method of wavelength calibration.

Warm plate blackbodies that illuminate the full detector array are used for flat fielding and radiometric calibration. Even though in slitless mode the zero order and first order images overlap for parts of the array, by knowing how each order is mapped onto the array we can radiometrically calibrate the entire detector array. These plates fill the entire field of view and are typically operated at temperatures of 300–360 K. The Leonid data was radiometrically calibrated using measurements of the warm plates made during each flight.

For guiding and visual comparison an intensified video camera was mounted on top of MIRIS and its output videotaped for the duration of the flight. The relative position of the fields was roughly calibrated by pre-flight observation of distant runway and hanger lights.

The MIRIS data files were time-stamped only once at the beginning of each file, which in retrospect was a poor choice. In examining the data subsequent to the mission it was necessary to get an event time off the video tape and then search the corresponding data file over an appropriate range of frames.

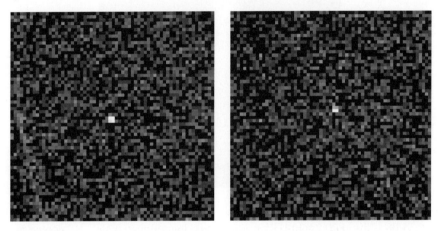

Figure 2. Enlargement of two Leonid meteor detections. Meteor from 17:54:08 UT on the left, and from 20:24:40: UT on the right, Nov. 17, 1998. Scale for each image is 0.064 radians on a side. Intensity is displayed in absolute numbers on a logarithmic scale in order to bring out possible wake. The detections are point sources, with no sign of persisting glow.

3. Results

The three mission video tapes, covering a six hour period, were each examined twice to obtain a record of the time of each meteor seen and an estimate its visible brightness, viz. faint, bright, or very bright. A total of 56 meteors were tabulated that included five very bright meteors including two that occurred while IR data were being recorded. After searching the MIRIS data files for possible detections, we identified one point source at the correct time at 17:54:08 UT. This source was seen in one frame and illuminated four pixels (Figure 2), consistent with the appearance of point sources when observed in zero order by this instrument. An integrated intensity of 9.0×10^{-12} W cm^{-2} in the 3.0–5.5 µm bandpass was determined for this meteor. Among the meteors comsidered bright for which we have examined the data, one other IR detection was made at 20:24:40 UT. This meteor was also only detected in one frame in zero order and illuminated four pixels with an integrated intensity of 5.0×10^{-12} W cm^{-2} in the 3.0–5.5 µm bandpass (Figure 2). The nominal error for the raw signal counts is 2–3 %. When taking into account the radiometric calibration uncertainty, the total uncertainty in measured brightness is 15%.

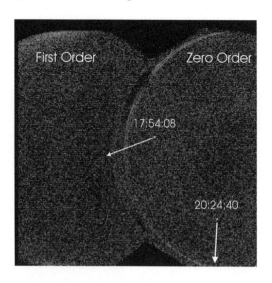

Figure 3. Composite MIRIS image of two Leonid meteor detections at 17:54:08 UT and 20:24:40 UT on a intensity logarithmic scale. By comparing this infrared image to visible images the position of each meteor along its trajectory at the time of the infrared observations was determined. The mid-wave IR detections are the point sources at the ends of the arrows that illustrate the directions of motion determined from the optical images of the meteors. The circular region on the right hand side of the image is the field of view of the instrument in zero order. Infrared radiation is dispersed to the left in first order. The orders overlap in the center of the detector array.

In order to confirm the detections we combined both MIRIS detections into one infrared image (Figure 3) and combined all video frames of the two meteor tracks in one visible image. These two composite images were then overlain adjusting for the different image scales which are both well determined. If the infrared sources are due to meteors then both detections must line up on their corresponding visible tracks in a unique orientation. Such was found to be the case, confirming the detections, determining the points along the trajectories where the infrared detections were made, and allowing a direct comparison of the infrared and visible brightnesses.

Figure 3 shows the composite infrared image and the direction of motion of each meteor determined from the visible trail.

Both meteors were also observed from the accompanying Electra aircraft, which provided multi-station support for stereoscopic measurements (Jenniskens and Butow, 1999). From the FISTA and Electra video records, the trajectories of both meteors were determined with good convergence angles. Both were shown to be Leonid meteors. The different angle of projection on the composite image (Figure 3) is due to different observing directions at the two times.

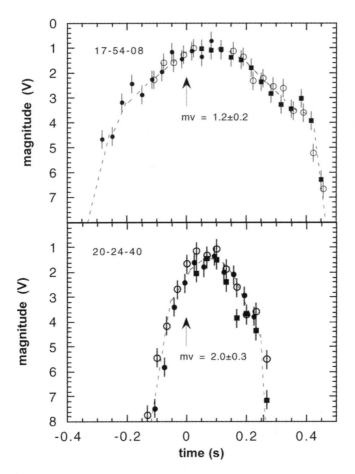

Figure 4. Visible light curves for two Leonid meteors at 17:54:08 UT and 20:24:40 UT detected in the infrared, as determined from visible imagery. Visible magnitude (mv) is plotted versus time, referenced to the time of the infrared observations of each meteor. The magnitude of each meteor at the time of the infrared detections is indicated. The circles, dots and filled squares are the meteor brightnesses measured using three different reference stars in the field of view.

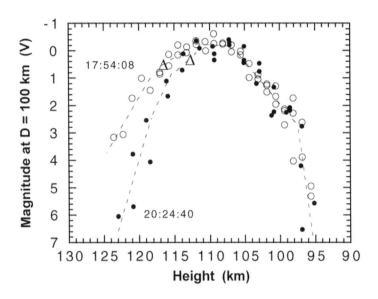

Figure 5. Visual magnitude vs. Height plot for two Leonid mid-IR meteors. Magnitude normalized to a distance of 100 km is plotted vs. height for each meteor. The meteor with the more shallow trajectory (17:54:08 UT) reached a higher brightness early in its trajectory. Below 112 km altitude both meteors were equivalent in absolute brightness vs. altitude. Open circles are the measured visible brightnesses for the 17:54:08 UT meteor while closed circles are for the 20:24:40 UT meteor. The two triangles show the points in the trajectory where the infrared detections were made.

The visible light curves for the two meteors was derived from the video record (Figure 4). The brighter of these two meteors had a visible brightness of +1.2 ± 0.2 magnitude at the time of the IR detection while the fainter meteor had a visible brightness of +2.0 ± 0.3 magnitude. This corresponds to an average 1.24 ± 0.25 x 10^{-16} W cm^{-2} Å^{-1} and 0.59 ± 0.13 x 10^{-16} W cm^{-2} Å^{-1} over the passband of the intensified cameras. The video spectral response is given in Jenniskens (1999), from which we adopt an effective band width of 3161 Å. Hence, meteors emitted 23 ± 5 and 27 ± 6 times more light in the the MIRIS 3.0–5.5 μm passband than at visible wavelengths at the point of detection.

The meteor that occurred later during the same night at 20:24:40 UT had a shorter trajectory mainly because it entered the atmosphere at a steeper trajectory. Both light curves had a similar shape with a relatively early maximum, a gradual decline that was followed by a rapid decrease of intensity at the end of the trajectory. These light curves resemble those for other Leonid meteors observed on Nov. 17, 1998 as part of the Leonid MAC campaign (Murray *et al.*, 1999). Using the stereoscopic data it is possible to calculate the light curve as a function of altitude (Figure 5).

The light curves are nearly identical below 112 km, but differ early in the trajectory where the more shallow meteor reaches a higher brightness sooner. Surprisingly, the infrared detection of these two meteors were made at a similar altitudes of 117 and 113 km (triangle symbols in Figure 5), just before the maximum in the visible light curve.

4. Discussion

4.1. SINGLE SOURCE DETECTION

The MIRIS frames for both meteors are shown in composite form in Figure 3, with the directions of motion for the two meteors indicated. Given the observed motions of the meteors on the visible videotape (corresponding to approximately 30 infrared pixels of motion per infrared frame), both meteors were still positioned within MIRIS's field of view on the infrared frames following the detection frames, although meteor 17:54:08 UT was only barely so. No hint of detections were seen on either of those frames. For meteor 17:54:08 UT, however, the expected location in the frame subsequent to the detection frame was near the transition from zero to first order where the noise is significantly greater, which may have masked detection in that frame.

If the time development of the infrared light curve followed the trend of the visible light curve then the 20:24:40 UT meteor might have been detected in at least one frame prior to the detection frame, while the 17:54:08 UT meteor might have been detected in at least two frames before being too faint. Tracing back the paths of the two meteors, no detections were seen on any of the preceding frames.

The observations of these meteors in only a single frame, we believe, is partially explained by the high speed motion of the meteors, the low data recording rate, and the particular read out method of the detector array. In the model 4256 camera, integration times shorter than the frame rate are implemented by reducing the number of pixels integrating at any time and progressively scanning across the detector array. For most of our observations the camera was set up with an integration time of about one half the frame time or less, with the meteor trails oriented roughly perpendicular to the detector rows. This means that for most of our observations less than one half the detectors were integrating at any time. In that situation a source moving across rows has a considerable chance of never – or only rarely – illuminating an integrating pixel. However, this is not true for a meteor that moves in a direction generally parallel to the rows, such as meteor 17:54:08 UT.

This results implies that the infrared light curves have faster rise and/or decay times than the visible light curves resulting in a briefer time interval over which meteors are bright enough to be observed in the mid-wave infrared with this instrument – perhaps in the range of 300 msec or less.

If the optical luminosity of meteors is uniform for all meteors with a typical value of 1% the total energy (Jacchia et al. 1967), our results imply that about 25% of the emitted intensity is released in the 3.0–5.5 micron passband at the point in the development of the meteors where we made our detection.

4.2. METEOR SPECTRUM

Why did we not detect the first order spectrum of these meteors?

The brighter of the two meteors was observed near the edge of the zero order field of view in the dispersion direction. For that position in the focal plane, the majority of the associated first order spectrum is located off the detector array with only the 3–4 μm wavelength region falling on the array. If the energy detected in zero order were evenly spread out over the 100 pixels that cover a 3–5.5 μm spectrum then a first order spectrum should have been observed at a signal-to-noise ratio of 3. Since we did not detect a signal in the 3–4 μm section of the spectrum we infer that most of the energy observed in zero order lies in the range 4–5.5 μm.

The second meteor we detected was located at a position in the field of view for which the associated first order spectrum was located on the array. For this fainter meteors the anticipated signal-to-noise ratio in first order would have been less than 2 and thus the failure to detect a spectrum is not surprising. Were the energy from these sources primarily in the form of line emission we would have expected a clear first order detection.

For the 4–5.5 μm emission to exceed the 3–4 μm emission, a blackbody temperature of 600 - 700 K is implied. For that temperature range, the IR-to-visible ratio is very much greater than the factor of 25–30 observed. The visible emission of meteors, however, is not due to blackbody emisison. Our observations are consistent with the IR emission coming from the ablating meteoroid itself and imply a solid object at a temperature of < 800 K at 115 km altitude but with the visible emission coming from the plasma in the meteor's head and wake.

For at least one of the two meteors we cannot attribute the lack of a detection in the pre-detection frame to the camera timing issue noted above. For the 17:54:08 UT meteor we attribute the lack of a detection in the pre-detection frame to the meteoroid emitting a fainter signal due to a lower meteoroid temperature. Based on the measured noise for the pre-detection frames we derive an upper limit of 450 K for the temperature of this meteoroid 60 msec prior to our detection when the object was at an altitude of 120 km.

5. Conclusions

The two broadband detections reported here are the first published radiometrically calibrated mid-wave infrared detections of meteors. These detections place constraints on the brightness, spectral characteristics, and time duration of the mid-wave infrared radiation from Leonid meteors. The mid-wave infrared light curve differs from that at optical wavelengths, with significant emission just prior to the peak of the visible light curve, where the luminosity is 25 ± 4 times that at optical wavelengths. From the lack of a first order spectral record in the data, we conclude that the 3.0–5.5 µm emission is primarily a broadband signal with the majority of the emission appearing longward of 4 µm and an upper limit of 800 K for the temperature of the solid body.

The two meteors detected first became visible on the intensified video camera at an altitude of 124 km. At an altitude of 120 km the temperature of at least one of these objects was less than 450 K. By the time the meteoroids reached an altitude of 115 km their temperatures increased to likely values of 600 - 700 K, and were certainly no greater than 800 K. At this point in their trajectories they were still solid objects. Sixty msec later, by which time their visible trails reached maximum brightness, neither meteor was detected in the infrared. We cannot determine if this was due to an actual decrease in the infrared emission from the meteoroids (which one might expect from a decrease in meteoroid surface area as the object ablates), location of the meteoroid signal on detectors where the background noise was higher than elsewhere masking its presence, or instrumental effects related to the detector array timing.

Study of the time development of the infrared signal of similar meteors will require continuous time coverage, preferably at higher frame rates than those used for these observations. The need for such observations was one motivation for the installation of a new Radiance HS camera head added to the system since the 1998 Leonid observations. This new camera head operates in a true snapshot mode, at higher frame rates, and integrates all pixels simultaneously for all integrations times.

Acknowledgments

This work has been supported at the Aerospace Corporation by the Mission-Oriented Investigations and Experiments program, and the Internal Research and Development program. Many Aerospace staff members assisted in making the success of these observations possible. Chief among these were R. Young and M. Ben-Ami who provided key technical support for deployment of the instrument on FISTA. J. Kristl at SDL and his team including T. Hudson, and S. Nierman from Hanscom AFB were generous in providing hardware and technical expertise. U.S. Air Force, MSgt. M. Padilla, Capt. K. Thompson (Edwards AFB) and TSgt. J. Turner (Kadena AFB) offered invaluable critical support. The '98 Leonid MAC was supported by NASA's Exobiology and Planetary Astronomy programs, the NASA Ames Research Center and the NASA

Advanced Missions and Technologies Program for Astrobiology. *Editorial handling*: Frans Rietmeijer.

References

Jacchia, G.J., Verniani, F., and Briggs, R.E.: 1967, *Smithson. Contrib. Astrophys.* **10**, 1–139.
Jenniskens, P.: 1999, *Meteoritics Planet. Sci.* **34**, 959–968.
Jenniskens, P. and Butow, S.J.: 1999, *Meteoritics Planet. Sci.* **34**, 933–943.
Murray, I.S., Hawkes, R.L., and Jenniskens, P.: 1999, *Meteoritics Planet. Sci.* **34**, 949–958.
Warren, D.W., Hackwell, J.A., Brames, B.J., and Skinner, W.J.: 1994, in D.L. Crawford (ed.), *Proc. SPIE, vol. 2198, Instrumentation in Astronomy VIII*, [SPIE, Bellingham, WA], pp. 479–486.
Warren, D.W., Hayhurst, T.L., Rossano, G.S., Hackwell, J.A., and Russell, R.W.: 1998, in A.M. Fowler (ed.), *Proc. SPIE, vol. 3354, Infrared Astronomical Instrumentation*, [SPIE, Bellingham, WA], pp. 168–177.

COMPUTATION OF ATMOSPHERIC ENTRY FLOW ABOUT A LEONID METEOROID

IAIN D. BOYD
Department of Aerospace Engineering
University of Michigan, Ann Arbor, Michigan, USA
E-mail:iainboyd@engin.umich.edu

(Received 15 May 2000; Accepted 7 August 2000)

Abstract. The flow field is computed around a 1 cm diameter Leonid meteoroid as it enters the Earth's atmosphere at an altitude of 95 km and a velocity of 72 km/s. These conditions correspond to a Knudsen number of 4 and a Mach number of 270. To accurately compute the gas flow, these extreme nonequilibrium conditions require application of a kinetic approach and the present work employs the direct simulation Monte Carlo method. A meteoroid ablation model is included in the computations and is found to play a significant role. The computational results predict that a large region of the flow field is affected by meteoroid ablation that produces an extended wake at high temperature in a state of thermal equilibrium. These findings are in qualitative agreement with spectroscopic observations of the 1998 Leonid meteoroid shower. The computations indicate that the results are sensitive to the material properties assumed for the meteoroid.

Keywords: Ablation, atmospheric entry, meteoroid, rarefied gas dynamics

1. Introduction

Annual meteoroid showers such as the Leonids and the Perseids are of interest for a variety of reasons. It has been proposed that comets and meteoroids entering the Earth's atmosphere carried with them extraterrestrial chemical elements that were required to develop life on Earth. In addition, the passage of high velocity meteoroids through the atmosphere is believed to affect the overall aerothermochemistry. Also, there has been concern over the possible interference of Earth orbiting satellites by the increased rate of meteoroid impacts on the spacecraft that occurs during the showers. It is estimated that the Leonid showers of 1998 and 1999 were at the peak of the 33 year cycle, and no adverse effects have been reported for any spacecraft. However, the showers have raised the general issue of the need to understand this type of potential hazard.

As part of the international scientific study of the Leonids, NASA and the United States Air Force, in a collaborative effort, undertook two airborne flight experiments (one each in 1998 and 1999) called the Leonid Multi-Instrument Aircraft Campaigns (Leonid MAC). Details of these experiments are summarized by Jenniskens and Butow (1999).

One of the instruments in the 1998 study was a high resolution slitless CCD spectrograph. The data from this instrument were surprising. The spectra were characterized by relatively high temperatures (about 4,300 K) based on both atomic lines and rotational band structure of molecules (Jenniskens et al., 2000). It was surprising under the nonequilibrium conditions associated with entry of a Leonid meteoroid that the flow should be characterized by an elevated temperature in a state of thermal equilibrium.

In an attempt to help to understand these spectral observations, it is the primary goal of this study to compute the two dimensional flow field about a typical Leonid meteoroid entering the Earth's atmosphere. It will be found that the flow conditions provide very rarefied, hypersonic flow, and so the numerical method employed is the direct simulation Monte Carlo (DSMC) technique (Bird, 1994). This is believed to be the first attempt to apply the DSMC technique to a computation of this type. In the paper, the flow conditions and the properties of the meteoroid are first described. Then a brief description of the DSMC method and code employed are provided. Results of three different computations are presented and discussed. Areas where further work is needed are also considered.

2. Model of A Leonid Meteoroid

The physical properties of meteoroids in the Leonid shower are not well known. Here, we consider a representative case of a spherical meteoroid of diameter 1 cm, entering the Earth's atmosphere at an altitude of 95 km at a speed of 72 km/s. The standard atmospheric conditions at this altitude are as follows: number density of $N_2 = 2.8 \times 10^{19}$ m^{-3}, number density of $O_2 = 7.0 \times 10^{18}$ m^{-3}, number density of O $= 4.9 \times 10^{17}$ m^{-3}, temperature = 176 K. These conditions give a free stream Knudsen number (the ratio of mean free path to body diameter) of about 4, and a Mach number (the ratio of meteoroid velocity to the speed of sound) of 270. The high Knudsen number indicates that the gas flow around the meteoroid is almost collisionless. The high Mach number (this is ten times higher than that of the Space Shuttle as it enters the atmosphere) indicates that any collisions (either gas-gas or gas-surface) are extremely energetic. This clearly represents a very strong nonequilibrium flow condition. It is to be emphasized that this set of parameters is taken to be representative. In the showers, there are variations in meteoroid size, altitude, and velocity. The present investigation represents an initial attempt to see if a computational analysis of the flow field can provide any useful information.

Table I

Meteoroid Type	M_{meteor} (kg/kg-mol)	Q (kJ/g)
H-Chondrite	23.6	6.3
Comet (Halley)	8.26	3.8

It will be found in the computations that including ablation of the meteoroid due to interaction with the atmosphere is significant. This process can be modeled in a simple way following the approach described by Bronhsten (1983). In this model, the number of surface material atoms evaporated due to impact by an air molecule is given by:

$$N_{meteor} = \frac{M_{air} V_{air}^2}{2 M_{meteor} Q} \quad (1)$$

where M_{air} and V_{air} are the mass and velocity of the impacting air molecule, respectively, M_{meteor} is the mass of the meteoroid particle evaporated, and Q is the heat of vaporization. The parameters M_{meteor} and Q depend on the meteoroid material. There is uncertainty concerning the material of Leonid meteoroids. In this study, two different types of meteoroid are considered: (1) an H-chondrite meteoroid; and (2) a comet-like meteoroid (data from comet Halley are employed here). Leonids are generally assumed to consist of comet-like material. However, small Leonids may be depleted of volatiles due to solar irradiation, and in this case their composition may be closer to the chondritic one. These two meteoroid materials differ in their chemical composition giving rise to differences in average particle mass and heat of vaporization. The values used in this study are obtained from Jaroscewich (1990) and Jessberger, et al. (1988), and are listed in Table I.

In the flow field computation (described in detail below), each time a particle of any species impacts on the surface of the meteoroid, a number of vaporized particles given by Eq. 1 is introduced into the flow field from the meteoroid surface. The meteoroid vapor is assumed to be a single species with the average mass of the material (as listed in Table I). The heats of vaporization given in Table I correspond to a material temperature of 2,500 K and this is used as the surface temperature of the meteoroid.

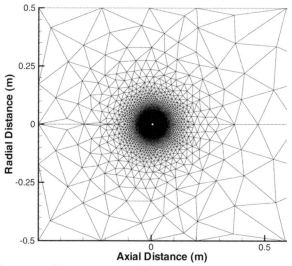

Figure 1. Close up view of the computational grid in the vicinity of the meteoroid.

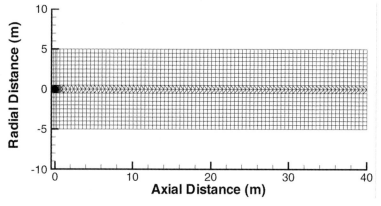

Figure 2. Computational grid for the complete flow field.

3. Model of the Flow Field

The highly rarefied nature of the flow under consideration suggests application of a kinetic based computational approach. In the present work, the direct simulation Monte Carlo method (DSMC) (Bird, 1994) is employed. This technique has been developed and applied to a variety of rarefied flows for many years. Some of its greatest successes have been in application to high altitude, hypersonic flows of spacecraft and missiles (Rault, 1994; Boyd, Phillips, and Levin, 1998). However, the meteoroid entry Mach number of 270 is at an unprecedentedly high level in comparison to these prior studies.

The DSMC technique uses model particles to emulate the motions and collisions of real molecules. Each particle represents a much larger number of actual atoms or molecules. The particles move through physical space over a time step that is smaller than the mean time between collisions. The particles are collected into computational cells that have dimensions of the order of the local mean free path. With such small cells, the assumption is made that the exact location of particles within the cell is negligible in deciding which particles in the cell may collide with one another. Therefore, pairs of particles are formed at random, and a probability of collision for each pair is computed based on the relative velocity of the pair, and the physical properties of the particles. Macroscopic properties are obtained from the microscopic particle properties by time averaging. A thorough description of the method and its application can be found in Bird (1994).

In the present work, a general DSMC code called MONACO (Dietrich and Boyd, 1996) is employed. MONACO offers great flexibility in terms of the computational grid through the use of unstructured, triangular cells. This is an important requirement for the computation of the flow fields around meteoroids due to the large dynamic range of density (and therefore mean free path, and therefore cell size) encountered in these flows. A close up view of the computational grid in the region close to the meteoroid is shown in Figure 1. The meteoroid is represented by the small circle near the middle of the image. The cells are tightly clustered around the meteoroid due to the fact that the total flow field density is found to increase significantly in this region. This is discussed further in the next section. The complete computational domain is shown in Figure 2. This grid was finalized after several low resolution computations.

The physical modeling for this initial study of meteoroid flows is kept to a relatively simple level. Air is represented by three chemical species (N_2, O_2, and O). The vapor ablated from the meteoroid surface is assumed to be a single chemical species with mass given by the values listed in Table I. Thermal relaxation is allowed between the translational, rotational, and vibrational energy modes. The Variable Hard Sphere (VHS) collision model is employed (Bird,1994) in which the collision cross section is given by:

$$\sigma = \sigma_{ref} \left(\frac{g}{g_{ref}} \right)^{1-2\omega} \quad (2)$$

where g is the relative velocity, a value of $\omega = 0.7$ is used for air, and σ_{ref} and g_{ref} are reference values. The reference collision diameter for these studies is 4 Å. No chemical reactions are considered. All species

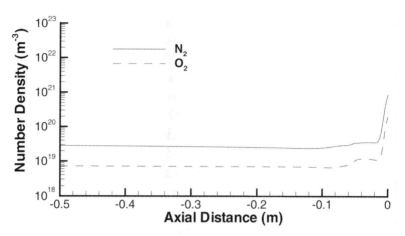

Figure 3. Profiles of number density along the stagnation streamline: no ablation.

are assumed to be in their ground electronic states. In reality, as will be seen, there are highly energetic collisions that occur immediately in front of the meteoroid that would lead to chemical reactions (in particular, molecular dissociation), electronic excitation, and ionization of the air species. Inclusion of these more detailed phenomena are left to future studies. It is expected that inclusion of chemical reactions will increase the number density of air species in the wake by close to a factor of two, assuming that all molecules are dissociated into atoms. The model for meteoroid ablation is described above. The properties of both the impacted particles after reflection and any ablated vapor particles are determined assuming diffuse reflection at the meteoroid surface temperature of 2,500 K.

4. Results and Discussion

Three different sets of computational results are presented. The first involves flow without any meteoroid ablation; and the second and third consider ablation of the two different types of meteoroid material listed in Table I. In all three cases the computational grid shown in Figure 2 is employed. The computations employ 2 million particles and are performed on four processors of an SGI Origin parallel computer in the Keck Computational Fluid Dynamics Laboratory at the University of Michigan. For the cases that include ablation, 30,000 iterations are required to reach steady state. The macroscopic results presented here are then obtained by sampling over a further 50,000 iterations. The total computation time for each case is about 18 hours.

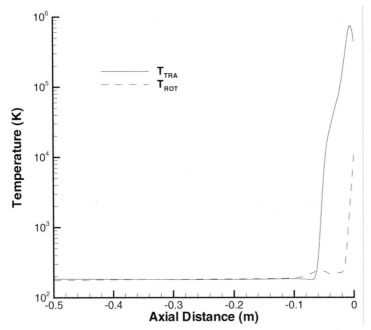

Figure 4. Profiles of temperature along the stagnation streamline: no ablation.

4.1. (1) NON-ABLATING METEOROID

In Figure 3, the number density profiles of molecular nitrogen and oxygen along the symmetry line in the region in front of the meteoroid (i.e. along the stagnation streamline) are shown. There is a significant rise in the densities of the air species immediately next to the body. This increase is due primarily to the fact that the body surface temperature is significantly smaller than the adiabatic stagnation temperature of this high speed flow. The profiles of translational and rotational temperature along this same line are shown in Figure 4. The translational temperature reaches a value of almost 1,000,000 K close to the surface. Note that the adiabatic stagnation temperature under these conditions is about 2,500,000 K. There are two important points to be made about this high temperature. The first is that under these strongly nonequilibrium conditions, the velocity distribution function will not have its equilibrium Maxwellian form. Therefore, the definition of temperature is unclear. In the DSMC method, the temperature computed represents a measure of the width of the overall velocity distribution function. Along the stagnation streamline, the velocity distribution function contains two distinct populations. One originates in the free stream, and is characterized by a low temperature (176 K) with a large off-set from zero (72 km/s). The second population consists of molecules

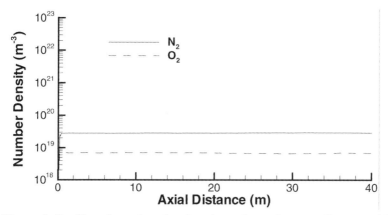

Figure 5. Profiles of number density along the wake centerline: no ablation.

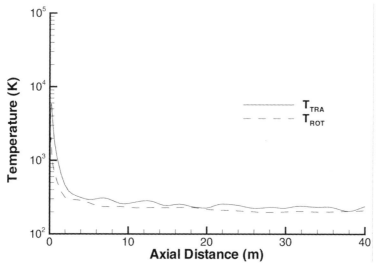

Figure 6. Profiles of temperature along the wake centerline: no ablation.

reflected from the meteoroid surface that are characterized by a higher temperature (2,500 K) and a lower, negative velocity (-1 km/s). The width of the overall distribution may be large, but almost no molecules are characterized by this temperature in the thermal sense. The second point is that these large temperatures may be somewhat reduced in the real flow through inelastic collision events such as dissociation, ionization, and electronic excitation.

In Figures 5 and 6 the species number densities and temperatures along the symmetry line behind the meteoroid (i.e. in the wake) are shown. The number densities are constant at their free stream values. The temperatures rapidly decrease to their free stream values from the

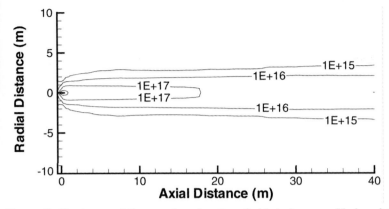

Figure 7. Contours of the number density of ablated vapor: H-chondrite meteoroid.

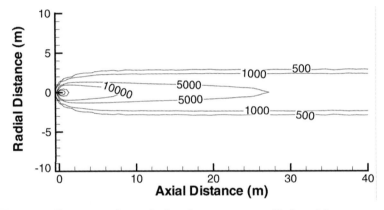

Figure 8. Contours of translational temperature: H-chondrite meteoroid.

elevated levels generated in front of the meteoroid. The translational and rotational modes are close to being in thermal equilibrium.

These results do not show the behavior found in the MAC spectroscopic investigations of the 1998 Leonid shower. This suggests that meteoroid ablation may provide an important mechanism in generating the high temperature, thermal equilibrium region that was observed.

4.2. (2) H-Chondrite Meteoroid

Before performing the computations, it is clear that a significant amount of ablation should occur in these flows using the simple evaporation model, Eq. 1. Each air molecule impacting at the free stream velocity on the H-chondrite meteoroid is predicted to ablate about 500 meteoroid particles. Of course, as ablation proceeds, a cloud of ablated vapor surrounds the meteoroid, and it becomes less likely for air molecules to strike the meteoroid directly. Therefore, a steady state is eventually

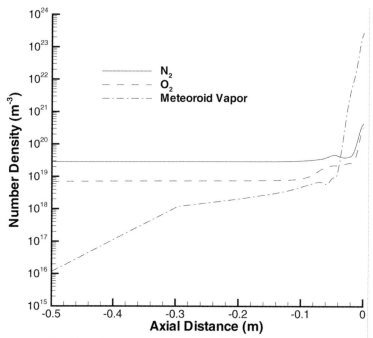

Figure 9. Profiles of number density along the stagnation streamline: H-chondrite meteoroid.

reached which forms a stable vapor cloud that expands rapidly away from the meteoroid. In Figure 7, contours of the number density (in m^{-3}) of meteoroid vapor are shown for the entire computational domain. A cylindrical wake with a waist diameter of about 6 m extends far behind the meteoroid. The corresponding translational temperature contours (in K) are shown in Figure 8. Due to the significantly higher densities of the air species within most of the meteoroid trail, these translational temperatures represent those of the air species. The meteoroid ablation leads to a large region of high temperature air in the wake of the meteoroid with values of several thousand degrees.

The species number densities and temperatures along the stagnation streamline are shown in Figures 9 and 10, respectively. The very high density cloud of meteoroid vapor in front of the body is clearly shown in Figure 9. The peak density of the vapor is more than two orders of magnitude higher than the peak density of the air species. The mean free path of the flow immediately adjacent to the meteoroid is about 10^{-5} m giving a local Knudsen number based on the meteoroid diameter of 0.001 which is in the near-continuum regime. The large range of Knudsen number encountered in the flow field requires use of sub-cells in the DSMC computation to adequately resolve the collisional behavior

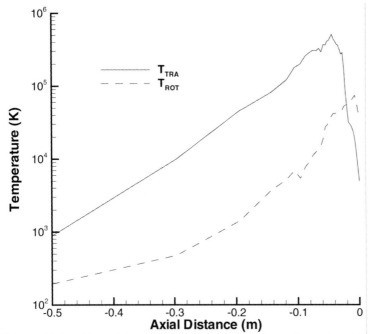

Figure 10. Profiles of temperature along the stagnation streamline: H-chondrite meteoroid.

close to the body. Far away from the meteoroid, the density of the ablated vapor decays rapidly. In comparison with the temperature profiles shown in Figure 4 for the case without ablation, the results provided in Figure 10 indicate that the rise in temperature occurs over a much larger spatial region. Again, caution is needed in the interpretation of translational temperature. The results in Figure 10 are an indication of the width of the velocity distribution function in a multi-species, strongly nonequilibrium environment.

The structure of the flow field behind the meteoroid is considered in Figsures 11 and 12 which show the number density and temperature profiles along the centerline in the wake flow downstream of the meteoroid. Here it is found that the air species densities are almost identical to the case without ablation, and the density of the ablated material slowly decays. The temperatures rapidly decay immediately behind the meteoroid and then slowly decrease with distance. The temperatures far behind the meteoroid are significantly higher than those computed for the no ablation case. It is significant to note that the computations including meteoroid ablation predict temperatures of several thousand degrees with the translational and rotational modes in thermal equilib-

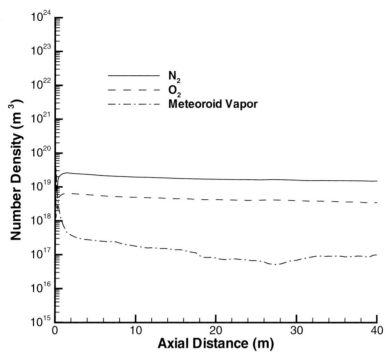

Figure 11. Profiles of number density along the wake centerline: H-chondrite meteoroid.

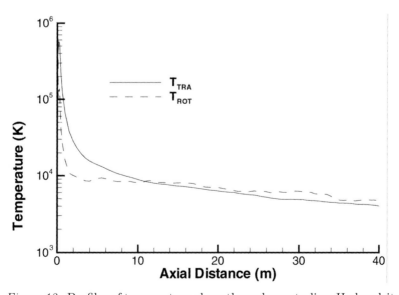

Figure 12. Profiles of temperature along the wake centerline: H-chondrite meteoroid.

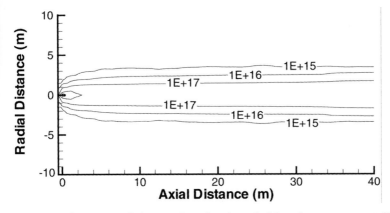

Figure 13. Contours of the number density of ablated vapor: comet-like meteoroid.

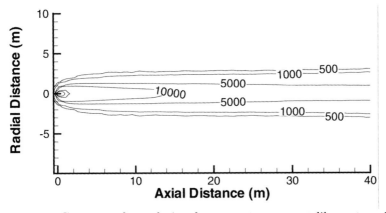

Figure 14. Contours of translational temperature: comet-like meteoroid.

rium. These characteristics of the results are in qualitative agreement with the airborne spectral measurements.

4.3. (3) COMET-LIKE METEOROID

To investigate the sensitivity of the results shown in Figures 7–12, the meteoroid material is changed to the comet-like substance with properties as listed in Table I. Flow field contours of the number density of the ablated meteoroid vapor and of translational temperature are shown in Figures 13 and 14, respectively. As might be expected, there are some changes to the wake structure. The properties of the comet-like material result in the ablation of about 2,350 meteoroid particles for each impact of a free stream air molecule on the surface. This increase in the number of ablated particles is in part due to the lighter mass of the comet-like material and this will also lead to a greater degree

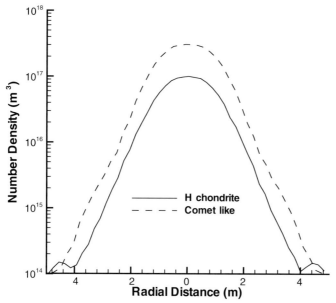

Figure 15. Radial profiles of ablated vapor number density at 20 m behind the meteoroid.

of diffusion of the meteoroid vapor through the surrounding air. These characteristics are seen in the flow field contours. The number density of ablated vapor in the wake is higher for the ablation of comet-like material and the translational temperature is also higher.

It is found that the profiles of the species densities and the temperatures of the different energy modes along the stagnation streamline and the wake streamline are similar for this case to those shown in Figures 9 through 12. Comparisons of the results for the two different meteoroid materials are compared in Figures 15 and 16, where radial profiles at a distance of 20 m behind the body are shown for ablated vapor number density and translational temperature. The density of ablated material is at least a factor of three higher than for the H-chondrite meteoroid for most of the profile. The translational temperature for the comet-like case is about 2,000 K higher on the centerline.

5. Conclusions

The direct simulation Monte Carlo method was employed to compute the two-dimensional flow field around an ablating meteoroid. The properties of the body were representative of a Leonid meteoroid. It was found that a simple ablation model led to the formation of an exten-

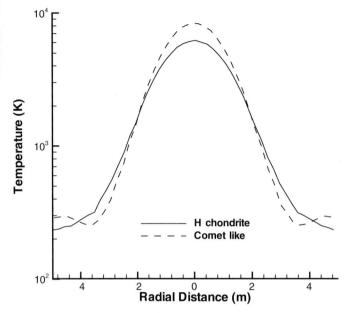

Figure 16. Radial profiles of translational temperature at 20 m behind the meteoroid.

ded, high temperature wake of several thousand degrees behind the meteoroid. Due to the build up of ablated vapor around the meteoroid, there was a sufficient number of intermolecular collisions to maintain the gas in the wake in a state of thermal equilibrium. These findings of the computational study were in qualitative agreement with airborne spectrographic measurements. It is therefore concluded that this type of computational analysis can aid greatly in the understanding of meteoroid ablation phenomena.

One of the most important predictions from the computations is the physical size of the meteoroid trail. In the present study, a waist diameter of 6 m is predicted. Observations of different meteoroids give a wide range of sizes. For example, radiometeor data indicate initial radii of meteor trains in the range of 2 to 7 m (Ceplecha, et al., 1998). Photographs of fast meteors show trail lengths of 50 to to 150 m (Babadzhanov and Kramer, 1968). Of course, the trail size will vary with meteoroid altitude, velocity, size, and composition. Assessment of the present computer predictions requires further experimental study to carefully characterize the luminous volume of meteoroid trails for which spectra are measured.

As stated at the beginning, this study is preliminary in nature. While the main goals of the investigation were achieved, it is clear that much work is required to improve the physical modeling included in the computations. For example, there is the need to include dissociation and ionization reactions of the air species colliding at very high energy with ablated meteoroid vapor. In addition, there is the need to improve the meteoroid ablation model to include more microscopic phenomena.

In addition to improving the physical modeling, there is a need to understand the sensitivity of the computational results to the flow conditions considered. In particular, computations should be performed for variations in the meteoroid size, altitude, and velocity. Attempts should be made to use the flow field results to compute synthetic spectra for direct comparison with the airborne data measured during the Leonid MAC.

Acknowledgements

The author expresses his gratitude to Peter Jenniskens, Olga Popova, and Christophe Laux for extremely helpful discussions during this study. Funding was provided in part by the NASA Exobiology program through a grant with the SETI Institute. The computations were performed on an SGI Origin 2000 funded by AFOSR grant F49620-99-1-0164 with Dr. Marc Jacobs as monitor. *Editorial handling:* Noah Brosch.

References

Babadžanov, P. B. and Kramer, E. N.: 1968, in L. Kresak and P.M. Millman (eds.), *Physics and Dynamics of Meteors*, p. 128–142.
Bird, G. A.: 1994, *Molecular Gas Dynamics and the Direct Simulation of Gas Flows*, Oxford University Press, 458 pp.
Boyd, I. D., Phillips, W.D., and Levin D.A.: 1998, *Journal of Thermophysics and Heat Transfer* **12**, 38–44.
Bronshten, V. A.: 1983, *Physics of Meteoric Phenomena*, D. Reidel Publ. Co., Dordrecht, 356 pp..
Ceplecha, Z., Borovička, J., Elford, W.G., ReVelle, D.O., Hawkes, R.L., Porubcan, V., and Simek, M.: 1998, *Space Science Reviews* **84**, 327-471.
Dietrich, S. and Boyd, I.D.: 1996, *Journal of Computational Physics* **126**, 328–342.
Jaroscewich, E.: 1990, *Meteoritics* **25**, 323–337.
Jenniskens, P. and Butow, S.J.: 1999, *Meteoritics Planet. Sci.* **34**, 933–943.
Jenniskens, P., Packan, D., Laux, C., Wilson, M., Boyd, I.D., Popova, O., Krueger, C.H., and Fonda, M.: 2000, *Earth, Moon and Planets* **82–83**, 57–70.
Jessberger, E. K., Christoforidis, A., and Kissel, J.: 1988, *Nature* **332**, 691–695.
Rault, D.F.G.: 1994, *Journal of Spacecraft and Rockets* **31**, 944–952.

SCREENING OF METEOROIDS BY ABLATION VAPOR IN HIGH-VELOCITY METEORS

OLGA P. POPOVA, SVETLANA N. SIDNEVA,
VALERY V. SHUVALOV, AND ALEXANDR S. STRELKOV

*Institute for Dynamics of Geospheres RAS, Leninsky pr.38,bld.6,
117979 Moscow, Russia
E-mail: olga@idg.chph.ras.ru*

(Received 1 June 2000; Accepted 18 August 2000)

Abstract. The ablation is calculated of non-fragmenting 10^{-2} to 10 cm sized meteoroids for typical Leonids with a velocity of 72 km/s at altitudes of about 90 - 110 km. At altitudes below the onset of intensive evaporation, a dense vapor cloud is formed around the body. This vapor cloud screens the meteoroid surface from direct impacts of air molecules. A particle beam model is developed that describes the air–meteoroid interaction. Based on this model, the physical parameters of the vapor cloud are calculated and compared with observations collected during the Leonid Multi-Instrument Aircraft Campaign. General agreement with measured temperatures is found. In addition, boundaries of the model's applicability to specific encounter conditions are given.

Keywords: Ablation, Leonids, meteoroids, meteors

1. Introduction

Meteors are considered a potential source of organic matter and metallic compounds at the time of the origin of life (Jenniskens and Butow, 1999), but surprisingly little is known about the physical conditions in ordinary meteors and the fate of ablated material in the Earth's atmosphere.

All processes connected with or initiated by meteoroid entry are essentially affected by fragmentation and ablation. Numerous studies have been carried out to show the influence of fragmentation on the meteor's light curve and deceleration, as a result of which different meteroid models (dustball, droplet solution, multi-component solid grain) have been proposed. Comprehensive reviews may be found in Bronshten (1983) and Ceplecha *et al.* (1998).

The intense meteor observations during the recent Leonid showers have provided a wealth of data on fast Leonid meteors, with an entry velocity of 72 km/s and meteor brightness in the range +3 to -10 magnitude, which corresponds to sizes in the range of 10^{-2} to 10 cm (Jenniskens *et al.*, 1998).

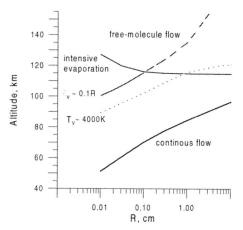

Figure 1. Boundaries of different flow regimes in the presence of a dense ablation vapor cloud in front of a Leonid meteor.

In this paper, we consider the ablation processes of such very fast meteoroids. The grain model of Hawkes and Jones (1975) appears to be the most suitable for the bodies considered here (Murray *et al.*, 1999). The model considers the meteoroid to fall apart quickly from the evaporation of a glue, after which the emission properties are representative for an ensemble of small solid grains. Our approach is to calculate the thermodynamic properties of the ablated material for a single solid body at different points along its trajectory. A dust-ball model can then be constructed by combining the effect on individual meteoroid fragments.

2. The ablation model

Although ablation can involve fragmentation and the generation of debris, here we will assume that the majority of the ablated material is released in vapor form (Ceplecha *et al.*, 1998). We consider meteoroid ablation as a process of evaporation (Bronshten, 1983). We do not consider the differential ablation of individual mineral compounds (McNeil *et al.*, 1998). Instead, the simultaneous evaporation of the mixture as a whole at the given temperature is assumed.

At given altitude, the behavior of meteoroids entering the atmosphere is determined by their size and velocity. The heating of bodies of sizes $R < 0.01$ cm is determined by the balance between incoming air energy flux and the thermal radiative cooling. The altitude where energy losses due to meteoroid heating and thermal radiative cooling become much smaller than the incoming air energy flux is called the height of intensive evaporation (Bronshten, 1983). This boundary may be roughly determined following Ceplecha et al. (1998), taking into account scarce information on the meteoroid physical properties. Figure 1 shows the corresponding altitude for a range of meteoroid sizes.

At the altitudes of 80-125 km, where most visual Leonids are observed (Betlem et al., 1999), the mean free path of molecules is bigger than the meteoroid size in this range. Hence, the flow is a free-molecular one. However, after the beginning of intensive evaporation the appearance of meteoric vapor affects that situation. The boundary of pure free-molecule regime will shift to higher altitudes. The same is valid for continuous flow. For the Leonid velocity of 72 km/s, that transition to continuous flow occurs about at the lower line in Figure 1. In that case the flow can be described in the framework of gasdynamics and hydrodynamic codes are applicable.

The very high entry velocity of Leonids provides a high evaporation rate of meteoroid. And the presence of vapor is essential in transition regime from free-molecule to continuous one. Simple estimates of the evaporation rate

$$\dot{m} \sim (q/Q) \sim \rho_a V^3/(2Q) \sim \rho_v c_v \qquad (1)$$

show that a vapor with density

$$\rho_v \sim \dot{m}/c_v \qquad (2)$$

is formed around the body. Here Q is the specific ablation energy, q is the energy flux at the meteoroid surface, ρ_a is the air density, c_v is the vapor sound velocity. The pressure of the vapor is proportional to the cube of meteoroid velocity:

$$p_v \sim \rho_a V^3/(2Qc_v) \qquad (3)$$

and for high velocities may be higher than aerodynamical loading, which is proportional only to square of the velocity:

$$p_a \sim \rho_a V^2 < p_v \qquad (4)$$

That results in the formation of a dense vapor cloud around the body, which will screen the meteoroid itself from direct impacts by the air molecules.

Roughly, the screening by the vapor is essential below the altitude of intensive evporation if the mean free path length in the vapor is smaller than the body size:

$$\ell_v \sim 0.1R \qquad (5)$$

This line is shown in Figure 1. In that case formed vapor cloud may be described hydrodynamically. For most of the Leonids considered here, the interaction with the atmosphere takes place in a transition regime between a pure free-molecular flow and a continuous flow regime (Figure 1).

Finally, we can also estimate the altitude where the temperature of the vapor exceeds 4,000 K, beyond which ionization rapidly increases and the vapor cloud is no longer optically thin for radiation. This boundary is marked as a dashed line in Figure 1. Below this line, vapor cloud temperatures are expected to be higher than 4,000 K.

3. Air beam model

There is an analogy between meteoroid evaporation under impacts of air molecules and the action of a high-energy ion beam on a target in vacuum in the inertial confinement of fusion experiments. The differences are mostly in the velocity of the impinging particles and in the energy flux. Impinging air molecules may be considered as the particle beam with energy flux:

$$q_0 = \rho_a V^3 / 2 \qquad (6)$$

whereas the vapor is best described in the framework of gasdynamics. The energy transfer during the penetration of air particles into the layer of evaporated molecules can be described similar to radiation transfer assuming some effective absorption coefficient. Impinging molecules need several collisions in order to be slowed down to the ambient velocities. Consideration of the particle dynamics of evaporating meteoroid atoms (see below) shows that the energy transfer takes place in about 10 collisions. The effective absorption coefficient (κ) for energy transfer in terms of specific mass may be written as:

$$\kappa = (n\ell_a \rho_v)^{-1} \qquad (7)$$

where ℓ_a is the free path length of air in the vapor, ρ_v is the vapor density, and $\kappa \sim 10^7$ cm^2/g for $n \sim 10$.

This approach to the air-meteoroid interaction in the transition regime, in the framework of an "air beam model", allows us to take into account evaporation, heating of vapor formation, screening of the meteoroid surface and radiative cooling of the heated vapor.

The absorption of incoming energy takes place in the layer with specific mass $m = 1/\kappa$. When this amount of mass has been evaporated, the formed vapor will screen the surface of the meteoroid and decrease the evaporation rate. When the vapor cloud's pressure exceeds the ambient air pressure, i.e. $p_v \gg p_a$, the vapor expands much like a hydrodynamic flow into a vacuum. The outer layers of the vapor cloud attain a supersonic velocity, whereby the density decrease in a 1D radial symmetry, or 2D geometry lateral expansion of the vapor cloud, is usually found to be faster than in the case of ablation from a plane surface. Hence, the outer layers become transparent more rapid for impinging air molecules, which can now penetrate deeper. Because of that, absorption of the energy of impinging air molecules occurs mainly in the subsonic part of the flow. For smaller vapor densities, screening decreases and ablation rate and vapor density increases. Similarly, for larger vapor densities, screening increases and vapor production and density decrease. This creates a balance, wherein a self-regulating regime of absorption, vapor heating, and expansion is realized (Nemtchinov, 1967; Maljavina and Nemtchinov, 1972).

It can be shown by analysis of a set of equations (Nemtchinov, 1967), as well as by direct numerical simulation (Maljavina and Nemtchinov, 1972), that in this self-regulating screening regime the density of the vapor in the sonic cross section (vapor velocity = sound speed) is such that:

$$\ell_a \approx R/2 \qquad (8)$$

This relationship permits us to define the vapor density ρ_v^* that balances the mass and energy equations to obtain all other parameters in this sonic section (Nemtchinov, 1967). Thus, vapor parameters in the total flow area can be derived directly from 1D hydrodynamic numerical simulations or by solving the ordinary differential equations that describe the 1D flow (Nemtchinov, 1967; Popova, 2000). Figure 2 (left) shows the vapor density in the sonic cross section for different body sizes. Different points are for different atmospheric loading.

When the sonic surface has formed, the flow outside this surface has little effect on the flow near the body. The boundaries of this regime can be estimated as follows. The flow becomes subsonic when the vapor pressure in the sonic section p_v^* becomes equal or smaller than aerodynamical loading. In that case, the incoming air pressure p_a can not be further neglected in the area close to the body. The higher the altitude and the smaller the meteoroid size, the bigger is the vapor pressure excess above p_a in the sonic cross-section. This is illustrated in Figure 2 (right), showing the relationship of vapor pressure to the atmosphere loading in the sonic cross-section for different body sizes. At the altitude of 114 km the condition $p_v^* \geq p_a$ holds for all bodies 0.01-10 cm in size. At the altitude of 86 km this condition is true only for $R \sim 0.01$-0.1 cm meteoroids. This estimate closely agrees with result of the 2D simulations described below.

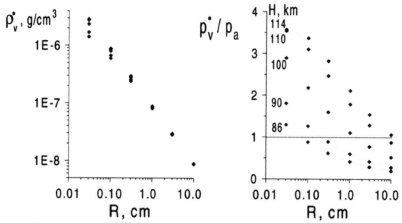

Figure 2. Vapor density (left) and the relationship of vapor pressure to the atmosphere loading (right) in the sonic cross-section for different body sizes R.

Aerodynamical loading also affects the supersonic part of the flow if the deceleration time for some mass element is smaller than the time of flight. Calculations show that the incoming momentum should be taken into account for big bodies at low altitudes. For a 1 cm body, this may be relevant below 105 km altitude. In that case the incoming momentum in the supersonic part of the flow should be taken into consideration approximately in the same area where the flow becomes subsonic. The air beam model with momentum absorption is useful both above and below this boundary.

4. Monte Carlo simulation

The air beam model was tested with Monte Carlo (MC) type calculations, in order to be certain that the air energy absorption is well described by the air beam model. The MC method has been extensively used to study free molecular and transition regime flows (Bird, 1976). Recently, the MC approach was used to describe the rarefied flow around a Leonid meteoroid with a simplified ablation model (Boyd, 2000). We simulated the energy transfer from air flow to meteoroid body, concentrating on the ablation process itself and the vapor cloud surrounding the meteoroid.

Our model is a 2D MC calculation of the action of impinging air molecules from one direction on a spherical vapor cloud with densities from 10^{-11} to 10^{-5} g/cm^3. The interaction of the impacting air particles with the meteoroid and vapor cloud is regarded as paired elastic collisions. The simulations are performed both for a constant cross-section

and for a cross-section depending on the impacting-particle velocity:
$$\sigma(v) = \sigma_0(v_{00}/v) \qquad (9)$$
where v_{00} is the particle velocity corresponding to a kinetic energy of 1 eV, because the real interaction cross section usually depends on velocity.

The initial kinetic energy of high velocity air molecules is much larger than the thermal energy of the vapor. The simulation process involves the construction of a trajectory for the impacting air particle and several generations of vapor recoil particles to the point of their "absorption". Absorption implies the decrease of particle kinetic energy to the value less than some threshold. It was proven by simulations that the main results do not depend in an important manner on the threshold value, for reasonable values. All further calculations are carried out for a threshold energy 0.5 eV.

The flow of primary air particles consists of molecules N_2 and O_2 in the ratio 0.8 and 0.2, respectively. Because the kinetic energy of the primary particles is much greater than the bond energy of these air molecules, it is assumed that molecular dissociation occurs during the first collision. One atom dissipates, and the other continues to move in the same direction to the point of the next interaction. It is supposed for simplicity that the meteoroid is Silicium gas with high density, $\rho_{si} = 2.3$ g/cm^3. Chemical interactions of unexcited molecules and atoms in the meteoroid vapor and air are neglected.

The particle trajectories are calculated in a volume consisting of the meteoroid body with radius R = 1 cm and a vapor cloud R_{cloud} = 19 cm. The density distribution in the vapor cloud is assumed to be almost uniform with constant density value ρ_c, in order to understand better the collision processes. Only at the boundaries the density changes rapidly, from $\rho_{si} = 2.3$ g/cm^3 at the inner boundary down to zero at the outer one. The value of ρ_c is a variable in the range from $1 \times \rho_{c0}$ to $10^{-6} \times \rho_{c0}$, where $\rho_{c0} \sim 10^{-5}$ g/cm^3. Hence, the radial specific mass of the vapor cloud m_v changes from about $7 \cdot 10^{-5}$ to $7 \cdot 10^{-11}$ g/cm^2. An important parameter is the ratio of the vapor cloud specific radial mass m_v and the nonthermal (air) radial mass (m_a) The adopted vapor densities cover the total range between "optically thick" $(m_a \ll m_v)$ and "optically thin" $(m_v \ll m_a)$ cases. The optical thickness of the vapor varies in the range $\tau = m_v/m_a \sim 10^{-2}$ to 10^4.

This size of the vapor cloud (R_{cloud} = 19 cm) was chosen from results obtained in the course of preliminary estimates. It overestimates the real size due to overestimation of momentum transfer. Nevertheless, the size of the cloud is not critical when considering the air particle interactions with a constant density vapor cloud.

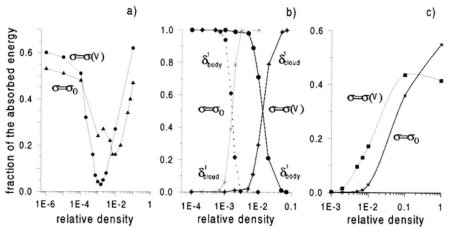

Figure 3. (a) - the absorbed energy fraction δ_{abs} versus vapors relative density; (b) - the energy fractions absorbed by meteoroid (δ^1_{body}) and by the vapor cloud(δ^1_{cloud}) versus vapors relative density. The air flow radius $R = 1$ cm; (c) - the absorbed energy fraction δ_{abs} versus vapors relative density if the air flow radius $R_{cloud} = 19$ cm.

4.1. AIR FLOW INTERACTION WITH METEOROID

We consider the flow of air particles impacting the cross section area of a meteoroid with radius $R = 1$ cm, surrounded by ablation vapor. The energy absorbed in the meteoroid-vapor system $E_{abs} = E_{body} + E_{cloud}$, and dimensionless values

$$\delta_{abs} = E_{abs}/E_{flow}$$

$$\delta^1_{body} = E_{body}/E_{abs}$$

$$\delta^1_{cloud} = E_{cloud}/E_{abs} \qquad (10)$$

are the main characteristic of the model and describe the action of air particles on the meteoroid and vapor cloud.

The value of δ_{abs} shows what part of air flow energy E_{flow} incoming through the meteoroid cross-section ($R = 1$ cm) is absorbed by cloud-body system. The fractions δ^1_{body} and δ^1_{cloud} demonstrate the distribution of absorbed energy inside the system. The sum $\delta^1_{body} + \delta^1_{cloud}$ is equal to unity.

In Figure 3 we show these fractions for a range of relative densities and two assumed interaction cross sections σ ($\sigma = \sigma_0$ and $\sigma = \sigma(v)$).

If the vapor cloud is transparent for impacting air particles, e.g.:

$$\rho_c = 10^{-6} \div 10^{-5} \rho_{c0} \qquad (11)$$

and $\tau \sim 0.1 \div 0.01$, then $\delta_{abs} \approx 0.52$, i.e. half of the incoming energy is absorbed by the system. The values obtained for $E_{abs} \approx E_{body}$ are twice as low than value obtained in the simplified ablation model by Boyd (2000). This decrease is probably caused by the contribution of inclined impacts (Pleshivtsev, 1968).

In Figure 3(a,b), there is a sharp decrease of values δ_{abs} (i.e. fraction of the energy absorbed by the system) and δ^1_{body} (i.e. fraction absorbed by body itself) for values of density $\rho_c \sim 10^{-4} \rho_{c0}$ ($\tau \sim 1$). The minimum of δ_{abs} is observed when $m_v \sim (1\text{-}2) \times 10^{-7}$ g/cm^2, which corresponds to $\tau \sim 10 \text{ - } 20$. With vapor density increase, the fraction of energy obtained by body itself decreases, while incoming flow energy absorbs mainly in the cloud (Figure 3b). The increase of δ_{abs} after this minimum is caused by vapor energy absorption. At high vapor densities, all energy absorption occurs in the cloud ($\delta^1_{cloud} = 1$). The dependence of particle interaction cross-section on velocity $\sigma = \sigma(v)$ leads to the beginning of screening under higher vapor density (Figure 3b).

The energy absorption efficiency δ_{abs} has a local maximum (Figure 3a), This is caused by the presence of two competing processes – the decrease of the total air particle number and the increase of the number of particles which reach the meteoroid. In some interval of density, the number of secondary particles increases and results in partial compensation of the decrease of the number of primary particles. However, as the vapor density increases, the layers from which recoil particles can reach up to the meteoroid begin to be screened by vapor external layers. The value of δ^1_{body} falls down to zero, i.e. a total screening of the meteoroid body occurs. Screening proceeds when the vapor optical thickness for the most penetrating primary particles achieves the value $\tau \sim 10$ (Figure 3).

4.2. Air flow interaction with the vapor cloud

We also consider the air interaction with a well developed vapor cloud around the meteoroid, assuming the same density distribution. The incoming flow of air particles is distributed uniformly over the cross-section of the vapor cloud (for $R_{cloud} = 19$ cm). The fraction of energy absorbed by the cloud-body system δ_{abs} is presented in the right graph of Figure 3. When the density of cloud is low, the absorption is determined by the meteoroid itself and only a small part of energy incoming through the large cross section is absorbed. The energy part absorbed in the vapor cloud increases monotonically when the vapor density increases. The energy is totally absorbed by the vapor in the cases ($\tau > 10 \text{ - } 100$):

$$\rho_c > 10^{-3} - 10^{-2} \rho_{c0} \tag{12}$$

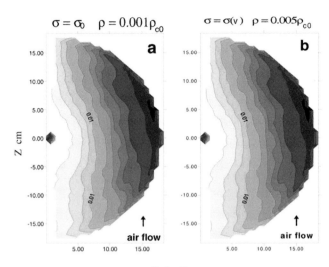

Figure 4. Absorbed energy spatial distribution for vapor of relatively low density.

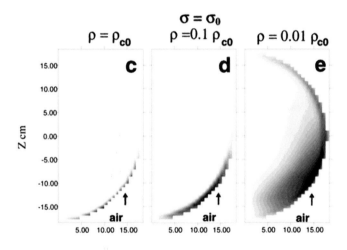

Figure 5. Absorbed energy spatial distribution for vapor of relatively high density.

While the impinging air particles initiate the collisions, further energy transfer is determined by the secondary vapor particles that carry the momentum. The spatial distribution of absorbed energy is determined mainly by the energy transport of several generations of vapor particles.

The energy distribution in the regime of partial screening is presented in Figure 4. The meteoroid is at the grid center, impinging air molecules are coming from below. Density contours show the transferred energy. The energy absorption spans the whole volume of the vapor cloud. The maximum energy absorption is at the volume surface near the symmetrical plane, perpendicular to the air flow. The captured air particles, stopped after all energy has been transferred, are distributed in the outer layers of the vapor cloud. The highest density of captured air molecules is towards the back end of the cloud.

As the vapor cloud density ρ_c increases, the region of highest energy absorption shifts to the volume periphery and becomes degraded to a thin layer located on the frontal surface of the cloud (see Figure 5).

These simulations show that the size of initial particle influence area is larger than the initial particle mean free path l_a by an order of magnitude and provide additional foundations to use air beam model for description of ablation.

5. 2D gasdynamic approximation

The vapor parameter estimates presented in Section 3 were obtained from a 1D radial geometry, whereas the vapor cloud surrounding a real meteoroid is three dimensional. However, a cylindrical geometry may be assumed and a good approximation may be obtained from a two dimensional approach. In order to reveal how the 2D approach changes the picture we perform 2D simulations for several specific cases. According to the air-beam model it is assumed that the energy release due to deceleration of incoming air molecules is determined from the equations:

$$\frac{\partial q}{\partial x} = -\rho_v \kappa q \qquad (13)$$

$$\frac{\partial e}{\partial t} = -\frac{1}{\rho_v}\frac{\partial q}{\partial x} \qquad (14)$$

where ρ_v and e are the vapor density and thermal energy, q is the energy flux, which is equal to q_0 at the outer boundary, and x is the coordinate along the trajectory. Some portion of the beam energy is absorbed by the dense vapor which results in the heating and expansion of the vapor. The part of the energy flux reaching the body surface causes evaporation.

There are four governing parameters in this problem: q_0, Q, κ and the meteoroid radius R. Let us introduce the dimensionless variables according to:

$$q = q/q_0, r = r/R, x = x/R, e = e/Q$$

$$u = uQ^{-1/2}, \rho = \rho_v q_0^{-1} Q^{1/3}, p = p_v q_0^{-1} Q^{1/2} \qquad (15)$$

Here, p_v and u are vapor pressure and velocity. The ambient air density (or pressure) is not a governing parameter because it is much less than the vapor density in the region where the gasdynamical approximation is valid. The effective adiabatic exponent γ for comet substance vapor varies in the range from 1.1 to 1.4 (Kosarev et al., 1996). For simplicity of the argument, and for facilitating a comparison with analytical estimates and 1D simulations, the vapor is assumed to be an ideal gas with specific heat ratio $\gamma = 1.4$. The system of gasdynamical equations and boundary conditions written using these dimensionless variables defines the flow, which depends on only one dimensionless parameter:

$$\eta = q_0 R Q^{-3/2} \kappa \qquad (16)$$

Following the results of the previous section it is assumed that the air molecule beam is absorbed at a distance of ten molecule free path ℓ_a:

$$\kappa = (10 \ell_a \rho_v)^{-1} = \sigma (10 \mu)^{-1} \qquad (17)$$

where σ is the gaskinetic cross-section and μ is the vapor molecule mass. In that case:

$$\eta = 0.05 \sigma \mu^{-1/2} R \rho_a V^3 Q^{-2/3} \qquad (18)$$

We have performed numerical simulations of this problem with the use of the SOVA hydrodynamic code (Shuvalov, 1999), for three values of η: 5, 50, 500, which correspond to a Leonid meteoroid altitudes of 110, 100 and 90 km respectively ($V = 72$ km/s, $Q = 6$ kJ/g, $R = 1$ cm). The falling body is considered to have a spherical shape. Density and energy distributions around the meteoroid at the altitude of 100 km are shown in Figures 6 and 7 (left panels).

The white lines in these figures show the boundary of the region where the molecule free path is significantly less (by a factor of 5 or more) than the meteoroid size. The falling body is surrounded by a dense vapor cloud, which can be described using the gasdynamic approximation. The molecular free path in the ambient air equals about 17 cm. Therefore, gasdynamics can not be applied to study the motion of a nonablating body at this altitude.

The results of the 2D calculation are rather close to those obtained from the 1D spherical problem. The sonic surface is also formed (it is marked by a black line), although it is not spherically shaped. The flow outside this surface does not influence the flow near the meteoroid. The main difference between the 1D and 2D results concerns the absorption of the beam energy. In the case of 1D spherical symmetry, a considerable portion of energy flux (about 50 %) is absorbed outside the sonic surface. In the 2D problem, this fraction is less.

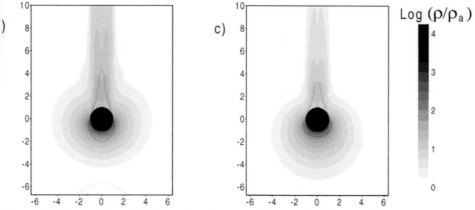

Figure 6. Distribution of relative density around a 1 cm body at 100 km altitude; left - only energy absorption is considered, right- momentum and mass absorption are also taken into account.

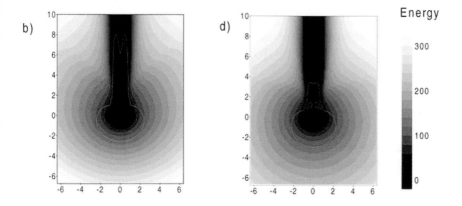

Figure 7. As Figure 6, but for dimensionless internal energy distribution.

The 2D modeling can be used as a starting point for introducing modifications in the 1D simulations, concerning details of energy, momentum, and mass transfer processes.

However, the beam causes not only energy transfer to the vapor, but also momentum and mass. The model of momentum transfer is essential in the determination of the vapor density. The assumption of $p_v \sim \rho_a V^2$ at the outer vapor boundary results in an overestimation of the vapor density. To estimate this effect in a second simulation, we have taken into account the momentum and mass release. It is assumed that the mass and momentum fluxes are absorbed in the same way as the energy flux and with the same value of the absorption coefficient. The results of the second run are shown in Figures 6 and 7(right panels).

The impulse absorption acts as a counter pressure. Its influence increases when the altitude decreases, and at the altitude of 90 km it causes the flow to be subsonic. This flow change is in a good correlation with 1D estimates (Figure 2, right). However, the model of momentum transfer should still be verified in direct MC simulations, because the absorption coefficient for momentum transfer may differ from that for energy.

6. Vapor cloud parameters and the influence of radiation

The dimensionless set of gasdynamic equations depends on one parameter η that determines how the vapor parameters change with body size and altitude. Internal energy and temperature increases with a decreasing altitude and an increasing size. The vapor cloud density is mainly determined by the absorption coefficient and is about 10^{-5} g/cm^3 near the surface, while rapidly decreasing with the distance from the body down to the ambient air density. The size of the vapor cloud increases with meteoroid size and increasing altitude. Roughly, the size of the dense vapor cloud is 5 - 10 times of body size. The rarefied vapor penetrates further from the body than that, but the number density of vapor particles is lower than the air density at these distances. The total size of vapor penetration is about 5-10ℓ_a (50 - 100 cm at altitudes 90 - 100 km), which is in a good agreement with results of the direct MC simulations (Boyd, 2000).

The 2D-modeling is performed for an ideal gas with $\gamma = 1.4$. The real equation of state (EOS) results in different temperatures values. 1D simulations with real EOS of comet substance (with an assumed composition similar to comet Halley, including water ice) and H-chondrites (Kosarev et al., 1996) show that the maximum vapor cloud temperatures are of order 4,000 - 40,000K for 0.1 - 1 cm bodies at 90 - 110 km altitude. The use of either H-chondrite or comet substance properties leads to similar dependencies of parameters on the size and altitude.

The high temperature of the vapor asks for the inclusion of vapor thermal radiation into the models. According to Borovička et al. (1999) and Jenniskens et al. (2000), the observed Leonid meteor radiation shows only a moderate deviation from thermal equilibrium. MC simulations by Boyd (2000) demonstrated that gas in the meteor wake is nearly equilibrated, although rotational and translational temperatures differ in the head. We do not consider here the possible effects of such deviations from Local Thermodynamic Equilibrium (LTE) and only estimate how the radiation may change the thermodynamical parameters.

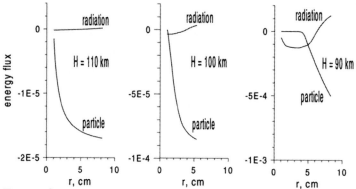

Figure 8. Energy fluxes in relative units for particle beam and radiation at different altitudes.

For that, 1D simulations, adjusted by results from the 2D modeling, are performed with a real EOS, including effects of radiation. Radiation transfer in a multigroup approximation (we assume 8 spectral groups) is taken into account according to the method of Svettsov (1994).

The total energy fluxes in relative units for particle beam and for the radiation are compared in Figure 8 for a 1 cm cometary body (including ice). The evaporation is caused by air molecule impacts at 110 km altitude. Below this altitude, the particle evaporation is accelerated by the UV radiation generated in the plasma. Moreover, at lower altitude the air molecules do not reach the meteoroid surface and heating of the meteoroid may be dominated by radiation rather than energy transport through the vapor. The lower the altitude, the bigger is the role of radiation in the ablation. However, it should be noted here that deviation from LTE can change the radiative flux.

In conclusion, the radiation has a pronounced effect on the thermo-dynamical parameters of the vapor. At all altitudes, it decreases the maximum temperature. However, at low altitudes it may also increase the vapor density and the vapor cloud size, as is shown in Figure 9 for the case of an altitude of 90 km.

7. The meteor wake

Our model predicts vapor parameters in front of the meteoroid. The total luminous area includes a region behind the body called the wake. The precise boundary between the head and the wake, as well as a shape of the luminous volume, is not well known. On instantaneous meteor photographs ($6 \cdot 10^{-4}$s exposure) the average image length of fast meteors is of about 50 - 150 m (Babadžanov and Kramer, 1968).

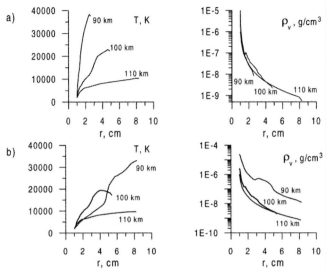

Figure 9. Influence of radiation on thermodynamical parameters for 1 cm comet body at different altitudes. Vapor parameters are found: (a) - without radiation; (b) - including radiation transfer.

The wake itself is described in the work by Boyd (2000). However, a crude estimate of the wake parameters near the meteoroid itself can be made using the air beam model. A cylindrical layer of vapor is stripped by the incoming air and is thrown into the nearby wake. The initial state of the spatial distribution of the stripped vapor layer is the result of conditions formed in the vapor cloud around the body. Its further evolution is determined by expansion into the undisturbed air and subsequent radiative cooling.

Simulations show that for a cometary body the peak temperature drops by a factor of 2 at a distance of only about 1 m behind the body. At the distance of about 5 meters, the vapor temperature is still about $\sim 5{,}000\text{-}6{,}000$K. Hence, the extended hot area far exceeds the meteoroid size itself. Our estimate agrees with result of MC simulations by Boyd (2000), where the formation of a large (10 - 40 m in length) wake with temperature of about 5,000-10,000K is calculated. And this identifies the wake as the source of the temperature measured from atmospheric line and band emission of Leonid meteors in the 1998 Leonid MAC (Jenniskens et al., 2000).

8. Spectra

Meteor spectra can be calculated from the thermodynamical parameters in the luminous volume and the opacities for radiation. In our

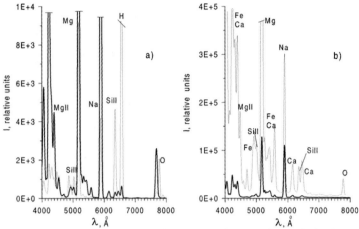

Figure 10. Example of spectra emitted by vapor layers of equal thickness for 1 cm body at 100 km altitude: a) for cometary composition (at a distances of 2 m (thin) and 4 m (thick) behind the meteor head); b) H-chondrite composition (at a distances of 0.2 m (thin) and 2.5 m (thick)).

calculations, the optical properties of comet material and H-chondrite vapor are those calculated by Kosarev *et al.* (1996). As it was mentioned above we do not consider the differential ablation in our modeling, the simultaneous evaporation of the mixture as a whole is assumed. Because of that, the chemical composition of the vapor is suggested to correspond to H-chondrite composition (no ice) or to the composition of comet Halley (with ice). These model spectra are very approximate, because we only have information on the assumed properties of the wake near the meteor head. Only vapor radiation is taken into account, and we do not estimate the radiation from the heated air.

Spectra from the vapor layers at different distances from the meteor head are shown in Figure 10. Registered meteor spectra consist of the total radiation of vapor layers and air radiation.

The spectra change with the distance from the meteoroid. Lines with high excitation energy (SiII, H, MgII, some O lines) originate mainly from the hot area of wake and meteor head and warmer portions of the wake. The lower excitation lines of MgI and NaI, on the other hand, are radiated from a more extended area in the wake (Figure 10).

The evidence of high temperature in the vapor are both predicted and observed in the Leonid spectra. Many Leonid spectra were observed during the Leonid MAC missions. Borovička *et al.* (1999) analyzed video spectra in the range 380-880 nm from millimeter – centimeter sized Leonids ($+2...-6^{mag}$). The altitudes of the observed spectra are not given. It was found that about 2/3 of radiated energy in the given wavelength range is due to atmospheric air emissions, about 1/6 of the energy is radiated as thermal continuum and another 1/6 is due to the line emissions of the meteoric vapor. The spectra were found to have two components of quite different temperature. The temperature of main component is estimated of about 4500 K, while the second component was about 10,000 K (Borovička, 1994). From this second component, Borovička et al.(1999) identified Mg^+ emission and find that the intensity increases for larger Leonids. Ionized silicium lines presented in theoretical spectra were identified in the photographically recorded Leonid meteors by Hirose et al. (1968) and in Leonid fireball video spectra by Borovička and Jenniskens (2000).

If we compare observations and theory, we find that the diameter of the vapor volume derived from the observations (Borovička et al., 1999; Jenniskens et al., 2000) appears to be one to two orders of magnitude greater than the meteoroid size (of about 1-10 cm for a 0^{mag} Leonid), which may confirm the effect of vapor accumulation around the body.

The estimate of vapor density ($\rho_v \sim 10^{-7}$ g/cm^3) in Borovička et al. (1999), proves to be similar to our (10^{-5} to 10^{-10} g/cm^3), whereas we predict higher temperatures in the area of air-vapor interaction.

Theoretical spectra depend on altitude and body size. The high resolution spectra recorded during Leonid MAC 1999 also demonstrate that spectra change with altitude (Abe et al., 2000). But the detail influence of altitude, size and composition on theoretical spectra should be considered in future.

9. Conclusions

The interaction of air molecules with the ablation vapor cloud of non-fragmenting 0.01 - 10 cm sized meteoroids with a velocity of 72 km/s at altitudes of about 90-110 km were calculated. It is found that below the altitude of intensive evaporation a dense vapor cloud is formed around the meteoroid which screens its surface for direct impacts of air molecules.

A particle beam model describes the air-meteoroid vapor interaction well. The results were compared to direct MC calculations for specific cases. The direct MC simulation estimates the fraction of energy transferred to the meteoroid and the conditions of meteoroid shielding. Initial air molecules penetrate into the vapor to a depth larger than the initial particle mean free path ℓ_a by an order of magnitude. The deposition length equals approximately 10 mean free paths of the most penetrating primary particles and of vapor recoil fast particles. As a whole, the results of the 2D simulations show that the differences between the 1D and 2D models are not substantial. Hence, the 1D approximation, supplemented with a more complete physical model, may be used to study this problem.

Preliminary comparison of theoretical predictions with observations shows both agreements and disagreements. Nevertheless the main features of ablation are evident: a big volume of evaporated mass travels with the meteor body substantially increasing its effective size; at high altitudes and for small bodies the flow is supersonic, the air loading is not an important parameter; for given body size the momentum transfer and vapor radiation should be taken into account with decreasing altitude; ablation is dominated by direct impacts at high altitude but radiation dominates the ablation at low altitudes. Calculated vapor spectra show general agreement with observations.

The present model is still rather crude. Especially, we did not take into account the electron thermal conductivity and did not study in detail the stripping of vapor cloud molecules into the wake. We considered only a relatively large particle size and only two meteoroid compositions. Clearly, this is only a first step. Further analysis of observational data and elaboration of theoretical model are needed to understand the physical processes that determine ablation of small meteoroids. Nevertheless, our consideration revealed the questions which should be cleared up and allow to obtain qualitative picture and estimates of the main flow characteristics.

Acknowledgements

This work was partially supported by Sandia National Laboratories under contract BC1842. We are grateful to Prof. I.V. Nemtchinov for his interest and fruitful discussions and Ms.L.M.Beletskaya for technical assistance. Olga Popova thanks Dr.Peter Jenniskens for helpful discussions and numerous corrections of the manuscript. *Editorial handling:* Noah Brosch.

References

Abe, Sh.,Yano, H., Ebizuka, N., and Watanabe J-I,: 2000, Meteoritics Planet. Sci, in press.
Babadžanov, P.B., and Kramer, E.N.: 1968, in L. Kresak and P. Millman (eds.), *Physics and Dynamics of Meteors*, 128–142.
Betlem, H., Jenniskens, P., van't Leven, J., ter Kuile, C., Johannink, C, Zhao, H., Lei,Ch., Li, G., Zhu, J., Evans, S., and Spurny, P.: 1999, *Meteorit. Planet. Sci.*, **34**, 979–986.

Bird, G.A.: 1976, *Molecular gas dynamics*, Oxford Engineering Sciences Series Clarendon Press. Oxford, 238 pp.
Borovička, J.: 1994, *Planet Space Sci.*, **42**, 145–150.
Borovička, J. and Jenniskens,P.: 2000, *Earth,Moon,Planets*, **82–83**, 399–428.
Borovička, J., Stork, R., and Bocek, J.: 1999, *Meteorit. Planet. Sci*, **34**, 987–994.
Boyd, I.D.: 2000, *AIAA paper* **2000-0583**, 38th Aerospace Sciences Meeting&Exhibit, 10-13 January, Reno, NV.
Bronshten, V.A.: 1983, *Physics of Meteoric Phenomena*. D. Reidel. Publ. Co. Dordrecht, 356p.
Ceplecha, Z., Borovička, J., Elford, W.G., ReVelle,D., Hawkes,R.L., Porubcan, V., and Simek, M.: 1998, *Space Sci. Rev.*, **84**, 327–471.
Hawkes, R.L., and Jones, J.: 1975, *Mon.Not.R.Ast.Soc.*, **173**, 339–356.
Hirose, H., Nagasawa, K. and Tomha, K.: 1968, in Kresak and Millman (eds.), *Physics and Dynamics of Meteors*, 105–118. I.A.U. XXX.
Jenniskens, P., de Lignie, M., Betlem, H., Borovička, J., Laux, C.O., Packan, D., and Kruger, C.H.: 1998, *Earth, Moon and Planets* **80**, 311–341.
Jenniskens, P. and Butow, S.J.: 1999, *Meteorit. Planet. Sci.*, **34**, 933–943.
Jenniskens, P., Wilson, M.A., Packan, D., Laux, C.O., Krueger C.H., Boyd, I.D., Popova, O.P., and Fonda, M.: 2000, *Earth, Moon and Planets* **82–83**, 57–70.
Kosarev, I.B., Loseva, T.V., and Nemtchinov, I.V.: 1996, *Solar System Research* **30**, 265–278.
McNeil, W.J., Lai, S.T., and Murad, E.: 1998, *J.Geophys.Res.* **103**, 899–911.
Maljavina, T.B. and Nemtchinov, I.V.: 1972, *Prikl.Mech. i Techn.Fizika* **5**, 59–75 (in Russian).
Murray, I., Hawkes, R.L., and Jenniskens, P.: 1999, *Meteorit. Planet. Sci.* **34**, 949–958.
Nemtchinov, I.V.: 1967, *Prikl. Matem. i Mech.* **31**, 300–319 (in Russian).
Pleshivtsev, N.V.: 1968, *Cathode Sputtering*, Atomizdat, Moscow (in Russian).
Popova, O.: 2000, *AIAA paper* **2000-0587**, 38th Aerospace Sciences Meeting&Exhibit, 10-13 January, Reno, NV.
Svettsov, V.V.: 1994, *Comp.Maths Math.Phys.* **34**, 365–376.
Shuvalov, V.V.: 1999, *Shock Waves* **9**, 381–390.

SEARCH FOR EXTRATERRESTRIAL ORIGIN OF ATMOSPHERIC TRACE MOLECULES – RADIO SUB-MM OBSERVATIONS DURING THE LEONIDS

DIDIER DESPOIS, PHILIPPE RICAUD, NICOLAS LAUTIÉ
NICOLA SCHNEIDER and THIERRY JACQ
Observatoire de Bordeaux, B.P. 89, F-33270 Floirac, France
E-mail: despois@observ.u-bordeaux.fr

NICOLAS BIVER
Observatoire de Paris-Meudon, 92195 MEUDON Cedex, France
and IfA, University of Hawaii, 2680 Woodlawn Drive, Honolulu, HI 96822 USA

DARIUSZ C. LIS, RICHARD A. CHAMBERLIN
and THOMAS G. PHILLIPS
California Institute of Technology, Downs Laboratory of Physics,
MS 320–47, Pasadena, CA 91125, USA

MARTIN MILLER
I. Physikalisches Institut, Universität zu Köln,
Zülpicher Str. 77, 50937 Köln, Germany

PETER JENNISKENS
SETI Institute, NASA ARC, MS 239–4, Moffett Field, CA 94035, USA

(Received: 9 August 2000; Accepted 2 September 2000)

Abstract. To identify the effect of meteor showers on the molecular content of the upper atmosphere of the Earth, we have carried out ground-based observations of atmospheric HCN. HCN radio observations at CSO (Hawaii) on Nov 18/19, 1999, the night after the second Leonid shower maximum, show unusually low HCN abundances above 45 km altitude, which are only recovered after sunrise. We also investigated UARS/HALOE satellite data on H_2O and O_3. No correlation appears of year round H_2O and O_3 around 55 km with annual meteor showers, nor with meteor activity at the time of the 1998 Leonid shower.

Keywords: Early Earth, H_2O, HCN, Leonids 1999, lower thermosphere, O_3, mesosphere, meteors, micro-wave, radio

1. Introduction

Earth and planets are not isolated from interplanetary space. Interplanetary Dust Particles (IDPs) enter continuously into the Earth's atmosphere as sporadic meteors or meteor showers up to a total estimated amount of 20–40,000 tons/year (Love and Brownlee, 1993;

Maurette, 1998). These particles were identified as being responsible for the atmospheric neutral atom metal layers (Na, Fe, ...) observed around 90 km altitude (e.g. Höffner et al., 1999; Chu et al., 2000). Recent work has shown that water ice IDPs of cometary origin may explain the presence of H_2O in the upper atmosphere of the giant planets Saturn, Uranus, Neptune (Feuchtgruber et al., 1997). Water ice may survive in the meteoroids while at Saturn's heliocentric distance, but evaporates rapidly at 1 A.U. from the Sun. However, these meteoroids also contain less volatile hydrogen containing organic matter. When hydrogen is released, it can recombine with atmospheric oxygen atoms and ozone, and may thus provide an alternative source for the high altitude atmospheric water ice of noctilucent clouds. The organic matter in meteoroids can also release other compounds, such as HCN or CN radicals when decomposed during ablation.

Looking back in time 4 Gyr ago, at the time of the origin of life, IDPs are believed to be a major source of carbon on the surface of the early Earth, when their input rate was supposed to be a hundred times higher than today (Chyba and Sagan, 1996). It is unclear how much organic carbon may have been delivered through ablation in the atmosphere. The details of the interactions between the entering particles and the atmosphere are still poorly understood, despite years of modelling attempts. In particular, it is not known to what extent molecules, not only atoms and ions, can be injected or produced by meteors in the atmosphere, a key issue in determining the prebiotic chemistry of the early Earth (Jenniskens et al., 2000a).

In order to study qualitatively and quantitatively the impact of meteors on the molecular content of the Earth's upper atmosphere, we started searching for time variations in molecular lines during intense meteor showers, using both existing satellite observations and new ground-based radio sub-mm observations. This provides the most straight forward way to link the presence of a given molecule to the extraterrestrial input. More difficult alternatives are the study of isotopic ratios variations (e.g. D/H), or the precise analysis of the vertical abundance profiles at high altitude.

We present here two such studies: 1) ground-based observations of HCN with special emphasis on the Leonid showers of November 1998 and 1999; and 2) an effort to see meteor related variations in the 1991-1999 O_3 and H_2O contents of the upper atmosphere as retrieved from the Halogen Occultation Experiment (HALOE) on board the Upper Atmosphere Research Satellite (UARS).

2. Search for HCN variations

2.1. INTRODUCTION

HCN is a minor constituent of the Earth atmosphere, with a typical volume mixing ratio around 10^{-10} HCN per air molecule. At present, the main source of HCN in the lower atmosphere is expected to be biomass burning (Crutzen and Andreae, 1990). The atmospheric HCN has been observed since 1981, first in the infrared, then at microwave radio frequencies (Coffey *et al.*, 1981; Carli *et al.*, 1982). Figure 1 gathers most of the available measurements of HCN in the atmosphere. These measurements come from a variety of techniques listed in the figure caption.

Globally, above 30 km, HCN measurements are in excess of model predictions based on standard photochemistry and biomass burning as the only HCN source (solid lines in Figure 1). This excess has been explained by 1) ion-catalyzed reactions in the entire stratosphere, involving CH_3CN as a precursor (Schneider et al., 1997) and/or 2) a high altitude source as a result of chemical production from the methyl radical CH_3 (Kopp, 1990), or from injection or production by meteors (Despois *et al.*, 1999). HCN is a minor constituent of cometary ices. HCN polymers or copolymers have been suggested as constituents of cometary refractory organic matter (Matthews, 1997, and refs. therein), and would thus be present in the incoming meteoroids, if these polymers survived their stay in interplanetary space after ejection. HCN may also be created from the CN radical decomposition product of organic carbon, after reaction with hydrogen-bearing molecules.

To test the hypothesis of HCN input by meteoroids or the formation in the upper atmosphere from meteoric ablation products, we decided to monitor the HCN submillimeter lines around a major shower: the Leonids.

2.2. OBSERVATIONS

In 1998, observations of the HCN radio lines were attempted at various radiotelescopes: IRAM 30m (Spain), Bordeaux 2.5m, Caltech Submillimeter Observatory CSO 10m and the JCMT 15m (Hawaii). Unfortunately, instrument problems (IRAM, Bordeaux) and bad weather (Hawaii) prevented obtaining a usefull sensitive limit.

In 1999, time was granted on two telescopes: the CSO 10m and KOSMA 2.5m (Switzerland) telescopes. Weather did not allow any useful observations at KOSMA. Observations at the Caltech Submillimeter Observatory took place from Nov. 16 to Nov 25, 1999, with unfortunately bad observing conditions at the time of the peak on November

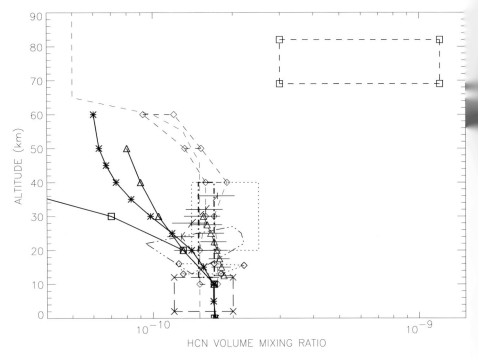

Figure 1. HCN in the atmosphere: models (thick black lines) and observations (others). Observations cover southern mid-latitudes (triangle; Zander *et al.*, 1988) to northern high latitudes (-...-; Spreng and Arnold, 1994), using mass spectrometry (- - - with square; Kopp, 1990), infra-red (+-+; Abbas *et al.*, 1987), and millimeter-wave techniques (- - - with diamond; Jaramillo *et al.*, 1988), with ground-,aircraft-based, balloon-,rocket-, and space-borne sensors. Photochemical models (thick black lines) are from Brasseur *et al.* (1985, with triangle for a photodissociation rate $J_{HCN}=0$, and with square for $J_{HCN}=J_{HCL}$) and Cicerone and Zellner (1983, with asterisk). A deficit in modeled HCN appears in the middle atmosphere, although tropospheric amounts agree reasonably well. Note the discrepancy between the photochemical models and the Kopp data around 75 km (dashed rectangle in the upper right corner of the plot). The other data are from Coffey *et al.* (-.-; 1981), Carli *et al.* (...; 1982), Rinsland *et al.* (x-x; 1982), and Schneider *et al.* (... with diamond; 1997). The dashed line without symbol represents the *a priori* HCN profile used in radiative transfer calculations.

18. Here, we report on the observations of the J = 3–2 rotational line of HCN at 265.88 GHz, which was observed in frequency switch mode on Nov. 19, from 15 UT until 21 UT.

In addition, atmospheric HCN observations in Hawaii were obtained serendipitously at JCMT from cometary observations at other periods of the year and can be used for reference. The frequency of the HCN lines observed are: HCN(4–3) = 354.51 GHz at JCMT, and HCN(3–2) = 265.88 GHz at CSO.

2.3. RESULTS

A typical example for a non-Leonid night observation at JCMT is shown in Figure 2. The HCN line is well separated from other emission lines. The other line we see on the spectrum is a CO line that comes from the image band (JCMT measurements are double-side band) and, due to that and to the reduction process of frequency-switched spectra, the line appears as a negative signal.

The CO line is narrower than the HCN line because CO (created in Earth's upper atmosphere from the UV photon dissociation of CO_2) is more abundant in the mesosphere than in the stratosphere, whereas HCN is more abundant at stratospheric altitudes (Figure 1). At lower altitudes, the spectral lines are pressure broadened, increasingly so for lower altitudes. Hence, the HCN line is more pressure-broadened than the CO line, except for the small feature on top of the profile that is from high altitude (> 60 km) HCN. The measurement technique (frequency-switch) removes most of the broad low-altitude wings of the HCN line, so that by choosing the right frequency switch, we end up with a line profile that gives information only above mid-stratosphere (above 35 km).

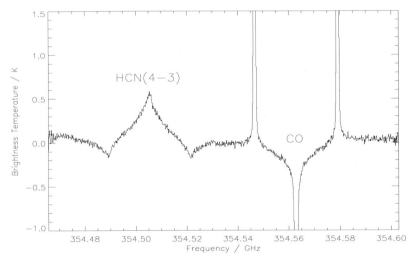

Figure 2. The HCN(4–3) line at 354.51 GHz (left part of the spectrum) observed at JCMT on a day without strong meteor shower activity (25th March, 1998). The frequency switch throw was 16.2 MHz and the integration time 1.5 h. The CO line on the right side of the spectrum comes from the image band of the radiometer and is therefore negative. The mesospheric CO line illustrates the expected width for high (> 70 km) altitude molecules.

The CSO measurements on November 19, 1999, were averaged over 1 hour intervals (2 hours for last scan), and shown in Figure 3. Sunrise in Hilo was at 16:31 UT that day.

We were expecting a small increase in the HCN content at high altitude from the direct release of molecules from ablated meteoroids. Instead, we saw a quite different behavior: in the night after the Leonid shower, the HCN concentration was much less than on other days, only recovering to previous line strength after sunrise (Figure 3). Such a night-to-day variation was not observed on the previous days leading up to the peak of the Leonid shower. Unfortunately, observing conditions prevented the acquisition of good data on the night of the peak itself.

2.4. Comparison with modeled spectra

The line profiles are directly related to the height dependence of the volume mixing ratio (abundance), because the atmospheric pressure broadening quickly decreases with altitude. Above 70 km altitude, Doppler broadening is the dominant broadening mechanism, and thus the line no longer contains specific information about the vertical abundance profile at high altitudes.

We applied the forward model part of the software called Microwave Odin Line Estimation REtrieval (MOLIERE) to simulate the measurements that were made at CSO and JCMT. MOLIERE was developed in the framework of the Swedish Space Corporation (SSC) Odin satellite mission to analyse the microwave measurements of the Sub-Millimeter Radiometer (SMR) onboard Odin (Baron, 1999; Baron et al., 2000). The forward model provides the atmospheric emission measured by radiometers, after taking into account the radiative transfer along the line of sight in a non-scattering atmosphere, the refraction of the signal through the atmosphere, and the satellite and radiometers characteristics. Temperature, pressure and molecular species concentration profiles are given by models or satellite data. Spectroscopic parameters are taken from the Jet Propulson Laboratory, Geisa, and HITRAN catalogues; hyperfine HCN line intensities and frequencies at 265.88 GHz are from Biver (1997). This code has been adapted for ground-based measurements and to the frequency-switch observing technique. It also contains an inversion procedure based on the Optimal Estimation Method (Rodgers, 1976), which was used in the next step of analysis to retrieve the vertical abundance profile of HCN from the spectra.

Figure 4 shows the HCN volume mixing ratio profile and the resulting line shape. The chosen HCN volume mixing ratio vertical profile uses the *a priori* MOLIERE profile from Figure 1. The result is char-

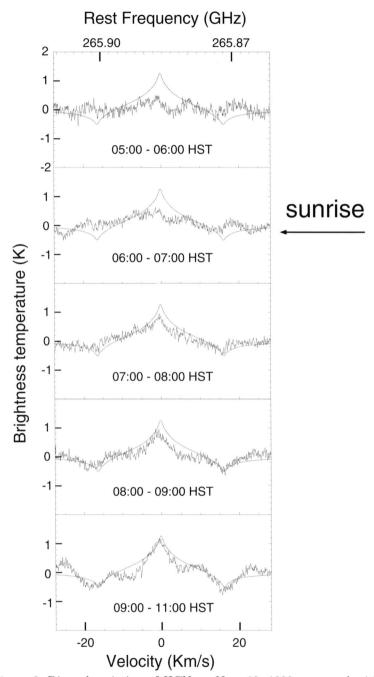

Figure 3. Diurnal variation of HCN on Nov. 19, 1999 measured with the Caltech Submillimeter Observatory (CSO). The J=3–2 rotational line of HCN at 265.88 GHz has been observed in frequency switch from 15 to 21 UT (05 to 11 HST Hawaiian Standard Time). Sunrise was at about 16:30 UT. The solid line shows a model spectrum of the HCN line computed with the MOLIERE code.

acteristic of our CSO observations on Nov. 19 during day time (Figure 3). It is also characteristic of HCN profiles measured at JCMT at other times in the year (Figure 2).

Figure 5 shows a modeled spectrum calculated for a volume mixing ratio that falls off rapidly above 45 km. This model spectrum reproduced the observed intensity. We conclude that during the night of Nov. 19, the mid altitude HCN concentration was significantly diminished.

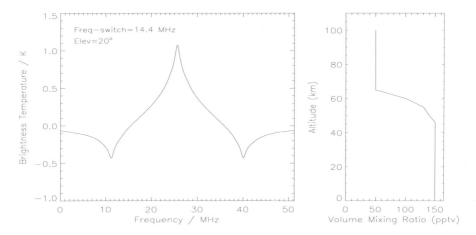

Figure 4. Simulation with MOLIERE of a CSO spectrum and the associated reference HCN volume mixing ratio profile (1pptv = 10^{-12}) used for the computation (here the same profile than the "MOLIERE: A priori" HCN profile plotted in Figure 1).

3. Discussion.

The present observations show no evidence for a direct input of HCN at mid and high altitudes, but it is possible that the meteoric ablation products affected the atmospheric chemistry in the mesosphere and lower thermosphere. Our HCN measurements were done at solar longitude 236.85-90 (J2000), in the night after the secondary maximum in the Leonid activity profile, which peaked over Hawaii at solar longitude 235.9 (Jenniskens *et al.*, 2000b). Any ablation products would have had time to disperse horizontally. However, it is not easy to understand how meteors can affect HCN concentration at such altitudes. HCN may have been chemically attacked by a reactive species such as ozone, atomic oxygen or electrons, known to be produced by the trails of the more massive meteoroids, which are ablated at these altitudes, or produced by lighter ones and transported from higher altitude. Alternatively, a coincidence with a purely atmospheric phenomenon is also

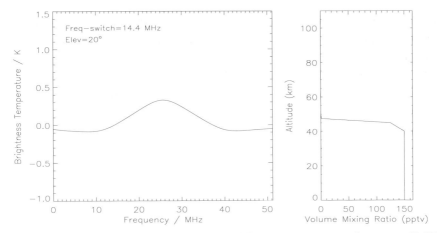

Figure 5. Simulation with MOLIERE of a CSO spectrum using a "truncated" HCN volume mixing ratio profile shown on the right of the picture. With this profile, which indicates no HCN above 45 km, we obtain a spectrum close (in intensity) to the one measured on Nov. 19, 1999 from 05:00 to 06:00 at CSO (see Figure 3).

not excluded, as nothing is known yet about the short-term behavior of this molecule due to gravity waves and tides. In order to establish the link with meteor shower activity, future observations will need to be combined with a characterization of the background airglow.

3.1. H_2O AND O_3 AT 55 KM

To investigate the possibility of ozone concentration variations, we investigated satellite data on this trace gas to search for any correlation with meteor activity. We also looked for H_2O as an indicator of hydrogen release.

Systematic measurements of O_3 and H_2O in the mesosphere have been performed with the HALogen Occultation Experiment (HALOE instrument) onboard the Upper Atmosphere Research Satellite (UARS, launched in September 1991). The HALOE instrument measures the absorption of solar radiation at sunrise and sunset by a variety of trace gases (Russell et al., 1993), sweeping from 80° South to 80° North in latitude approximately every month.

The data covers the years 1991 to 1999, with an altitude coverage up to 80 km. We created averages over various latitude bands and present in Figure 6 the results for the latitude band going from 30° South to 30° North across the equator, for an altitude characterized by a pressure of 0.46 hPa, which corresponds to 55 km according to the standard atmosphere. This altitude is reached by bright and slow meteors, hence a direct influence could be expected from annual shower activity. It is also possible that the ozone and water may be affected by meteoric input at higher altitudes. Ozone and water vapor data at higher altitudes (between 60 and 80 km, by approximately 3 km step) have also been observed by HALOE, which is more relevant for this study, but the signal to noise ratio in the data is much less.

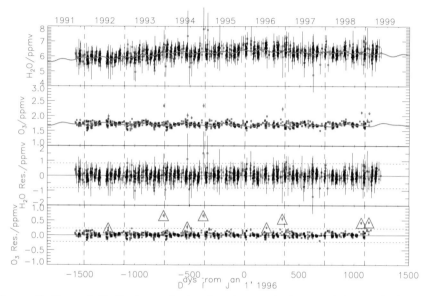

Figure 6. O_3 and H_2O atmospheric abundances (volume mixing ratio in ppmv; 1ppmv = 10^{-6}) at 55 km (1991-1999) from UARS/HALOE satellite measurements between 30° South and 30° North. Upper two curves show the rough data for H_2O and O_3 respectively, with the regression model (sinusoidal curve) superimposed. The lower two curves display the residuals (measurements minus model) after having removed the expected long term variations. Dotted lines indicate the 3σ level. Data points above this level are shown by triangles.

In the analysis of the data, we took into account semi-annual, annual, quasi-biennal, solar cycle oscillations (amplitudes and phases), and linear trend over the 9-year period. Using a linear regression model, we removed these oscillations in order to search for a correlation of any residual variation with a meteor shower.

At some periods, the O_3 abundance is more than 3σ above the regression model values (triangles in Fig. 6). However, no correlation has been found at present with annual meteor showers.

3.2. FUTURE WORK

With an eye on the 2001 and 2002 encounters, we note that radio sub-millimeter observations are very suitable for airborne application in a future Leonid Multi-Instrument Aircraft Campaign, in combination with existing airglow measurements. A good control of sensitivity and baseline irregularities can improve the quality of the observations. Other atmospheric minor constituents can be observed in a similar manner: for example, H_2CO and CH_3CN. In addition, satellite data are expected from the Microwave Limb Sounder instrument (MLS) onboard the UARS satellite (measurements of CH_3CN) and from the soon to be launched (December 2000) Odin Astronomy/Aeronomy satellite.

Acknowledgements

We thank referees Steve Charnley and Frans Rietmeijer for helpfull comments. This work was supported by grants from CNRS/INSU (G.D.R Exobiologie and Action Spécifique "Grands Télescopes Etrangers") and CNES, which also provided financial support to Nicolas Lautié. Didier Despois and Philippe Ricaud thank the CSO for providing excellent working conditions, very valuable technical help, and free accomodation at the telescope. CSO is supported by the National Science Foundation grant AST-9980846. *Editorial handling:* M. Fonda.

References

Abbas, M. M., Guo, J., Carli, B., Mencaraglia, F., Carlotti, M., and Nolt, I. G.: 1987, *Geophys. Res. Lett.* **14**, 531–534.

Baron, Ph.: 1999, *Développement et validation du code MOLIERE: Chaîne de traitement des mesures micro-ondes du Satellite Odin*, PhD thesis, Université de Bordeaux, France.

Baron, Ph., Ricaud, Ph., de La Noë, J., Eriksson, J. E. P., Merino, F., Ridal, M., and Murtagh, D.: 2000, *Can. Journ. of Phys.*, in press.

Biver, N.: 1997, *Molécules mères cometaires: observation et modelisation*, Ph.D. Thesis, Université de Paris VII, France.

Brasseur, G., Zellner, R., de Rudder, A., and Arijs, E.: 1985, *Geophys. Res. Lett.* **12**, 117–120.

Carli, B., Mencaraglia, F., and Bonetti, A.: 1982, *Int. J. Infrared Millimeter Waves* **3**, 385–394.

Chu, X., Pan, W., Papen, G., Gardner, C.S., Swenson, G., and Jenniskens, P.: 2000, *Geophys. Res. Lett* **27**, 1807–1810.

Chyba, C. F., and Sagan, C.: 1996, in P. J. Thomas, C. F. Chyba, C. P. McKay (eds.), *Comets and the Origin and Evolution of Life*, Springer Verlag, p. 147–174.

Cicerone, R. J., and Zellner, R.: 1983, *J. Geophys. Res.* **88**, 10689–10696.

Coffey, M. T., Mankin, W. G., and Cicerone, R. J.: 1981, *Science* **214**, 333–335.

Crutzen, P.J. and Andreae, M.O.: 1990, *Science* **250**, 1669–1678.

Despois, D., Biver, N., Ricaud, Ph., Dobrijevic, M., Baron, Ph., Kieken, J., Selsis, F., Jacq, T., Billebaud, F., Lis, D., Crovisier, J., Paubert, G., Schneider, N., and

Matthews,H.: 1999, Poster Abstract at *Asteroids, Comets and Meteors 1999*, Ithaca, USA, July 1999.

Feuchtgruber, H., Lellouch, E., de Graauw, T., Bézard, B., Encrenaz, T., and Griffin, M.: 1997, *Nature* **389**, 159–162.

Höffner, J., von Zahn, U., McNeil, W. J., and Murad, E.: 1999, *J. Geophys. Res.* **104**, 2633–2643.

Jaramillo, M., de Zafra, R. L., Barrett, J. W., Parrish, A., and Solomon, P. M.: 1988, *Geophys. Res. Lett.* **3**, 265–268.

Jenniskens, P., Packan, D., Laux, C., Wilson, M., Boyd, I.D., Popova, O., Krueger, C.H., and Fonda, M.: 2000a, *Earth, Moon and Planets*, **82–83**, 57–70.

Jenniskens, P., Crawford, C., Butow, S.J., Nugent, D., Koop, M., Holman, D., Houston, D., Jobse, K., Kronk, G., and Beatty, K.: 2000b, *Earth, Moon and Planets* **82–83**, 191–208.

Kopp, E.: 1990, *J. Geophys. Res.* **95**, 5613–5630.

Love, S.G. and Brownlee, D.E.: 1993, *Science* **262**, 550–553.

Matthews, C. N., Pesce-Rodriguez, R. A., and Liebman, S. A.: 1997, in C. B. Cosmovici, S. Bowyer, and D.Werthimer (eds.), *Astronomical and Biochemical Origins and the Search for Life in the Universe*, IAU Coll. 161, Editrice Compositori, p. 179–187.

Maurette,M.: 1998, in André Brack (ed.), *The Molecular Origins of Life: assembling pieces of the puzzle*, Cambridge Univ. Press, p. 147–186.

Rinsland, C. P., Smith, M. A. H., Rinsland, P. L., Goldman, A., Brault, J. W., and Stokes, G. M.: 1982, *J. Geophys. Res.* **87**, 11119–11125.

Rodgers, C. D.: 1976, *Rev. Geophys. Space Phys.* **14**, 609–624.

Russell, J. M., III, Gordley, L. L., Park, J. H., Drayson,S. R., Hesketh, W. D., Cicerone, R. J., Tuck, A. F., Frederick, J. E., Harries, J. E., and Crutzen, P. J.: 1993, *J. Geophys. Res.* **98**, 10777–10797.

Schneider, J., Bürger, V., and Arnold, F.: 1997, *J. Geophys. Res.* **102**, 25501–25506.

Spreng, S. and Arnold, F.: 1994, *Geophys. Res. Lett.* **21**, 1251–1254.

Zander, R., Rinsland, C. P., Farmer, C. B., Namkung, J., Norton, R. H., and Russell, J. M.: 1988, *J. Geophys. Res.* **93**,1669–1678.

ORGANIC MATTER IN DUST OF COMET 21P/GIACOBINI-ZINNER AND THE DRACONID METEOROIDS

NIKOLAI N. KISELEV
*Astronomical Observatory of Kharkov National University,
Sumskaya Str. 35, 61022 Kharkov, Ukraine
E-mail: kiselev@astron.kharkov.ua*

KLAUS JOCKERS
Max-Planck-Institut fur Aeronomie, Katlenburg-Lindau, Germany

and

VERA K. ROSENBUSH
Main Astronomical Observatory of National Academy of Sciences, Kiev, Ukraine

(Received 30 May 2000; Accepted 24 July 2000)

Abstract. Unusual behaviour of the polarization of scattered light on dust particles in comet 21P/Giacobini-Zinner is discussed. Contrary to all other comets, the wavelength gradient of the polarization of the continuum is negative. It can be associated with unusual properties of the dust particles. In particular, one of possible interpretations is connected with a high abundance of organic matter in its dust. The Draconid meteors associated with this comet support the presence of organic matter in the large dust grains of comet 21P/Giacobini-Zinner, which is depleted in carbon species in the gas phase.

Keywords: Comets--individual: 21P/Giacobini-Zinner, Draconids, meteoroid, polarimetry

1. Introduction

At present time, the association of meteor showers with periodic comets is well established by the similarities of their orbital parameters, but the correlation between the physical characteristics of dust in comets and meteoroids is less studied. Photometric and spectral data indicate that the chemical composition of most comets is very similar despite the fact that the gas and dust production rates may vary widely (A'Hearn *et al.*, 1995). There are exceptions. Comet 21P/Giacobini-Zinner (hereafter 21P/G.-Z.) is unusual in its chemical composition. It belongs to the group of "depleted" comets characterized by low abundances of carbon in the gas phase. That organics may be in the dust phase. Most dusty comets display similar polarimetric properties of their dust

(Levasseur-Regourd *et al.*, 1996; Jockers, 1997) but it is not yet clear whether the polarimetric properties of "depleted" comets differ from those of normal comets. Thus, the purpose of our observations of comet 21P/G.-Z. was the study of polarimetric features of a "depleted" low-carbon comet. We find that comet 21P/G.-Z. showed unusual polarimetric properties that can be constrained by the properties of the Draconid meteoroids associated with this comet.

2. Observations

CCD imaging polarimetry of comet 21P/G.-Z. was conducted at the 2-m telescope of the Pik Terskol Observatory (Northern Caucasus) on November 20, 1998 and January 25, 1999 using a two-channel focal reducer of the Max-Planck-Institute for Aeronomy (Germany) (Jockers *et al.*, 1997). It was possible with this polarimeter to carry out polarimetric measurements in two wavelength intervals simultaneously. Narrow-band continuum filters centered at 4430/44 Å (blue continuum) and at 6420/26 Å (red continuum) were employed. We also made similar observations of comet C/1998 U5 (Linear). The CCD images were processed using the Münich Image Data Analysis System (MIDAS) reduction package. The image processing included bias correction, flat field division, removal of cosmic ray events, the determination of the polarization parameters of the instrumental system, the alignment of subimages with different polarization direction, and the construction of polarimetric images. After correction for instrumental polarization the observed polarization degree of the polarized standard stars deviates on average from the catalog values by 0.3 ± 0.2 % and the position angle of the plane of polarization by 1 degree.

The mean degree of polarization of comet 21P/G.-Z. in an area 35,300 x 32,200 km^2 (November 20, 1998) and in an area 15,050 x 15,050 km^2 (January 25, 1999) are given in Figure 1. Comet 21P/G.-Z. was also observed with the 70-cm reflector at the Observatory of Kharkov National University on November 19, 21 and 22, 1998. We used a one-channel photopolarimeter with a fast-rotating polaroid. The comet was observed at large air masses and was somewhat faint for this telescope. Thus, we only used the wide-band red (7228/1140Å) filter and the 88 arcsec diaphragm (54,500 km at comet). Apart from November 22, the observation conditions were poor. The instrumental polarization and zero-point of the polarization plane were determined from observations of standard stars with zero and large polarization, respectively. Instrumental polarization, which does not exceed 0.1%, was subtracted from the cometary data. The observations and data reduction are described in more detail in the paper by Kiselev *et al.* (2000).

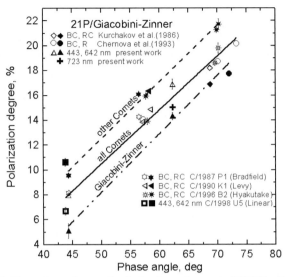

Figure 1. Phase dependence of polarization of comets 21P/Giacobini-Zinner, C/1987 P1 (Bradfield), C/1990 K1 (Levy), C/1996 B2 (Hyakutake) and C/1998 U5 (Linear) for the blue (open symbols) and red (filled symbols) continuum filters. All dusty comets display the same phase dependence of polarization in the blue spectral range (solid line). The polarization of comet 21P/Giacobini-Zinner in the red spectral range (dash-dotted line) is 2% less than in the blue range and 4 % less than the polarization of the other comets in the red range (dashed line).

3. Results

Figure 1 shows our data on the phase dependence of polarization of comet 21P/G.-Z. compared to those obtained by Kurchakov *et al.* (1986) and Chernova *et al.* (1993) for the same comet. We also show similar data for comets C/1987 P1 (Bradfield), C/1990 K1 (Levy) (Chernova *et al.*, 1993), C/1996 B2 (Hyakutake) (Kiselev and Velichko, 1998) and C/1998 U5 (Linear) that were observed at the same phase angles.

One can see that the degree of polarization of the dust in comet 21P/G.-Z. in the blue continuum is in a good agreement with that of dust in the comets Bradfield, Levy, Hyakutake, and Linear U5. At the same time the polarization of comet 21P/G.-Z. in the red continuum is consistently lower by about 2% than in the blue continuum. Comets Bradfield, Levy, Hyakutake, and Linear U5 display the commonly observed increase in the polarization with increasing wavelength for dusty comets. We can clearly see the anomalous wavelength dependence of polarization in comet 21P/G.-Z. No other comet observed previously showed a similar wavelength dependence of polarization.

4. Discussion

The NH_2 emission at $\lambda 6408$ Å and a weak unidentified emission at $\lambda 6428$ Å fall within the red continuum filter used for the imaging polarimetric observations (Kiselev et al., 1999). The depolarization effect depends on the relative contribution of emissions with respect to the continuum. This gas contamination may be very critical for comets with extremely low continuum and a normal abundance of NH_2 species. Konno and Wyckoff (1988) and Beaver et al. (1990) have found that NH_2 was depleted in comet 21P/G.-Z. and that its abundance was 5 times less than that in normal comets. At the same time, Sanzovo et al. (1996) pointed out that comet 21P/G.-Z. has an intermediate dust-to-gas mass ratio and consequently has intermediate strength of continuum with respect to normal comets. The NH_2 (0,8,0) $\lambda 6335$ Å band is the strongest transition in the discussed wavelength region. Using the fluxes of the (0,8,0) band of NH_2 and of a continuum window at 6250 Å of width 36 Å by Fink and Hicks (1996), we find the depolarizing influence of NH_2 when the observations were carried out in a band pass at $\lambda 6335$ Å of 36 Å width. The maximal depolarization in the NH_2 (0,8,0) band is only 1.7 %. For comet 21P/G.-Z. all NH_2 bands, which pass through the filters $\lambda 6420/28$ Å, RC, R and $\lambda 7228/1140$ Å are much weaker than the NH_2 (0,8,0) emission band (see Figure 4 in Fink and Hicks, 1996). According to Korsun (private communication) the flux ratio for the NH_2 (0,8,0) transitions to the NH_2 emission at $\lambda 6408$ Å is about 30 for comet C/1996 Q1 (Tabur). Consequently, the reduced polarization of comet 21P/G.-Z. in the red filters cannot be explained by the NH_2 contamination. Kiselev et al. (2000) described the procedure to account for the depolarization effect of gas emissions in more detail. We also checked if an inversion of the polarization color in the local parts of the coma can produce a negative spectral gradient of polarization for the whole coma of comet 21P/G.-Z.. We made cuts along the comet-Sun line through the red and blue polarimetric images obtained on 25 January. We found that polarization in the cut for the blue image is permanently larger than that in the red image.

Thus, it seems reasonable to assume that comet 21P/G.-Z.'s observed anomalous spectral dependence of polarization is caused by the properties of its dust particles. There is other observational evidence that comet 21P/G.-Z. differs from other comets. This "depleted" comet has low C_2, C_3, NH_2, and NH emissions (Cochran and Barker, 1987; Schleicher et al., 1987; Konno and Wyckoff, 1988; Beaver et al., 1990) and is also depleted in C_2H_6 (Weaver et al., 1999). The average color of comet 21P/G.-Z. is redder (BC-RC = 0.24^m) than that of comets Bradfield (0.07^m) and Levy (0.13^m) (Kolokolova et al., 1997), and Hyakutake (0.18^m) (Kiselev and Velichko, 1998). In addition, IR-photometry data by Hanner et al. (1992) showed that comet 21P/G.-Z. has a weak 10-μm silicate feature supporting a low abundance of submicron silicate grains. It should be pointed out that all comets with a low abundance of submicron silicate grains belong to a group of comets characterized by low degrees of polarization (10-15 %) for both the blue and the red domains of spectrum at large phase angles. Again we find comet 21P/G.-Z. to be an exception to this rule.

According to Kolokolova and Jockers (1997), the wavelength dependence of the degree of polarization may help to determine the composition of the cometary dust. These authors have shown that the spectral dependence of polarization is mainly determined by the spectral dependence of the absorption index of the materials that are present in the dust. They have performed Mie calculations of the wavelength dependence of intensity and polarization of particles consisting of different materials (see Fig. 4 in Kolokolova and Jockers, 1997). The power-low size distribution $n(r) \sim r^{-a}$ with index $a = 3$ was used. All calculations were carried out for phase angle close to that of maximum polarization, namely $\alpha = 90°$. Kolokolova and Jockers (1997) found that in the visible wavelength domain only the astronomical silicate shows increased polarization with increasing wavelength while the organic materials and particles of mixed astronomical silicate + organics provide a negative gradient of polarization. The size distribution of particles is another factor that determins the wavelength gradient of polarization. An abundance of large particles in the coma may also produce a redder colour as well as an anomalous wavelength dependence of polarization. The redder dust colour in comet 21P/G.-Z. can also be explained by a decrease in the imaginary part of refractive index with wavelength. Thus, we could explain the negative wavelength gradient of polarization seen in comet 21P/G.-Z. by a higher content of organic materials in its dust, a very high abundance of large particles, or a combination of both.

The observed relatively low abundance of C_2, C_3 and C_2H_6 in comet 21P/G.-Z. could mean that the volatile carbon-chain molecules are depleted but that other carbon-bearing molecules in this comet may not be depleted (Weaver et al., 1999). Therefore, we find no inconsistency between the reduced contents of carbon and presence organic matter in comet 21P/G.-Z. The presence of organic substances in the dust particles of comet 21P/G.-Z. may be inferred from properties of its associated meteor stream. It is well known (Jacchia et al., 1950) that the Draconids have short atmospheric trajectories (~10 km) that disappear at great heights at about 100 km. This behavior is consistent with meteoroids that are very soft, fragile and are quickly evaporated. According to Ceplecha (1977), meteoroids of the Draconid stream are composed of material with an extremely low density of about 0.2 g cm^{-3}. Wu and Williams (1995) analysed the data on observed showers of the Draconid meteor stream. They provided a computer simulation of the dynamic behavior of five hundred large meteoroids (0.25 cm) and five hundred small meteoroids (0.03 cm) which were ejected at each perihelion passage of comet 21P/G.-Z. from 1900 to 1985. They concluded that most of the observed features of the

Draconids can be explained if these meteoroids are very soft and have a very short lifetime.

Lebedinets (1987, 1991) has shown that the evaporation of the stony and iron meteorites begins with a temperature of T ≈ 2,300 K but that the upper temperature limit of evaporation of Draconid-type meteoroids is less than 1,000 K. Thus, Lebedinets (1987, 1991) has suggested that the Draconid meteors consist of the organic materials. If meteor velocities are 20 km sec^{-1}, the maximum altitudes of the beginning heights of the ablation trails of the Draconids should be 94 km. However, according to Jacchia *et al.* (1950), 73 % of meteors appeared at more than 94 km altitudes, while 29% of these meteor trails had ending heights above 94 km. It follows that from ~30% to ~70% of the Draconid meteors may be dominated by organic materials.

The presence of organic carbon in comets is beyond any doubt. Greenberg (1982, 2000) hypothesized that the complex organic molecules should be formed in core-mantle grains due to ultraviolet photoprocessing of the ices. The presence of organic materials in comets is confirmed by spectral feature in the 3.3–3.7 μm IR range. (Knacke *et al.*, 1987; Tokunaga *et al.*, 1987). Large amount of so-called CHON grains (up to 30 %) which associated with organics were first detected in comet 1P/Halley (Jessberger *et al.*, 1988). It is also obvious that the relative abundances of the organic substances may vary among comets. It is possible that comet 21P/G.-Z. has a relative large abundance of organic materials compared to all other comets. It is common knowledge that meteoroids of different streams have a different density, which vary in the range from 0.1 to 8 g/cm^3 (Babadzhanov, 1994). Meteoroids of different types ranging from iron to very rich in volatile materials occur in the streams of the Perseids, Taurids and α-Capricornids with parent comets 109P/Swift-Tuttle, 2P/Encke and 45P/Honda-Mrkos-Pajdusáková, respectively. Meteoroids with very volatile compounds, in addition to the Draconids, also occur in the Orionids and η-Aquarids which parent body is comet 1P/Halley. The Leonid meteoroids may be rich in carbonaceous materials (Jenniskens *et al.*, 2000). Its parent body is comet 55P/Tempel-Tuttle. The Geminids, Virginids, δ-Aquarids and τ-Aquarids have dense meteoroids. Future work should look for similarities between the physical characteristics of meteoroids of these streams and their parent bodies.

5. Conclusion

The results of CCD imaging and photoelectric polarimetry of comet 21P/G.-Z. have shown that the polarization of this comet in the red spectral range is 2% less than in the blue range and 4% less than the polarization of other comets in the red range. The unusual spectral trend of polarization of comet 21P/G.-Z. may be caused by a large content of organic substance in its dust and/or an overabundance of large particles. Presence of organic matter in dust of comet 21P/G.-Z.

is confirmed by the properties of the Draconid meteoroids associated with this comet.

Acknowledgements

We thank F.J.M. Rietmeijer for numerous suggestions that helped improve the paper, and A.C. Levasseur-Regourd and an anonymous referee for reviews. N.K. is grateful for the financial support from the Local Organizing Committee of the Leonid MAC-99 Workshop at the University of Tel Aviv (Israel). *Editorial handling:* Frans Rietmeijer.

References

A'Hearn, M.F., Millis, R.L., Schleicher, D.G., Osip, D.J., and Birch, P.V.: 1995, *Icarus* **118**, 223–227.
Babadzhanov, P.: 1994, in A. Milani, M. Di Martino, A. Cellino (eds.), *Asteroids, comets, meteors 1993*, Kluwer Academic Publishers, Dordrecht, pp. 45–54.
Beaver, J.E., Wagner, R.M., Schleicher, D.G., and Lutz, B.L.: 1990, *Astrophys. J.* **360**, 696–701.
Ceplecha, Z.: 1977, in A.H. Delsemme (ed.), *Comets, Asteroids, Meteorites, Interrelations, Evolution and Origin*. University Toledo Press, Toledo, 143–152.
Chernova, G.P., Kiselev, N.N., and Jockers, K.: 1993, *Icarus* **103**, 144–158.
Cochran, A.L. and Barker, E.: 1987, *Astron. J.* **92**, No.1, 239–243.
Fink, U. and Hicks, M.D.: 1996, *Astrophys. J.* **459**, 729–743.
Greenberg, J.M.: 1982, In L.L.Wilkening, (ed.), *Comets*, University of Arizona Press, Tucson, Arizona, pp. 131–163.
Greenberg, J.M.: 2000, *Earth, Moon and Planets* **82–83**, 313–324.
Hanner, M.S., Veeder, G. J., and Tokunaga, A.T.: 1992, *Astron. J.* **104**, No.1, 386–393.
Jacchia, L.G., Kopal, Z., and Millman, P.M.: 1950, *Astrophys. J.* **111** 104–133.
Jenniskens, P., Wilson, M.A., Packan, D., Laux, C.O., Krüger, C.H., Boyd, I.D., Popova, O., and Fonda, M.: 2000, *Earth, Moon and Planets* **82–83**, 57–70.
Jessberger, E.K., Christoforidis, A., and Kissel, J.: 1988, *Nature* **332**, 691–695.
Jockers, K.: 1997, *Earth, Moon, and Planets* **79**, 221–245.
Jockers, K., Rosenbush, V.K., Bonev, T., and Credner, T.: 1997, *Earth, Moon, and Planets* **78**, 373–379.
Kiselev, N.N., Jockers, K., Rosenbush, V.K., and Korsun, P.P.: 1999, ACM'99, Poster, Ithaca, NY, USA.
Kiselev, N.N. and Velichko, F.P.: 1998, *Icarus* **133**, 286–292.
Kiselev, N.N., Jockers K., Rosenbush V.K., Velichko F.P., Bonev T., and Karpov, N.: 2000, *Planet. Space Sci.*, in press.
Knacke, R.F., Brooke, T.Y., and Joyce, R.R.: 1987, *Astron. Astrophys.* **187** 625–628.
Kolokolova, L. and Jockers, K.: 1997, *Planet. Space Sci.* **45** 1543–1550.
Konno, I. and Wyckoff, S.: 1988, *Adv. Space. Rev.* **9**, 163–168.
Kurchakov, A.V., Nosov, I.V., Rspaev, F.K., and Churyumov, K.I.: 1986, *Komet. Tsir.* **No. 350**, 4.

Lebedinets, V.N.: 1987, *Astron. Vestn.* **21**, 262–271.
Lebedinets, V.N.: 1991, *Sol. Syst. Res.* **25**, 49–53.
Levasseur-Regourd, A.C., Hadamcik, E., and Renard, J.B.: 1996, *Astron. Astrophys.* **313**, 327–333.
Sanzovo, G.C., Singh, P.D., and Huebner, W.H.: 1996, *Astron. Astrophys. Suppl. Ser.* **120**, 301–311.
Schleicher, D.G., Millis, R.L., and Birch, P.V.: 1987, *Astron. Astrophys.* **187**, 531–538.
Tokunaga, A.T., Nagata, T., and Smith, R.G.: 1987, *Astron. Astrophys.* **187**, 519–522.
Weaver, H.A., Chin, G., Bockelee-Morvan, D., Crovisier, J., Brooke, T.Y., Cruikshank, D.P., Geballe, T.R., Kim, S.J., and Meier, R.: 1999, *Icarus* **142**, 482–497.
Wu, Z. and Williams, I.P.: 1995, *Planet. Space. Sci.* **43**, 727–731.

PREDICTING THE STRENGTH OF LEONID OUTBURSTS

ESKO J. LYYTINEN
Kehäkukantie 3 B, 00720 Helsinki, Finland
E-mail:esko.lyytinen@minedu.fi

and

TOM VAN FLANDERN
Meta Research, 6327 Western Ave, NW, Washington, DC 20015, USA
E-mail: tvf@mindspring.com

(Received 27 June 2000; Accepted 18 August 2000)

Abstract. A simple model is described that predicts the time of occurrences and peak activity of Leonid shower outbursts. It is assumed that the ejection speeds of escaping particles at each return of the parent comet near perihelion are very small, but solar radiation pressure acting differently on different particles causing a spread of particles into different period orbits. Earlier papers predicted the position of the resulting dust trails. This paper sets forth to better predict the strength of the expected outbursts by considering the role of non-isotropic effects in the interaction with the solar radiation on the dispersion of particles away from the dust trail center. This paper determines the approximate magnitude of the relevant effects. Predictions for the next few years are presented that include such considerations, for reasonable assumptions of particle properties. For example, earlier predictions for the 1999 storm of ZHR = 6,000–7,000 are now reduced by a factor of two, which is in better agreement with the observed ZHR ~ 4,000. The success of the technique, when applied to historic meteor storms and outbursts without need of additional free parameters, lends confidence to the soundness of the underlying model and to its application for future predictions. We predict that the best encounters of this return of the parent-comet will occur in the years 2001 and 2002.

Keywords: Comet, comet ejection, Leonids, meteor, meteor shower, meteoroid, model predictions, orbital dynamics, radiation pressure

1. Introduction

This work was initiated in early 1999 after the 1998 fireball outbursts arrived unexpectedly and far "off-schedule" (Lyytinen, 1999). Its premise, low ejection velocities, is based on the satellite model of comets, which was developed earlier (Van Flandern, 1981; 1999). In that model, a comet is supposed to have a nucleus with an orbiting debris cloud system. The mass of the debris cloud can be comparable to, or in excess of, the mass of the primary nucleus. It is postulated that at each perihelion there is gravitational escape of particles through the L1 and L2 Lagrangian points. Smaller particles, i.e., most meteor size particles, are actually stable only in inner orbits around cometary nuclei, and are driven out from orbits around the cometary nucleus mainly by radiation pressure. Slightly different, but regarding the formation and course of trails, essentially similar assumptions have been made in earlier work by Kondrat'eva and Reznikov (1985) and Kondrat'eva et al. (1997) and recently in the work by McNaught and Asher (1999). The first to suggest the importance of radiation pressure was Kresak (1976). The first to realize the formation of trains due to different orbital periods was Pavel (1955).

If the ejection speeds are very small with little dispersion and the orbital dispersion is caused mainly by solar radiation pressure, then the radiation pressure acts as if solar gravitation were decreased, thereby increasing the semi major axis, thus also the period of revolution, according to the particle size. That increase is proportional to the radiation pressure/mass ratio. As a result of differences in orbital period, the particles form trails after one revolution.

Transverse spread is caused by the pressure of non-symmetric absorption and re-radiation of solar radiation. It is the spread along the comet orbit and the transverse spread that are responsible for the observed intensity of the shower. This is usually expressed in terms of Zenith Hourly Rate (ZHR). Our predicted peak rate prior to the 1999 storm was ZHR = 6,500 ± 500 hr^{-1} (Lyytinen, 1999). After the observations of the storm suggested a peak activity of ZHR ~4,000 hr^{-1} (Arlt et al., 1999; Jenniskens et al., 2000), we re-addressed earlier concerns on such non-isotropic effects of the interaction of the meteoroids with sunlight. In this paper, we will introduce a better treatment of the radiation pressure effects to arrive at more reliable estimates of the peak activity of future meteor storms. Some other refinements of the model are also introduced.

2. Method

The backward integration of the parent comet was conducted as in Yeomans et al. (1996), adjusted to agree with observed returns since 1366. From each perihelion a number of particles (typically about 140) with different radiation pressure for each were integrated forward. The radiation pressure introduces an effective decrease of the gravitational pull for each particle. We recorded the time, radius vector and longitude at the descending node crossings of the 1950.0 ecliptic plane of each particle. Figure 1 gives a graphic presentation of the trails around the present return of the parent comet. The data was manually checked for encounters with the Earth.

Corrections for the true (mean) Earth orbital plane of the date were made for longitude and for the radius vector. Our earlier work did not include the radius correction (Lyytinen, 1999). The heliocentric distance of ecliptic crossing r_D is affected in the fifth decimal. In comparison to our earlier model, the perihelion distance q of the parent comet needs a reduction of 0.000077 AU according to the most recent orbital elements by Nakano (1998) This brings the adopted orbit close to the value used in McNaught and Asher (1999a). So each r_D value in the earlier results was reduced by this amount.

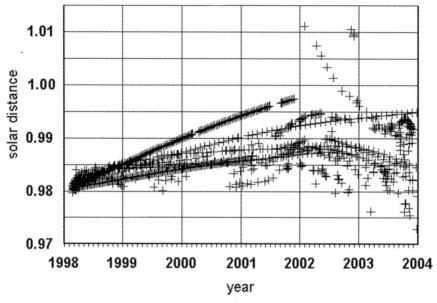

Figure 1. Distance from the Sun of modeled trail particles at the descending node. The vertical lines are at the start of January. The Earth-Sun distance is at about 0.988 to 0.989 AU. A concentration of points near this value in mid-November (as in the year 2001) indicates a likely meteor storm.

3. Treatment of radiation pressure

What is usually called simply radiation pressure is here called "regular radial pressure". It is a repulsive force, which is inversely proportional to solar distance, effectively lowering the gravitational force. Other radiation pressure effects can be non-radial and can have a different dependence on solar distance. Notably, variations in radiation force as a result of the particle's surface to mass ratio, rate of rotation, etc., will lead to dispersion of particles, effectively widening the trail perpendicular to the Earth's orbit and in the Earth's orbital plane. For the purpose of calculating the shower's activity, one can group the various effects according to their influence on the dispersion of the stream. The general term radiation force is used to cover all the effects that describe the particle's response to solar radiation.

3.1. THE REGULAR RADIAL PRESSURE

The radiation force on a particle consists of three parts: due to absorption of solar radiation energy, due to re-radiation of the thermal energy, and due to reflection (scattering). The radiation pressure force on a totally absorbing particle of diameter d, neglecting the other two effects, at a distance r from the Sun is:

$$F_r = \pi \, (d/2)^2 \, S_0 \, c^{-1} \, r^{-2} \qquad (1)$$

The constant S_0 is the solar radiation constant at Earth's distance and c is the light speed.

In the ideal case where there is no absorption and all the energy would be reflected directly backwards (to the Sun) like a flat mirror put normal to the solar radius, the total effect would be twice that given by (1). Similarly, in the ideal case that all the energy would be re-radiated as thermal radiation directly backwards toward the Sun, the force by re-radiation would be the same as that of scattering. Hence, if the particle is non-rotating, spherical, non-conducting and homogenous on its surface, we can assume that the total force due to absorption, re-radiation, and scattering is directed away from the Sun and obeys an inverse square law. Particles affected by solar radiation pressure in this manner are effectively in a weaker gravitational field. The motion of the particle without further perturbations is a Keplerian ellipse. Because the particles are assumed to escape into their own orbits near the perihelion of the comet's orbit, radiation pressure has the effect of creating a slightly wider elliptic orbit, with a slightly bigger semi-major axis, and longer orbital period.

The ratio of radiation pressure to gravity (β) for similar particles is inversely proportional to the particle diameter. It is mostly affected by the surface-to-mass ratio of the particles. So different-sized particles get into elliptic orbits with different orbital periods, thereby forming elongated trails without much tendency to spread perpendicular to the orbit.

Non-spherical particles can have different average surface-to-mass ratio, so actually it is expected that to some degree different mass particles will exist even with similar orbital periods. This also applies to particles with different densities.

The net effect of the radial component of the radiation pressure is to distribute the particles along the comet orbit and delay the population of particles relative to the comet after even a single orbit. Subsequent orbits cause this dispersion to increase by about the same amount each return. In the case of Leonid meteoroids, a radiation pressure of one thousandth of the gravity of the Sun will give a period of revolution of 380 days longer than that of the parent comet (which has an orbital period of about 33.3 years). The force of radiation pressure as described by (1) on a one–millimeter-diameter sphere with density 1 g cm^{-3} is about this big.

Key to the predicting encounter times for Leonid outbursts is the point where the so formed dust trail crosses the ecliptic plane.

3.2. PERTURBATIONS

The inverse-square and radial force law of Equation 1 is the starting approximation. All departures from this law are perturbations, that are discussed in the next sections. Such perturbations affect the shape of the orbit and the ecliptic plane crossing distances away from the Sun. Because the Leonid meteoroids are observed near the perihelion of their orbits, perturbations on the total angular moment arising far away from perihelion are of this type. The net effect is a dispersion of particles giving rise to the shape of the Zenith Hourly Rate (ZHR) activity curve of the meteor storm.

The purpose of this section is to show that there exist mechanisms that can quantitatively explain the scatter of particles, to the extent that it is sensible to construct a ZHR model.

3.2.1. The "A2 effect"

A non-graviational induced period change, over the regular radiation pressure and with little effect otherwise, is called here the "A2" effect, because of an analogy with non gravitational effects in comet orbital dynamics. The "A2" effect is in particular caused by perturbations on the speed of the particles due to a perturbing component along the plane and normal to solar radius that arises near perihelion. Such perturbation mainly affects the orbital period but hardly at all the ecliptic plane crossing radius. The Poynting-Robertson effect is an example of such a perturbation, even though very small. In the Poynting-Robertson effect, radiation is emitted with higher momentum towards the direction of motion, effectively slowing down the grain. The "A2 effect" by the other mechanisms described in this paper can be to either direction, at least over a short time span.

These perturbations do not directly affect the meteor outburst predictions because the particles that would encounter the Earth are replaced by others with similar ecliptic plane crossing radii and nodes. Only the orbital period determines whether particles reach the ecliptic when the Earth is nearby. The true effect of such perturbations is to change somewhat the encounter conditions and cause slightly altered differential planetary perturbations over its orbit. The magnitude of such perturbations is comparable with the radial effect for old trails and cannot be separated in a straight-forward manner. The spread of particles by this effect can be modeled by computer simulations that include planetary perturbations. The A2 effect mainly affects encounters with old trails.

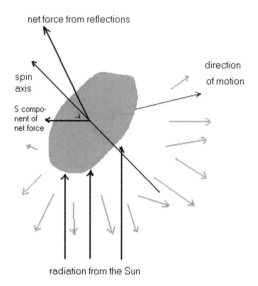

Figure 2. Non-isotropic scattering may give force components that change the orbit and, among others, affect the orbital period, i.e. produce an A2 effect.

3.2.2. Non-isotropic scattering of radiation

Consider the perturbing forces from radiation scattered non-isotropically from a dust grain in three directions: R, along the radius vector away from the Sun; W, normal to the orbital plane; and S, in the orbital plane towards the direction of motion and normal to the previous directions (Figure 2a). For both the components R and S, there appears a $\sin(2*(v+\text{constant}))$ type of dependency on the direction of the force (where v is the true anomaly; with a different constant for each particle and force component). This effect mostly concerns flattened particles, but can arise even for spherical particles with non homogenous surfaces. An imaginary extreme example would be a two-sided mirror such as a solar sail. This mechanism perturbs the period, but during one orbit the net is zero. This type of perturbation does not cause a net change of orbital period if there is no further dependency of solar distance r than the normal $1/r^2$. The conclusion of no orbital period change with this type of perturbation was derived from a simple program integrating the effect of the proposed mechanism during one orbit, running the program for various different spin axis directions of the particle. A test particle with the regular radiation pressure of one thousandth of solar gravity ($\beta = 0.001$) was used. It may be that the true dependency of these components is more complex than in the studied example and may cause some A2 effect as well.

This effect, even in the pure form, does cause a dispersion. This arises mostly from perturbations near aphelion, even though $1/r^2$ reaches its minimum there. Assuming that the component S is one hundredth of the total radiation pressure at aphelion, and that the test particle further obeys the "double sine law" sin(2*v+constant) (or in this case cos(2*v)), we find that our test particle above gives a shift in q of ± 0.00016 AU during one revolution. This applies only to a high albedo particle. If the particle albedo is small, this effect may be too small to explain much of the observed particle dispersion in meteor storms.

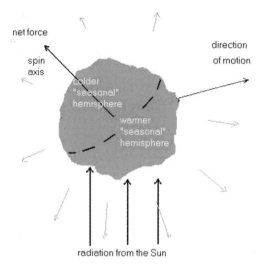

Figure 3. The mechanism for seasonal Yarkovsky is illustrated here. This effect can perturb the orbital motion of a particle much like the effect illustrated in Figure 2.

3.2.3. Non regular thermal re-radiation: the seasonal Yarkovsky effect

Previous perturbations do not consider rotation of the particle and a delay of re-emission, causing the (diurnal) Yarkovsky effect. This effect arises from the fact that the there is a lag in the temperature of the surface during one revolution causing the temperature in the "afternoon" to be larger than at a similar phase during the "morning". Because the rotation of a small body is quite fast, the diurnal Yarkovsky force is rather small (Olsson-Steel, 1987). The effect depends on thermal lag and unknown rotation properties.

Instead, we will consider what is now called the "seasonal Yarkovsky", which was recognized by Rubincam (1995, 1998) and Farinella *et al.* (1998) (e.g. see Rubincam, 1995; Vokrouhlicky and Broz, 1999). Their treatment is for asteroid sized objects, down to 0.1 m. The seasonal Yarkovsky effect arises from the tilt of the spin axis, which puts one hemisphere towards the Sun and the other away from the Sun. The thermal re-radiation from the "summer" hemisphere causes an effective force in the direction of the "winter" hemisphere, in a direction different from the radial direction towards the Sun. The effect depends on conveying thermal energy effectively to the night side within the diurnal period, by rotation. The effect does not depend on rotation properties (within a reasonable range of spin rates), but on distributions of the spin axis tilt. The strength of the seasonal Yarkovsky depends on a particle's thermal conductivity, as well as its size. For small grains, the seasonal effect is several orders of magnitude larger than the diurnal variant. The effect for asteroids is a modest semi major axis reduction (Rubincam, 1995; Vokrouchlicky and Broz, 1999).

We derived an approximate formula for this force in the case of spherical particles with a distribution of spin angles. We assume that the Sun's radiation is totally absorbed and calculate the difference in re-emission from the day and night time sides. The radiation from the shadowed side is equal to the thermal energy going by conduction through the particle. The thermal conduction is proportional to the difference in temperature of the two hemispheres. We find for a particle whose spin axis is pointed towards the Sun:

$$F = J S_o d^2 (c r^2)^{-1} (1 + L (K/d) r^{3/2})^{-1} \qquad (2)$$

where J is an (adjusting) factor, whose effect will get introduced in the numerical coefficient in equation (4), K is the thermal conductivity and:

$$L = 2^{-1/4} S_o^{-3/4} \sigma^{-1/4} \qquad (3)$$

where σ the Stefan Bolzmann coefficient. Note that as thermal conductivity K approaches zero, this will give the distance dependency $1/r^2$. Hence, the force from thermal re-radiation given by the seasonal Yarkovsky effect, in the case of a black body and axis directed to the Sun, is proportional to the gravity at the distance in question. Equation 2 can be re-written as:

$$F_{SY} = 0.18 \beta (1+L (K/d) r^{3/2})^{-1} \qquad (4)$$

If the body is not black, the resulting forces must be multiplied by the absorption coefficient; i.e., with (1–albedo). The force components vary with sin(2 x angle).

For the purpose of checking the accuracy of the approximate solution of non linear equations, we put equation (4) in the general form:

$$F_{SY} = 0.18 \beta (1+H)^{-1} \qquad (5)$$

where we treat $H = L (K/d) r^{3/2}$ as one variable. The maximum proportional error is only about 4 per cent. The H range with this size error is around 0.05 – 0.4, while it is smaller outside this range. This size error arises from the approximations in the solution of nonlinear equations. Approximations in the spherical shape geometry will introduce further errors, but the particles are not expected to be spherical in reality.

The ecliptic crossing of the Leonids is close to the perihelion of the orbit. So it was numerically computed how the Equation 4 force affects q. We assumed that the spin axis is situated in the orbital plane, keeping its orientation during one orbit and varying the direction in the plane with different runs to get some maximal values for the variability of q by this effect. Within the same computer runs, the change in semi major axis was computed. This gives the possible change of the period of revolution in one orbit; i.e., the A2 effect.

With near-zero thermal coefficient K, the maximum change of q during one revolution is about 0.003 au. The A2 effect is zero. When K increases, the change of q decreases quite rapidly while the A2 effect increases before reaching a maximum. At its peak, the orbital period change is about two weeks over one orbital revolution. Unfortunately, this is in the parameter range for H (0.05 – 0.4) where the assumptions that go into Equation 4 may break down. Olsson-Steel (1987) gives the value of K for terrestrial silicates as K = 3.5 W m^{-1} K^{-1}. The value of K for returned lunar samples is only about one percent of this. This should be further reduced by about one and a half orders of magnitude to get into the region of the biggest A2 effect (for a one mm particle). With the values of the lunar samples, an A2 effect a bit less than a day is reached. With this value, the change of q is only about 0.00001 au. A change in q essentially translates directly to a change in the radial ecliptic plane crossing distance of the particle, r_D. This is too little to explain the direct spread of r_D inferred from observations. Hence, the seasonal Yarkovsky effect will either change q (and r_D), or create a larger dispersion along the orbit.

We have not studied the seasonal Yarkovsky effect normal to the plane, but the perturbing normal component has the dependency (at least approximate) of sin(angle-along-plane). This angle is from the node of the "particle equator" in the orbital plane. So this clearly has an effect on the plane and causes some scatter normal to the orbital plane.

In the case of a very flat particle, the mean surface to mass ratio is now bigger than with spherical particles having a given value of β resulting in a bigger mass particle than a spherical particle with the same β value. Further, for a sheet-like particle, the range of total radiation pressure force varies from its maximum value to practically zero. If the Sun shines normal to the particle spin axis, then the temperature of the particle falls. This actually gives a further reduction of the re-radiation force in the direction of the spin axis.

3.2.4. *Effects due to precession of the particle spin axis*

Olsson-Steel (1987) considered the effects due to precession of the particle spin axis for a particle with 1 mm radius, whereas we have it at

0.5mm. So it is quite close to this particle size. For an assumed spin rate, Olsson-Steel calculated the angular change of axis by precession as approximately 10^{-6} rad s^{-1} (at 1 AU). Although Olsson-Steel noticed that this would imply that the spin axis revolves around the Sun-meteoroid line several times in each orbit, we find a lower precession rate. Olsson-Steel takes as the force for torque the whole radiation pressure, which is clearly too coarse an approximation. Actually, for a spherical particle, the torque is zero both for absorption and thermal re-radiation. This applies also for flattened elliptic particles, if we neglect diurnal effects and the Poynting-Robertson type effects are also neglected in all cases. For scattering, it is also zero for a sphere, but non-zero for elliptic flattened particles. With flattened particles, it may be reasonable to expect the precession rate to be about two orders of magnitude smaller than the value above. This would imply that the spin axis can remain more or less fixed in space for several orbital revolutions.

For the prediction of meteor storms, the assumption of fixed orientation is reasonable. However, at the extreme Δa range (large initial difference in orbital period) as in the outburst of 1969, this is no longer true, but the model and predictions have not been altered for this. As a consequence, the predicted 2006 and maybe 2007 outbursts will have perhaps twice the ZHR given in Table I below.

4. The meteor stream model

From the previous considerations, we have constructed a stream model that is partly empirical by aiming to comply with past Leonid shower observations, but derives some properties from the assumed radiation pressure spread in section 3. For the most part, the form of the equations have a theoretical basis.

The spread into different orbital periods by the assumed regular radiation pressure is already embedded within the trail computations. These computations will also give the spread along the orbit that is expressed by the *mean anomaly factor* f_M, as introduced in McNaught and Asher (1999a).

The radial dispersion from the trail center occurs by the mechanism of non-isotropic scattering or thermal re-radiation of solar radiation, as discussed in Section 3.

It is assumed that the dispersion affects all three dimensions, one of which appears in f_M. The spread in the two other dimensions is assumed proportional to orbit number. This is because the rotation axis is expected to maintain its rough direction for several orbits giving cumulative spread effects. Further it is proportional to original Δa (difference of original semi major axis from that of the comet at ejection time). This comes from the fact that the non-regular radiation effects can be assumed to be proportional to the regular force. This is especially true for scattering, but less so for the seasonal Yarkovsky effect. The direct radial scatter is expected to be largely due to the scattering.

The radial density ρ_r of the trails is assumed to be of the form:

$$\rho_r \sim (1 + \Delta r^2)^{-0.5p} \quad (6)$$

where Δr is the difference in radial ecliptic plane crossing distances of Earth and meteoroid $r_E - r_D$, scaled according to:

$$\Delta r = (r_E - r_D) / (0.00059 \times n \times \Delta a) \quad (7)$$

where n is the number of orbits since ejection and $(r_E - r_D)$ and Δa are in astronomical units. Equation 6 was derived from the observed activity profiles of past meteor storms with the assumption, that the profile shape radially (from Sun) is the same as normal to the orbital plane. In equation 7, the terms n and Δa arise, because the trail width is expected to widen directly with n and also with an increase of Δa, practically resulting in a widening with increasing distance from the comet. From published graphs of the 1966 storm (Jenniskens, 1995; Mason 1995; Brown 1999), the free parameter p was derived as $p = 2.7 \pm 0.7$. The scaling factor 0.0002 in Equation 7 was derived from the 1966 storm case, which had $n = 2$ and $\Delta a = 0.169$. In addition, the densities are assumed proportional to the distance from the center raised to the power -2.7 far away from the trailet center.

We now introduce the empirical function $fn(\Delta a)$, which is the ZHR for a one-revolution central encounter having the original Δa. Noticing further the mentioned spread (and dilution) directly in two dimensions with n and further with f_M, we get this expression for ZHR:

$$ZHR = fn(\Delta a) \ (f_M / n^2) \ (1 + \Delta r^2)^{-0.5p} \quad (8)$$

In principle, there may be two ways to improve the Δr scaling Equation 7. One is to compare the dispersion scaling radial and normal to the orbital plane. The other is to get a mutual fit for $fn(\Delta a)$ for central and non central passes (for example comparing storms in 1833 and 1966). A trial with the first would change the above from $\Delta r = 0.0002$ to 0.00025 AU. However, the parameter was left as originally assumed.

An approximation of $fn(\Delta a)$ was derived from a fit of observed outbursts. The peak ZHR observations used are the same as used by McNaught and Asher (1999a). For one of the outbursts this was actually the only source available, while for others there was no clear reason to change those values. The course of this function with points derived from observations is shown in Figure 4.

The resulting predictions are given in Table I. The table lists the year of return, the number of revolutions since ejection (n), the difference in radial ecliptic plane crossing distances (the miss distance), the function f_M, the difference in semi-major axis, the predicted Zenith Hourly Rate, the predicted time of the peak (in solar longitude, J2000), the corresponding date, and some comments on what mechanisms may affect the characteristics of the shower.

TABLE I

Year	Rev.	rE-rD	fM	delta a	Pred. ZHR	Sol.long. 2000.0	Date Nov.	Time UT	Comments
2000	2	-0.0012	0.55	0.30	215	235.270	17	07:50	mostly faint
2000	4	0.00080	0.135	0.116	700	236.279	18	07:50	
2000	8	0.00080	0.250	0.065	700	236.103	18	03:40	fM locally bigger, but strongly affected by A2 ef
2001	4	0.00025	0.135	0.144	6100	236.467	18	18:22	
2001	5	0.00178	0.114	0.095	60	236.29	18	14:10	
2001	6	0.00135	0.123	0.080	110	236.20	18	12:00	
2001	7	-0.00043	0.140	0.081	2000	236.115	18	09:58	
2001	9	0.0001	0.260	0.043	1500-2000	236.433	rev:s 9,10,11		strongly affected by A2 ef
2001	10	0.0006	0.160	0.030	600	236.423	ZHR comb.		
2001	11	0.0004	0.160	0.026	260	236.425	18	ab. 2500 to 3000 ab. 17:30	
2002	4	-0.00004	0.148	0.174	7400	236.894	19	10:44	
2002	5	0.00148	0.115	0.120	160	236.72	19	06:45	
2002	7	-0.00013	0.130	0.114	4500	236.612	19	04:02	
2003	15/14 co/me	-0.001 to -0.003	1 - 2.5 extr.conf.!	0.28	250	230.69 to230.78+	13 !!!	13 to19	very special also uncertain
2006	2	-0.0002	0.470	0.94	50	236.618	19	04:48	maybe strong teles-
2007	2	-0.0004	0.560	1.06	30	236.107	19	22:55	copic outbursts

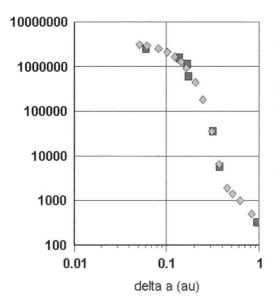

Figure 4. The graph shows the approximation of fn(Δa) from seven observed peak Zenith Hourly Rates. Squares are derived from observations of 1866, 1999, 1966, 1833, 1869, 1867, and 1969. Diamonds show the derived numerical approximation.

5. Discussion

5.1. THE A2 EFFECT

For young trails, the A2 effect is expected to have a relatively small influence on the predictions. Even for not quite so old trails (n ~ 4–6), this effect is probably dominated by planetary perturbations at large radial distances. For old trails, the A2 effect typically makes the ZHR smaller (by increasing the dispersion in the shower), but can also increase the rates by decreasing the miss distance (r_E–r_D). The inclusion of this in the ZHR-formula was tested by assuming that it mostly affected the radial direction. However, there appear to be cases where it mostly shifts the longitude of the ecliptic crossing. Further, the effect is hard to scale, and frequently the situation seems to be very irregular, for example f_M changing quite strongly with the applied effect. The 11 rev. 1903 outburst ZHR as compared to the prediction is very suggestive of the existence of this effect and in principle may be used for scaling. This was an encounter with small miss distance. The ZHR may have been a few hundred (Jenniskens, 1995). The model gives a post-prediction value of around 3600 without assuming an A2 effect.

Even though this effect was not included in the ZHR formula, this has been to some degree considered in the predictions. For older trails (7 revolutions and up) this effect is in some cases assumed to reduce the ZHR, and this may increase the ZHR in the n = 10 case in 2001. This effect also changes the timing, maybe up to hours for distant encounters. The effect on timing is not included in the table data, except in the special case in 2003, where this effect may help to bring particles near the Earth orbit a few hours after the nodal crossing. The suggested timing is stretched a few hours past the nodal crossing.

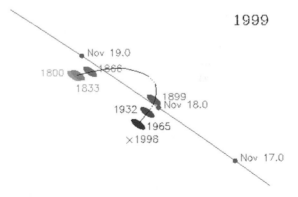

Figure 5. Drawing a smooth curve trough the trails in the order of n may be a practical means of timing far encounters such as with the 1866 trail. Original Figure by David Asher, reproduced and edited with permission.

5.2. POSSIBLE PRACTICAL MEANS OF TIMING DISTANT ENCOUNTER OUTBURSTS

While looking at the David Asher ecliptic plot for the year 1999, we recognized a special behavior in the timing. Even though the region around 16 to 20 UT Nov 18 is not very well mapped by observations, it appears as if the distant encounter with the 5 rev trail would have happened some hours prior to the nodal encounter (Jenniskens *et al.*, 2000). A treatment as in Figure 5 may explain the timing. It is as if there were some kind of sheets existing between the trails. This is probably explained by A2 type effects. This may suggest that some particles are more affected than would otherwise be anticipated.

5.3. CONSIDERATIONS ON THE WIDTH (DURATION) OF AN OUTBURST

In principle, this model should also give the width of outbursts at least for close encounters. Some first exploration of parameter space was made, but not a thorough treatment to present here. For young trails, spreading in length almost uniformly with orbit number, the width should be about proportional to the time interval from the passage of the parent comet. This relation seems not to hold for the observed 1969 outburst with large original Δa.

The comparison of observations with the model is greatly complicated by the effect of the miss distance. Before the 1999 storm, we assumed that the profile of the trails in all directions of cross section is the same. This would lead to a purely mathematical means of deriving the change of the width according to the miss distance. The observed duration of the 1999 storm shows that this assumption led to an overly large duration prediction. The conclusion was that the assumptions on the cross section are not true. Actually, the observed behavior may give a further clue to the cause of the dispersion. Further modeling of these radiation pressure phenomena is suggested, by considering, simultaneously, the dispersion near the descending node radially and normal to plane. With this modeling, the observed relative widths radially and normal to orbital plane should also be in agreement. A comparison of observed duration (further changed to a normal–to-the-orbit-plane value) with the radial dependency according to the model, gives the radial spread to be about 2.5 times the normal for young trails. Because of the inclination of the cometary orbit, the scatter happens to be roughly equal in the ecliptic plane both radially and along the direction of Earth's motion.

Jenniskens *et al.* (2000) deal with the profile shape and width of the 1999 Leonid storm as measured very precisely from airborne observations during the Leonid Multi-Instrument Aircraft Campaign. They find a good fit with a Lorentz curve shape. Interestingly, the distribution in Equation 6 does also have a Lorentz shape when $p = 2$ (we have it 2.7 ± 0.7). This paper's ZHR model could be easily transformed to use this Lorentz form. Of course, the $fn(\Delta a)$ would need to be determined accordingly. The effect on the predicted ZHRs would probably not be too much if the observations are concentrated in the same Δr range as the predicted cases.

In addition, Jenniskens *et al.* (2000) find that the whole trailet pattern is shifted by about 0.0003 AU from the assumed dependency and plot of observed meteor storm profile widths. This conclusion depends on the assumption that the Lorentz shape applies to the radial distribution as it does to the normal distribution. As pointed out by Jenniskens *et al.*, this can not be proven yet due to lack of data. For the widening not to happen as the simple assumption in Jenniskens *et al.*, requires a more rapid decline at angles between radial and normal from the trails. Interestingly, our model finds that the profile can have this Lorentz shape in both radial and normal directions to the plane, and yet not widen as expected. For example, if y represents the radial distance from trail center and x that normal to the orbital plane, then these are already scaled according to the mutual central half-widths (to get a simplified expression). If the ZHR is of the type:

$$ZHR(x,y) = ZHR_{max} / (1 + x^2 + y^2) \qquad (9)$$

then the observed Lorentzian broadening in the x-direction (with increasing pass distance y) does arise. Here, ZHR_{max} is the maximum corresponding to the center of the trail, and not the observed maximum. However, the same broadening will not arise when the dust density in the trailet has the form:

$$ZHR(x,y) = ZHR_{max} / ((1+x^2)(1+y^2))$$

$$= ZHR_{max} / (1+x^2+y^2+x^2y^2) \qquad (10)$$

although the distribution is of Lorentz form both in the x and y directions, but not at intermediate cross sections. The last form shows that with this, the decline is more rapid into other than axial directions. In our model (Equation 8), the expressions in parentheses, either $(1+x^2+y^2)$ or both $(1+x^2)$ and $(1+y^2)$, should be raised to the power $p/2 = 2.7 \pm 0.7 / 2 = 1.35 \pm 0.35$.

If there is scatter in q caused "somewhere", and the scatter normal to the plane is caused somewhere else, and these are sufficiently independent from each other, then there doesn't arise the otherwise expected dependency of outburst duration with miss distance. The observations are suggestive that there is no or little dependency of duration on the miss distance to somewhat out of the trail centers, but there is dependency at larger miss distances.

It is expected that one key factor with the dependency of the width is that the ecliptic crossing happens near perihelion. So angular momentum changes normal to the plane near aphelion, where the efficiency of non-isotropic scattering of solar radiation is high, do not efficiently cause scatter normal to the orbital plane (near perihelion), whereas in-plane effects there do affect the change of q. The seasonal Yarkovsky effects at a distance of roughly 2 AU may be mainly responsible for the scatter normal to the plane. Both these depend on the direction of the spin axis. However, it is not known without special modeling if these effects are independent enough to explain the observations.

5.4. UPCOMMING ENCOUNTERS

Let us examine the predictions in Table I. As noticed before (Kondrat'eva and Reznikov, 1985; McNaught and Asher, 1999; Lyytinen, 1999), very strong meteor storm encounters are still to come. Especially in 2001 and 2002 rates can go up to ZHR = 6,000 or almost twice the level of activity of the 1999 meteor storm. That event has now ZHR = 3,200 in our model, not too different from the observed value of ZHR = 3,700 (Arlt et al., 1999). The ZHR value of 2,500 to 3,000 in the line 2001 and 11 rev. means the combined value of trails with 9, 10 and 11 revolutions.

The first encounter ahead is that of the year 2000. In both the n = 4 and 8 rev. cases, the trail centers will pass inside the Earth's orbit. It has been the other way (outside Earth's orbit) with all the observed outbursts, and we caution that this supposed symmetry (as in the model) in r_E-r_D may not be true. Also, the 2000 encounter with the n = 8 rev. trail is actually with a "piece" of trailet cut on both sides by perturbations of earlier passes near the Earth. By purely gravitational solution, the mean anomaly factor seems to be locally quite high, but it is expected that the A2 effect smoothens the trail fragment into a longer piece in space. In that case, the A2 effect does not seem to help (applied in either way) bring the particles closer to the Earth. In the prediction table, the mean anomaly factor has been dropped into about one third of the purely gravitational local value, but this is not much more than guessing (except that it does need reduction).

Further ahead, we expect that the encounters with young trails, mainly in 2006 and 2007, will be vastly more abundant in smaller particles (about one fourth of a mm in diameter), which will lower the rates in less perfect conditions. The special case in 2003 may also belong to this group.

The closest pass of the one-rev trail from 1965 (not included in the prediction table) happens in 2005 at about 0.0006 AU, whereby the trailet is passed inside the Earth's orbit. This may give practically no visual meteors, but may give a brief outburst of 0.1mm particles instead that are a target of telescopic observations. The miss distance from the trail center is only about 2.5 times the geosynchronous distance. The geocentric nodal crossing is at solar longitude 235.569. This is on Nov 17 at about 21:52 UT (2005). The encounter with geosynchronous satellites on the day side will be up to about a half-hour later, depending on the exact location of the satellite.

Looking back in time, the 1998 "storm component" demonstrates a large dispersion of dust perpendicular to Earth's orbit. Considering the miss distances for various trailets, it would seem that the "storm component" was most likely caused by the three revolution trail (1899). This trail would have passed at about 0.004 AU from the Earth's orbit.

On the other hand, this trail had a gap (caused by perturbations of a near pass by Earth in 1965) of about one month in just the position of the encounter in 1998. This region of the trail should have been empty. The trail from 1932 would have passed at 0.005 AU, and that from 1965 at almost at 0.007 AU. Because of the near pass of Jupiter around the 1899 return, the nodes of earlier trails don't fit at all. On the other hand, the presence of particles in this region can be explained if there exists quite a large A2 effect, a non-gravitational period change.

6. Summary

Considering the very good post-predictions of historic Leonids storms and the successful prediction of the 1999 storm, with the trail models as in Kondrat'eva and Reznikov (1985), McNaught and Asher (1999), and Lyytinen (1999), it is reasonable to expect that good timing predictions for future outbursts can now be made. The strength of outbursts is more difficult to predict. In this paper, we addressed various aspects that can affect the dust density (as reflected in the Zenith Hourly Rate). A semi-empirical model was built for the dust density in the trails that gives a good fit to past observations. It is therefore expected that predictions for future meteor storm encounters will also not be too far off. The main uncertainty is the assumed symmetry in the radial dispersion to both sides of the trails. Also, there are no observations yet of encounters with old trails from five to ten revolutions ago. The radiation pressure effects are found to be strong enough to explain the total observed dispersion away from the trail centers. A better understanding of these effects needs further attention.

Acknowledgments

We thank David Asher for support of the research by enabling direct comparisons of integrations, and discussions on the treatment of radiation pressure effects. This work was partially sponsored by Meta Research. We thank the editor for numerous improvements of the manuscript. This work furthers the goals of the Pro-Amat Working Group of IAU C22. *Editorial handling*: Peter Jenniskens.

References

Arlt, R., Bellot Rubio, L., Brown, P., and Gyssens, M.: 1999, *WGN, Journal of the IMO* **27**, 286–295.
Brown, P.: 1999, *Icarus* **138**, 287–308.
Jenniskens, P.: 1995, *Astron. Astrophys.* **295**, 206–235.
Jenniskens, P., Crawford, C., Butow, S., Nugent, D., Koop, M., Holman, D., Houston, K., Kronk, G., and Beatty, K.: 2000. *Earth, Moon and Planets*, **82–83**, 191–208.
Kondrat'eva, E. D. and Reznikov, E. A.: 1985, *Sol. Syst. Res.* **19**, 96–101.

Kondrat'eva, E. D., Murav'eve, I. N., and Reznikov, E. D.: 1997, *Sol. Syst. Res.* **31**, 489–492.
Kresak, L.: 1976, Bull. *Astron. Inst. of Czechoslovakia* **27**, 35–46.
Lyytinen, E.: 1999, *Meta Research Bulletin* **8**, 33–40.
Mason, J.W.: 1995, *Journal of The British Astronomical association* **105**, 219–235.
McNaught, R.H. and Asher, D.J.: 1999a, *WGN, Journal of the IMO* **27**, 85–102.
McNaught, R.H. and Asher, D.J.: 1999b, *Meteoritics Planet. Sci.* **34**, 975–978.
Nakano, S.: 1998, *Minor Planet Circular* 31070
Olsson-Steel, D.: 1987, *MNRAS* **226**, 1–17.
Pavel, M.: 1955, J. *Atmos. Terr. Phys.* **2**, 168–177.
Rubincam, D. P.: 1995, J. Geophys. Res. **100**, E1, 1585–1594.
Van Flandern, T.: 1981, *Icarus* **47**, 480–486.
Van Flandern, T.: 1999, *Dark Matter Missing Planets & New Comets*, North Atlantic Books, Berkeley, CA, 2nd edition, chapter 10.
Vokrouhlicky, D. and Broz, M.: 1999, *Astron. Astrophys.* **350**, 1079–1084.
Yeomans, D.K., Yau, K.K., and Weissman, P.R.: 1996: *Icarus* **124**, 407–413.

GLOBAL GROUND-BASED ELECTRO-OPTICAL AND RADAR OBSERVATIONS OF THE 1999 LEONID SHOWER: FIRST RESULTS

P. BROWN[1], M.D. CAMPBELL[1], K.J. ELLIS[2], R.L. HAWKES[3,*],
J. JONES[1], P. GURAL[4], D. BABCOCK[3,5], C. BARNBAUM[6],
R.K. BARTLETT[7], M. BEDARD[8], J. BEDIENT[9], M. BEECH[10],
N. BROSCH[11], S. CLIFTON[12], M. CONNORS[13], B. COOKE[12],
P. GOETZ[14], J. K. GAINES[7], L. GRAMER[15], J. GRAY[1],
A.R. HILDEBRAND[16], D. JEWELL[17], A. JONES[1], M. LEAKE[6],
A.G. LEBLANC[3, 18], J.K. LOOPER[6], B.A. MCINTOSH[19],
T. MONTAGUE[20], M. J. MORROW[9], I.S. MURRAY[3, 21],
S. NIKOLOVA[1], J. ROBICHAUD[1], R. SPONDOR[22], J. TALARICO[17],
C. THEIJSMEIJER[1], B. TILTON[23], M. TREU[24], C. VACHON[22],
A.R. WEBSTER[1], R. WERYK[1], AND S.P. WORDEN[25]

* Physics Department, Mount Allison University, 67 York St., Sackville, NB,
Canada E4L 1E6 E-mail: rhawkes@mta.ca

[1]University of Western Ontario, London, ON, Canada
[2]Communications Research Centre, Ottawa, ON, Canada
[3]Mount Allison University, Sackville, NB, Canada
[4]Science Applications International Corporation, Arlington, VA, USA
[5] present address: Centre for Research in Earth and Space Science, York University, Toronto, ON, Canada
[6]Valdosta State University, Valdosta, GA, USA
[7]United States Air Force, Pentagon, Washington, DC, USA
[8]HQ United States Air Force Directorate of Weather, Washington, DC, USA
[9]Meteor Group Hawaii, Hawaii, USA
[10]Campion College, University of Regina, Regina, SK, Canada
[11]Wise Observatory and Tel Aviv University, Tel Aviv, Israel
[12]NASA Marshall Space Flight Center, Huntsville, AB, USA
[13]Athabasca University, Athabasca, AB, Canada
[14]United States Air Force, HQ AFSPC/DORW Colorado Springs, CO, USA
[15]North American Meteor Network, Florida, USA
[16]University of Calgary, Calgary, AB, Canada
[17]HQ USAF Space Command, Colorado Springs, CO, USA
[18]present address: Saint Marys University, Halifax, NS, Canada
[19]London, ON, Canada
[20]United States Air Force, Washington, DC, USA
[21]present address: University of Regina, Regina, SK, Canada
[22]Department of National Defence, Ottawa, ON, Canada
[23]United States Air Force, HQ AFSPC/XPXY, Colorado Springs, CO, USA
[24]Space and Reconnaissance Requirements Div., United States Air Force, Washington DC, USA
[25]HQ United States Air Force, Pentagon, Washington, DC, USA

(Received 12 June 2000; Accepted 17 August 2000)

Abstract. A total of 18 image intensified CCD detectors were deployed at 6 locations (two in Negev Desert, Israel, and one in each of the Canary Islands, Long Key in Florida, Haleakala in Hawaii, and the Kwajalein Atoll) to provide a real-time reporting system, as well as data for subsequent detailed analysis, for the 1999 Leonid shower. Fields of view ranged from 9 to 34 degrees, with apparent limiting stellar magnitudes from about +7 to +9. In addition, a dual frequency (29.850 and 38.15 MHz) automated meteor radar with directional determination capability was located in the Canadian Arctic at Alert, Nunavut and provided continuous monitoring of the shower from a location where the radiant was constantly above the horizon. Both the radar and electro-optical systems successfully recorded the activity of the shower in real time, and typical real-time activity plots are presented. Post-event analysis has concentrated on the Israel electro-optical wide field cameras and the time interval centered around the peak of the storm. About 2700 meteors have been digitized, with 680 measured for this analysis. Of these 371 were well enough determined to permit a single-station technique to yield approximate heights. Light curves and photometric masses were computed for these 371 Leonids which form the basis of the preliminary results reported in this paper. These cameras recorded Leonid meteors with peak luminosity in the magnitude range –3 to +5, corresponding to the photometric mass range 10^{-4} to 10^{-7} kg. A regression plot of photometric mass with magnitude did not indicate any change in light curve shape over the interval studied here. The peak flux as determined by the electro-optical observations was 1.6 ± 0.1 Leonid meteors of magnitude +6.5 or brighter falling on a one square kilometer area (oriented perpendicular to the Leonid radiant) per hour. This peak flux occurred at approximately 2:07 ±: 06 UT on Nov 18 1999, corresponding to solar longitude λ_o = 235.248 (epoch 2000.0). The radar results were consistent with this maximum flux rate and time. There was not a strong change in mass distribution over the few hours around maximum, although there is some indication that the peak interval was stronger in fainter meteors. Height histograms are provided for beginning, maximum luminosity and ending heights. It was found that maximum luminosity and ending heights were completely independent of mass, consistent with a dustball model in which the meteoroids are fragmented into constituent grains prior to ablation of the grains. However, the beginning height increases sharply (9.1 km per decade of photometric mass change) with increasing mass. This is possibly indicative of a volatile component which ablates early in the atmospheric trajectory.

Keywords: Leonids 1999, meteor, meteor flux, meteor shower, meteoroids, satellite impact hazard

1. Introduction

A series of distributed ground based stations can offer a cost-effective means of monitoring meteor showers. This paper will describe a network of six electro-optical locations and one multi-frequency radar location which were used to provide near real-time data on the 1999 Leonid shower to an operations control center. The main real-time goal was to provide mass-dependent flux information to satellite operators in near real time because of concern regarding the potential Leonid hazard to satellites (Beech *et al.*, 1995, 1997; Cevolani and Foschini, 1998). A secondary goal of the global ground-based campaign was to provide information on the mass distribution, atmospheric trajectories, orbits and physical structure of the Leonid meteoroids. This report provides some preliminary results from those analyses.

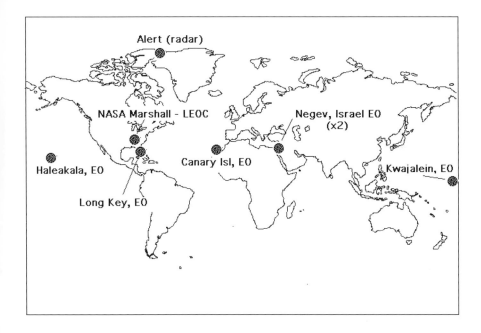

Figure 1. Map showing the location of the six electro-optical (EO) sites around the world, and the location of the dual-frequency radar in Alert, Canada, and the LEOC (LEonid Operations Control center) at NASA Marshall Space Flight Center in Huntsville, AB.

2. The network

We show in Figure 1 the observing locations used in this campaign. The central data node was located at NASA Marshall Space Flight Center in Huntsville, AL, USA. At this location, all data from the radar and electro-optical stations were processed in near real-time, with activity and predictions then forwarded to satellite operators (Treu et al., 2000). The electro-optical (EO) stations were located at six different locations, all of which were somewhat near 23° north latitude, so that the radiant would be high in the sky near dawn, and were selected to provide good longitude coverage.

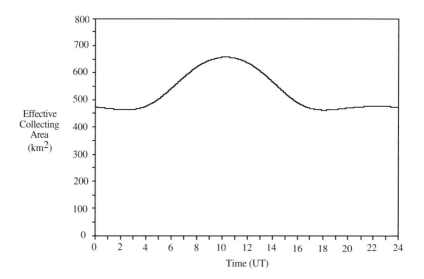

Figure 2. Effective radar collecting area for the Leonid radiant on November 18, 2000 from Alert.

The radar station was located at Alert, Nunavut, Canada (83° N, 68° W). From this location, the Leonid radiant is low in elevation but always above the horizon thus allowing for continuous observation. At this latitude, the specular reflection line for the radiant reaches almost overhead - a nearly optimum condition for radio based meteor studies. As a result, the effective echo collecting area for the radar, calculated as described in Brown and Jones (1995), varies only marginally throughout the course of a day (Figure 2). The radar system itself consists of two identical Skiymet radars that were operated in synchronism at frequencies of 29.850 and 38.150 MHz. A summary of the operational parameters of these systems is given in Table I. To obtain directional information, identical five-antenna interferometer arrays were used at each frequency with antenna spacings of 2 λ and 2.5 λ as illustrated in Figure 3. Using this configuration, the position of the specular reflection point could be compared with each echo direction and used to discriminate against non-Leonid meteors. As the radars were equipped with fully automated meteor detection software (see report by Hocking in this volume) real-time reporting of echo rates and other quantities was possible. It should be noted that an added benefit of this system is that dual-frequency simultaneous echoes can be used to derive an echo rate correction factor that can then be used to compensate for initial trail radius effects, since the loss in echo strength will be strongly wavelength dependent.

TABLE I

Parameter	Radar 1	Radar 2
Freq. (MHz)	29.850	38.150
Pulse Width (μs)	13.3	13.3
PRF (Hz)	2144	2144
GR (dBi)	6.2	6.2
GT (dBi)	7.0	7.0
PT (kW)	3.78	6.13
# Receivers	5	5

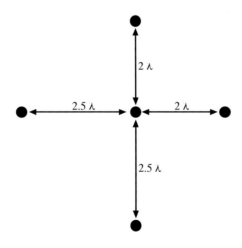

Figure 3. Interferometer spacing used in the receiving antennae.

The primary electro-optical location was in the Negev Desert in Israel. The predicted maximum of the storm would occur at about 4 am local time from this location. Two stations were set up, one at the Wise Observatory, near Mitzpe Ramon (30° 35' 45" N, 34° 45' 48" E) and the other near Revivim Kibbutz (31° 1' 45" N, 34° 42' 30" E). These were used in a triangulation mode (baseline 48.5 km) with four microchannel plate (MCP) image intensified CCD cameras at each location. Two image intensifiers at each site were generation II devices and two were generation III. The total spectral response extends from about 340 to 870 nm, although the generation III response is stronger in the red and near infrared than for the generation II systems which employ a near visual S-20 photocathode. Different focal length lenses were used at each site, resulting in different fields of view and limiting sensitivities, in order to enhance the determination of the mass distribution index. C-mount objective lenses with focal lengths from 25 to 75 mm were used, producing fields of view ranging from 35° to 9°, and a maximum limiting stellar magnitude on the most sensitive systems of nearly $+9^m$. The limiting meteoroid mass for the most sensitive cameras was approximately 2×10^{-8} kg for Leonid meteors. All CCD cameras used in the campaign were Cohu model 4910 scientific monochrome units operated at NTSC video frame rates (30 fps, with two interlaced video fields per frame). Three of the wide field cameras were equipped with MeteorScan real-time meteor detection software (Section 3).

At La Palma, in the Canary Islands (approx. 29N 16W) was located a secondary electro-optical site, with two gen II image intensified CCD cameras and two gen III image intensified CCD cameras. Each type of camera had one medium field of view (50 mm focal length lens) and one wide field (25 mm focal length) lens. Two of these cameras were used with MeteorScan automated detection software. At Long Key, FL, USA (approx. 24° 48' 48" N 80° 50' 0" W) were located two wide field gen II intensified CCD cameras, with one connected to MeteorScan automated detection. At Haleakala, HI, USA (approx. 20 N 156 W) was an electro-optical site which employed two wide angle generation III image intensified CCD cameras, with one connected to MeteorScan automated detection. At Kwajalein Atoll, Marshall Islands (approx. 9 N 167 W) was the final electo-optical site, with two generation III image intensified cameras, with one connected to MeteorScan automated detection. Unfortunately, on peak night weather conditions were poor on the

Canary Islands and at Kwajalein Atoll, and variable in Key West. Conditions were excellent at the Israel and Hawaii locations.

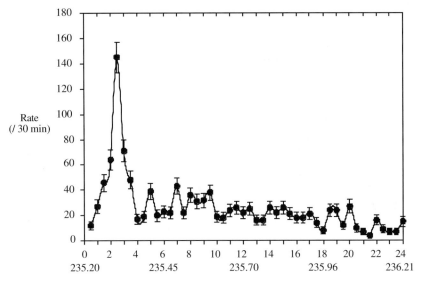

UT Nov 18 1999 on top; solar longitude (epoch 2000.0) on bottom

Figure 4. Automated detection results for the 38 MHz radar system for the day of Nov. 18 UT 1999. The rates are given as number of Leonids detected per 30 minute interval.

3. Real-Time Performance

Both the radar and the electro-optical systems employed real-time detection and characterization software. The output for the UT day Nov. 18 1999 for the 38 MHz radar unit is shown in Figure 4. The peak was clearly characterized by this real time data. Various real time displays could be remotely downloaded from the radar computers, including plots of radiant distribution, height distribution, flux and the echo profiles for individual events.

The electro-optical systems employed MeteorScan v2.1, developed by Pete Gural of Science Applications International Corporation (SAIC). The software runs under G3/G4 Macintosh systems, utilizing SCION LG-3 video digitizing cards. Full video frames (640x480 pixels) are digitized at 30 fps. After photometric and positional calibration, and determination of noise statistics for the CCD output, the software first performs frame subtraction to look for transient events. A noise map of the camera is performed (and updated) so that the mean and standard deviation of pixel brightnesses are known and used to determine significant transient enhancements in pixel brightness. A Hough transform algorithm is applied to significant transient bright points in order to search for linear segments (meteors). The user has control of three settings which can be used to set the sensitivity/false alarm performance of the detection system. When a linear segment is detected, the software determines the pointing direction and angular velocity, and does a check against possible meteor showers. Statistics were reported in 15 minute intervals during the Leonid campaign, providing number of members of each shower, as well as the number of unassociated meteors. One improvement to this version of MeteorScan was the addition of a routine which integrated the light intensity of the meteor trail to obtain a measure of the meteor magnitude (after comparison with stellar data obtained in the calibration segment). Magnitude distribution information was provided by MeteorScan in real time, although these real-time magnitudes were not consistently accurate. All magnitudes reported in this paper are based on post-campaign analysis, and not on the real-time MeteorScan values.

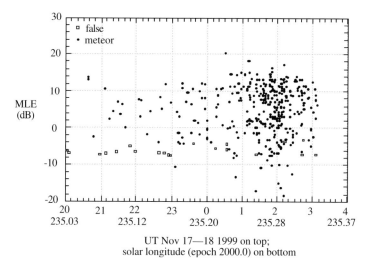

Figure 5: Plot of MeteorScan detections for one wide field camera located in Israel. The closed circles are real meteors (subsequently confirmed) while the open squares are false detections. MLE is the maximum likelihood estimator expressed in dB. It can be seen that there were very few false detections, and none of them were associated as Leonid meteors by the software.

In Figure 5 is shown the data collected from MeteorScan on peak night from one of the Israel wide field (25 mm focal length lens) cameras. The maximum likelihood estimator (MLE) in dB is plotted for each detection, with time on the x-axis. The higher the MLE the more certain the detection is a true event. After subsequent confirmation by a human observer, a different symbol was applied to meteors (circles) and false detections (open squares) in the plot. Clearly the performance of MeteorScan was outstanding, with very few false detections. None of the false detections which were recorded were incorrectly associated with the Leonids in the shower association segment of the software, so the real false detection rate of Leonids for this camera was 0%. This represented a significant improvement in MeteorScan performance over the software employed in Australia and Mongolia in the 1998 Leonid ground based campaign.

While there was some loss of MeteorScan detection efficiency during the peak of the shower, nevertheless the raw data as reported in real time accurately recorded the relative peak of the shower. In Figure 6 we show a plot of the MeteorScan output for the three Israel cameras which used real time detection, as a raw rate per 15 minute reporting interval. The maximum of the shower at approximately 2:00 to 2:10 UT Nov. 18, 1999 is evident even in these real-time data. The automated real-time reporting from both the radar and the electro-optical sites was successful.

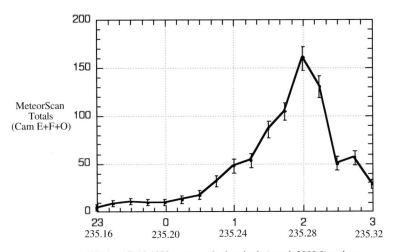

UT Nov 17–18 1999 on top; solar longitude (epoch 2000.0) on bottom

Figure 6: Real-time output of the MeteorScan automated meteor detection software on three of the wide field of view image intensified CCD cameras from Israel, reported in 15 minute intervals.

4. Post-Campaign Electro-Optical Analysis

We will now describe the post-campaign analsysis of a subset of the data collected from the electro-optical cameras. At the time of writing, the post-observation data analysis has concentrated on the cameras based in Israel, which had excellent weather during the peak of the Leonid shower. All peak-time (00–04 UT Nov 18, 1999) meteors from the Israel cameras have been digitized (~2,700 meteors, with the vast majority being Leonids). About 680 meteors have been measured, and a single station trajectory analysis (see below) performed on 371 well-measured Leonids. Magnitudes and photometric masses were determined for these 371 meteors. These data form the basis of the preliminary results presented here. It should also be mentioned that the data reported here are from the wide field of view (25 mm focal length) cameras which had a field of view of about 25° x 34° and which had an apparent limiting stellar sensitivity of about +7.5^m. The narrower field of view cameras had apparent limiting stellar sensitivity values of about +9.0^m. In this paper only the wide field results are analyzed, leaving the results of the narrow field cameras for a later paper. In the post-campaign re-analysis all meteors are redigitized (640x480x8 bit) and all meteor occurrences are human confirmed.

For positional work we employed the approach described by Hawkes *et al.* (1993). The pixel coordinates of a number of reference stars are measured, and then one applies a least squares fit between these coordinates and what an ideal (no distortion) camera would see (this is the *plate constants* approach described in detail by Wray (1967), Smart (1977), and Marsden (1982). The image measurement system provides a planar coordinate pair (x,y) for each point of interest (i.e. reference stars and meteor points). We then map (x,y) to a second planar coordinate system (ξ,η) which corresponds to what an ideal (zero distortion) system would observe. The η axis corresponds to motion along a great circle pointing in the direction of the north celestial pole, while the ξ axis points in the direction of increasing right ascension. The origin of the (ξ,η) system should be the plate center (or the reference star nearest to the plate center). Typically 15 reference stars were used in determining the positional coordinates.

Next the single station technique uses a three step process to assign shower associations. Both the direction of the meteor path and the angular velocity must be consistent with the shower. In the case of multiple shower possibilities, the best fit is assigned. Following shower association, the single-station analysis technique uses an assumed radiant and the apparent angular velocity to determine an approximate range to the meteor, and from that atmospheric trajectory information. The single station analysis technique is best suited for meteors with a relatively large number of frames, and covering a significant total angle. Under these conditions height errors may be as low as one to two km. However, in the case of meteors with only a few frames, height errors using this single station technique may be 15 km or worse. Precise dual station triangulation will be performed on meteors observed from both stations in Israel, but that data is not available for this report. Additional details on the single station analysis technique are provided by Hawkes et al. (1998), and it is essentially based on the pioneering work of Duffy et al. (1987).

For meteor photometry the method developed by Hawkes et al. (1993), Fleming et al. (1993), Campbell et al. (1999), and Murray et al. (1999) was used to determine the astronomical apparent magnitudes of these meteors. Essentially this procedure as follows. First, an area of interest (AOI) is drawn around a number of reference stars, and then a summation is obtained of the pixel intensity values over the AOI. A similar summation is obtained for background areas near the reference stars (the area just outside the star is used), and is subtracted from the reference star summation. A regression is found between the logarithm of this corrected pixel intensity summation and the astronomical magnitude, using a set of reference stars (done at the same time as the positional calibration). A similar AOI and background intensity summation is performed for the meteor points, and magnitudes obtained using the regression. Previous research on slewed stellar sources have validated this procedure (Fleming et al., 1993; Campbell et al., 1999).

The photometric mass is determined by integrating the meteor intensity equation (see e.g. McKinley, 1961; Ceplecha et al., 1998) over the entire luminous trail of the meteor in order to obtain the following relationship for the photometric mass:

$$m_p = \frac{2}{\tau_I v^2} \int I \, dt \qquad (1)$$

We have taken the velocity term out of the integral based on the assumption there is very little deceleration over the part of the trajectory when most of the light is emitted for meteors in the size range under consideration. It is natural to expect that higher velocity meteors will be more effective in producing light. Based upon experiments at moderately low meteor velocities, it is usually assumed that the luminous efficiency factor varies linearly with velocity. If we make this assumption we obtain the following result for the photometric mass:

$$\tau_I = \tau_0 v \qquad (2)$$

$$m_p = \frac{2}{\tau_0 v^3} \int I \, dt \qquad (3)$$

For those meteors with part of the light trail outside the field of view a correction should be applied to the apparent photometric mass. We have not applied such corrections here, but we did not include photometric analyses for those meteors deemed to have significant parts of the light curve undetected. The value of the luminous efficiency parameter above is based on the work of Verniani (1965) and has the value of 1.0×10^{-10} if one places all quantities in SI units and the resulting intensity is expressed in terms of number of 0 magnitude stars (McKinley, 1961 quotes a relationship by which a 0 magnitude meteor radiates 514 W in the visual range of the spectrum). It should be stressed that the luminous efficiency factor for fast meteors, such as the Leonids, is highly uncertain, since laboratory experiments cannot be performed at these speeds. These uncertainties would influence the absolute photometric masses, but not the relative masses important for scaling mass trends.

4.1. PHOTOMETRIC MASS AS A FUNCTION OF LEONID ASTRONOMICAL MAGNITUDE

For the data from one of the cameras (E) we plot in Figure 7 the logarithm (base 10) of the photometric mass (in kg) versus the astronomical magnitude at peak luminosity. It can be seen that with this wide field intensified CCD camera the Leonid meteors had astronomical magnitudes between -3 and $+5$, and photometric masses between 10^{-4} kg and 10^{-7} kg. The following regression relationship is obtained between photometric mass m, limiting astronomical magnitude at peak brightness M and the zenith angle z:

$$\log m \text{ (kg)} = -4.98 \pm 0.02 - 0.43 \pm 0.01 \, M - 0.07 \pm 0.26 \log(\cos z) \quad (4)$$

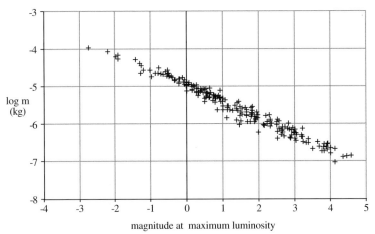

Figure 7. Plot of log photometric mass (in kg) versus meteor magnitude at peak luminosity for Leonids detected by one of the wide field intensified CCD systems.

The cos(z) dependence does not produce a statistically significant result since the meteors shown in this plot were all recorded over a very short time interval near Leonid maximum and there was insufficient variation in zenith angle. This relationship can be used to convert magnitude to photometric mass values for Leonid meteors in the magnitude range −3 to +5. If there was no change in shape of the meteor light curve with mass, one would expect a value of −0.40 for the magnitude term, which is approximated by our results. If one had a change in ablation mechanism, which resulted in changes in light curve shape, over the region investigated then the graph of log(m) versus M should not be linear. It is clear from Figure 7 that there is no indication of a dramatic change in light curve over the interval reported here. Since the relationship used to convert meteor shower population indices r to mass distribution indices s assume no change in shape of the light curve, it is important to establish the validity of this assumption.

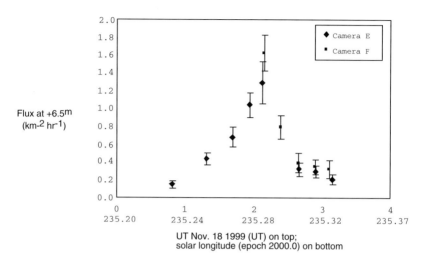

Figure 8. Meteor flux, expressed as number of Leonid meteors brighter than +6.5 magnitude which would impact on a one square kilometer surface perpendicular to the Leonid radiant plotted versus UT on Nov. 18, 1999. This is a compilation of data from two wide-field image intensified CCD cameras, with different symbols for each camera.

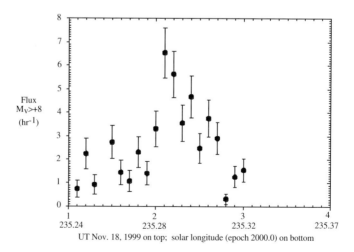

Figure 9. Leonid meteor radar flux divided into 6 minute time intervals for the 38 MHz meteor radar at Alert. The flux is to a limiting equivalent stellar magnitude of +8.

4.2. LEONID FLUX

One of the main goals of the campaign was to establish the Leonid flux in order to assess the hazard posed to satellites. While it is still early in the data analysis, we present in Figure 8 (based on the electro-optical data) a plot of the flux of Leonid meteors brighter than +6.5 astronomical magnitude versus universal time on Nov. 18, 1999. The flux is expressed in terms of number of meteors which would impact a one square kilometer surface which was oriented perpendicular to the Leonid radiant in one hour. The peak flux is 1.6 ± 0.1 Leonids brighter than +6.5^m per square kilometer per hour, and occurs at a time of 2:07 ± 6^{min} UT Nov 18 1999, which would correspond to a solar longitude value of λ_o = 235.248 (epoch 2000.0). This is consistent, within the precision of the small sample size, with the value obtained by an analysis of data from visual observers (Arlt et al., 1999) which found a peak flux of 1.4 Leonid meteors brighter than +6.5^m per square kilometer per hour, and occurs at a time of about 2:02 UT Nov 18 1999, which would correspond to a solar longitude value of λ_o=235.285 (epoch 2000.0). A detailed plot of the flux, as deduced from the 38 MHz radar, is shown in Figure 9. Here the flux is given to the radar effective limiting magnitude of +8.0. Only echoes which were within 3 degrees of the specular point from the Leonid radiant are included. We have used the radar data on November 14 (treated in the same way) as a control and subtracted these values in the same intervals to account for sporadic contamination. We have also corrected for initial trail radius effects, having computed that only 11% of all Leonid echoes are actually detected at 38 MHz, Assuming a Gaussian height distribution. This supports the idea that the maximum flux corresponded to a time of about 2:07 UT.

One interesting question is whether the mass distribution changes during the time of the intense peak. We plot in Figure 10 the photometric mass of each meteor recorded with wide field cameras E and F versus the universal time. There is no clear indication of a mass shift as the maximum is approached. Another method of looking at the data is presented in Figure 11, in which the hourly rate is reported for two different limiting photometric mass values. There is a slight hint that the central hour of most intense activity was mainly an enhancement of the fainter meteors.

182 GLOBAL ELECTRO-OPTICAL AND RADAR OBSERVATIONS

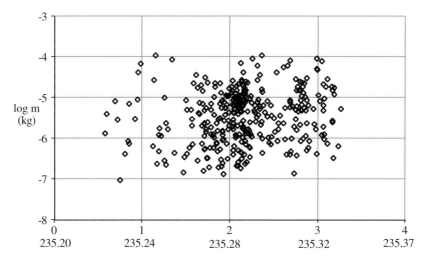

Figure 10. Plot of detections from two cameras around the time of the peak of the 1999 Leonid shower, with log photometric mass (in kg) plotted against universal time.

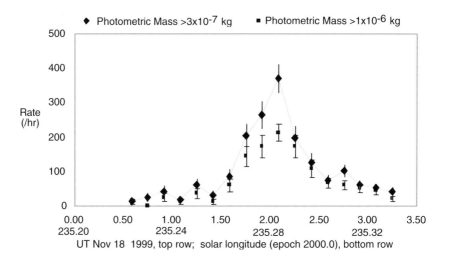

Figure 11. Meteor rate versus time (based on the electro-optical results) plotted for two limiting magnitudes (all meteors larger than 3×10^{-7} kg, and only those meteors larger than 1×10^{-6} kg). There is some hint that the peak of the shower was enhanced mainly in smaller meteors.

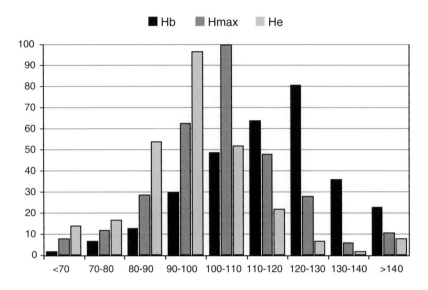

Figure 12. Histograms for beginning, maximum luminosity and ending heights for Leonid meteors. Note that uncertainties on some individual Leonids may be more than 10 km in height.

4.3. HEIGHT DISTRIBUTION

Heights of meteor atmospheric ablation profiles are one of the best indicators of the physical structure and chemical composition of meteoroids. For example, a composition rich in volatile organics would be expected to have beginning heights which are much higher than a meteoroid rich in less volatile metals. In-flight fragmentation would be expected to shorten the length of the light curves. Precision heights will await our future publication with the triangulation analysis. However, the single station procedure outlined earlier provided approximate heights. Figure 12 shows histograms of the beginning, maximum luminosity and ending heights for the meteors observed with two of the wide field of view cameras in the interval near the Leonid maximum. These results are consistent with the higher precision triangulation based heights which were measured for the 1998 Leonids (Campbell *et al.*, 2000), which indicated a mean beginning height of 114.1 km and a mean ending height of 97.1 km.

For comparison, Figure 13 shows the height distribution of Leonid echoes for all of Nov 18 at 38 MHz. Using the exact same selection criteria, the height distribution of echoes from Nov 14 are also shown. The average height increases by almost 8 km on the night of the Leonid maximum to 98 km, with a pronounced plateau above 110 km. Indeed, two maxima are visible; one just below 100 km associated with the "sporadic" component of the selected echoes (as is also visible on November 14) and a pronounced maximum at 110 km almost certainly due to the shower. The strong drop-off above 110 km is entirely due to the initial trail radius effect, suggesting strongly that the peak in the height distribution is higher still.

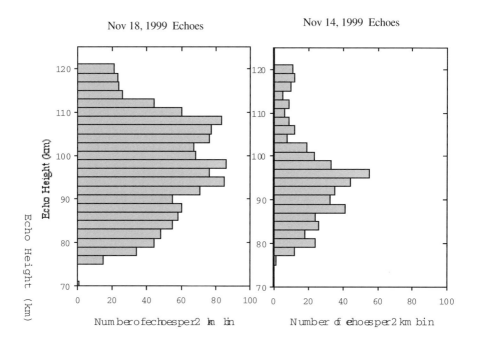

Figure 13. Radar height distribution at 38 MHz on Nov 18, 1999 (left) and on the control day Nov 14, 1999 (right). The average error in height is of order 2 km, the size of the binning used. Only echoes which were within 3 degrees of specular to the Leonid radiant are shown.

A good indicator of physical structure is how the heights vary with meteoroid mass (Hawkes and Jones, 1975). We show in Figure 14 a plot of beginning height as a function of mass, while Figure 15 shows the ending height as a function of mass. It should be kept in mind that these are low precision single-station meteor heights, so much of the scatter is simply due to the lack of precision in the technique for individual meteors.

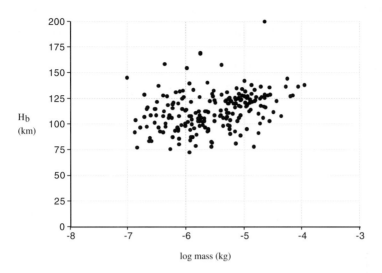

Figure 14. Plot of beginning height versus logarithm of the photometric mass of the Leonid meteoroid (in kg) based on the electro-optical observations. The best regression line shows that the beginning height increases by 9.1±1.9 km per decade mass increase.

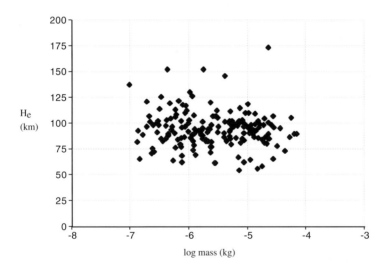

Figure 15. Plot of ending height versus logarithm of the photometric mass of the Leonid meteoroid (in kg) based on the electro-optical observations. There is no statistically significant change in ending height with meteoroid mass. A similar lack of significant change was found for the height of maximum luminosity (not shown).

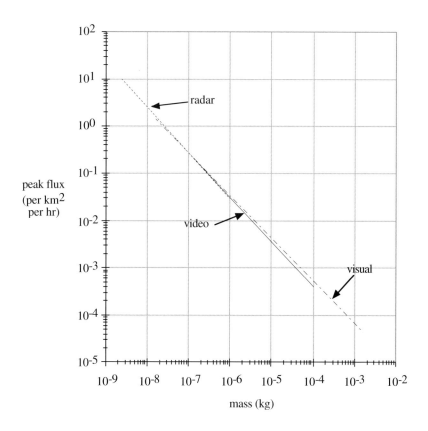

Figure 16. Measured peak flux of Leonids by radar (dotted line), video (solid line) and visual (dashed line) techniques. The flux is expressed in terms of cumulative number of meteors per square kilometer per hour which are larger than the mass value (in kg) plotted on the x-axis. Note that we have used the mass distribution indices determined by each method (radar s = 1.97; video s = 1.95; visual s = 1.90) to extend the curve over the approximate range of observations which contributed to the flux for each technique. The mass-magnitude relationship derived in this paper was used for converting magnitudes to masses in all cases.

The most interesting result is that both the maximum luminosity and ending heights showed no statistically significant dependence of heights on mass. There is some correlation of beginning height with mass ($R^2 = 0.11$) and the regression indicates that the mean beginning height rises by 9.1 ± 1.9 km for each increase of a factor of 10 in mass. Indeed an extrapolation of the beginning heights to the mass regime of the bright fireballs observed by Fujiwara *et al.* (1998) would be consistent with the high heights they observe. The reason that the beginning heights should vary so strongly with mass is not obvious. It should be remembered that the difference between the limiting sensitivity of the system and the meteor increases as one goes to brighter meteors. The most exciting interpretation would be that some volatile component, which is relatively weak in light production, is ablating at great heights.

4. Discussion

Real-time flux determinations from both ground-based, globally distributed electro-optical devices and an automated dual frequency radar system were successfully demonstrated. Both techniques yielded flux values and peak times which are consistent with other observers, within the precision of the data samples. Using the measured mass distribution indices for each technique the flux as a function of mass is shown in Figure 16. For radar data the number – amplitude distribution of underdense echoes was used to determine an average value of $s = 1.97 \pm 0.02$. Visual data are from Arlt (1999) where $s = 1.90 \pm 0.05$ and TV data are from this work, where the average value for the mass index was found to be $s = 1.95 \pm 0.05$. These mass distribution indices were used to extend the measured flux over the main mass range of the observations for each of the three techniques. The mass-magnitude relationship derived earlier in this paper was used to convert observed magnitudes to masses in all cases. This suggests that the flux curve was constant over a large mass range and that no obvious turnover in the mass distribution at smaller sizes occurred before $m \sim 2 \times 10^{-9}$ kg. The three methods are in excellent agreement regarding the actual flux value.

We have provided a relationship between astronomical magnitude and photometric mass for faint Leonids. There is no indication of a dramatic change in light curve shape (which would affect this mass-magnitude regression) over the interval −3 to +5 astronomical magnitude. There is some hint that the shower was richer in faint meteors near the time of the most intense shower activity, although more data is needed to establish this conclusively.

The heights reported here indicate that the majority of the 1999 Leonids (in the size range studied here) started between 100 and 130 km in height, had maximum luminosity points in the interval from 90 to 120 km and ended in the interval 80 to 110 km. It is interesting that there was no statistically significant dependence of ending or maximum luminosity heights on mass. This would be consistent with a physical structure (see Hawkes and Jones, 1975; also Beech, 1984) in which the meteoroid is almost completely fragmented into fundamental grains prior to the main grain ablation region of the atmosphere. However, the beginning heights did increase significantly with increasing mass. One interpretation is that a volatile component, as proposed by Elford *et al.* (1997) and Steel (1998), is ablating high in the atmosphere but producing relatively little light. For larger meteoroids the light is enough to be seen above the limiting sensitivity. The evidence for differential chemical ablation in Leonid meteors (Borovicka, 1999; von Zahn *et al.*, 1999) may be supported by this beginning height dependence on mass. Betlem *et al.* (1999) has provided precise trajectories of larger Leonids, including some very high ones, which are possibly indicative of early trail ablation of a volatile component.

One interesting question is how the height-mass dependence observed here compares with studies of sporadic meteors and those from other showers. Hawkes *et al.* (1984) and Sarma and Jones (1985) provide regression results for how heights depend on mass, velocity and other factors. They find that the beginning height does increase with increasing mass, while there is no statistically significant dependence of the height of maximum luminosity. The ending height has a slight dependence on mass, becoming lower for larger meteoroids.

Acknowledgements

The "Leonid 99" campaign reported here was funded by the US Air Force, NASA, the National Reconnaissance Office, the Canadian Space Agency, the European Space Agency, the Canadian Department of National Defence, the Defence Research Establishment Ottawa, as well as through research grants to J. Jones and R.L. Hawkes by NSERC (Natural Sciences and Engineering Research Council of Canada). The Wise Observatory and the University of Tel Aviv provided support for the campaign in Israel, the Isaac Newton Group of telescopes provided support for the campaign on La Palma. Additional support for the electro-optical sites in Maui were kindly provided by the 1st Space Survelliance Squadron, Air Force Space Command while those at the

Kwajalein Missile Range were made available through Army Space and Missile Command. *Editorial handling:* Peter Jenniskens.

References

Arlt, R.: 1999, *WGN: Journal of the IMO* **27**, 286–295.
Beech, M.: 1984, *MNRAS* **211**, 617–620.
Beech, M., Brown, P., and Jones, J.: 1995, *Q. J. R. astr. Soc.* **36**, 127—152.
Beech, M., Brown, P., Jones, J., and Webster, A.R.: 1997, *Adv. Space Res.* **20**, 1509–1512.
Betlem, H., Jenniskens, P., van Leven, J., ter Kuile, C., Johannink, C., Zhao, H., Lei, C., Li, G., Zhu, J., Evans, S., and Spurny, P.: 1999, *Meteoritics Planet. Sci.*, **34**, 979–986.
Brown, P. and Jones, J.: 1995, *Earth, Moon and Planets* **68**, 223–245.
Borovicka, J., Stork, R., and Bocek, J.: 1999, *Meteoritics Planet. Sci.* **34**, 987–994.
Campbell, M.D.: 1998, *Light Curves of Faint Meteors: Implications for Physical Structure* (B.Sc. Honours Thesis, Mount Allison University, Sackville, NB, Canada.
Campbell, M.D., Hawkes, R.L., and Babcock, D.D.: 1999, in V. Porubcan and W.J. Baggaley (eds.), *Meteoroids 1998*, Astron. Institute, Slovak Academy of Sciences, Bratislava, p. 363–366.
Campbell, M.D., Brown, P.G., LeBlanc, A.G., Hawkes, R.L., Jones, J., Worden, S.P., and Correll, R.R.: 2000, *Meteoritics Planet. Sci.*, submitted.
Cevolani, G. and Foschini, L.: 1998, *Planet. Space Sci.* **46**, 1597–1604.
Duffy, A.G., Hawkes, R.L., and Jones, J.: 1987, *MNRAS* **228**, 55–75.
Elford, W., Steel, D., and Taylor, A.: 1997, *Advances in Space Research*, **20**, 1501–1504.
Fleming, D.E.B., Hawkes, R.L., and Jones, J.: 1993, in J. Stohl and I.P. Williams (eds.), *Meteoroids and Their Parent Bodies*, 261–264.
Fujiwara, V., Ueda, M., Shiba, Y., Sugimoto, M., Kinoshita, M., Shimoda, C., and Nakamura, T.: 1998,. *Geophys. Res. Lett.* **25**, 285–288.
Hawkes, R.L. and Jones, J.: 1975, *MNRAS* **173**, 339–356.
Hawkes, R.L. and Jones, J.: 1984, *Bull. Astr. Inst. Czechosl.* **35**, 46–64.
Hawkes, R.L., Mason, K.I., Fleming, D.E.B., and Stultz, C.T.:1993, in D. Ocenas and D., Zimnikoval (eds)., *International Meteor Conference 1992*, International Meteor Organization, Antwerp, 28–43.
Hawkes, R.L., Babcock, D.D., and Campbell, M.D.: 1998, *Analysis Procedures and Final Electro-Optical Results*, 31 July 1998 CRESTech Contract Report, 58 pgs.
Marsden, B.G.: 1982, *Sky & Telescope* **64**, 284.
McKinley, D.W.R.: 1961, *Meteor Science and Engineering* (New York: McGraw-Hill).
Murray, I.S., Hawkes, R.L., and Jenniskens, P.: 1999, *Meteoritics Planet. Sci.*, **34**, 949–958.
Sarma, T. and Jones, J.: 1985, *Bull. Astr. Inst. Czechosl.* **36**, 9–24.
Steel, D.:1998, *Astron. Geophys* **39**, 24–26.

Treu, M., Worden, S.P., Bedard, M.G., and Bartlett, R.K.: 2000, *Earth, Moon and Planets* **82–83**, 27–38.
Verniani, F.: 1965, *Smithson. Contr. Astrophys.* **9**, 141–172.
von Zahn, U., Gerding, M., Hoffner, J., McNeil, W.J., and Murad, E.: 1999, *Meteoritics Planet. Sci.*, **34**, 1017–1027.
Wray, J.D.: 1967, *The Computation of Orbits of Doubly Photographed Meteors* (Univ. New Mexico Press, Albuquerque, U.S.A.).

LORENTZ SHAPED COMET DUST TRAIL CROSS SECTION FROM NEW HYBRID VISUAL AND VIDEO METEOR COUNTING TECHNIQUE – IMPLICATIONS FOR FUTURE LEONID STORM ENCOUNTERS

PETER JENNISKENS, CHRIS CRAWFORD[1], STEVEN J. BUTOW, AND DAVID NUGENT

SETI Institute, NASA/Ames Research Center, Mail Stop 239-4, Moffett Field, CA 94035, USA; [1]) 2349 Sterling Creek Road, Jacksonville, OR 97530, USA
E-mail: pjenniskens@mail.arc.nasa.gov

MIKE KOOP, DAVID HOLMAN, AND JANE HOUSTON

California Meteor Society, 1037 Wunderlich Drive, San Jose, CA 95129-3159, USA

KLAAS JOBSE

Dutch Meteor Society, Lederkarper 4, 2318 NB Leiden, The Netherlands

GARY KRONK

American Meteor Society, North American Meteor Network, 1117 Troy-O'fallon Road, Troy, Il 62294, USA

and

KELLY BEATTY

Sky & Telescope, 49 Bay State Road, Cambridge, MA 02138-1200, USA

(Received 4 June 2000; Accepted 18 August 2000)

Abstract. A new hybrid technique of visual and video meteor observations was developed to provide high precision near real-time flux measurements for satellite operators from airborne platforms. A total of 33,000 Leonids, recorded on video during the 1999 Leonid storm, were watched by a team of visual observers using a video head display and an automatic counting tool. The counts reveal that the activity profile of the Leonid storm is a Lorentz profile. By assuming a radial profile for the dust trail that is also a Lorentzian, we make predictions for future encounters. If that assumption is correct, we passed 0.0003 AU deeper into the 1899 trailet than expected during the storm of 1999 and future encounters with the 1866 trailet will be less intense than predicted elsewhere.

Keywords: Comet, comet: 55P/Tempel-Tuttle, dust trail, flux, Leonids 1999, Lorentz profile, meteor, meteor storm, predictions, satellite impact hazard, observing techniques

Earth, Moon and Planets **82–83**: 191–208, 2000.
©2000 *Kluwer Academic Publishers. Printed in the Netherlands.*

1. Introduction

The requirement for near-real time flux measurements from aircraft (Jenniskens and Butow, 1999) has led to the development of a hybrid technique of visual and video meteor observations. The method has a team of visual meteor observers view the video output of intensified cameras using video head displays (Figure 1). The cameras make it possible to conveniently observe part of the sky with a well defined field of view. The method proves particularly successful for airborne applications. The cameras are mounted behind optical windows in the aircraft and pointed at relatively low altitude, which achieves 3-4 times higher counts than from the ground (Jenniskens, 1999). We further boost the meteor count by visually inspecting the tapes rather than using automatic meteor detection software programs. The results enable a precise analysis of the 1999 Leonid storm rate profile. Of particular interest are the shape of the profile and possible deviations from a smooth mean behavior, which provide information about the ejection mechanism and shower dynamics in the planetary environment.

2. The Method

During the 1999 Leonid MAC mission, a team of eight visual observers first demonstrated this new approach on-board the "*Advanced Ranging and Instrumentation Aircraft* (ARIA)", operated by the USAF/452nd Flight Test Squadron. Details of the flight path and observing conditions are given in Jenniskens *et al.* (2000).

A counting tool was developed that records the detection of Leonid shower or sporadic meteors with the click of a mouse button. The tool has six entrance ports, which recorded the counts from one of six different intensified cameras. The four cameras considered here had a field of view of 39° x 29° and where mounted at an elevation of about 22° behind BK7 optical glass windows.

Each observer was assigned a mouse bearing a unique machine-readable identification number; each camera had its own designated computer port. The mice were chosen for their ergonomic design and their light-response buttons. The observer began each observing session by plugging the mouse into the computer port corresponding to the camera being used by the observer; the mouse was unplugged at the end of each viewing session. This permitted the computer to identify the starting and ending times of each viewing session, and determine which observer was watching from what camera at all times. Rotating the observer/camera pairings enabled calculation of individual observer and camera coefficients of perception from systematic differences in the counts.

Figure 1. Observer Jane Houston with video head display.

During the 1999 Leonid meteor storm, ARIA flew from the UK to Israel, from Israel to the Azores, and from the Azores to Florida on three consecutive nights. The peak of the storm occurred while enroute from Greece to Italy. Near-real time flux measurements were automatically transferred to a communication station on-board the aircraft, where the counts were sent to NASA/Ames Research Center by e-mail, telephone or direct internet access using INMARSAT satellite telephone lines. From NASA/ARC, the counts were further distributed to operation centers, such as the NASA and USAF sponsored LEOC at Marshall Space Flight Center and ESA's orbital debris center at ESOC, Darmstadt.

Shortly after the mission, several observers gathered at NASA/Ames Research Center to view, in the same manner, the video tapes that were recorded by four similar intensified cameras on-board the twin "*Flying Infrared Signature Technology Aircraft* (FISTA)". FISTA was about 150 km north from ARIA and the bulk of meteors are independent records.

3. Results

3.1. THE STORM PEAK

A total of 33,000 video Leonids were recorded in this manner, which account for about 3/4 of all Leonids on video. This compares with 277,172 Leonids that were observed by 434 visual observers worldwide and gathered by the International Meteor Organisation (Arlt et al.1999). Both data sets will be discussed together. The video data will be shown by black points, the previously published visual data by open squares. Although the number of video meteors is 8 times less than the visual record, the measurements are performed under much better controlled conditions, from which a more precise result can be expected.

Figure 2 shows the peak of the storm. No smoothing was applied. Individual points are 1-minute intervals. Each interval is an independent measurement. The video data are very smooth. The curve is featureless. A small depression at the peak can not be trusted because it is not present in the ARIA and FISTA data in the same way. We suspect that muscle fatigue in the button-pressing fingers started to become a problem at about that time. In hindsight, it appears that the technique works well for rates between ZHR = 5 and 5,000, but the technique will need modifications to conveniently cope with higher rates.

In this paper, our video rates are scaled to the visual Zenith Hourly Rates calculated by Arlt et al. (1999). The rates of Arlt et al. represent independent intervals of 2.8 minutes. We are not concerned with the absolute values, but with the shape of the curve. Hence, all data are plotted on a logarithmic scale, so that any scaling is a mere shift in the graph.

It is a compliment to the ground-based amateur visual observers and the airborne video observers to see how well both datasets agree! The time of the peak is confirmed at solar longitude $\lambda_o = 235.285 \pm 0.001°$, or t = 02:00.8 ± 1.5 min. UT (with $\Delta t = -0.5$ min topocentric correction following McNaught and Asher, 1999a). Also, the slopes of the activity profile are much the same. The visual data show more scatter around the mean, despite the higher number of meteors in each count. The profile is very smooth. We do not confirm the "additional clear enhancements" (Arlt et al., 1999), which were thought to be features in shower models.

These are probably the result of imperfect corrections for observer perception, observing conditions or other factors that affect visual observations. For the same reason, such features in the profiles from individual locations can not be trusted. In the remainder of this paper, we will concentrate on the gross features of the curves that are confirmed by both video and visual results.

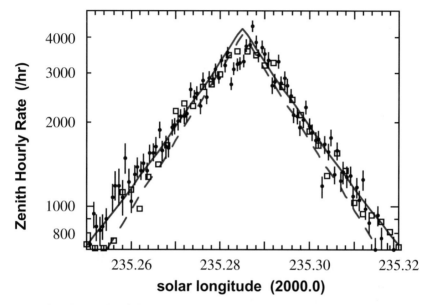

Figure 2. The peak of the 1999 Leonid storm. Solid points are our 1-minute counts (no smoothing applied). Open squares are data from Arlt *et al.* (1999). The dashed line shows the storm component (main peak), while the solid line is the best fit of all components together.

When plotted on a logarithmic scale, as in Figure 2, it is clear that the slopes of the storm peak are linear and well represented by an exponential equation like (Jenniskens, 1995):

$$ZHR = ZHR_{max} \, 10^{-B|\lambda_o - \lambda_o^{max}|} \qquad (1)$$

From a least squares fit, we find $B = 24 \pm 2$ per degree solar longitude for ZHR larger than 700. A slightly larger $B = 25 \pm 1$ value (and $ZHR_{max} = 4,100$ per hour) results when a composite of such curves is fitted to the profile that also accounts for other more shallow features. Note that this value is slightly less than the B = 30 ± 3 derived from the 1866, 1867, 1966 and 1969 Leonid storm profiles (Jenniskens, 1995), when Earth crossed deeper into the respective trailet.

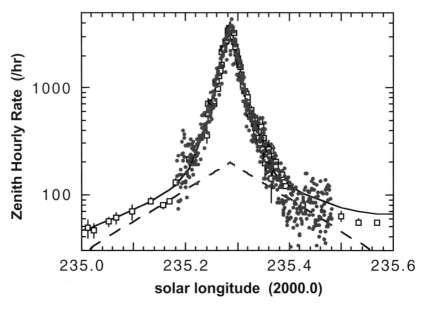

Figure 3. Wings of the profile, with background to the main peak indicated by a dashed line. Symbols as in Figure 2.

Above solar longitude $\lambda_{o_-} = 235.38$ (and below 235.20), rates level off significantly in both video and visual data (Figure 3). A similar background structure to the main peak was observed in the 1866 and 1867 profiles (Jenniskens, 1995). The slopes are near linear again on a logarithmic scale, with B = 2.5 ± 0.2. Combined with the annual Leonid background, we have B = 3.0 ± 0.3 for this component, slightly less than found before (B = 4–6), but typically from a smaller part of the activity profile. This structure appears to be centered within 0.01° from the center of the storm peak. When we assume the same peak time, we have a peak rate of $ZHR_{max} = 200 \pm 10$.

From the visual data (Arlt *et al.*, 1999), we conclude that the magnitude distribution index does not seem to change over the peak. This implies that the magnitude distribution index of the background component and main peak are the same (as we surmised earlier from the

1866 and 1966 profiles - Jenniskens, 1995). And that suggests strongly that both components are caused by the same physical processes, with no intrinsic merit to make a distinction between the two components. We expect to be able to verify from the video record that the two components can not be discriminated on ground of the magnitude distribution index, but will take this as a task for a future paper.

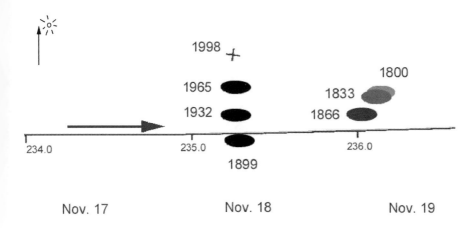

Figure 4. The Earth's path by debris trails ejected at various epochs during the return of 1999. Courtesy: David Asher, Armagh Observatory.

3.2. THE 1866 TRAILET

This material is though to have been ejected in 1899 (Kondrateva and Reznikov, 1985; McNaught and Asher, 1999b; Lyytinen, 1999). In Asher's diagrams of the path of Earth through the meteor shower, reproduced in Figure 4, the Earth approaches dust trails from 1866, 1833 and 1800 shortly after passing the 1899 and 1932 trails. Earlier during the 1998 return, we observed a peak in activity when Earth passed rather far from the calculated center of the 1899 debris trail (Jenniskens, 1999) Hence, we anticipated a second peak of activity just after solar longitude 236.0. Based on observations from Hawaii, Japan and China, Arlt *et al.* (1999) show this second maximum peak at solar longitude 235.87 ± 0.04. Leonid MAC observations in the night after the main peak show enhanced rates that appear to trace the declining branch of this component, showing a relatively fast decline (Figure 5). A curve with B = 1.6 would best fit the descending branch. However, a symmetric curve with this B value would raise the saddle between the two peaks and doesn't make a good fit to the ascending part.

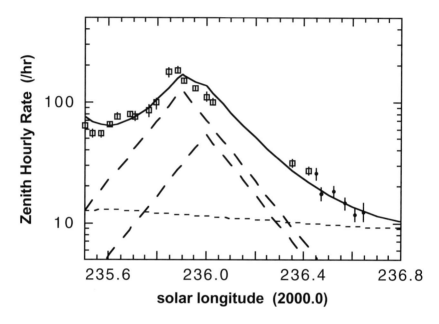

Figure 5. The second peak in the activity curve related to the 1866, 1833 and 1800 trails. Symbols as in Figure 2. The horizontal dashed line with narrow spacing shows the annual Leonid shower activity. Two other dashed lines show the contributions from the 1866 (left) and 1833/1800 (right) trailets. The solid line is the combination of all components.

In light of Figure 4, we interpret this peak as a composite of the result of several trailets. Hence, we fitted two sets of curves with B = 3.0 to the data, peaking at solar longitude of 235.90 (ZHR_{max} = 125) and 236.00 (ZHR_{max} = 50). That separation was taken to reflect the calculations by McNaught and Asher (1999b). However, the second peak occurred 0.14 degrees earlier than predicted, if the nearest point in Earth's orbit to the trailet center at the point of ecliptic plane crossing is considered. It follows that the current model does not precisely describe the position of the trails, at least not for later revolutions or trails at some distance from Earth's path.

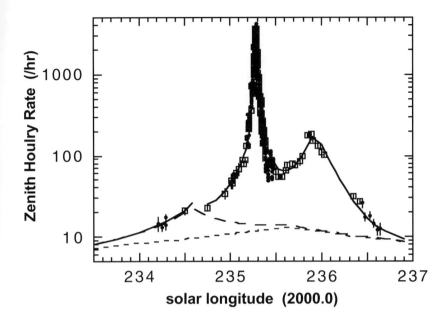

Figure 6. Possible presence of the Leonid Filament component (solid line). Symbols as in Figures 2 and 5.

3.3. THE LEONID FILAMENT?

In prior years, another dust component called the Leonid Filament was responsible for fireball showers in all years from 1994 until 1998 (Jenniskens, 1996). Its characteristic feature is the width, with $B = 1.1 \pm 0.1$ in all years, and its low magnitude distribution index. Jenniskens and Betlem (2000) predicted a return of this component at a lower level than in 1998, assuming that the Filament was the accumulation of many years of dust ejecta. Asher (1999), on the other hand, predicted no activity at all if the Filament was due to ejecta of the return of 1333 only.

The 1999 profile does not show a clear broad component that is readily defined as the Leonid Filament. This appears to confirm Asher's prediction. However, Leonid MAC observations in the night prior to the peak night (at solar longitude 234.5) show a significant enhancement of rates above expected levels that may in fact be caused by the Filament. The expected level being a mere extrapolation of the contribution from the annual Leonid shower, the main and background storm peak, and the second 1866/1833 component (Figure 6). This is consistent with few data shown in Arlt *et al.* (1999), which together trace a broader structure some time before the onset of the storm. Assuming that this is a profile with B = 1.1, that is caused by the Filament, then it has to peak at solar longitude 234.6 ± 0.2 with a peak rate of ZHR = 13 ± 2. The profile can not extend much more to higher solar longitude, because that would create a hump near the peak of the profile. In that case, the component is weaker than the ZHR ~ 120 expected (Jenniskens and Betlem, 2000) and peaked earlier than the expected ~ 235.1, but not far from the time of the 1998 peak.

Typically, this component has a significantly lower magnitude distribution index than the other shower components. We measured r ~2.1 on Nov. 16/17, which is the value expected. However, this value was based on meteor magnitudes called out by visual observers and represents only a very small fraction of observed meteors. A more complete magnitude analysis of the data can reveal if the proposed shower components are correct.

4. Discussion

In the past, shower profiles have been described in terms of Gaussian and exponential shapes (Jenniskens, 1995; Brown *et al.*, 1997). Now, we find that the Lorentz profile, known from damped oscillators, has a shape very similar to the peak and background combined:

$$ZHR = ZHR_{max} \frac{(W/2)^2}{(\lambda_o - \lambda_o^{max})^2 + (W/2)^2} \qquad (2)$$

W is the classical width of the profile at half the peak intensity (in degrees). Indeed, the main peak above ZHR = 300 is best fitted with a Lorentz profile of width W = 0.036±0.002° and ZHR_{max} = 3300±100, the line shown in Figure 7. The width is the only parameter that describes the shape of the curve. Indeed, the tail of the curve falls right on when the peak is fitted even if we ignore the background component.

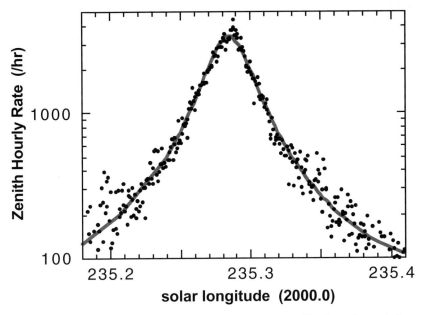

Figure 7. Fit of a Lorentz profile to the meteor storm profile. In order to bring out the small dispersion, error bars are not shown.

We find that data from past meteor storms show a similar good fit (Table I), which implies that each dust trailet itself has a Lorentzian cross section. This condition is necessary to account for the fact that we passed the dust trailets at different distances from the center in 1999, 1966 and 1866. If the dust distribution in a trailet (index "t") follows a Lorentz function as a function of r = distance from trailet center, then:

$$ZHR(r) = ZHR^t_{max} \frac{(W_t/2)^2}{r^2 + (W_t/2)^2} \qquad (3)$$

In that case, the cross section is also Lorentzian if we pass the center of the trailet along the Earth's orbit in a direction $X = \lambda_o$ (now in AU, with roughly 2π AU = 360 degrees neglecting curvature of the Earth's path) at a distance $Y = Y_o$ (measured in a direction perpendicular to Earth's orbit). Because, by substituting $r^2 = Y_o^2 + (X-X_o)^2$:

$$ZHR(X) = ZHR_{max} \frac{Y_o^2 + (W_t/2)^2}{(X - X_o)^2 + Y_o^2 + (W_t/2)^2} \quad (4)$$

which has a similar form as Equation 3. In that case, the width of the dust trailet equals:

$$(W_t/2)^2 = (W/2)^2 - Y_o^2 \quad (5)$$

and the peak rate in the trailet is:

$$ZHR^t_{max} = ZHR_{max} \frac{(W/2)^2}{(W/2)^2 - Y_o^2} \quad (6)$$

The peak of a Lorentzian is in fact well represented by the exponential curves used before (Equation 1). The new representation only adds a tail to the distribution, which is assumed to be a natural consequence of the dispersion mechanism.

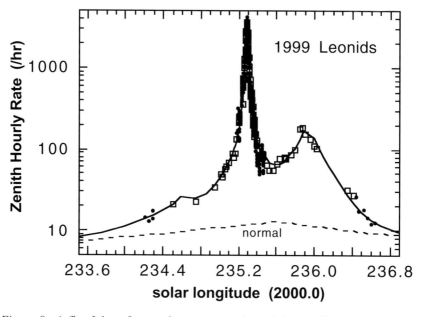

Figure 8. A fit of three Lorentzian curves to the activity profile, which describe the 1899-1932 trails (storm), the 1866-1800 trails (2nd peak) and the Filament. Symbols as in Figures 2 and 5.

It is not clear, at present, what physical mechanism is responsible for this tail in the distribution. Lorentzian distributions are characteristic for damped oscillators, and perhaps a natural consequence of the orbital evolution in the three body system meteoroid-Sun-Jupiter.

Given the good representation in the case of the 1999 Leonid storm, we applied the Lorentzian fit to other shower components (Table I). We find that this year's shower profile is well represented with three Lorentz curves representing storm, 1866 peak, and Filament (Figure 8).

TABLE I

Year	From	Δ C-E * (AU)	M_o ** (°)	W/2 (AU)	ZHR_{max} (hr^{-1})	λ_o^{max} (°)
observed:						
1999	1899	-0.0007	18.4	0.00031±0.00002	3,300±100	235.285
1998	1899	(+0.0044)#	16.6	0.00087±0.00008	70±20	(235.28)#
1966	1899	-0.0001	16.6	0.00024±0.00005	15,000±3000	235.166
1965	1899	+0.0017	5.8	0.0024±0.0007	100±50	235.40
1969	1932	+0.0000	49.1	0.00026±0.00005	200±50	235.265
1999	1866	+0.0016	19.0	0.0027±0.0003	130±15	235.95
1998	1866	+0.0040	16.6	0.003±0.002	10±10	236.0
1866	1733	-0.0004	9.1	0.00024±0.00007	14,000±2000	233.323
1867	1833	-0.0002	19.9	0.00024±0.00012	5,000±1000	233.411
predicted:						
2000	1866	+0.0008	29.5	(0.0011)	(70)	236.28 *
2000	1932	-0.0012	29.5	(0.0009)	(207)	235.29 *
2001	1866	+0.0002	40.3	(0.00053)	(72)	236.46 *
2002	1866	+0.0000	51.1	(0.00035)	(38)	236.86 *
2002	1966	+0.0018	51.1	(0.0021)	(4)	235.27 *

*) Minimum distance between Earth and Comet orbit, from McNaught & Asher (1999b)
**) Mean anomaly of trail particles
#) Large uncertainty because of perturbation by Earth in earlier encounter

From older data, we note that especially the 1966 profile as calculated by Brown *et al.* (1997) is a perfect Lorentzian, and not a Gaussian as proposed there. The "storm" peak of 1998 (Jenniskens, 1999) is an exception. That profile was clearly assymetric, which differs from a Lorentzian profile. A sum of two profiles could fit that curve, but it is not easy to assign the components to the debris of a particular return. Perhaps, this debris was disturbed by prior close encounter with the planets, as proposed in McNaught and Asher (1999b). We do recognize an enhancement that can be associated with the passage of the 1866/1833/1800 trailets, in order to account for relatively high rates in the night after the maximum (Arlt and Brown, 1998; Jenniskens, 1999).

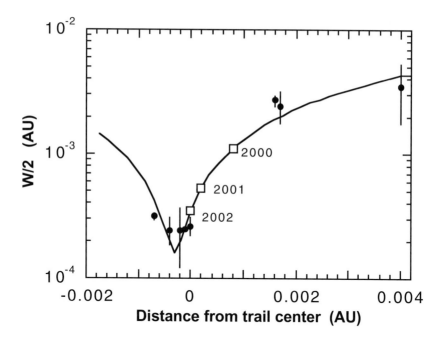

Figure 9. The width of the profile as a function of distance from the center of the trailet. The solid points are observations (Table I), the open squares are predicted values based on the fitted Lorentz curve (solid line).

The width of the profile is expected to gradually increase if the Earth passes further away from the center of the trailet. Near the center is a core with a steep slope, which has a more shallow tail further out. The core is typical for the 1866, 1867, 1966, 1969 and 1999 profiles, while the profiles of 1998, 1965 and the second peak of 1999 are cases of further out. If we plot the width versus the distance to the trailet center (Y_o), as calculated by McNaught and Asher (1999b), then we find that Equation 5 (solid line in Figure 9) indeed does fit the result, allowing for at least ± 0.0001 AU uncertainty in the calculated trailet positions. The intrinsic width of the dust trailet is calculated at $W_t = 0.00032 ± 0.00008$ AU.

However, the fit is good only if the calculated trailet pattern (together making up the comet dust trail) is shifted outward by about +0.0003 AU. The curve in Figure 9 should center on zero. We conclude that the Earth crossed about 0.0003 AU deeper into the debris trail ejected in 1899 than predicted. Unfortunately, that means that the Earth will not cross quite as deep into the 1866 epoch trailet in 2001 and 2002.

On top of that are two more factors that influence the peak rate in future years: 1) the rate of decrease of dust density away from the comet for a pristine trailet of 1 revolution, and 2) the decay of dust density with each subsequent revolution.
Regarding the decay of dust density with subsequent revolutions, we assume that the dust density falls off inversely with the number of revolutions (N), which is expected if the spreading is mainly due to differences in orbital period of the particles in the dust trailet. Here, we ignore the fact that the peak of the particle density also shifts progressively along the comet orbit in time. In that case, the peak dust density at a given position after 1 revolution is:

$$ZHR^t_{max} (1 \text{ rev.}) = ZHR^t_{max} \times N \qquad (7)$$

Figure 10 shows the density of dust in the center of the trailet after one revolution, calculated from the observed peak ZHR value. This value was corrected to a center-of-trailet value for a Lorentz distribution with adopted $W_t = 0.00032$ AU perpendicular to Earth's orbit centered on the trail centers calculated by McNaught and Asher (1999b), and by taking Equation 7 into account. Unfortunately, only one data point (the return of 1969) is available to constrain the slope of the dashed line in Figure 10. All other observations fall in a rather narrow range of mean anomaly. Any error in the 1969 result will bear heavily on the assumed dependence on mean anomaly and the predictions that follow from it.

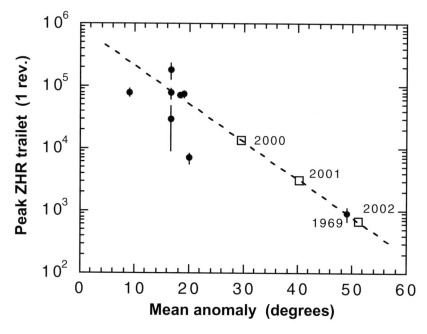

Figure 10. Peak dust density in the trailet after 1 revolution, as derived from the flux profiles of past meteor storms and outbursts. Three lower points all refer to pre 1899 trailets.

From Figure 10, it is possible to predict the peak activity in 2000–2002. The time since perihelion passage for each return is marked on the dashed line with an open square. The predicted peak rate follows from this by corrections according to Equation 7 and Equation 6. We find ZHR = 70 in 2000 (1866 trailet), ZHR = 210 in 2000 (1932 trailet), ZHR = 70 in 2001 (1866) and ZHR = 40 in 2002 (1866), whereby the width of the profiles should gradually decrease from W = 0.0022 AU (1866) and 0.0018 AU (1932) in 2000, to W = 0.0011 AU in 2001 and 0.0007AU in 2002.

These predictions are somewhat disappointing given the high predicted rates by McNaught and Asher (1999b) of 10-35,000 and 25,000 for the 1866 trailet encounter of 2001 and 2002, respectively. However, one should keep in mind that the observations on which these predictions are based are very limited. Also, the expected rates are higher if we would include the anticipated progressive shift of the peak of the particle density along the comet orbit in time.

4.1. FUTURE OBSERVATIONS

The video record of the 1999 Leonid meteor storm is a treasure trove of information that can be further analyzed. Unlike the hybrid visual-video observation technique, such in-depth analysis is time consuming. Some preliminary results are presented in Gural and Jenniskens (2000).

After full analysis of the 1999 shower, the big unknown will still be the dispersion perpendicular to Earth's path and the exact position of the dust trail center. Only the year 2000 encounter can shed light on this. The predicted distances to the trailet centers are small enough to get significant increased rates and recognize the component from other shower components. Also, the distance to the trailet center is not as small as in 2001 and 2002, when we are on the steep slope of the Lorentz profile. Small natural shifts in the trail center can cause great variations in rates that are can not easily be interpreted in terms of the width of the dust trail perpendicular to the Earth's path. On the other hand, while the year 2000 provides a 2-dimensional picture of what is now only a 1-dimensional view of dust trails, the years 2001 and 2002 will provide a three dimensional perspective by providing important clues to how quickly the dust density falls off away from the comet position.

The method described in this paper promises a detailed picture of the dust density in comet dust trailets by combining theory and observations of future Leonid showers. Observations in future years will test the assumptions that go into the model, such as the cylindrical geometry and the position of the trail. Each future encounter will be a strong test for refining the theoretical multi-trailet model of comet dust trails.

Acknowledgements

We thank ESA's Michael Schmidhuber for assisting in the "flux measurement team" and Morris Jones and Pete Gural who assisted in the visual examination of the FISTA tapes after the flight. The flux measurements in the Leonid MAC 1999 mission were supported by grants from USAF/XOR and the NASA Planetary Astronomy and Suborbital MITM programs. *Editorial handling:* Noah Brosch.

References

Asher, D.: 1999, *MNRAS* **307**, 919–924.
Arlt, R. and Brown, P.: 1999, *WGN, Journal of the IMO* **27**, 267–285.
Arlt, R., Bellot-Rubio, L., Brown, P., and Gijssens, M.: 1999, *WGN, Journal of the IMO* **27**, 286–295.
Brown, P., Simek, M., and Jones, J.: 1997, *Astron. Astrophys.* **322**, 687–695.
Gural, P. and Jenniskens, P.: 2000, *Earth, Moon and Planets* **82–83**, 221–248.
Jenniskens, P.: 1995, *Astron. Astrophys.* **295**, 206–235.
Jenniskens, P.: 1996, *Meteoritics Planet. Sci.* **31**, 177–184.
Jenniskens, P.: 1999, *Meteoritics Planet. Sci.* **34**, 959–968.
Jenniskens, P. and Butow, S.J.: 1999, *Meteoritics Planet. Sci.* **34**, 933–943.
Jenniskens, P. and Betlem, H.: 2000, *Astrophys. J.* **531**, 1161–1167.
Jenniskens, P., Butow, S.J., and Fonda, M.: 2000, *Earth, Moon and Planets* **82-83**, 1–26.
Kondrat'eva, E.D. and Reznikov, E.A.: 1985, *Sol. Syst. Res.* **19**, 96–101.
Lyytinen, E.: 1999, *Meta Research Bulletin* **8**, 33–40.
McNaught, R.H. and Asher, D.J.: 1999a, Meteoritics Planet. Sci. **34**, 975–978.
McNaught, R.H., and Asher, D.J.: 1999b, WGN, Journal of the IMO **27**, 85–102.

COMPARING METEOR NUMBER FLUXES FROM GROUND-BASED AND AIRPLANE-BASED VIDEO OBSERVATIONS

DETLEF KOSCHNY AND JOE ZENDER

European Space Agency, Space Science Department, Keplerlaan 1, NL-2200
Noordwijk, The Netherlands
E-mail: dkoschny@estec.esa.nl

(Received 31 May 2000; Accepted 7 August 2000)

Abstract. We operated identical Low-Light level TV cameras to observe the Leonid 1999 meteor storm, both from a ground-based station in Southern Spain and from the ARIA airplane of the Leonid Multi-Aircraft Campaign. The ground-based camera was pointed to a fixed position about 50° from the zenith, the airborne camera was pointed through a window at 75° from the zenith. During the peak of the Leonid storm, the two cameras were located between 10° and 20° apart in geographical longitude. The recorded meteor numbers differed by a factor 5.3 ± 0.4, the airborne camera recording the higher rates. This is much more than what could be expected from the geographical separation of the cameras. The different elevation angles and altitudes of the cameras can explain this. Pointing the camera low to the horizon results in a much larger volume in the atmosphere which is observed, resulting in higher meteor count rates. However, the meteors are on average much further away than when observing high above the horizon. The atmospheric extinction reduces the brightness of the meteors, effectively reducing the count rates. For two ground-based cameras, these two effects are expected to compensate. Due to the high altitude of the airborne camera, the increasing effect of the count rate dominates. We set up an atmospheric extinction model taking into account Rayleigh scattering that quantitatively explains the number difference. Using the same model, we predict number differences for cameras observing from the same location, but pointed at different elevation angles. For typical observing conditions, neglecting this effect can result in differences up to a factor of 10 in the derived meteor numbers.

Key words: Leonid MAC, meteor, meteor flux, Zenithal Hourly Rate

1. Introduction

The Space Science Department of the European Space Agency, ESA/SSD, organized a ground-based observing effort of the Leonid meteor shower in November 1999 from two sites in Southern Spain, namely Calar Alto Observatory (CAHA) and the Observatory Sierra Nevada (OSN). Several science goals were addressed, mainly using Low-Level TV cameras and CCD imagers: High-resolution imaging of meteors for precise position and magnitude determination, double-station observations for orbit determination, and simple number determinations. ESA/SSD also participated in the Leonid MAC campaign (Jenniskens *et al.*, 2000), providing one camera on an airborne platform to the flux monitoring effort. This camera was identical to one of the cameras used from the ground and allows a direct comparison of the measured meteor numbers.

2. The setup

We built a small series of identical Low-Light Level TV cameras at ESA/SSD, called ICC1 to ICC5 (ICC = Intensified CCD Camera), a schematics of which is shown in Figure 1. The camera uses a 50 mm f/0.75 Rayxar lens (Old Delft), a 2nd Generation MCP image intensifier with fiber input window, model no. XX1700DB from DEP (highest sensity between 400 nm and 600 nm), a Sony XC-77CE video camera with a 2/3" CCD, coupled via a fiber taper to the intensifier producing 756 (H) x 581 (V) pixels resolution. The video signal was routed via a time inserter (Ingenieurbüro Cuno) and recorded on standard VHS (airplane) and S-VHS (ground) video recorders. The resulting sky image was 9 ° by 12 ° in size, the faintest visible stars were around 8.5 mag.

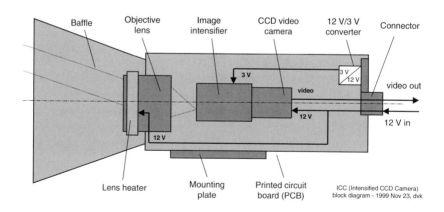

Figure 1. Block diagram of the LLTV camera used both from ground and from the airplane.

On ground, the camera was set up at the Observatory Sierra Nevada (OSN) outside the building on a tripod. The VCR, time inserter, and a control monitor were set up inside the building. The camera was pointed roughly at the pole star, i.e. due North and an angle of 50° away from the zenith.

The airborne camera was mounted in the ARIA airplane on an optical bench, looking out through a window to the left side at 90° to the flight direction and at 75° to the airplane zenith. By identifying the star fields during the observing times taken into account here, we verified that the airplane zenith corresponded to real zenith in these times. The window is of BK7 optical glass and does not affect the recorded image of the meteors in the spectral region of interest. The location and altitude of the cameras are given in Table I.

3. Data evaluation and results

No automated meteor detection software was used for the determination of the number fluxes, since tests of automated software systems showed that these could suffer from selection effects based on short path lengths of meteors and low detection probability near the noise limit of the imagery. Thus, the videotapes of the two cameras were visually inspected. The time of a meteor was determined by reading off the time inserter display. The stream association and the maximum magnitude were estimated by path orientation and visual comparison to the surrounding stars, respectively. The ground-based camera (ICC5), recorded 248 meteors in 03h25m. The airborne camera (ICC1), recorded so many meteors that a complete inspection of the tape was not performed. Rather, we inspected three intervals of 5 minutes each. A total of 220 meteors were recorded in the total 15 minutes. See Table I for details.

TABLE I

	Time interval	Raw no.	Latitude	Longitude	Elevation
Airborne camera	18 Nov, 01h46m - 01h51m	106	35.95 °N	22.08 °E	11.0 km
	18 Nov, 02h11m - 02h16m	104	37.17 °N	19.55 °E	11.0 km
	18 Nov, 03h28m - 03h33m	10	38.74 °N	10.87 °E	11.0 km
Ground-based camera	17/18 Nov, 23h40m - 03h23m (split in intervals)	223	37.06 °N	3.38 °W	3.0 km

From these raw counts, we determine the effective ZHR according to the following formula:

$$eZHR = \frac{n_{met}}{T_{eff}} \cdot F \cdot r^{6.5-lm} \cdot \frac{1}{\sin z} \qquad (1)$$

where $eZHR$ is the effective Zenithal Hourly Rate of the video camera, n_{met} is the observed number of meteors, T_{eff} is the effective observing time for the respective time interval, F a correction factor for cloud cover, r the population index of the meteor stream, z the elevation of the radiant at the actual observing position (changing over time for the airplane).

The limiting magnitude of the two cameras was identical (8.5 mag), the population index r was assumed to be $r = 2.0$ (Molau et al., 1999). There were no clouds during the observation, thus $F = 1$.

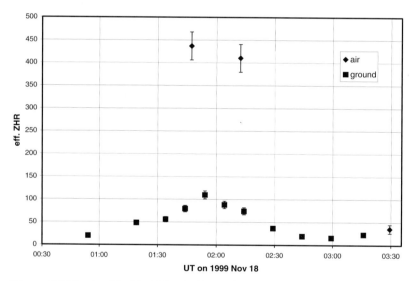

Figure 2. Effective Zenithal Hourly Rates (eZHR) of the airborne camera (diamonds) and the ground-based camera (squares). The error bars are the square root of the raw number counts, as typically used for visual observations. Note that the eZHR is much smaller than the ZHR as derived by visual observations due to the small field of view of the cameras.

The resulting numbers are shown graphically for both cameras in Figure 2. The time intervals for the ground-based camera were chosen between 10 and 60 minutes, depending on the absolute number of meteors. The maximum was observed shortly before 02h00m UT, in the interval from 01h50m to 02h00m. This is consistent with the result of other groups, *e.g.* Molau *et al.* (1999) gives 01:58 and 02:11 UT as the times of two prominent peaks in video data. For visual observations, Arlt *et al.* (1999) derive 02:02 +/- 2 min, which is somewhat later than our maximum. Note, however, that the absolute numbers in the intervals of the ground-based camera are only between 15 and 35. It is expected that a detailed analysis of the airborne data will yield a more accurate determination of the peak time due to the much higher absolute numbers.

It is apparent that there is a significant number difference between the observed meteor counts. Comparing the effective ZHR at 01:48 and 02:13 UT, the center times of the 5-min-intervals counted for the airborne camera, with the interpolated values at the same times for the ground-based camera yields a ratio of 4.9 and 5.6, respectively.

In the following sections, this difference will be explained by the difference in atmospheric extinction, the different apparent velocities of the meteors, and the difference in the observed volume in the atmosphere due to the different camera elevations.

4. The atmospheric extinction model

The light emitted by the meteor will be partially absorbed and scattered by the atmospheric molecules and by aerosols. The main effects are: (a) Rayleigh scattering by atmospheric molecules; (b) Mie scattering by aerosol particles; and (c) absorption by atmospheric molecules.

Rayleigh scattering will be dominant in all wavelengths and is quite effective. Star light passing through the atmosphere will loose about 40 % of its intensity at a wavelength of about 600 nm (Houghton, 1986). This effect is the most dominant one in visual wavelengths and will be the one and only modelled here.

Mie scattering at aerosols will also be important in visual wavelengths if many aerosol particles are present. This will play a role for ground-based observations at low elevation angles above the horizon. Here, we compare observations from an aeroplane with observations from a high-altitude site on the ground at high elevation angles. In both cases, not many aerosols will be present. Therefore, the contribution of Mie scattering will be neglected. For a ground observer at sea level, Mie scattering would be an issue.

The absorption by atmospheric molecules plays a large role in the ultraviolet and the infrared wavelength regime and will also be neglected, since the spectral sensitivity of the camera is limited to the visible spectral range.

The scattering along the line of sight follows a functionality similar to absorption and is known as Lambert's law. The scattering which occurs when radiation of intensity *I* traverses an elementary slab of atmosphere of thickness dz is proportional to the mass of absorber $\rho\, dz$ in unit cross-section of the slab. ρ is the density of the absorber. The scattering is also proportional to the intensity of the radiation:

$$dI = -I\sigma_R \rho\, dz \qquad (2)$$

where σ_R is the Rayleigh scattering coefficient. Integrating (2) results in

$$I = I_0 \exp\left(-\int \sigma_R \rho\, dz\right) \qquad (3)$$

The Rayleigh scattering coefficient is defined as:

$$\sigma_R = \frac{32\pi^3}{3N_0 \lambda^4 \rho_0}(n-1)^2 \qquad (4)$$

where N_0 is the number of molecules per unit volume, ρ_0 the density and n the refractive index, all at conditions of standard temperature and pressure.

We set up a numerical model to determine the reduction in light due to scattering for different observer altitudes above the ground and different elevation angles of the viewing direction.

Figure 3 shows the geometrical assumptions. The observer is assumed to be at the altitude h_{obs} above the Earth's surface. The observer looks into the direction δ away from the zenith. The meteor is modelled as a light source with intensity I_0 in Wm^{-2}ster^{-1} at altitude h_{met}. From the geometry, the initial distance between the light source and the observer is calculated. Then, the light attenuation following the path s is numerically calculated using (3) and considering the reduction of light intensity due to the quadratic distance law. The density ρ in (3) is taken from a look-up table derived from the Standard International Reference Atmosphere (COSPAR 1972). The Rayleigh scattering coefficient is calculated by assuming a wavelength $\lambda = 600$ nm, $N_0 = 2.547 \cdot 10^{25}$ m^{-3} and $\rho_0 = 1.16$ kg/m^3 (Houghton 1986).

The refractive index n of dry air is calculated using Elden's formula (Houghton, 1986):

$$(n(\lambda) - 1) \times 10^6 = 64.328 + 29{,}498.1\,(146 - \lambda^{-2})^{-1} + 255.4\,(41 - \lambda^{-2})^{-1} \quad (5)$$

The light intensity is calculated in step sizes of 1 m for the complete distance between the source and the observer. The calculated reduction in intensity, expressed as a magnitude difference Δm for the two camera altitudes and elevation of the line of sight are given in Table II, where the extinction Δm in magnitudo is given for different combinations of zenith angle δ and meteor altitude h_{met}. A zenith angle of 50° corresponds to an observer altitude of 3.0 km (ground-based observer), a zenith angle of 75° to 11.0 km (airborne observer). Δm_{scat} is the magnitude loss by scattering, Δm_{dist} the loss by distance, and Δm_{val} the loss by apparent velocity. The last column gives the total magnitude loss.

TABLE II

δ	h_{obs} km	h_{met} km	s_i km	Δm_{scat} mag	Δm_{dist} mag	Δm_{vel} mag	Δm mag
50° grnd	3.0	80	123	0.48	0.46	2.1	3.04
		100	154	0.48	0.94	2.1	3.52
		120	184	0.47	1.33	2.1	3.90
75° air	11.	80	286	0.37	2.28	1.6	4.25
		100	352	0.38	2.73	1.6	4.71
		120	416	0.38	3.09	1.6	5.07

5. Apparent velocity of the meteor

Another effect that needs to be taken into account is the magnitude loss due to the different apparent velocities of the meteors in the two camera systems. The faster a meteor moves the higher the number of pixels over which the light is distributed. The loss is given in Jenniskens *et al.* (1998) as:

$$\Delta m = 2.5 \log (\omega_{meteor} / \omega_0) \quad (6)$$

where ω_0 is the subtended angular size of the sky imaged by one pixel divided by the integration time of the pixel. In real cameras the light of point sources is not imaged onto single pixels but is gaussian distributed over a certain number of pixels due to the point-spread function of the optics. Therefore the loss formula given above is an upper limit. Rather than setting ω_0 to the angular size of one pixel, we set it to the full width at half maximum (FWHM) of the point spread function for a typical meteor magnitude of 6 mag, in our case 4 pixels or 4.4'. The movement of a meteor for the airborne camera was measured to be 17 pixels, the light was thus distributed to approximately 17 x 4 = 68 pixels. For the ground-based camera the light was distributed to 28 x 4 pixels = 112 pixels. This yields a magnitude reduction relative to the FWHM of 2.5 log (68/16) = 1.6 mag for the airborne camera, 2.5 log (112/16) = 2.1 mag for the ground-based camera.

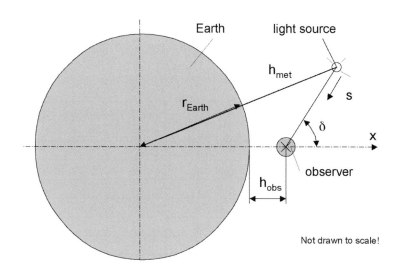

Figure 3. The geometrical assumptions for the model. Details see text.

6. Observed volume in the atmosphere

It is assumed that all meteors appear in the altitude range between 80 and 120 km above the Earth's surface. Thus, the sampled volume of the atmosphere gets larger as the zenith angle of the observing camera gets larger, see Figure 4. To simplify the determination of the sampled volume, we assume that the cameras have 10° circular field of view.

Thus, the area which a camera sees at a distance s_i can be calculated by:

$$A_i = \pi \, (s_i \, \tan(5°))^2 \qquad (7)$$

We calculate the sampled atmospheric volume by calculating the volume of the cone

$$V = (s_{120} - s_{80}) \cdot (A_{s80} + A_{s120})/2 \qquad (8)$$

where A_{s80} the area according to (6) for s_i as given in Table I for the meteor altitude of 80 km, A_{s120} the area for s_i for the altitude of 120 km, s_{80} the distance of the meteor at an altitude of 80 km, s_{120} the distance for 120 km altitude. The resulting volumes are:

$$V_{air} = 490367 \text{ km}^3 \text{ for } \delta = 75°$$
$$V_{ground} = 44006 \text{ km}^3 \text{ for } \delta = 50°$$

and the ratio of the two is:

$$V_{air} / V_{ground} = 11.1 \qquad (9)$$

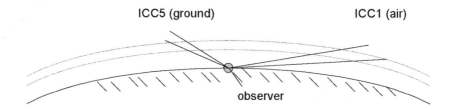

Figure 4. Qualitative illustration of the sampled atmospheric volume for different zenith angles.

7. Application of the model

The cumulative number of meteors seen for different limiting magnitudes is described by (see *e.g.* Rendtel *et al.*, 1995):

$$\frac{n_{mag1}}{n_{mag2}} = r^{mag1-mag2} \quad (10)$$

or, in our case:

$$\frac{n_{mag\_ground}}{n_{mag\_air}} = r^{\Delta m\_ground - \Delta m\_air} \quad (11)$$

From Molau *et al.* (1999) we take $r = 2.0 \pm 0.3$. To keep the calculation simple, we assume for each of the two cameras the extinction value at a meteor altitude of 100 km, *i.e.* $\Delta m\_air = 4.71$ mag, $\Delta m\_ground = 3.52$ mag from Table II. The number ratio for the two cameras can thus be calculated to

$$\frac{n_{mag\_ground}}{n_{mag\_air}} = (2.0 \pm 0.3)^{3.52-4.71} = 0.45 \pm 0.08 \quad (12)$$

The total number ratio is the product of the ratio between the two volumes (8) and the difference due to the extinction and the different apparent velocities (11):

$$\frac{n_{ground}}{n_{aur}} = \frac{V_{ground}}{V_{air}} \cdot \frac{n_{mag\_ground}}{n_{mag\_air}} = 5.0 \pm 0.9 \quad (13)$$

8. Results and Conclusion

By comparing the observed meteor numbers between the two cameras, we found that the airborne camera observes 4.9 and 5.6 times more meteors than the ground-based one. Using a scattering model to calculate the reduction in light intensity and considering the difference in observed volume due to the different elevations of the camera field of views we arrive at a predicted number difference of 5.0 ± 0.9, with the error due to the uncertainty of the population index. Thus, our model is in good agreement with the observations. There seems to be a slight tendency to underestimate the difference. The reason is assumed to be the fact that we neglect Mie scattering, which results from aerosols and dust in the atmosphere. This effect would further reduce the number of meteors observable on ground, thus raising the difference with respect to our prediction.

We neglected effects of changes in population index due to the different location on the Earth. This, however, is expected to be small since the airplane was reasonably close to the ground-based observing site.

With the presented model it can be shown that for video observations with a small field of view the elevation of the camera above the horizon cannot be neglected in the determination of effective Zenithal Hourly Rates of meteors.

To assess the importance of using an airplane as a platform for a meteor camera, we plot the difference in meteor numbers with respect to an observation in the zenith for a ground-based observer assumed to be at 3 km altitude and an airplane at 11 km altitude. The result is shown in Figure 5 (note that the effect of the difference in apparent angular velocity is not shown, since it depends on the pixel scale of the cameras used)ρ

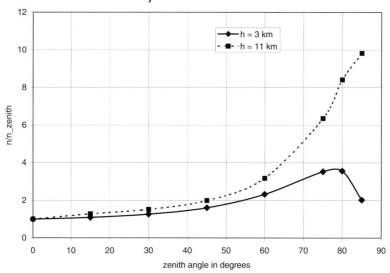

Figure 5. Number of meteors as seen for two different observer altitudes as a function of camera elevation, normalized to the number of meteors seen at the zenith from the ground-based observer. The field of view of the camera was assumed to be circular with 10 deg diameter, the population coefficient is $r = 2$.

It can be seen that increasing the zenith angle (reducing the elevation angle) increases the number of meteors that are observed with respect to the number seen at the zenith. For a ground-based observer, however, this trend reverses for zenith angles larger than about 75°. There, the distance to the meteor gets so large that the light scattering overcompensates the increase in observed volume. For an observer at

11 km altitude, this reversal cannot be seen. In addition, especially at larger zenith angles, the number increase is significantly higher than that observed from the ground.

The fact that the Mie scattering is strongly reduced at higher altitudes further enlarges this difference. We conclude that using airplane platforms for meteor cameras is highly recommendable, especially when observing at high zenith angles.

Acknowledgements

Thanks to Michael Schmidhuber for flying on the airplane and operating the camera for us. We also thank Felicitas Mokler for visually inspecting the video tapes and her support in analysing the data. Alex Jeanes provided organisational support. The cameras were partially funded by contract ESA/ESOC No. 13807/99/D/CS. We thank two anonymous reviewers for their valuable comments. *Editorial handling:* Noah Brosch.

References

Arlt, R., Bellot Rubio, L., Brown, P., and Gyssens, M.: 1999, *WGN, Journal of the IMO* **27**, 286--295.
COSPAR: 1972, *International Reference Atmosphere*, Akademie-Verlag Berlin, Berlin.
Houghton, J. T.: 1986, *The Physics of Atmospheres*, Cambridge University Press, Cambridge, 271 pp.
Jenniskens, P., de Lignie M., Betlem H., Borvicka J., Laux C. O., Packan D., and Krüger C. H.: 1998, *Earth, Moon, and Planets* **80**, 311--341.
Jenniskens, P., Crawford, C., Butow, S.J., Nugent, D., Koop, M., Holman, D., Houston, J., Jobse, K., Kronk, G., and Beatty, K.: 2000, *Earth, Moon and Planets* **82-83**, 191–208.
Molau, S., Rendtel, J., and Nitschke, M.: 1999, *WGN, Journal of the IMO* **27**, 296--300.
Rendtel, J., Arlt, R., and McBeath, A.: 1995, *Handbook for Visual Meteor Observers*, International Meteor Organisation, Mechelen, 310 pp.

LEONID STORM FLUX ANALYSIS FROM ONE LEONID MAC VIDEO AL50R

PETER S. GURAL

Science Applications International Corporation, 4001 N. Fairfax Drive, Suite 500, Arlington, Virginia 22203, USA
E-mail: peter.s.gural@saic.com

and

PETER JENNISKENS

The SETI Institute, NASA Ames Research Center, Mail Stop 239-4, Moffett Field, California 94035, USA
E-mail: pjenniskens@mail.arc.nasa.gov

(Received 31 May 2000; Accepted 16 August 2000)

Abstract. A detailed meteor flux analysis is presented of a seventeen-minute portion of one videotape, collected on November 18, 1999, during the Leonid Multi-instrument Aircraft Campaign. The data was recorded around the peak of the Leonid meteor storm using an intensified CCD camera pointed towards the low southern horizon. Positions of meteors on the sky were measured. These measured meteor distributions were compared to a Monte Carlo simulation, which is a new approach to parameter estimation for mass ratio and flux. Comparison of simulated flux versus observed flux levels, seen between 1:50:00 and 2:06:41 UT, indicate a magnitude population index of $r = 1.8 \pm 0.1$ and mass ratio of $s = 1.64 \pm 0.06$. The average spatial density of the material contributing to the Leonid storm peak is measured at 0.82 ± 0.19 particles per square kilometer per hour for particles of at least absolute visual magnitude +6.5. Clustering analysis of the arrival times of Leonids impacting the earth's atmosphere over the total observing interval shows no enhancement or clumping down to time scales of the video frame rate. This indicates a uniformly random temporal distribution of particles in the stream encountered during the 1999 epoch. Based on the observed distribution of meteors on the sky and the model distribution, recommendations are made for the optimal pointing directions for video camera meteor counts during future ground and airborne missions.

Key Words: Clustering, flux, fragmentation, Leonids 1999, meteor, meteoroid, optimal video pointing, satellite impact hazard, simulation

1. Introduction

The spatial number density of particles in the Leonid meteor stream is of paramount concern to the U.S. Air Force and NASA, as well as other satellite operators. The 1999 Leonid shower was anticipated to cause several orders of magnitude increase in the population of fast meteoroids for satellites exposed to the shower (Jenniskens *et al.*, 1998). Near real-time reporting systems were set up globally to provide immediate access to the severity of the meteor storm during the period of the shower (Brown *et al.*, 2000). The NASA and USAF sponsored Leonid Multi-Instrument-Aircraft Campaign participated in this effort by providing near real-time meteor counts to satellite operators worldwide (Jenniskens *et al.*, 2000a).

An accurate calibration of fluxes and the measurement of the meteoroid size distributions, however, demands an after-the-fact analysis, involving the data processing of video tapes collected from eight cameras taken on three consecutive nights of data gathering (Jenniskens *et al.*, 2000b). Such work is also needed to reveal size-dependent dispersions in the shower and possible non-statistical fluctuations that can be traced back to particles breaking up in space (Jenniskens and Butow, 1999).

This paper presents an analysis of a small time interval from just one of the tapes recorded. This analysis includes a comparison to a meteor simulation model in order to derive key meteor shower parameters such as mass index and spatial number density. In comparison, such parameters have in the past been determined through measured counts and magnitude binning with correction factors applied to account for various geometric effects (Koschack, 1990a; 1990b). Using a simulation to attempt to match observed flux levels for the derivation of shower characteristics, is an alternative approach and one that is new for the meteor community. Monte Carlo techniques have been used in the past to study the observational influences on both the magnitude distribution index and zenith hourly rate corrections (e.g. Van der Veen, 1986; Arlt, 1998). Our work significantly expands on the technique by directly estimating meteor shower parameters via simulation and measurement comparisons and by including higher fidelity modeling. The advantages are that in the process of improving the fidelity of the simulation to match observations, magnitude loss models and instrument characteristics are refined that help to characterize the visibility and flux of the shower under different viewing conditions. Based on these models, recommendations for best viewing directions can be made for future data collection missions using similar techniques. The principal goal of this work, however, is to estimate the actual spatial number density of meteoroids during the peak flux of the 1999 Leonid storm.

2. Measured Data Set

The Leonid MAC mission involved two aircraft, ARIA and FISTA, flown in tandem with multiple instruments onboard for collecting data on meteor tracks and trains (Jenniskens *et al.*, 2000). The video record studied here was taken onboard the Advanced Ranging and Instrumentation Aircraft (ARIA), looking due south at low elevation. It was designated AL50R (ARIA Low elevation 50mm Right). The flight path for the night of maximum, November 18, 1999, took the aircraft on a east-northeasterly path from Israel to the Azores over a period of eight hours. The second tape from the night of November 18 was studied in detail because it contained the imagery during the time of peak storm flux around 2:00 UT. This part was also documented by visual inspection of the video record, which provided a baseline for subsequent analysis.

The imaging camera consisted of a low f-number objective of 50mm focal length f/1.4 coupled to an AEG multi-channel plate second-generation image intensifier from standard military use "night-goggles". The output of the intensifier was recorded with a Sony CCD-TRV65 Handycam Hi-8 camcorder by macro focusing on the back phosphor plate of the intensifier. The camera was pointed through one of the ARIA's BK7 optical quality glass windows whose transmission characteristics did not affect the sensitivity of the camera in the spectral region of interest. The camera performed similarly to the systems described in Jenniskens (1999a), with slightly worse coupling of the higher resolution intensified star images with the relatively low camcorder image resolution and a slightly different distribution of noise levels across the intensifier.

3. Data Analysis Procedure

The data were recorded on Hi-8mm videotape and required digitization prior to performing the various analyses necessary for flux estimation. A Scion Corporation LG-3 video frame grabber installed in a Macintosh G3/400 MHz computer formed the basis of the imagery analysis hardware. This system was capable of capturing every frame in the video record at full resolution in real time.

To obtain an accurate estimation of measured flux above the imaging system's limiting magnitude it was decided to avoid using the automated meteor detection features of the software "MeteorScan" written by one of the authors (Gural, 1999a). Although this software typically provides detection efficiencies approaching 80% of the meteors on a single pass of videotape, the software can suffer from selection effects based on short path lengths of meteors and low detection probability near the noise limit of the imagery. For the imager chosen, many of the lowest elevation meteors have very short streaks which are difficult to identify as meteors for the line detection and moving target detection algorithms in the software. Thus an interactive human detector was employed which has been found to be very efficient at motion detection of very faint objects in video.

The "MeteorScan" software was modified to first digitize several minutes of imagery and then play it back in one second movie loops for visual analysis. This required some special purpose software just to digitize the data since writing video streams to disk was found to be slower than real time. Thus four separate passes of the tape were required with 25% time overlap of one-second sequences. These were later synched up, via a spatial image correlation, on the changes in the seconds character of the superimposed time stamp.

Each second of video was replayed continuously so the user had many looks at the imagery to determine if a meteor existed. In addition to the raw imagery display, the user could also examine the difference frames, where a moving track stands out more clearly above the noise. Once a track was identified, the user would position a cursor over the beginning and ending points, mouse clicking to define the meteor start and stop positions. The track would be marked on the screen so that during the next one second of time overlapped video, the user would not repeat the count of the same meteor. After sequencing through roughly fifty seconds of video or one CD's worth, the selected tracks would be reprocessed in higher fidelity to refine estimations of begin and end points, apparent angular speed, integrated intensity, and frame of peak brightness. The refinement was done by applying a matched filter along the meteor's path and varying four track parameters in a downhill simplex search for the maximum likelihood ratio. This provided the best fit of a propagating line to the video sequence of images.

Only those meteors that had an end point within the camera's field of view were counted thus defining the effective field of view (Jenniskens, 1999a). It is necessary to calibrate the field of view to convert pixel coordinates for the end points to azimuth and elevation in an earth inertial coordinate system centered on the imager. This was done at the beginning of each period of fifty seconds using several known stars in the field and applying the astrometry algorithms of Steyaert (1990). Inputs to this process were the ARIA's position in latitude, longitude, and height obtained off recorded logs from the Global Positioning System. For the AL50R, the field of view was found to be stable to within plus or minus one degree for the time interval studied. The azimuth and elevation of the image center (Figure 1 and 2) wavered little except for a aircraft turn of three degrees at 1:54 UT. Thus the elevation was deemed stable enough to be able to bin meteor counts in elevation without having to calibrate each individual image frame.

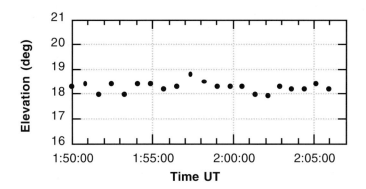

Figure 1. Image center elevation as a function of time. Note the relatively stable imager pointing during the data analysis interval.

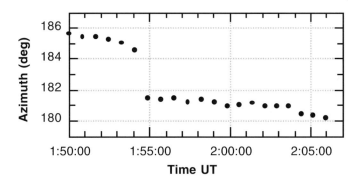

Figure 2. Image center azimuth as a function of time. The ARIA aircraft executed an azimuth adjustment of three degrees at 1:54 UT.

4. Results

4.1. SPATIAL DISTRIBUTION

In total there were 1512 meteors identified on the AL50R videotape from 1:50:00–2:06:41 UT of which 99.2% were Leonids. A plot of the end point positions in azimuth and elevation (Figure 3) after the aircraft turn shows the elevated counts near the horizon as one looks through a larger atmospheric volume. This high count of meteors at low elevation is a key advantage to flying at high altitude. This was first recognized and came as a surprise during the 1998 Leonid MAC mission (Jenniskens, 1999b).

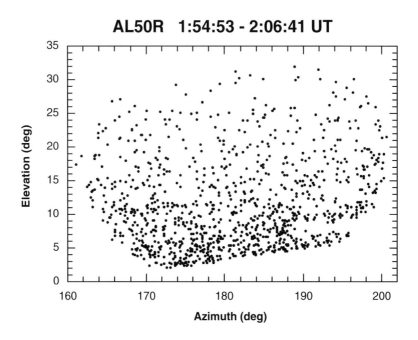

Figure 3. End point positions of AL50R meteors recorded between 1:54:53 UT and 2:06:41 UT.

The response of the imager is fairly uniform across in azimuth over almost the entire field of view. This bodes well for not having to correct for imager sensitivity or lens vignetting in the later analysis. A lack of meteors seen in the lower right corner can be accounted for by the interference from the superimposed time stamp that blocked the visibility of meteors from the user in that portion of the field of view. The sloped edge of the field of view at the lowest elevation angles is a result of the camera mounting alignment and pitch up of the aircraft relative to the local horizon.

4.2. TEMPORAL DISTRIBUTION

Examination of the raw video counts versus time during the time of the storm's peak, shows a fairly flat flux profile (Figure 4) with activity in the AL50R field of view averaging 80 meteors per minute with several bursts of activity as high as 110 meteors per minute. At the ten second resolution scale (Figure 5), it is apparent that a relatively constant background level seems to be punctuated with flux rates that can nearly double over very short observation times. These possible waves of meteors deserve attention in future analysis of the remaining video records from the other cameras. Please note that the flux estimation made in the latter part of this paper will represent an average level over the period of time analyzed and should include the caveat that waves of meteoroids can easily raise the spatial number density on a short time basis.

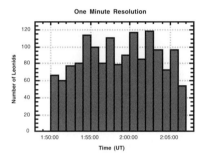

Figure 4. Raw Leonid video counts versus time at a one minute resolution.

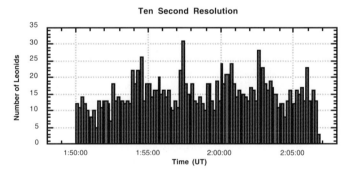

Figure 5. Raw Leonid video counts versus time at a ten second resolution.

4.3. METEOR TRAIL PROPERTIES

It was first thought that the larger distances to those Leonids near the horizon would result in the detection of a smaller part of the meteor trajectory. The issue is that the meteor's light curve has a brightening and fading temporal response and the dimmer begin and end points would be invisible due to extinction and distance losses. Thus we would expect the trail lengths to be more foreshortened than theoretically predicted. We examined the observed trail lengths and compared them to theoretical values based on typical Leonid beginning and end heights. The heights chosen were based on extrapolation of photographic beginning heights suggesting a beginning height of only 108 km. whereas the typical end point of Leonids have been found to have a height of 95 km (Jenniskens *et al.*, 1998; Brown *et al.*, 2000). Of course, beginning and end heights are derived from meteors observed at short range (high elevation angle) so relating the low elevation observations to the shorter-range measurements of the past can be problematic.

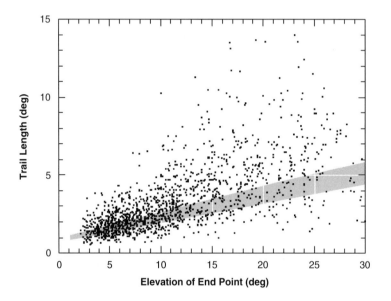

Figure 6. Visible meteor trail lengths versus elevation compared to typical Leonids with beginning heights of 108 km. and ending heights of 95 km. for the AL50R at the time of peak flux. The dots are measured trail lengths from the video and the gray area represents expected values given average Leonid parameters over the entire field of view.

From the measured positions of the meteor begin and end points in image coordinates and the calibration of the field of view in stellar coordinates, one can determine radiant distance and the measured visible trail length presented to the low elevation imager (Figure 6). A predicted set of trail lengths can be defined by using an average beginning height of 108 km, an end height of 95 km, and solving for the geometric lengths presented to all positions in the AL50R field of view for the appropriate Leonid radiant position. This was done for the ARIA's position at 2:00 UT on November 18, 1999. It can be seen that the measured meteor trail lengths (dots) fall in and around the average predicted Leonid values (gray area). The spread in measured values is caused by both a distribution of begin heights and missing components of the fainter tracks due to distance.

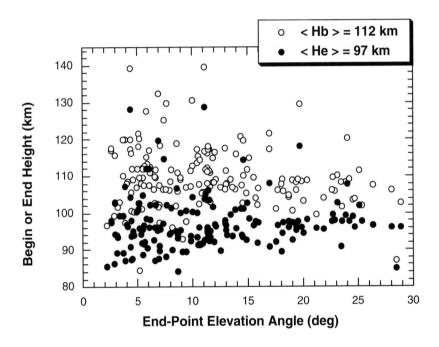

Figure 7. Leonid beginning and ending heights based on single station estimates showing similarity to published values and no apparent trend with elevation angle. Shown is a set of 172 meteors whose angular velocities were measured more precisely representing all meteors observed between 1:59:48 and 2:01:19 UT.

To double check the beginning and end heights chosen, one can use a single station formulation for H_e and H_b, since we know the entry velocity for Leonids (71 km/sec), the sensor altitude (11.0 km), the distance of the track from the radiant, and the apparent angular velocity ω. We first re-evaluated the angular velocity more accurately for all meteors over a period of one and a half minutes centered near 2:00 UT, and computed the heights using a ground based formula (Gural, 1999b) corrected for high altitudes. R is the earth's radius and H_{sensor} the sensor altitude.

$$H = [(R + H_{sensor})^2 + 2 (R + H_{sensor}) |q| \sin(\text{Elevation}) + |q|^2]^{1/2} - R \quad (1)$$

where:

$$|q| = V_\infty \sin(\text{Radiant Distance}) / \omega(\text{rad/sec}) \quad (2)$$

The formula is valid for any point along the meteor's track but suffers from larger errors as the elevation decreases (± 2 km at 25 degrees versus ± 5 km at 5 degrees) for the given angular velocity accuracy (± 0.0065 rad/sec limited by the video resolution). Nevertheless, the mean begin and end heights match very closely to the values suggested from other published results (Figure 7). There is also no obvious trend in the data as a function of elevation angle, although the errors associated the low elevation single station measurements make this somewhat inconclusive.

5. Flux Modeling and Simulation

A Monte Carlo meteor influx simulation was developed for comparison with the observations. The objective was to estimate the spatial density of meteors from the 1999 Leonid storm and to determine the optimal camera pointing directions for future data collection missions.

The technique relates the raw observed meteor counts to a statistically derived sample of counts based on a few parameterized shower variables. The statistical flux levels are obtained by ray tracing simulated meteor paths through realizable geometry and magnitude losses, and counting them over a restricted field of view. This method avoids some of the pitfalls inherent in adjusting measured meteor counts for perception and low elevation observations that have been used until now. However, the method requires reasonably good models for magnitude losses from atmospheric extinction, distance fading, and apparent angular speed. Although the ARIA and FISTA low elevation video data were collected at an altitude with little to no extinction loss, the simulation includes such losses to better model observations that could be potentially collected at any altitude.

The simulation is initialized with a uniformly distributed set of randomly positioned meteoroids in a three dimensional cylinder (Figure 8). The region in space containing the meteoroids is aligned with its long axis parallel to and centered on the observer's radiant vector **r** with the orthogonal dimensions representing the cross-track orientation. All the particles are assumed to move in parallel along the direction **r** with entry velocity V_∞. The cross-section of the particle swarm is taken large enough to encompass every look direction of an all-sky sensor that intersects a sphere of radius beginning height plus Earth's radius. This defines a meniscus that extends to the observer's zero elevation horizon in all azimuth directions, and is assured to be intersected by meteoroids in all possible places independent of radiant elevation. The meteoroids are assumed to travel in straight lines without deceleration along the atmospheric path. Zenith attraction is not accounted for but can be neglected in the case of the Leonids with high entry velocities at medium radiant elevations. Another factor ignored for this simulation was that the magnitude is dependent on the entry angle of the meteor. However, for radiant elevations above ten degrees the curvature of the atmospheric cap is so slight as to cause little variation in entry angle across the sky at a given point in time. Since the period of analysis covered only a degree change in radiant elevation angle it was deemed unnecessary to model the entry angle effect at this point.

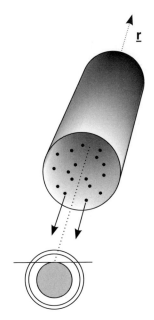

Figure 8. Geometric model of the meteor simulation developed for the analysis of observed flux and prediction of best pointing of image intensified video cameras. In the lower portion of the figure, the inner circle represents the earth's surface surrounded by two concentric circles at the meteor begin and end heights. The cylinder is a three-dimensional volume in space containing randomly distributed particles whose central axis is aligned with the observer's radiant vector **r** (dotted line). Each particle moves with the same velocity in parallel paths indicated by the arrows pointing towards the Earth.

The meteoroids are randomly assigned a magnitude referenced to a particle ablating at the zenith 100 kilometers above the earth's surface. The magnitude distribution follows from a user-specified magnitude distribution index (r) and the classic number density formula,

$$n(\text{magnitude}) = N \, r^{(\text{magnitude} - lm)} \qquad (3)$$

such that lm is defined as the limiting magnitude and N as the total number of meteoroids of magnitude lm. Each meteoroid is ray traced to the intersection with the atmosphere and assigned a random magnitude from the distribution before any losses are accounted for. The value of limiting magnitude used in the simulation was based on the sensor's sensitivity and the r-factor was allowed to vary to find the best fit to the observations.

The meteoroid must pass two visibility criteria before being counted as "seen" by the simulated imaging system. The first criteria, is that the meteor must be above the observer's horizon at some point along the visible portion of the meteor's path. For the elevation count analysis defined in the previous section, only those meteors were counted that had their endpoints visible in the camera's field of view. Thus for the purpose of the simulation, the end height was used as the defining position for the first visibility test. As there can be two solutions for that point in space, entry and exit from the atmospheric cap, the position of first entry into the atmosphere is used in the test.

The second criteria required the meteor's magnitude to exceed the limiting magnitude, after taking into account losses for extinction, distance, and the meteor's apparent angular velocity. The extinction loss, as a function of air mass "X", was based on a quadratic fit to the tables found in Roth (1994, Appendix B). For a ground based observer the expression used for visual magnitudes was:

$$\Delta m_v = -0.003 \, X^2 + 0.228 \, X - 0.225 \qquad X < 35 \text{ air masses} \quad (4)$$

Alternative formulations for extinction have been published, such as the model developed by Koschny and Zender (2000), and could easily be incorporated into future simulations

To obtain the extinction at higher observer altitudes, it was decided to calculate the air mass at the new observer's height for each elevation angle and apply the ground-based extinction for that air mass. The air mass is given by integrating the density of the 1962 U.S. Standard Atmosphere along an elevation line from the observer position to a point 50 kilometers in altitude. This is normalized by the integral of air density from the ground to the zenith. Thus given the elevation and sensor height, the air mass X was computed and the magnitude loss due to extinction Δm_v

determined. Extinction losses are normalized to airmass = 1.0 (zenith). The airmass is in fact less than one for zenith observations at altitude. The loss as a function of elevation for ground based, mountaintop, and airborne altitudes shows the advantage of an airborne sensor (Figure 9). Clearly the extinction losses are far lower when imaging at the high altitude of an airborne mission and is a major contributor to the high meteor counts seen by the aircraft's low elevation cameras during the storm.

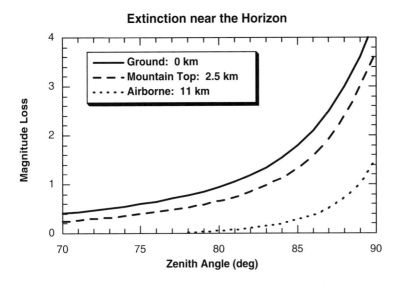

Figure 9. Extinction as a function of zenith angle for various sensor altitudes.

Magnitude losses for the meteor's distance from the observer beyond 100 kilometers is accounted for with the expression:

$$\Delta m = 5.0 \log_{10} (\text{Distance in km.} / 100) \quad (5)$$

The distance loss is normalized for a distance of 100 km, hence there is again a small magnitude gain near the zenith where the aircraft is less than 100 km from a meteor positioned directly overhead.

The final magnitude loss expression includes the effect of the apparent angular velocity "ω" of the meteor and the short time span it spends on a given pixel in the CCD camera. The loss is taken as (Jenniskens *et al.*, 1998):

$$\Delta m = 2.5 \log_{10} (\omega_{meteor} / \omega_o) \quad (\omega_{meteor} > \omega_o) \quad (6)$$
$$\Delta m = 0 \quad (\omega_{meteor} < \omega_o)$$

where ω_o is the subtended angular size of the sky imaged by the pixel divided by the integration time of the pixel. The losses are taken as zero when the meteor stays within a single pixel during the entire integration period of a single video frame ($1/60^{th}$ of a second for interlaced video), otherwise there would be a net magnitude gain from blind application of the formula. The effect of angular velocity loss is to lower the counts of higher elevation meteors since they have a faster apparent angular velocity than those near the horizon.

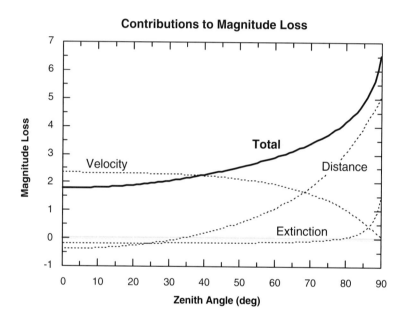

Figure 10. Total magnitude loss versus zenith angle for azimuth 180° and Leonid radiant position as seen from ARIA at 2:00 UT November 18, 1999. The losses include the effects of extinction, distance, and meteor angular velocity.

The combination of these three loss terms (Figure 10) is added to the simulated meteor's initially assigned magnitude with the result compared to the limiting magnitude. If a simulated meteor is brighter than the limiting magnitude and appears above the observer's horizon it is declared as having been detected. Here, it is assumed the sensing system has unity response across the field of view. For a human observer it would be necessary to add a detection efficiency factor as a function of the off axis angle from the line-of-sight direction, and to consider detection along the entire luminous track of the meteor rather than just at the end point. For the AL50R imager, the response across the field of view will be considered flat but adjustments will be made to the sensor area due to clipping of the image in the corners and tilt of the imager with the horizon.

When all visibility criteria are passed, the meteor is counted and binned into an azimuth / elevation table with one degree resolution. One billion simulated meteors are examined producing an all-sky map of observable flux for a given set of r-factor, radiant coordinates, entry velocity, end height, limiting magnitude, ω_o, and observer altitude. A final correction to normalize every bin to "meteors per square degree" is made by dividing each counting bin by the cosine of the elevation angle.

6. Comparison of the Simulation with Observations

6.1. ESTIMATING THE SHOWER PARAMETERS

To make an estimate of the spatial number density of meteoroids during the Leonid storm, it is necessary to determine an r-factor, the magnitude distribution index, which best matches the observed measurements. For the 1999 Leonids, results thus far for published visual data (Arlt et al., 1999), and a preliminary estimate for radar and video data (Brown et al., 2000), indicate an r-factor of 2.3 on the night of November 18, 1999. These all have been based on binning magnitude counts and either computing the mass ratio or magnitude distribution index.

The attempt in this paper is to determine the r-factor independently of magnitude estimates and rely on a simulation tool to match the observations. If the camera can be assumed to respond equally well over the entire field of view, then the variation in observed elevation counts can be compared to simulated levels by varying the r-factor until a good match is obtained. It was verified that the limiting magnitude did not vary below 20 degrees elevation by examining the sky background in the lower half of the image. The upper half did show a darker sky background, but the critical area for the fits to slope in the elevation counts is below 20 degrees so the uniform response assumption is valid in that region.

A plot of meteor counts (Figure 11) at each elevation for the AL50R observations was determined by averaging the counts across in azimuth as both the measurements and simulation show a weak flux dependence on azimuth for the southerly look direction. Aircraft pitch and roll were stable enough that angle-pointing variations of only plus or minus one degree occurred over the time period of interest. The averages were taken over three degree elevation swaths to smooth out the effects of sample size and platform pointing instability. The comparison of the measured points (dots) to the simulated curves (solid and dashed) for various r-factors are shown. Based on the simulated results, the best r-factor found was r =1.8 ± 0.1.

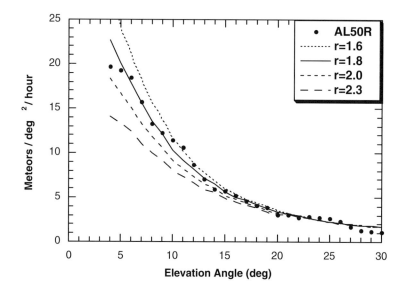

Figure 11. Measured meteor counts per square degree per hour compared to simulated counts for various r-factors. All curves have been normalized to equivalent flux at thirty degrees elevation.

The discrepancy in the r-factor with other published results is puzzling. We checked if the measurements on the ground may be affected by observing geometry and conditions. Using r = 1.8 and determining the actual magnitude ratios that would have been observed by a video camera on the ground using the simulation tool, yielded an r-factor of 1.8 to 1.9, depending on pointing. Thus the model shows that there is no distortion of the r-factor observed on the ground relative to the initial particle distribution in space. This eliminates one possible cause of the discrepancy. It is assumed that the other published results have corrected for meteor distances. If not, this could overestimate the number of faint meteors and raise the r-factor. Another possibility is that the magnitude loss factors are not modeled correctly in the simulation for apparent angular velocity and some camera slew tests through stellar fields would help quantify the actual loss numbers.

The model also produces a total meteoroid count in a given volume of space, which can be scaled to the actual counts seen by the camera and applied to all elevations. This in turn leads to a spatial density estimate of meteoroids above a given mass. The mass limit is set by the limiting magnitude of the sensing system, which was found to be +6.5 in the raw video imagery. Although imaging processing of the video through multi-frame averaging, mean removal and contrast enhancement can lower this to at best +7.5, the higher limiting magnitude was adopted for the simulation as the user processed the raw imagery without the benefit of image enhancement.

Scaling the initial simulation meteoroid count of one billion with r-factor of 1.8 to obtain the counts per square degree observed, resulted in a simulation density of 1,000,000 particles over the period of observation. Since the period of observation covered 1:50:00-2:06:41 UT or 1,001 seconds, a Leonid simulation with particle velocities of 71 km/sec would have a long axis length of 71,071 kilometers. The simulated cross section of the spatial volume was 4.36×10^6 km^2, and thus the *average* spatial density of meteoroids that produced the observed flux was found to be 3.2 ± 0.8 particles per million cubic kilometers.

Thus the average flux of the Leonids at storm peak was 0.80 ± 0.20 particles per square kilometer per hour. If we adopt the Lorentzian profile derived from the meteor counts in near real-time (Jenniskens *et al.*, 1999; 2000b), it is possible to relate the mean flux to the flux at the peak. In that case, the *peak* flux would then be 0.80 x 1.03 = 0.82 ± 0.19 particles per square kilometer per hour. This flux level is for particles with intrinsic magnitude brighter than m_v = +6.5, which is thought to be equivalent to a mass of greater than 22 micrograms (Verniani, 1973; Hughes, 1987).

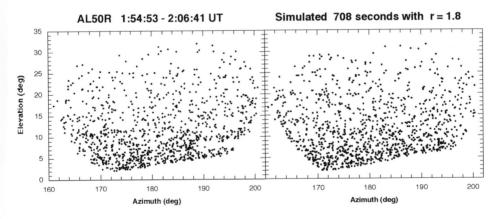

Figure 12. Simulated end point positions for the AL50R field of view (right) for a time duration that matches the measured data plot of figure 3 (left).

6.2. CLUSTERING ANALYSIS

The calculated flux is computed as a mean over the observed area and several minutes of time. Examination of the videotape gives the impression there are waves meteors on short time scales and raises the question: is there any spatial or temporal correlation that show significant deviations from the mean flux?

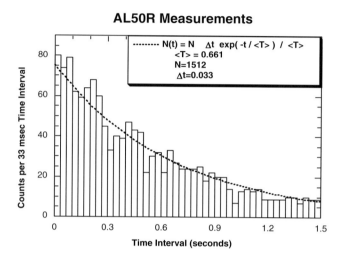

Figure 13. Histogram of time of arrival differences between successive meteors for the measured Leonids and a curve of the theoretical result for a spatially random meteoroid distribution.

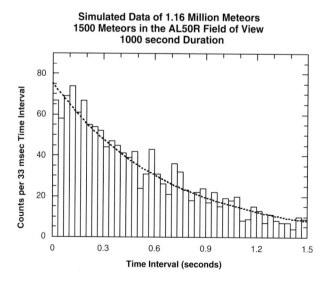

Figure 14. Histogram of time of arrival differences between successive meteors using the meteor simulation.

The simulated flux level for the AL50R look direction of due south near the horizon compares favorably with the observed count distribution as seen by the AL50R camera (Figure 12). The simulation correctly predicts the strongest flux measurements will occur at very low elevation angles due to the lack of any appreciable atmospheric extinction at high sensor altitudes.

There is some apparent spatial correlation in the observed data that deserves further attention. The observed pattern (Figure 12, left) seems to show clusters and filaments on a spatial scale of about 50 km that are not apparent in the random simulated data (Figure 12, right). A more definitive answer on spatial clumping will need to await the analysis of more of the video record and a three dimension clustering evaluation.

Such spatial correlation can result from larger meteoroids that disintegrated at some point in the past. If such breakup would occur on approach of the Earth, one would expect temporal clustering of meteors on time scales less than 1 second. These are smaller time intervals than have been looked at before (e.g. Porubcan, 1968; Ofek, 1999). To address this issue, a histogram was made of the time difference between the brightest point in the temporally adjacent meteor tracks using all meteors in this analyzed video segment. Because of to the large range in meteor distances, the times of arrival were adjusted to reflect the change in position of the meteor end point along the radiant vector for the associated azimuth and elevation angles of each meteor.

Comparison of the video measurements to the simulated case and also to the theoretical curve for a random distribution following Porubcan (1968) and Ofek (1999), all show no deviation from a uniformly random distribution of meteoroids (Figure 13 and 14). At the temporal resolution of 66 milliseconds in the video imagery (5 kilometers spatially for the Leonids) there appears to be no clumping of meteors. We conclude that meteoroids do not tend to break up on approach to the Earth in the interplanetary medium or in the Earth's magnetosphere. Rather, the spatial correlation, if real, must result from breakup during a prior return of the comet. Even after a single evolution, each cluster of particles will tend to disperse like a dust trail and spread out rapidly prior to the time of arrival at the Earth. It might even be possible that such breakup occurs in the comet atmosphere shortly after ejection. In that case, temporal variations such as found in Figure 5, may signify the breakup of larger grains, and density correlations may exist over relatively large spatial scales.

7. Optimal Pointing in Flux Measurements

The remainder of this paper is concerned with the optimum viewing conditions for flux measurements that follow from the current simulation model, as well as variations in flux measurements that can occur as a result of pointing choices of camera fields. This discussion improves on earlier results in Jenniskens *et al.* (1998;1999a).

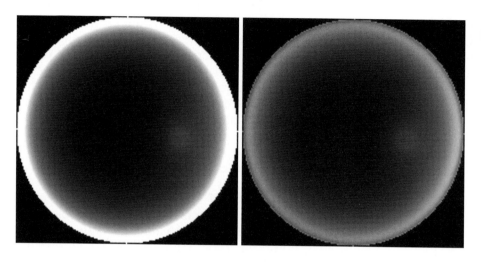

Figure 15 (left). Simulated all-sky flux levels for an airborne imager and r = 1.8.
Figure 16 (right). As Figure 15, for a ground based imager.

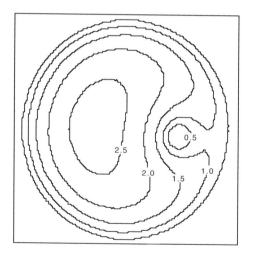

Figure 17. Magnitude losses due to the apparent velocity impact on integration time.

An illustration of the all-sky simulated flux levels for the night of the Leonid storm maximum is shown (Figure 15 and 16). The images are oriented with azimuth zero up, with azimuth measured positive in a clockwise fashion. The zenith is located dead center in the image with the horizon defined by the outer edge of the circular region. The radiant is located right of center. The simulated results are based on parameters at the time of peak flux for November 18, 1999 at 2h UT. For the ARIA's position, eleven kilometers high over the Mediterranean Sea just south of Greece, the Leonid radiant was at an elevation of 48° and an azimuth of 97°. The Leonids were assumed to have an end height of 95 kilometers. The camera's resolution of 3.7 arc minutes per pixel with 1/60 second integration time was determined from an earlier field of view calibration of the star imagery which also yielded a limiting magnitude of 6.5.

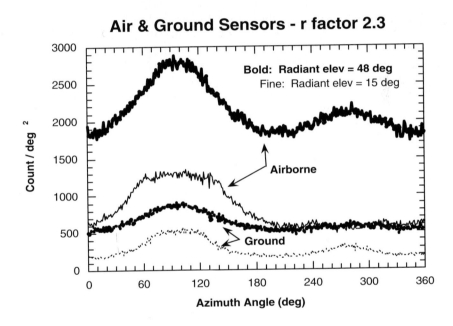

Figure 18. Azimuth cut of flux for air and ground based sensors for two different radiant elevation angles. These cuts were taken at the peak flux elevation of 3-4 degrees in the airborne case and 7-8 degrees in the ground case.

Comparing the airborne to ground flux simulation for a given radiant elevation and r-factor of $r = 1.8$, there is at least a three-fold improvement in the number of meteors seen by flying at high altitudes and looking at low elevation angles (Figure 15 and 16). What also is interesting is that ground based CCD imagers are better off pointing at a low elevation towards the azimuth of the radiant to obtain the maximum number of meteors recorded. This is because the angular velocity is minimized (Figure 17). Another interesting feature in the all-sky flux plot is a hot spot of activity located at the radiant position. This feature is due to a one to two magnitude gain in detection of faint meteors from their slow apparent velocity in and around the radiant position of the otherwise fast moving Leonids. However, it has been remarked before that the very slow moving meteors near the radiant are not easily detected by a visual observer due to their highly foreshortened tracks, leading to a typical loss in counts for that look direction (Jenniskens, 1999a).

Next, several curves are presented to show the impact of radiant elevation and azimuth, sensor altitude, and limiting magnitude on a set of simulated flux counts (Figure 18 through 23). An elevation cut through the radiant and an azimuth cut near the maximum count elevation angle are presented in these figures extracted from an all-sky simulation result similar to figures 15 and 16. Unless otherwise indicated the results are for a meteor storm with r-factor of $r = 2.3$, a 71 km/sec entry velocity, 95 km end height, radiant azimuth near 97 degrees, and a billion initial particles passing through an area 4.36×10^6 km^2. The higher r-value is chosen to reflect values reported by ground-based observers (Brown et al., 2000).

Figure 18 illustrates the effect of azimuth of the pointing direction relative to the azimuth of the radiant. These cuts are for very low elevations representative of the peak counts computed. For the airborne case the elevation angles selected are a few degrees lower than the ground's peak elevation due to extinction effects. Note that the counts are integrated over one square degree and sensors with a wide field of view will have a different total count response when integrated over their associated viewing angles. In the southward direction of the AL50R sensor, the flux is relatively insensitive to azimuth but is at its lowest value because the angular velocity is at its highest (Figure 17). This was advantageous for our averaging the flux across the imager field of view but resulted in a loss of nearly 30% of the potentially observable meteors. In our model, the ground based system captures only 40% of those seen from the air for the same pointing direction.

Examining the elevation cuts passing through the radiant and 90 degrees from the radiant (Figure 19 and 20) shows the advantage again of pointing in the direction of the radiant's azimuth. The airborne based sensor clearly shows at least a three-fold improvement over ground based measurements in observed flux levels in the low elevation directions. This gain is because below the radiant, the angular velocity is lower than elsewhere. The angular velocity is very low near the radiant also, but that does not necessary lead to higher meteor counts (Jenniskens et al., 1998). Typically, the point meteors are not easily identified in the noisy background of the intensified video cameras. The effect is clearer in photographic data, where the radiant is an efficient area for detecting meteors. For flux measurements, the radiant is not a good location because the angular velocity changes rapidly with location in the field of view. That makes the result extremely sensitive to detector properties.

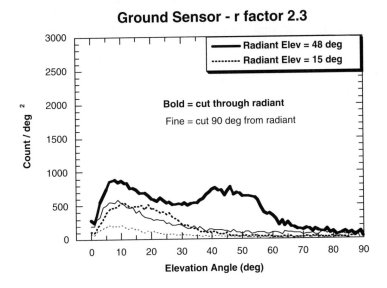

Figure 19. Elevation cut of flux for a ground-based sensor for two different radiant elevation angles. The bold lines at elevation cuts that pass through the radiant. The fine lines are elevation cuts for an azimuth ninety degrees from the azimuth of the radiant.

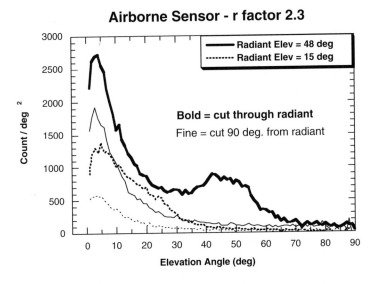

Figure 20. Elevation cut of flux for an airborne-based sensor for two different radiant elevation angles. The bold lines at elevation cuts that pass through the radiant. The fine lines are elevation cuts for an azimuth ninety degrees from the azimuth of the radiant.

The impact of r-factor on the air and ground results shows similarity with one another (Figures 21 and 22). Above 40 degrees elevation there is no r-factor dependence on total flux observed. Only below 40 degrees can the combined effects of extinction, distance, and speed begin to spread the curves as more of the fainter parts of the distribution are missed due to the increasing magnitude losses. Finally not until very low elevation angles does the airborne flux levels soar due the dominating effects of extinction on the ground. If the shower is dominated by fainter particles and a high r-factor then it appears prudent to point the imager at somewhat higher elevation, as noted before (Jenniskens *et al.*, 1998).

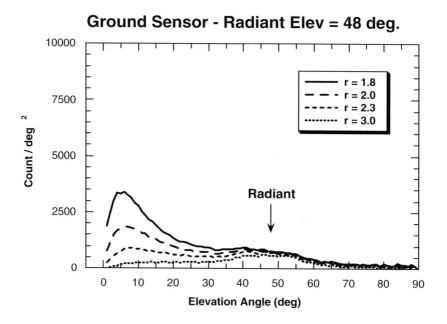

Figure 21. Elevation cut of flux for a ground sensor for different r factors.

Finally, the impact of limiting magnitude was examined, because for the 2000 and 2002 Leonids, there will be the interfering effects of moonlight (Figure 23). For the airborne case, the presumption is that the limiting magnitude drop may amount to no more than 0.2 magn., assuming a look direction away from the moon's position in the sky. For the ground case however, the losses would be 0.5 magn. or perhaps as high as 1.0 magn. for non-ideal conditions. The impact is to lower the apparent ground measured fluxes by 30 to 60 percent. Without better understanding of the effects of light scattering on limiting magnitudes at various elevations, the effect in the model is simply a scale factor on the flux in a given direction.

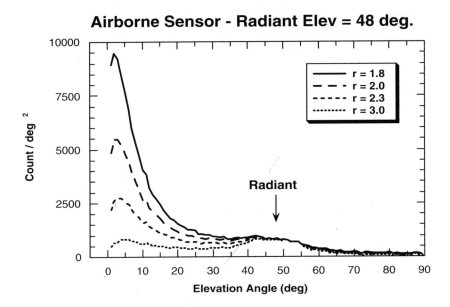

Figure 22. Elevation cut of flux for an airborne sensor for different r factors.

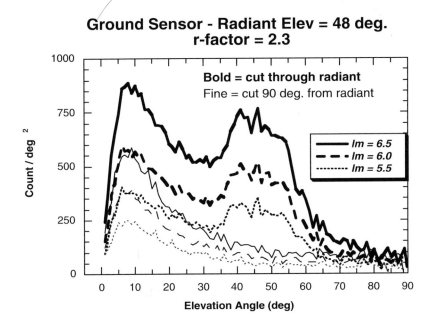

Figure 23. Elevation cut of flux for a ground-based sensor for different limiting magnitudes.

In conclusion, overall pointing recommendations for video sensors in order to maximize measured flux seem equivalent for air and ground based systems. Perpendicular to the radiant at low elevation provides highest counts and uniform conditions across the field of view. Yet higher rates are observed towards the azimuth of the radiant at low elevation angles. However, poor detectability of short meteor trails and a rapidly changing angular velocity with location in the field of view, makes the results from such observations extremely sensitive to detector properties.

In order to improve our understanding of the instrument properties in relation to observing geometry and observing conditions, it is best to do simultaneous measurements with multiple cameras at different azimuth and elevation angles and compare the results with a model as described in this paper. Koschny and Zender (2000) have published the first effort at this approach.

Until now, only a small fraction of videotaped data has been analyzed. Future work will concentrate on the possible spatial and temporal correlation detected in this paper. Of particular interest is the information that may be retrieved about the ejection and subsequent breakup of large meteoroids. In addition, the concentration of meteors near the horizon is still a surprise in our opinion. An unlikely low r-value was needed to provide a realistic distribution of meteors in the simulation model. Hence, the spatial model of meteor rates on the sky needs to be further improved to account for the observed increase of meteor rates near the horizon and improve the absolute calibration of the meteor influx.

Acknowledgments

During the meteor storm, Mr. Gural was a member of the global ground flux monitoring team lead by Dr. Peter Brown, but unfortunately was clouded out in the Canary Islands during the peak flux. He is most grateful for having been able to witness and contribute to the analysis of this once in a lifetime meteoric event from videotape. This furthers the goals of the Pro-Amat Working Group of IAU Commission 22. The work for the 1999 Leonid Multi-Instrument Aircraft Campaign was sponsored by NASA's Planetary Astronomy and Atmospheres Program. Leonid MAC was sponsored by NASA's Planetary Astronomy, Exobiology, Sub-orbital MITM, and Advanced Missions and Technologies for Astrobiology programs, NASA Ames Research Center, and the U.S. Air Force/XOR. The ARIA aircraft's operational support was provided by the USAF/452nd Flight Test Squadron. *Editorial handling:* Noah Brosch.

References

Arlt, R.: 1998, *WGN, Journal of the IMO* **26**, 239--248
Arlt, R., Rubio, L.B., Brown, P., and Gijssens, M.: 1999, *WGN, Journal of the IMO* **27**, 286--295.
Brown, P., Campbell, M.D., Ellis, K.J., Hawkes, R.L., Jones, J., Gural, P., Babcock, D., Barnbaum, C., Bartlett, R.K., Bedard, M., Bedient, J., Beech, M., Brosch, N., Clifton, S., Connors, M., Cook, B., Goetz, P., Gaines, J.K., Gramer, L., Gray, J., Hildebrand, A.R., Jewell, D., Jones, A., Leake, M., LeBlanc, A.G., Looper, J.K., McIntosh, B.A., Montague, T., Morrow, M.J., Murray, I.S., Nikolova, S., Robichaud, J., Spondor, R., Talarico, J., Theijsmeijer, C., Tilton, B., Treu, M., Vachon, C., Webster, A.R., Weryk, R., and Worden, S.P.: 2000, *Earth, Moon and Planets* **82–83**, 167–190.
Gural, P.: 1999a, *MeteorScan Documentation and User's Guide, Version 2.3*, Published by Peter Gural, Sterling, Virginia, 120 pp.
Gural, P.: 1999b, *WGN, Journal of the IMO* **27**, 111--114.
Hughes, D.W.: 1987, *Astron. Astrophys.* **187**, 879--888.
Jenniskens, P.: 1999a, *Meteoritics Planet. Sci.* **34**, 959--968.
Jenniskens, P.: 1999b, Adv. Space Res. **23**, 137--147.
Jenniskens, P. and Butow S.J.: 1999, *Meteoritics Planet. Sci.* **34**, 933--943.
Jenniskens, P., de Lignie, M., Betlem, H., Borovicka, J., Laux, C.O., Packan, D., and Krüger, C.H.: 1998, *Earth Moon and Planets* **80**, 311--341.
Jenniskens, P., Crawford, C., and Butow, S.: 1999, *WGN, Journal of the IMO* **28**, 58--36.
Jenniskens, P., Butow, S.J., and Fonda, M.: 2000a, *Earth, Moon and Planets* **82–83**, 1–26.
Jenniskens, P., Crawford, C., Butow, S.J., Nugent, D., Koop, M., Holman, D., Houston, J., Jobse, K., Kronk, G., and Beatty, K.: 2000b, *Earth, Moon and Planets* **82–83**, 191–208.
Koschack, R. and Rendtel, J.: 1990a, *WGN, Journal of the IMO* **18**, 44--58.
Koschack, R. and Rendtel, J.: 1990b, *WGN, Journal of the IMO* **18**, 119--140.
Koschny, D. and Zender, J.: 2000, *Earth, Moon and Planets* **82–83**, 209–220.
Ofek, E.O.: 1999, *WGN, Journal of the IMO* **27**, 172--176.
Porubcan, V.: 1968, *Bull. Astron. Inst. Czechoslov.* **19**, 316--323.
Roth, G. D.: 1994, *Compendium of Practical Astronomy, Volume 3, Stars and Stellar Systems*, Springer-Verlag, Berlin, 321 pp.
Steyaert, C: 1990, *IMO Monograph No. 1, Photographic Astrometry*, Published by the Computer Commission of the International Meteor Organization, 59 pp.
Van der Veen, P.: 1986b, *Radiant, Journal of the Dutch Meteor Society* **6**, 41--45.
Verniani, F.: 1973, *J. Geophys. Res.* **78**, 8429--8462.

THE LEONID METEORS AND SPACE SHUTTLE RISK ASSESSMENT

JAMES F. PAWLOWSKI

NASA Johnson Space Center, Mail Stop SN3, Houston, TX 77059
E-mail: james.f.pawlowski1@jsc.nasa.gov

and

THOMAS T. HEBERT

Dept. of Electrical and Computer Engineering, The University of Houston,
4800 Calhoun Rd., Houston, TX 77204-4793
E-mail: thebert@uh.edu

(Received 31 July 2000; Accepted 18 August 2000)

Abstract. The November 1999 Leonid meteor shower was videotaped with a low light level camera from the grounds of the National Aeronautics and Space Administration (NASA) Johnson Space Center (JSC) in Houston, Texas. Additionally, observations of the Leonids were recorded both digitally with the Liquid Mirror Telescope (LMT) and with a low light level camera at the JSC Observatory near Cloudcroft, New Mexico. The tapes were analyzed using a computer-automated meteor analysis system developed at JSC. These results were used to form estimates of the Leonid mass-distribution. These estimates were compared to a Leonid mass distribution model used by NASA in risk assessment calculations associated with Space Shuttle missions. The observed data agrees favorably with the NASA model in the 0.002 to 0.02 milligram range (based upon the LMT observations) and in the 0.02 to 0.2 gram range (based upon the low light level camera observations). This comparison supports the continued use of this model.

Keywords: Flux, Leonids 1999, Liquid Mirror Telescope, meteor, meteor shower, satellite impact hazard

1. Introduction

The primary interest in the Leonid meteors at JSC lies in their mass distribution (Pawlowski, 1999). The mass distribution is critical to the determination of potential damage to the Space Shuttle and to the International Space Station (ISS). Currently, JSC uses a Leonid mass distribution model that defines a simple time-varying mass-dependent flux.

The idea behind this model is to use predictions such as the maximum Leonid Zenith Hourly Rate (ZHR_{MAX}) to compute a probable penetrating flux on a spacecraft. The current flux model $9 \times 10^{-15} (ZHR_{MAX}) m^{-1.3}$ per meter^2 where the mass m, in units of milligrams, is based on theoretical and empirical data accumulated to date (Matney and Prior, 1999), and 1-s = -1.3 The chief parameter of interest is the Leonid mass distribution index of s = 2.3. To assess the risk to each Space Shuttle mission, this Leonid mass distribution flux model is used in conjunction with orbital debris and other meteoroid models (Christiansen, 1998). If the risk to the Space Shuttle is too high, flight designers modify the mission or postpone it. In 1997, 1998, and 1999, no Space Shuttle missions were conducted during the Leonid activity. Typical measurements of the mass distribution index of the Leonids has ranged from 1.4 to 2.2 (Murray et al., 1999; Correll et al., 1999; Jenniskens, 1999). In this paper, we perform a data-based evaluation of the use of the conservative NASA mass distribution index of s = 2.3 in this model.

The Leonid mass distribution is also important to the design and operation of the ISS wherein its vulnerable components are shielded to protect against meteoroids and orbital debris. Extra vehicular activity (EVA) from the ISS will also be planned in accordance with the assessed risk from meteor showers. Therefore the primary purpose in videotaping the 1999 Leonid meteor shower and analyzing the data was to derive a measurement comparison with the existing Leonid mass distribution model.

2. Observations

The effort by JSC to videotape the Leonid meteor shower of 1999 was performed on the grounds of JSC in Houston, Texas and at the JSC Observatory at Cloudcroft, New Mexico. The implicit assumption by using these sites is that observations from these sites would lead to mass distributions representative of the Leonid meteoroid population. A low light level video camera with a field of view (FOV) of 35° was operated at both locations, and the JSC Liquid Mirror Telescope (LMT) with a FOV of 0.278° was also operated at Cloudcroft. Table I shows the characteristics of these instruments: Location, camera, focal length, aperture, image intensifier, field of view (FOV), and star limiting magnitude (Star LM). The potential viewing period for the 1999 Leonids was midnight to dawn on November 17, 18, and 19. This translates to 06:00 to 12:00 UT in Houston and 07:00 to 13:00 UT at Cloudcroft. Clear skies in Houston allowed video to be recorded during this complete period. Viewing conditions at Cloudcroft only allowed video to be recorded from 07:00 to 10:20 UT on November 18 and from 07:00 to 13:00 UT on November 17 and 19. During these times all three instruments were pointing vertically.

TABLE I

Location	Camera	Focal Length	Aperture	Image Intensifier	FOV	Star LM
JSC	Marshall	45 mm	f/2.0	25 mm, Gen II	35	5
Cloudcroft	Sony	50 mm	f/1.4	25 mm, Gen II	35	6
Cloudcroft	LMT	5.157 m	f/1.72	25 mm, Gen III	0.276	18

2.1. ANALYSIS

Video from the low light level cameras was screened visually to detect all meteors. Video sequences of all candidate Leonid meteors were digitized and transferred to PC. A semi-automated meteor analysis software was used to measure the direction of these meteors. The radiant at the time of each detection was projected onto a viewing plane normal to the vertical camera viewing direction. A comparison between this angle and that of each observed meteor was made. An acceptance criterion of $< 8.0°$ from the radiant angle was used for the first half of the viewing period. During the second half of the viewing period the radiant was higher, introducing greater measurement variance. An acceptance criterion of $< 9.0°$ was used for this period.

Digital video tapes recorded from the LMT (0.278° FOV) were transferred to computer and processed by an automated meteor-detection software that requires approximately 12 hours of Pentium-II CPU time to process 1 hour of digital video. To detect meteors, this software applies a library of matched filters to each frame of the digital video. Each matched filter is optimal for a particular sample in the three dimensional space (relative meteor velocity, FOV crossing angle, position in the FOV). This meteor detection software has been shown to outperform the combined efforts of three trained reviewers in detecting meteors, satellites, and orbital debris (Hebert et al., 1999). All meteors, sporadics and Leonids, captured by the LMT were cataloged by this software and archived onto CD-R. Leonids were also distinguished from sporadics using a directional criterion. Since the LMT detected extremely faint meteors, we modeled the true radiant of these Leonids as having a radius of 5 degrees under the assumption that smaller meteors were ejected with higher relative velocities resulting in a greater spread in orbital parameters. Combined with an average uncertainty in measured angle of 1.1° and an uncertainty in angular alignment of the image frame with the Hubble Guide Star catalog of 2.0°, an average acceptance criterion of $< 8.1°$ was used. Additionally, meteors satisfying the directional criterion were subjected to a velocity criterion that removed candidate Leonids whose field-interval trail-lengths were completely observed within the LMT FOV, indicating that the meteor's angular velocity was well below 0.276 deg/frame. The expected Leonid angular velocities ranged from 0.28 deg/frame to 1.23 deg/frame during the observation period. Approximately 3% of all meteors detected by the LMT were categorized as Leonids.

Leonid meteor counts per hour were generated for each instrument and are displayed in Table II. The 80 Leonids detected by the 35° low light level camera used at Cloudcroft was more than the 40 Leonids detected in Houston by a similar camera. This difference is attributable to the better viewing conditions at Cloudcroft. Additionally, in comparison to the low light level cameras, the LMT detected significantly more Leonids (151) during the same viewing period. This is due to the higher sensitivity of the LMT as well as the higher detection sensitivity of the automated detection software.

Computer-assisted measurements of the absolute magnitudes of meteors satisfying the direction and velocity criterions listed above were computed using a meteor analysis system developed at JSC. To operate this software, the operator views a video sequence of the meteor frame-by-frame and visually selects the frame in which the meteor exhibits its peak brightness. Stars from the Hubble Guide Star Catalog that are visible in the frame are automatically overlaid in red upon the video frame. The operator outlines the meteor and the automated software computes an integrated intensity after subtracting an estimate of the sky background. The apparent intensities of the three visible stars closest in integrated intensity to the meteor are computed and their corresponding magnitudes are accessed from the Hubble Guide Star Catalog. Using the equation

$$M = -2.5 \log_{10} I_M + B \qquad (1)$$

depicting the relationship between absolute magnitude and intensity, where M = magnitude and I_M = intensity of a star of apparent magnitude M, the constant B is solved using a least squares fit. For meteors detected with the low light level cameras, the intensity of the meteor is computed and its absolute magnitude is calculated from Equation (1).

Computation of the absolute magnitude M of Leonid meteors detected by the LMT is somewhat more involved according to the formula

$$M = -2.5 \log_{10} (2 \alpha I_M) + B - 0.73 \qquad (2)$$

To understand Equation (2), we note that the Sony DV format in which the Leonids were recorded divides each 1/30 sec frame into two 1/60 sec fields. Given the 0.278° FOV of the LMT, an average altitude of 110 km at detection, and an average 71.7 km/sec velocity, the light from detected Leonid meteors is integrated only over one field-interval of 1/60 sec, rather than the complete frame-interval of 1/30 sec. The star intensities, on the other hand, are integrated over the complete frame-interval. This introduces the factor of 2 multiplying I_M in Equation (2). Further, the narrow FOV captures only a measurable fraction of the Leonid field-interval trail-length. The portion of the trail-length captured in the LMT FOV varies according to the radiant direction and according to whether the particular Leonid meteor passes directly through the center of the circular FOV or nearer the edge. For each detected Leonid, the term α in Equation (2) is formed by dividing the field-interval trail-length w_f at the corresponding radiant by the length w_o of that portion of the trail that is within the FOV of the LMT. The last term, 0.73 magnitudes, in Equation (2) is also due to the narrow FOV of the LMT. Each Leonid meteor is seen in only a single video frame. Therefore, it is uncertain whether the meteor was observed at the beginning, the peak, or the end of its light curve. Using the shape of light curves from Murray *et al.* (1999), an average correction factor of 0.73 magnitudes was applied.

TABLE II

TIME (UT)	6-7	7-8	8-9	9-10	10-11	11-12	12-13
JSC							
11/17/99	1	1	1	2	2	0	-.-
11/18/99	0	3	8	5	4	5	-.-
11/19/99	1	0	1	2	3	1	-.-
Cloudcroft							
11/17/99	-.-	0	6	4	13	9	1
11/18/99	-.-	5	7	12	3	CLOUDY	
11/19/99	-.-	0	4	4	7	3	2
Cloudcroft/LMT							
11/17/99	-.-	5	7	10	9	32	2
11/18/99	-.-	3	5	7	1	CLOUDY	
11/19/99	-.-	7	9	8	18	20	8

Given the absolute magnitude M of a Leonid, the meteor's mass m is then estimated using the equation (Sarma and Jones, 1985):

$$\log_{10} m = 0.495 \, [\, 9.88 - 7.17 \log_{10} (71.7 \text{ km/s}) + 0.1 \log_{10} (\cos Z) - M \,] \quad (3)$$

where m = mass in grams, M = magnitude, and Z = zenith angle (90° minus the elevation angle of the Leonid radiant).

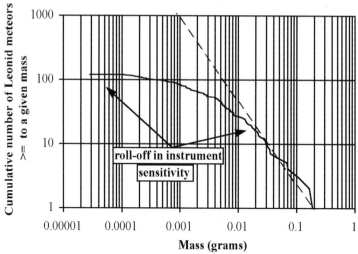

Figure 1. The 1999 Leonid Meteor Mass Distribution from the low light level cameras at Houston and Cloudcroft (solid line). Also shown is the NASA Leonid mass distribution model (dashed line).

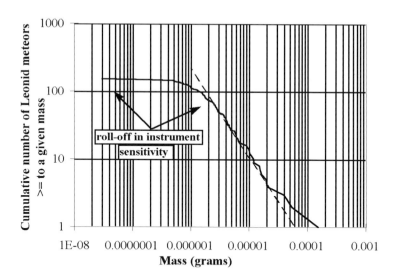

Figure 2. The Leonid mass distribution obtained from the 1999 Liquid Mirror Telescope observations at Cloudcroft (solid line). Also shown is the NASA Leonid mass distribution model (dashed line).

None of the Leonids observed by the low light level camera at Cloudcroft passed through the 0.278° FOV of the LMT at that same location. By chance, all the Leonid meteors detected by the LMT were very faint, so much so that no Leonids detected by the LMT were of sufficient brightness as to be observable in the images from the low light level camera at Cloudcroft.

4. Results and Conclusions

Figure 1 shows the Leonid mass distribution obtained from the 1999 low light level camera observations at both JSC and Cloudcroft. Also shown is the NASA model for the Leonid mass distribution normalized to the number of detected Leonids. Figure 2 shows the Leonid mass distribution obtained from the 1999 Liquid Mirror Telescope observations at Cloudcroft and the normalized NASA Leonid mass distribution model. Visually, the comparison between the data and the model looks very good in the center region of Figures 1 and 2. A least-squares fit to the measured mass distribution from the low light level cameras in the mass range of 0.02 to 0.2 grams in Figure1 yields a least-squares estimate of the mass distribution index of $s = 2.07 \pm 0.66$. This is a deviation of 10 % from the NASA mass distribution model within this mass range. A least-squares fit to the measured mass distribution from the LMT in the mass range of 0.002 to 0.02 milligrams in Figure 2 yields a mass distribution index of $s = 2.20 \pm 0.27$. This is a deviation of 4 % from the NASA mass distribution model within this mass range. The mass ranges selected for comparison correspond to the regions where the Leonid count does not suffer from decrease in instrument sensitivity or lack of sufficient counting statistics.

The results of the 1998 Leonid shower exhibited a large difference in slope from the model (Pawlowski,1999). This difference has since been attributed to the abundance of bright meteors in the 1998 shower which flattened the slope in the range of the heavy meteoroids. The 1999 shower did not show this unusual characteristic. As shown in the results above, there is excellent agreement between the NASA model and the LMT data within the corresponding range of small Leonids and an acceptable agreement between the model and the low light level data in the larger size regime for the 1999 Leonids. This agreement supports the continued use of the model.

Acknowledgments

The authors wish to acknowledge the fine telescope observations performed by Anna Scott (Aura) at Cloudcroft, the technical guidance provided by Dr. Mark Matney (Lockheed-Martin), and the video data reduction performed by Kandy Jarvis and Tracy Thumm both employed by Lockheed-Martin. *Editorial handling:* Noah Brosch.

References

Correll, R., Campbell, M., LeBlanc, A., Worden, S., Hawkes, R., Montague, T., and Brown, P.: 1999, in D. Lynch *et al.* (eds.), *Proc. 1999 AIAA Leonid Threat Conf*, 12 pp.

Hebert, T., Africano, J., and Stansbery, E.: 1999, *Automated detection of orbital debris in digital video data from a telescope*, International Astronautical Congress, Amsterdam, the Netherlands, October, 8 pp.

Jenniskens, P.: 1999, *Meteoritics Planet. Sci.* **34**, 959–968.

Matney, M. and Prior, T.: 1999, *Leonid Mass Distribution Model*, NASA internal communication, Houston,TX, 2 pp.

Murray, I., Hawkes, R., and Jenniskens, P.: 1999, *Meteoritics Planet. Sci* **34**, 949–958.

Pawlowski, J.: 1999, *Meteoritics Planet. Sci.* **34**, 945–947.

Sarma, T. and Jones, J.: 1985, *Bull. of the Astron. Inst. of Czechoslovakia* **36**, 9–24.

Christiansen, E.: 1998, *Orbiter Meteoroid/Orbital Debris Impacts*, JSC-28033, NASA Houston,TX, 100 pp.

VISUAL OBSERVATIONS OF THE 1998 AND 1999 LEONIDS IN ISRAEL

ALEX MIKISHEV AND ANNA LEVINA

Israeli Astronomical Association, Astronomia Esperanto-Klubo, str. Amazia Ha-Melech 20/1, IL-74484, Ashdod, Israel
E-mail: mikisal@inverness.co.il

(Received 1 June 2000; Accepted 17 August 2000)

Abstract. Results from visual observations of the Leonid showers in 1998 and 1999 in Israel are presented. They were processed by the method which was developed at Engelhardt Astronomical Observatory (Kazan, Russia). The values of the mass index and the Zenithal hourly rate (ZHR) are estimated for both showers. The peak of the 1999 shower was over the Middle East and the conditions for observation were optimal. The calculated ZHR for 1999 Leonid shower was up to ZHR ~ 3,700 based on 2-minute intervals, while the peak time was at solar longitude λ_o = 235.281 ± 0.001 (J2000), which corresponds to $1^h 57^m$ ± 2^m UT November 18. The shower width was 0.032 ± 0.002 ° solar longitude. The population index was higher during the storm of 1999 than in 1998.

Keywords: Dust trail, Leonids 1999, meteors, meteor shower

1. Introduction and observations

The Israeli Astronomical Association (IAA) organized collaborate observations of the Leonid showers in Israel during the last two years. A large number of amateur astronomers participated in these events. They performed visual, photographic and video-assisted observations. The main purpose of this paper is to analyze the visual observations.

The observations of the Leonid outburst in Nov. 1998 were carried out by two groups of observers. The first group was located in the Ashdod area (31°49' N, 34°39' E), 35 km south from Tel-Aviv, the other in Mitzpe Ramon (in the center of the Negev desert; 30°35' N, 34°45' E). The observers in Mitzpe Ramon, however, mainly carried out photographic work. The Ashdod group totaled 14.6 hours of observational time on the night of maximum activity of the meteor shower, counting to 1056 Leonids. The 1998 Leonids shower was characterized by bright fireballs with long luminous trajectories with persistent trains of up to 10 min duration.

The 1999 Leonid storm was observed from Mitzpe Ramon, Neve Shalom (30 km east from Tel-Aviv), Avichayil (31 km north from Tel-Aviv), and Holon (southern suburb of Tel-Aviv). Excellent observing conditions resulted in a total of 13,405 Leonids being counted in 38.17 hours of observing time. Rates increased to 5 meteors per second. Preliminary analysis of the data was performed as part of a global analysis of visual observations by the International Meteor Organisation (Arlt, 1998; 1999). Here, we apply a different analysis method, which was developed at Engelhardt Astronomical Observatory (Kazan, Russia). Results are in good agreement.

2. Processing method

The method for visual rate analysis developed by the meteor department of the Engelhardt Astronomical Observatory (Belkovich *et al.*, 1995) is easy to use. The reduced zenithal hourly rate (ZHR) is proportional to the meteoroid flux density with masses greater than a certain value corresponding to the meteors with visual magnitude +3 for given velocity. The method reduces the observed number of meteors to the number of meteors brighter than +3 magn. By contrast, the IMO method increases the rate to include all meteors up to magnitude +6.5. This introduces uncertainty, especially for inexperienced observers. However, both methods correlated well in the description of the main characteristics of meteor showers (Belkovich and Ishmukhametova, 1995).

Firstly the population index r was calculated from the linear part of the logarithm of meteor number, N, versus meteor magnitude M (Figure 1). It was obtained from the slope of a best least squares fit regression line through the logarithmic true meteor numbers, see for example Figure 1. We assume that the meteor population index ($r = N(m+1)/N(m)$) corresponds to the meteoroid mass index (S) by the equation (Koschak and Rendtel, 1990):

$$S = 1 + 2.5 \log r \qquad (1)$$

Then all S values are averaged in certain intervals of chosen solar longitude. Averaged S are used for reductions of observed meteor numbers. The zenith hourly rate, using the parameter S, is calculated from (Andreev *et al.*, 1983):

$$ZHR = \frac{N \times k}{T(\cos Z)^S} \qquad (2)$$

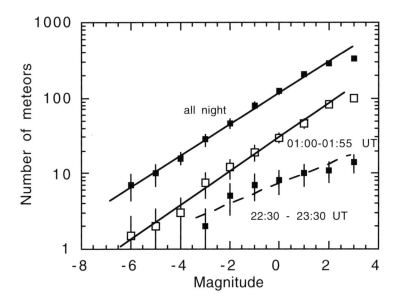

Figure 1. Distribution of meteor magnitudes from 3 experienced observers on the night of November 16/17, 1998.

Here N is the observed number of meteors, k is a correction to account for meteors of brightness less or equal $+3^m$ that are not seen, T is the effective time of the observation and Z is the zenith distance of the meteor radiant. Several corrections are not taken into account, for example, the detection probability as a function of meteor path length and angular velocity, and the limiting magnitude that may be affected by moon light. These depend on the characteristics of the human eye and can not be derived analytically. In principle, these corrections can be found by a correlative analysis, but here we apply the formula without corrections. First of all, because the observing conditions were perfect during the night, with no Moon interference. Secondly, the time period covered is relatively short, with little variations in meteor path length or angular velocity.

The threshold of the magnitude (+ 3) was chosen because it is close to the effective limiting meteor magnitude. The correction for this threshold is relatively small, albeit not negligible (Jenniskens, 1994). This constraint minimizes subjective factors for the observers (like eye sensitivity). It is a powerful tool when rates are high and numbers are sufficient to determine linear parts in the log-magnitude distribution for small time intervals.

3. Results

3.1. THE 1998 LEONIDS

Let us first consider the observations from the Leonid return of 1998. The night of the shower maximum, November 16-17, was divided into two time intervals. Figure 1 shows the meteor magnitude distribution for each time interval. The first hour of observation is not well characterized, with few Leonids observed. However, after 01^h00^m UT the luminosity function shows the expected exponential behavior. The averaged value of S for solar longitudes from 234.41 until 234.60 equals to $S = 1.45$ (or $r = 1.5 \pm 0.2$). The result is in good agreement with other estimates (Arlt, 1998). The ZHR at the time of the Israeli observations increased from 190 until 300 for solar longitude 234.46–234.55 (J2000), again in good agreement with Arlt (1998).

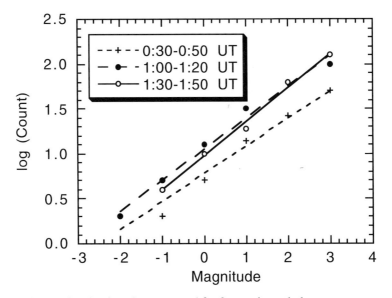

Figure 2. Luminosity function averaged for 2 experienced observers at three time intervals on the night of November 18, 1999.

3.2. THE 1999 LEONIDS

The mass index S during the 1999 Leonid storm was determined in three time intervals leading up to the peak (Figure 2). The value of S

changed from 1.77 at 00:40 UT, to 1.87 at 1:10 UT and 1.95 at 1:40 UT (or r increasing from 2.04 to 2.39). After that, rates increased so much that magnitude estimates were no longer complete. Note that Arlt et al. (1999) found a constant r = 2.2 over this time interval.

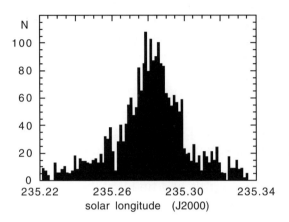

Figure 3a. Two-minute time distribution of meteors counted by Anna Levina near the activity peak on November 18, 1999.

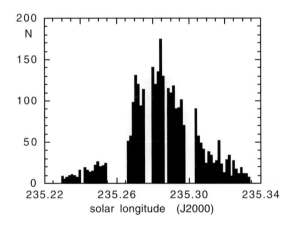

Figure 3b. As Figure 3a for observer Shlomi Eini.

Figure 3a and 3b depict the rate of Leonid meteors from two experienced observers during the night of peak activity on Nov. 17/18, 1999. The temporal resolution is two minutes, with no smoothing applied. Observer Levina finds a fairly smooth curve, with no clear evidence for filamentary structure. Observer Eini has some breaks during the observations and typically has 30% higher rates than Levina during the rise and descent of the activity, with most notably a peak at solar longitude 235.27 (or 1^h43^m UT), about 20 minutes prior to the real maximum. Arlt *et al.* (1999) reported a similar pre-maximum peak. Such peaks might be indicative of small scale ejection processes. However, note that the peak is not confirmed by the first observer and may be an effect of a changing concentration of the observers. It was not confirmed from observations onboard the Leonid Multi-Instrument Campaign (Jenniskens *et al.*, 2000). The ZHR reached 3,700 ± 200 for Levina and ZHR ~ 5,300 for Shlomi. Arlt *et al.* (1999) have a peak ZHR of 3,700. This suggests that Levina has a perception coefficient close to 1.0. After scaling the counts of Shlomi to those of Levina, we derive the mean ZHR curve in Figure 4. Following Jenniskens *et al.* (2000), we fitted a Lorentz profile and find that the center of the profile is at solar longitude 235.282±001, which corresponds to 1^h57^m UT November 18. The topocentric adjustment (McNaught and Asher, 1999a) would be only -0.5^m. The observed time of the meteor storm activity is in agreement with the predictions by Kondrat'eva and Reznikov (1985) and by Asher and McNaught (1999b). The width of the profile is 0.032±0.002° solar longitude, which compares to 0.036±0.002 derived from video observations during the Leonid Multi-Instrument Aircraft Campaign.

In comparision, we derived an average ZHR_{max} = 12 ± 3 for Nov. 16/17 and ZHR_{max} = 49 ± 9 for Nov. 18/19.

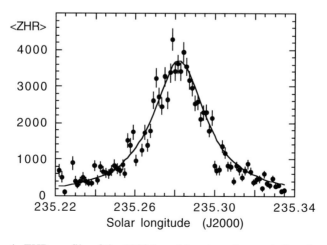

Figure 4. ZHR-profile of the 1999 Leonid meteor shower in Israel.

Acknowledgements

We thank the referees R. Arlt and M. Langbroek for their constructive and useful comments which have improved the quality of the paper. David Asher commented on the small scale filamentary structure of Leonids outbursts. We are very much indebted to all observers for their great work in collecting the data presented in this paper. The 1998 Leonids were observed by (effective observing time between brackets): Ofer Gabzo (GABOF, 0.25h), Orly Gnat (GNAOR, 0.17h), Dimitriy Kalayda (KALDI, 3.33h), Anna Levina (LEVAN, 7.21h), Semion Levin (LEVSE, 3.66h), and Eran Ofek (OFEER, 3.47h). The 1999 Leonids were observed by: Ronny Barry (BARRO, 1.77h), Sara Bordowitz (BORSA, 1.33h), Ofer Gabzo (GABOF, 3.07h), Orly Gnat (GNAOR, 1.40h), Vered Grinberg (GRIVE, 3.02h), Shlomi Eini (EINSH, 2.06h), Albert Kalifa (KALAL, 1.33h), Alexander Mikishev (MIKAL, 3.1h), Anna Levina (LEVAN, 13.01h), Eran Ofek (OFEER, 6.18h), Ronen Radayn (RADRO, 0.67h), and Gay Raviv (RAVGA, 1.29h). *Editorial handling:* P. Jenniskens

References

Andreev, V., Belkovich, O., and Tokhtasev, V.: 1984, *Solar System Res.* **17**, 244–248.
Arlt, R.: 1998, *WGN, Journal of the IMO* **26**, 267–285.
Arlt, R, Bellot Rubio L., Brown, P., and Gijssens, M.: 1999, *WGN, Journal of the IMO* **27**, 286–295.
Asher, D.J., Bailey, M.E., and Emel'yanenko, V.V.: 1999, *MNRAS* **304**, L53–L56.
Belkovich, O., Ishmukhametova, M., and Suleymanov, N.: 1995, *WGN, Journal of the IMO* **23**, 117–119.
Belkovich, O. and Ishmukhametova, M.: 1997, *WGN, Journal of the IMO* **25**, 79 — 84.
Jenniskens, P.: 1994, *Astron. Astrophys.* **287**, 990–1013.
Jenniskens, P.: 1999, *Meteorit. Planet. Sci.* **34**, 959–968.
Jenniskens, P., Crawford, C., and Butow, S.: 2000, *WGN, Journal of the IMO* **28**, 58–63.
Kondrat'eva, E.D., and Reznikov, E.A.: 1985, *Solar System Res.* **19**, 96–101.
Koschack, R. and Rendtel, J.: 1990, *WGN, Journal of the IMO* **18**, 44 — 58.
McNaught, R.H. and Asher, D.J.: 1999a, *Meteoritics Planet. Sci.* **34**, 975–978.
McNaught, R.H. and Asher, D.J.: 1999b, *WGN, Journal of the IMO* **27**, 85–102.

OBSERVATIONS OF THE LEONID METEOROID STREAM BY A MULTISTATION FS RADIO SYSTEM

GIORDANO CEVOLANI AND GIUSEPPE PUPILLO
Istituto ISAO-CNR, via Gobetti 101, 40129 Bologna
E-mail: G.Cevolani@isao.bo.cnr.it

and

ANTON HAJDUK AND VLADIMIR PORUBČAN
Astronomical Institute, Dubravska 9, 84 228 Bratislava (Sk)

(Received 31 May 2000; Accepted 2 August 2000)

Abstract. Results of joint campaigns of the Leonid meteoroid stream performed throughout the 1995–1999 years by the BLM (Bologna-Lecce-Modra) forward scatter (FS) radar, are shown and compared with visual observations. In 1998 and 1999 at both the stations of Lecce and Modra, the total reflection time is shown to give a better indication of the Leonid activity. For the two years the trends of the reflection time and of the overdense echoes (T > 8 sec duration) exhibit multiple peaks just at the maximum of the shower activity. In 1999, strong evidence of a short and extremely intense activity at the nodal longitude of the parent comet (November 18, 02h UT), is deduced from the radio data. The Leonid display is shown to follow a Gaussian activity profile and the particle density/stream width relationship is found to match observations of IRAS dust trails of comparable short-period comets. The mass distribution exponent shows a representation of an extended component of larger particles (brighter Leonids) in 1996 and especially in 1998. Average mass and population indices of radio and visual data show a similar trend in the 1995–1999 period.

Keywords: Flux, Leonids 1999, meteor, meteor shower, radar, radio forward meteor scatter

1. Introduction

Leonids are again on the rise as there is interest in the stream after the recent passage of the parent comet P/55 Tempel-Tuttle which periodically visits the inner solar system once every 33.25 years. Since a strong meteor shower activity was expected in 1998–1999, a systematic monitoring of the Leonids was highly requested in these years in order to obtain as much information on the flux and orbit of the stream as possible. The characteristics of an observed meteor display vary according to the Earth-meteoroid stream intersection geometry, the distribution of the meteoroids along the stream and the physical composition of the parent body. Times of enhanced meteor activity can be expected from every meteoroid stream. Roughly every 33 years, the Leonids generate usually a magnificent storm, when thousands of them illuminate the night sky. Nearly half of all the meteor storms reported during the last millennium are associated with the Leonids (Brown *et al.*, 1997). Kresak (1993a) has adopted a ZHR of one meteor every second, as the minimum ZHR to constitute a storm. Although the incoming particles are small (from dust to pebble), the Leonids glow brightly because they are the fastest of all the meteors. A typical Leonid meteor with a speed of about 71 km/s (more than 200 times faster than a rifle bullet), will start to glow at about 150 km. From a practical standpoint, it is also of interest to know that space platforms in Earth orbit could be at risk mainly during meteor storms. Despite the fact that several programmes have been funded to assess the risk coming from collisions with space debris, the threat to space platform integrity during times of meteor storms has only recently been investigated (Beech *et al.*, 1995; Cevolani and Foschini, 1998).

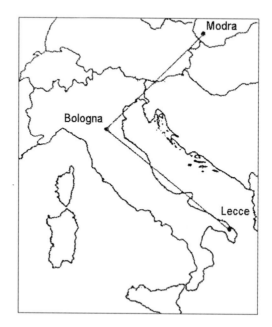

Figure 1. The geometry of the forward scatter BLM (Bologna-Lecce-Modra) radar.

2. Method

Since 1995, a new forward-scatter system for meteor observation has been operating in Italy, and starting from 1996, transmitting signals are emitted simultaneously along two mutually almost rectangular baselines with the transmitter at Budrio (44°.6N, 11°.5E) near Bologna and receivers at Lecce (40°.3N, 18°.2E) in Southern Italy and Modra (48°.4N, 17°.3E), Slovakia. The baseline distances Bologna-Lecce and Bologna-Modra are about 700 and 600 km, respectively. Figure 1 shows the geometry of the forward scatter BLM radar. The system was built for systematic monitoring of shower and sporadic meteor activity and comparing meteor flux data of a shower radiant at different baseline positions. The equipment utilises a continuous wave transmitting frequency at 42.7 MHz and a 0.5 kW mean power transmitted in the direction of both receiving stations. The transmitting and receiving antennas are horizontally polarised with an elevation angle of 15° (Cevolani et al., 1995).

3. Observations and activity of the Leonids in 1995-1999

3.1. SHOWER ACTIVITY

Radio observations of Leonids were carried out throughout 2 weeks (normally on November 10–25) each year of the 1995–99 period, but in 1995 the system was operating along the baseline Bologna-Lecce only. The observed data (all echoes, i.e. shower and sporadic) for three echo duration groups (all, > 1 s and > 8s) were recorded at Lecce and Modra. A correct derivation of a meteor shower activity curve and corresponding maxima involved a correction of the observed rates for the observability function of the utilized equipment (Cevolani et al., 1999a). The observability function describes the efficiency of the radar in detecting meteor echoes and depends on radar specific terms such as the gain pattern of the antenna, the power of the transmitter, the wavelength used and the range of the echo. Moreover, an influence is obtained by other factors such as the radiant elevation and mass distribution of the meteoroids.

From the raw hourly counts of radioechoes collected at the receiving stations, mean rates of the sporadic background were subtracted throughout the period of activity of the shower to produce shower rates. In 1995–97 shower counts exhibit multiple peaks at both the stations, also cited in other radio observations in Europe (Cevolani et al., 1999b). The 1997 Leonid activity showed a slow increase in the activity till late in the morning indicating that the Leonid maximum appeared later, not observed by the BLM system (the shower radiant being below the horizon at that time).

In 1998–99, the large number of long duration echoes made saturation effects important near the time of the peak of the shower. The power radar record was almost completely saturated at the time of the peak. Since in this period the activity and flux of Leonids was difficult to obtain, the reflection time of the radioechoes was found to give a better indication of the shower activity.

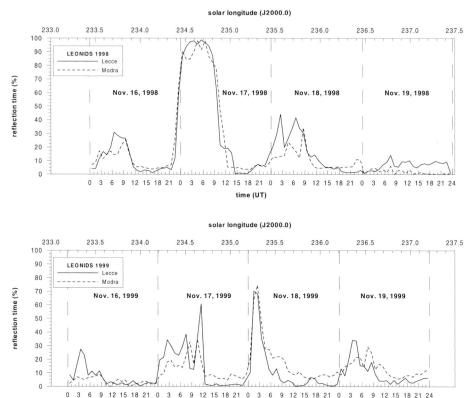

Figure 2. Reflection time (in percentage) of the radioechoes at the two stations of Lecce (*continuous line*) and Modra (*dashed line*) on November 16–19, 1998 (*upper part*) and November 16–19, 1999 (*lower part*), during the maximum activity of Leonids.

Figure 2 shows the reflection time (in percentage) of the radioechoes at the two stations of Lecce (*continuous line*) and Modra (*dashed line*) on November 16–19, 1998 (*upper part*) and November 16–19,

1999 (*lower part*), during the maximum activity of Leonids. In 1998, the reflection time of the radio signals in the 00–10 UT interval of November 17, was 91% and 88%, respectively at the two quoted stations causing obscuration of underdense trails by the numerous persistent echoes. These percentages of the reflection time represent the highest values recorded so far at our meteor radar stations during 25-year observations of meteor showers.

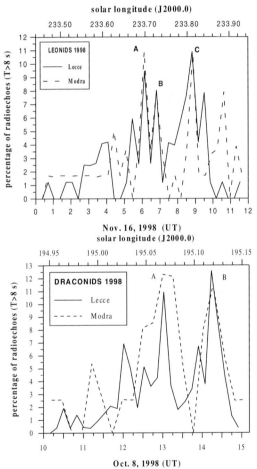

Figure 3. Number of long-duration (T>8 sec) radioechoes (in percentage) recorded at the stations of Lecce and Modra on November 16, 1998 during the Leonid display (*upper part*) and on October 8, 1998, during the Draconid display (*lower part*).

In 1999, there is strong evidence of a short and extremely intense activity at the nodal longitude of the parent comet. It is important to notice that a storm was observed visually from different locations exactly at the same solar longitude (235°.285, equinox 2000.0), corresponding to November 18, 1999, 02h UT, with a peak equivalent ZHR of 3,700 ± 100 based on about 3-minute intervals, whereby only observations with limiting magnitudes between +6 and +7 were taken into account to produce the ZHR-profile (Arlt et al., 1999).The behaviour observed when Earth encounters the meteoroid storm cloud depends critically upon: (i) the orbital position of the comet when the material is ejected; (ii) the initial ejection velocity; (iii) the influence of planetary perturbation; and (iv) the outgassing of residual volatiles from the parent comet. Over long-time scales, radiation pressure and the Poynting-Robertson effect will also have an important influence on the stream evolution. Sykes and Walker (1992) infer that cometary dust trails observed by the Infrared Astronomical Satellite (IRAS) and meteor storms are two phenomena of the same nature and the trail phenomenon is general to all short-period comets which are active members of the Jupiter family of comets.

Figure 3 exhibits the number of long-duration (T > 8 sec) radioechoes (in percentage) recorded at the stations of Lecce and Modra on November 17, 1998 during the Leonid display (*upper part*) and on October 8, 1998 during the Draconid display (*lower part*). In the case of the 1998 Leonids, the full width at half maximum of the common peaks (A, B, and C) recorded at both the stations is 0°.02 – 0°.03 corresponding to trail widths of about 20,000–25,000 km if an angle of about 163° with respect to the Earth motion is taken for Leonids. Similar results can be obtained in the case of the impressive outburst of Draconids in 1998. The width at half maximum of the common peaks (A and B) recorded at both the stations is 0°.03 – 0°.04 and corresponds to trail widths of about 25,000–30,000 km if an angle of about 32° relative to the Earth motion is taken for Draconids. These values are in good agreement with sizes of the dust trails ejected by short-period comets at heliocentric distances of 1 AU, discovered by the IRAS satellite. Kresak (1993b) indicates a proportionality of the trail width to the heliocentric distance and a typical width of 30,000 ± 10,000 km at 1 AU distances. According the same author, the IRAS dust trails are identical with meteor streams in which the particle density exceeds 400 ± 200 times that of the sporadic background at 1 AU.

Figure 4 shows the number of long-duration (T > 8 sec) and reflection time of radioechoes (in percentage) vs the solar longitude, recorded at the stations of Lecce and Modra on November 18, 1999 during the Leonid maximum. The full width at half maximum of the common peak recorded at both the stations appears more spread in solar longitude and at a first evaluation, the extent of the region perpendicular to the direction of the dust trail is 35,000–45,000. From visual data, Arlt *et al.* (1999) infers that the extent of the storm component is about 23,000 km. It is possible, too, that in 1999 the radar system observed effectively meteor particles having different masses with respect to those reported by visual data.

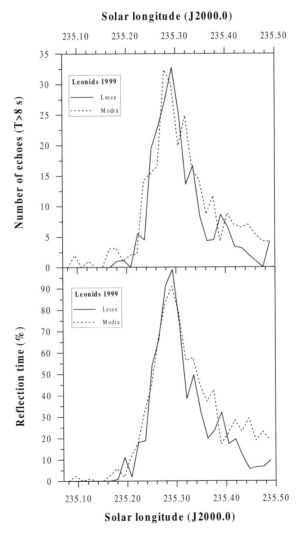

Figure 4. Number of long duration (T > 8 sec) and reflection time of radioechoes (in percentage) *vs* the solar longitude, recorded at the stations of Lecce (continuous line) and Modra (dashed line) during the maximum of the Leonids 1999.

3.2. MASS DISTRIBUTION

The radio data from Lecce and Modra enable us to derive the mass distribution index s and the population index r (s = 1 + 2.5 log r) from the observed echo durations of the shower and sporadic meteor echoes. The mass-distribution is usually determined from the echo duration distribution which is a much more sensitive function of mass distribution (see $N \propto m^{1-s}$, where N and m are number and mass of meteoroids, respectively). If diffusion is the dominant process of an echo decay, the distribution of the cumulative numbers of echo duration makes it possible to derive the exponent s from the relation (Kaiser and Closs, 1952):

$$Nc \propto T_D^{(-3/4)(s-1)} \tag{1}$$

in the form:

$$\log N_c = (-3/4)(s-1) \log T_D + \text{const} \tag{2}$$

where N_c is the cumulative number of echoes with the duration equal to and greater than T_D.

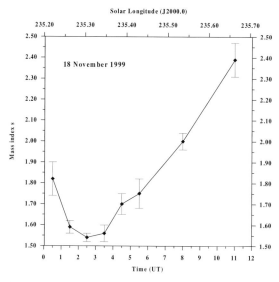

Figure 5. Variations of the mass index s during November 17–18, 1999 (Leonids + Sporadics), from combined Lecce and Modra data.

Figure 5 shows the variations of the mass index s during November 17–18, 1999 by using the combined Lecce and Modra data. The data were treated for each station separately and then combined together.

The Leonid shower is responsible for the low s values at the time of the storm, while sporadic meteors contribute to the higher s values after the storm, when the Leonid radiant set to the horizon.

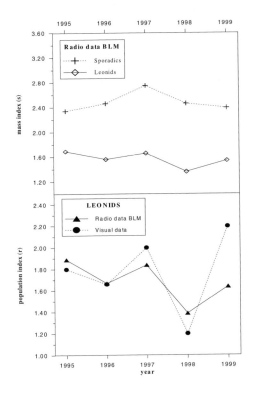

Figure 6. Variations of mass indices s of the Leonids and sporadics in 1995–99 deduced from the BLM radio data (*upper part*), and population indices of radio and visual data of Leonids in the same period (*lower part*).

Tables I and II show the mean mass indices (s) of the Leonids at different solar longitude intervals obtained for shower echoes (duration T > 1 sec) recorded in 1998 and 1999 during the peak hours by the BLM radar.

Finally, Figure 6 shows the variations of mass indices (s) of the Leonids and sporadics in 1995 – 99 deduced from the BLM radio data (*upper part*), and population indices of radio and visual data of Leonids in the same period (*lower part*).

TABLE I

Day	UT	Solar longitude (2000.0)	s
Nov. 16, 1998	21–24	234.32 – 234.45	1.54± 0.01
Nov. 17, 1998	00–03	234.45 – 234.58	1.31± 0.01
Nov. 17, 1998	03–06	234.58 – 234.70	1.29± 0.01
Nov. 17, 1998	06–09	234.70 – 234.83	1.39± 0.03
Nov. 17, 1998	09–12	234.83 – 234.95	1.46± 0.02
Nov. 17, 1998	21–24	235.33 – 235.46	2.64± 0.09

TABLE II

Day	UT	Solar longitude (2000.0)	s
Nov. 18, 1999	00–01	235.20 – 235.24	1.82 ± 0.08
Nov. 18, 1999	01–02	235.24 – 235.28	1.59 ± 0.03
Nov. 18, 1999	02–03	235.28 – 235.32	1.54 ± 0.02
Nov. 18, 1999	03–04	235.32 – 235.36	1.56 ± 0.04
Nov. 18, 1999	04–05	235.36 – 235.40	1.70 ± 0.05
Nov. 18, 1999	05–06	235.40 – 235.44	1.75 ± 0.07
Nov. 18, 1999	06–09	235.44 – 235.57	2.00 ± 0.04
Nov. 18, 1999	09–12	235.57 – 235.69	2.39 ± 0.08

TABLE III

	s - sporadics	s - shower	r	r_{vis}
1995	2.34	1.69	1.89	1.80
1996	2.46	1.56	1.67	1.66
1997	2.75	1.66	1.84	2.00
1998	2.46	1.36	1.39	1.20
1999	2.39	1.54	1.64	2.20

Table III shows the mean mass indices (s) of the Leonids and sporadics in the period 1995–1999. The final two columns show the magnitude population index (r) from radio forward meteor scatter and from visual observations r_v as reported in the literature.

The average population indices from radio and visual data for Leonids in 1995–99 are in good agreement except for the 1999 data. This discrepancy could be due to the shortness of the meteor storm and to the different bins of width in solar longitude used in radio and visual observations. On the other hand, a population index of 2.2–2.3 ($s = 1.9$) was found for visual data at the 1999 Leonid peak, as derived by a regression line method in the magnitude range -1 to $+3$ (Arlt et al., 1999). A value of $r = 2.3$ was adopted by these authors in correspondence to observations at the peak with only limiting magnitudes between +6 and +7, since the derivation of a population index turned out to be almost impossible. In this context, population indices derived from radio and visual data appear to be uncertain at the occurrence of the Leonid meteor stom, even if both the observational methods reveal a percentage lack of very faint and bright meteors in the storm component. This would justify in part the relatively low mass indices observed also a few hours after the Leonid 1999 peak.

4. Conclusions

Trends of long duration echoes and variations of reflection time obtained by the BLM radar exhibit a multiple peak activity, which is seen in connection with a filamentary structure of the Leonid meteoroid stream. The particle density/stream width relationship is found to match observations of IRAS dust trails of comparable short-period comets.

An extended component of larger particles (brighter Leonids) is found in 1996 and mainly in 1998 (very low population index for radio data). The average population indices of Leonids in 1995-99 deduced from radio and visual data are in good agreement except for the 1999 data. The derivation of indices from radio and visual data turned out to be uncertain only at the occurrence of the short-time Leonid meteor storm, even if both the observational methods revealed a percentage lack of very faint and bright meteors in the storm component.

Acknowledgements

The authors are indebted to G.Bortolotti, C.Franceschi and G.Trivellone of the ISAO-CNR in Bologna, and to E.Sbenaglia of the ISIATA-CNR in Lecce for their technical support during the observational campaigns. The financial support from the Agenzia Spaziale Italiana (ASI) is also acknowledged. *Editorial handling:* Noah Brosch.

References

Arlt, R., Rubio, L.B., Brown, P., and Gyssens M.: 1999, *WGN, Journal of the IMO* **27**, 286–295.
Beech, M., Brown, P., and Jones, J.: 1995,*Q. J.R. astr. Soc.* **36**, 127–152.
Brown, P., Simek, M., and Jones, J.: 1997, *Astron.Astrophys.* **322**, 687–695.
Cevolani, G., Bortolotti, G., Franceschi, C., Grassi, G., Trivellone, G., Hajduk A., and Kingsley S.P.: 1995, *Planet. Space Sci.* **43**, 765–772.
Cevolani, G. and Foschini, L: 1998, *Planet. Space.Sci.* **46**, 1597–1604.
Cevolani, G., Pupillo, G., Sbenaglia, E., and Trivellone, G.: 1999a, *Il Nuovo Cimento* **22C**, 747–754
Cevolani, G., Porubcan, V., Grassi, G., Hajduk, A., and Trivellone, G.: 1999b, in V.Porubcan and J.Baggaley (eds)., *Meteoroids 1998*, Astron. Institute, Slovak Academy of Sciences, Bratislava 1999, 406pp.
Kaiser, T.R. and Closs, R.L.: 1952, *Phil.Mag.* **43**, 1–32.
Kresak, L.: 1993A, in J.Stohl and I.P.Williams (eds.), *Meteoroids and their Parent Bodies*, Proceedings of the International Astronomical Symposium, Smolenice, Slovakia, July 6–12, 1992, 148–156.
Kresák, L.: 1993B, *Astron. Astrophys.* **279**, 646–660.
Sykes, M.V. and Walker, R.G.: 1992, *Icarus* **95**, 180–210.

PRECISE TRAJECTORIES AND ORBITS OF METEOROIDS FROM THE 1999 LEONID METEOR STORM

HANS BETLEM
Dutch Meteor Society, Lederkarper 4, 2318 NB Leiden, The Netherlands
E-mail: betlem@strw.leidenuniv.nl

PETER JENNISKENS
SETI Institute, NASA ARC, Mail Stop 239-4, Moffett Field, CA 94035
E-mail: pjenniskens@mail.arc.nasa.gov

PAVEL SPURNÝ
Astronomical Institute, Ondrejov Observatory, 251 65 Ondrejov, Czech Republic

GUUS DOCTERS VAN LEEUWEN, KOEN MISKOTTE, AND CASPER R. TER KUILE
Dutch Meteor Society, Lederkarper 4, 2318 NB Leiden, The Netherlands

and

PETER ZARUBIN AND CHRIS ANGELOS
Fremont Peak Observatory Association, P.O. Box 1110, San Juan Bautista, CA 95045

(Received 6 August 2000, Accepted 17 August 2000)

Abstract. Photographic multi-station observations of 47 Leonid meteors are presented that were obtained from two ground locations in Spain during the 1999 meteor storm. We find an unresolved compact cluster of radiants at $\alpha = 153.67 \pm 0.05$ and $\delta = 21.70 \pm 0.05$ for a mean solar longitude of 235.282 (J2000). The position is identical to that of the Nov. 17/18 outburst of 1998, which implies that both are due to comet 55P/Tempel-Tuttle's ejecta from 1899. We also find a halo which contains about 28% of all meteors. The spatial distribution of radiant positions appears to be Lorentzian, with a similar fraction of meteors in the profile wings as the meteor storm activity curve.

Keywords: Comet dust trial, dispersion, Leonids 1999, meteor, meteor orbit, meteor trajectory, orbital dynamics

1. Introduction

Precise orbits and trajectories of meteors provide insight into the orbital dynamics of meteoroids and the physical properties of the meteoroids during interaction with the atmosphere. Many open questions regarding the density and fragmentation of large meteoroids and the physical processes that determine the dynamics of meteoroid streams are summarized in the reviews by Ceplecha *et al.* (1998) and Jenniskens (1998).

The Leonid showers offer an unusual opportunity to study the formation and early evolution of a meteoroid stream. The first Leonid outburst was observed in 1994, amongst others from locations in California and Spain (Jenniskens *et al.*, 1996).

The discovery of enhanced Leonid activity led to the mounting of a multi-station photographic campaign at these two sites during the Leonids of 1995. Measurements of 23 precisely reduced Leonid orbits in Nov. 1995 showed a compact radiant defined by 7 orbits that were the first orbital elements of outburst Leonids (Betlem *et al.*, 1997). Similar measurements of seven orbits during a campaign from California in 1997 provided the first evidence for a systematic displacement of that compact radiant (Jenniskens and Betlem, 2000), which was confirmed by measurements from two sites in China during the outburst of 1998 in the night of Nov. 16/17 (Betlem *et al.*, 1999). Although this radiant was much more compact than that of the annual Leonid shower, we measured significant spread in the radiant positions of this component (Jenniskens and Betlem, 2000). Together with the relatively large nodal dispersion of this outburst component, called the Leonid Filament, this pointed towards a relatively high age. Indeed, numerical modeling shows the Leonid Filament to be the result of accumulation of debris in orbital resonances over the past 1000 years (Asher *et al.*, 1999; Jenniskens and Betlem, 2000).

Meteor storms are the result of more recent ejecta. From the tables of McNaught and Asher (1999) we conclude that the second peak in the night of Nov. 17/18, 1998, may have been caused by ejecta from 1899 or from 1932. We measured a significant dispersion in declination which, together with an asymmetric activity profile (Jenniskens, 1999), is evidence for planetary perturbations of the 1899 ejecta found by McNaught and Asher (1999) from numerical modeling.

Hence, we were eager to measure the radiant distribution for an unperturbed meteor storm, expecting the radiant to be more compact and the orbits of the meteoroids to be more similar. This would be a good test for our error estimates and an invitation to push the capability of the technique to the limit. Observations of the storm were performed from Spain, while similar observations from California served for reference. Only part of the data have been reduced at this time. Here, we report on the first data measured in Spain during the 1999 Leonid meteor storm.

2. Methods

The measurements were made from two observing sites at Punto Alto (38° 22' 45".25 356° 58' 1".88; height 772m) and Casa Nueva (39° 07' 01".30 357° 16' 39".90; height 785m) in Spain. Time exposures were made by fixed cameras, while an all-sky intensified video camera recorded the time of occurrence of the bright meteors. We deployed the same clusters of small (35 mm format) cameras with 50 mm f/1.8 optics and crystal controlled rotating shutters as in Betlem et al. (1999). The negatives were developed, scanned on Kodak Photo CD, and analyzed using interactive Astroscan software in the normal manner (Betlem et al., 1997; 1998). The typical astrometric accuracy is 0.003°.

3. Results

Around 1100 meteors were photographed from Punto Alto and about 700 were photographed from Casa Nueva on the night of Nov. 17/18, 1999. The first 65 precisely reduced orbits are presented in this paper.

Of the first 65 precisely reduced orbits, 47 turn out to have convergence angles larger than 20 degrees. These provide the most accurate results and are listed in Tables I and II. The convergence angle is the angle between the two planes defined by each observing site and the meteor trajectory. Table I gives the trajectories of 47 Leonids that are part of the 1999 Leonid meteor storm. The columns list apparent visual magnitude (Mv), beginning and end height (km), heliocentric velocity (Vh), geocentric velocity (Vg), geocentric radiant coordinates (α, δ), and convergence angle (Q). The heights represent the lowest and highest point of the recorded trajectory, respectively, which does not take into account that some meteors may have entered the field of view, while others may have left, without their actual beginning or end point being recorded. Table II lists the corresponding orbital elements of these 47

280 BETLEM *ET AL.*

Leonid storm meteors: perihelion distance (q), semi major axis (a), inclination (i), argument of perihelion (ω) and ascending Node (Ω).

TABLE I

Code	Day UT	Mv	H beg	H end	V h	V g	± V	α Geo	± α	δ Geo	± δ	Q max
1999008	18.01383	-5	118.72	92.19	41.39	70.69	0.16	153.62	0.06	21.67	0.03	20.71
1999010	18.03427	-1	117.28	107.67	41.25	70.54	0.06	153.61	0.04	21.74	0.04	29.81
1999011	18.03427	-1	117.26	108.31	41.33	70.63	0.15	153.61	0.16	21.65	0.04	27.45
1999013	18.04760	0	120.78	106.42	41.03	70.31	0.18	153.69	0.19	21.78	0.17	27.37
1999014	18.05128	-1	114.46	103.04	41.91	71.15	0.11	153.83	0.08	22.00	0.08	27.19
1999015	18.05166	-1	114.65	103.26	41.56	70.86	0.06	153.64	0.24	21.71	0.22	24.63
1999016	18.05387	0	113.75	104.70	41.69	71.00	0.12	153.80	0.12	21.56	0.11	31.83
1999017	18.05821	-3	114.18	94.85	41.49	70.78	0.03	153.63	0.00	21.73	0.00	38.30
1999018	18.05838	0	113.08	103.53	40.91	70.20	0.14	153.52	0.09	21.74	0.08	28.18
1999026	18.06705	-1	113.24	102.57	41.26	70.57	0.12	153.66	0.14	21.66	0.13	31.44
1999028	18.06954	0	111.74	100.69	41.54	70.82	0.14	153.67	0.07	21.79	0.06	33.26
1999029	18.06977	0	112.99	104.40	41.27	70.56	0.23	153.71	0.14	21.69	0.13	27.77
1999030	18.07054	0	113.34	101.66	41.61	70.90	0.08	153.75	0.06	21.70	0.06	27.89
1999031	18.07112	0	111.05	104.22	41.05	70.29	0.48	153.24	0.14	22.31	0.13	31.22
1999032	18.07148	-1	113.08	100.93	41.53	70.84	0.05	153.60	0.02	21.68	0.01	33.44
1999033	18.01383	-1	118.72	92.19	41.39	70.69	0.24	153.62	0.06	21.67	0.03	20.62
1999036	18.07451	-1	113.11	100.21	41.23	70.53	0.07	153.60	0.07	21.71	0.03	36.11
1999037	18.07529	0	115.05	102.11	40.94	70.22	0.25	153.76	0.07	21.71	0.07	77.71
1999041	18.07688	0	113.76	103.09	41.39	70.71	0.23	153.70	0.13	21.49	0.12	27.06
1999043	18.07907	-1	112.65	98.94	41.15	70.44	0.08	153.54	0.00	21.74	0.00	20.51
1999044	18.07940	-4	119.16	94.99	41.28	70.57	0.06	153.64	0.03	21.72	0.03	28.25
1999045	18.07948	-1	112.18	101.39	41.38	70.66	0.12	154.03	0.13	21.59	0.12	23.77
1999046	18.07979	0	114.00	105.73	41.31	70.65	0.27	153.33	0.22	21.55	0.21	25.74
1999047	18.08043	-1	113.17	101.67	41.25	70.60	0.07	152.57	0.12	21.85	0.11	27.82
1999050	18.08253	-1	112.65	100.59	41.50	70.79	0.10	153.70	0.01	21.73	0.00	32.01
1999057	18.08580	0	112.87	103.57	41.55	70.85	0.32	153.63	0.03	21.67	0.02	32.37
1999058	18.08630	0	114.87	103.03	41.55	70.82	0.12	153.68	0.00	21.82	0.00	33.11
1999059	18.09192	-1	114.47	99.26	41.30	70.60	0.11	153.65	0.04	21.69	0.04	70.05
1999061	18.08779	-1	118.14	97.71	41.45	70.76	0.14	153.50	0.08	21.70	0.07	76.81
1999064	18.08905	-1	113.69	96.81	41.36	70.60	0.05	154.39	0.01	21.71	0.00	35.47
1999065	18.08919	0	113.49	99.30	41.33	70.63	0.19	153.68	0.13	21.67	0.07	28.67
1999066	18.08936	0	112.92	104.85	40.60	69.88	0.18	153.31	0.11	21.96	0.04	26.54
1999067	18.08968	-1	110.83	94.94	41.06	70.33	0.08	153.84	0.11	21.78	0.10	21.83
1999071	18.09102	0	112.94	105.33	40.75	70.02	0.17	153.92	0.08	21.68	0.04	27.15
1999074	18.09296	0	114.02	99.25	41.04	70.33	0.15	153.66	0.05	21.72	0.04	77.33
1999079	18.10307	-1	113.12	100.91	41.10	70.38	0.11	153.92	0.17	21.62	0.16	19.85
1999081	18.09657	0	114.20	104.24	41.42	70.72	0.16	153.57	0.22	21.70	0.20	25.73
1999085	18.09917	-1	114.39	99.19	41.24	70.52	0.10	153.73	0.01	21.70	0.01	34.88
1999087	18.09788	0	112.99	101.69	41.61	70.89	0.26	153.73	0.09	21.78	0.08	47.05
1999088	18.10159	-4	115.76	89.90	41.53	70.83	0.02	153.71	0.02	21.64	0.02	30.23
1999105	18.11101	0	112.21	103.51	41.71	70.99	0.24	153.40	0.50	21.95	0.47	25.17
1999106	18.11076	-3	120.58	96.32	41.13	70.42	0.07	153.69	0.04	21.68	0.02	35.35
1999112	18.10750	1	111.15	99.28	41.47	70.73	0.31	154.14	0.12	21.71	0.06	25.70
1999114	18.11946	0	111.98	98.45	41.07	70.35	0.13	153.84	0.22	21.70	0.13	22.62
1999115	18.12153	-1	117.54	98.66	41.00	70.30	0.08	153.67	0.13	21.59	0.07	30.34
1999117	18.12295	-1	111.27	95.97	41.49	70.79	0.06	153.49	0.03	21.70	0.01	24.18
1999129	18.14267	-3	117.09	87.29	41.35	70.65	0.05	153.71	0.05	21.65	0.03	40.67
Mean			114.35	100.40	41.31	70.60	0.14	153.66	0.10	21.72	0.08	
Stand. Dev.			2.49	4.52	0.26	0.26		0.25		0.13		

TABLE II

Code	q	± q	1/a	± 1/a	i	± i	ω	± ω	Ω	± Ω
1999008	0.9843	0.0002	0.092	0.022	162.62	0.07	172.24	0.23	235.21500	0.00000
1999010	0.9844	0.0002	0.105	0.001	162.49	0.07	172.35	0.16	235.23561	0.00000
1999011	0.9843	0.0005	0.098	0.014	162.65	0.11	172.28	0.50	235.23561	0.00000
1999013	0.9842	0.0007	0.125	0.017	162.34	0.30	172.09	0.65	235.24906	0.00000
1999014	0.9843	0.0003	0.043	0.010	162.06	0.13	172.38	0.27	235.25277	0.00000
1999015	0.9844	0.0009	0.076	0.006	162.58	0.38	172.41	0.80	235.25314	0.00000
1999016	0.9837	0.0005	0.064	0.012	162.74	0.18	171.83	0.39	235.25536	0.00000
1999017	0.9845	0.0000	0.082	0.003	162.53	0.01	172.47	0.02	235.25974	0.00000
1999018	0.9847	0.0003	0.137	0.013	162.47	0.15	172.56	0.32	235.25992	0.00000
1999026	0.9850	0.0005	0.105	0.012	162.73	0.23	172.93	0.49	235.26866	0.00000
1999028	0.9845	0.0003	0.078	0.013	162.42	0.11	172.51	0.23	235.27117	0.00000
1999029	0.9842	0.0005	0.103	0.021	162.51	0.22	172.11	0.49	235.27140	0.00000
1999030	0.9841	0.0002	0.071	0.008	162.53	0.11	172.16	0.20	235.27218	0.00000
1999031	0.9863	0.0004	0.123	0.045	161.76	0.24	174.22	0.52	235.27277	0.00000
1999032	0.9846	0.0046	0.078	0.005	162.64	0.02	172.56	0.07	235.27313	0.00000
1999033	0.9843	0.0002	0.092	0.022	162.62	0.07	172.24	0.23	235.21500	0.00000
1999036	0.9846	0.0002	0.107	0.006	162.54	0.06	172.49	0.21	235.27618	0.00000
1999037	0.9839	0.0003	0.134	0.023	162.39	0.12	171.85	0.28	235.27697	0.00000
1999041	0.9839	0.0005	0.092	0.022	162.86	0.21	171.95	0.45	235.27856	0.00000
1999043	0.9848	0.0000	0.115	0.117	162.51	0.01	172.69	0.04	235.28078	0.00000
1999044	0.9845	0.0001	0.102	0.005	162.51	0.05	172.40	0.11	235.28111	0.00000
1999045	0.9829	0.0005	0.093	0.011	162.50	0.20	171.08	0.43	235.28119	0.00000
1999046	0.9852	0.0007	0.100	0.026	162.97	0.36	173.16	0.77	235.28150	0.00000
1999047	0.9874	0.0002	0.105	0.007	162.93	0.19	175.93	0.40	235.28214	0.00000
1999050	0.9844	0.0000	0.082	0.010	162.49	0.02	172.34	0.06	235.28427	0.00000
1999057	0.9845	0.0002	0.077	0.030	162.65	0.07	172.52	0.18	235.28756	0.00000
1999058	0.9846	0.0000	0.078	0.011	162.38	0.02	172.56	0.06	235.28807	0.00000
1999059	0.9845	0.0002	0.100	0.010	162.56	0.06	172.40	0.14	235.29373	0.00000
1999061	0.9850	0.0003	0.086	0.013	162.66	0.13	172.91	0.27	235.28957	0.00000
1999064	0.9817	0.0000	0.095	0.004	162.08	0.01	170.16	0.04	235.29085	0.00000
1999065	0.9843	0.0003	0.097	0.018	162.58	0.13	172.27	0.29	235.58202	0.00000
1999066	0.9857	0.0003	0.165	0.017	162.20	0.10	173.53	0.37	235.29116	0.00000
1999067	0.9838	0.0004	0.122	0.008	162.25	0.17	171.78	0.37	235.29149	0.00000
1999071	0.9833	0.0003	0.151	0.016	162.31	0.09	171.22	0.29	235.29283	0.00000
1999074	0.9844	0.0002	0.124	0.014	162.46	0.08	172.27	0.17	235.29479	0.00000
1999079	0.9833	0.0007	0.119	0.011	162.47	0.27	171.36	0.58	235.29797	0.00000
1999081	0.9848	0.0008	0.089	0.015	162.61	0.35	172.74	0.74	235.29842	0.00000
1999085	0.9842	0.0001	0.106	0.009	162.47	0.02	172.15	0.07	235.30104	0.00000
1999087	0.9844	0.0003	0.072	0.025	162.42	0.15	172.41	0.32	235.29975	0.00000
1999088	0.9843	0.0001	0.079	0.002	162.65	0.03	172.26	0.06	235.30348	0.00000
1999105	0.9857	0.0015	0.062	0.023	162.37	0.79	173.75	1.65	235.31298	0.00001
1999106	0.9843	0.0001	0.116	0.006	162.52	0.04	172.22	0.13	235.31273	0.00000
1999112	0.9828	0.0005	0.084	0.029	162.26	0.13	171.06	0.41	235.30944	0.00000
1999114	0.9838	0.0008	0.122	0.012	162.37	0.16	171.80	0.69	235.32150	0.00000
1999115	0.9843	0.0003	0.129	0.007	162.65	0.13	172.14	0.27	235.32358	0.00000
1999117	0.9851	0.0001	0.083	0.006	162.67	0.03	173.10	0.11	235.32502	0.00000
1999129	0.9844	0.0002	0.095	0.004	162.59	0.06	172.36	0.17	235.34490	0.00000
Mean	0.9844	0.0004	0.099	0.016	162.50	0.14	172.39	0.33		
Stand. Dev.	0.0009		0.024		0.22		0.86			

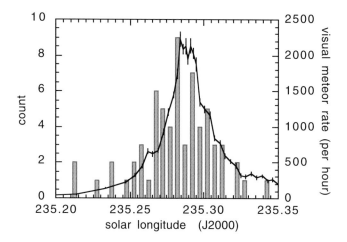

Figure 1. Rate of reduced multi-station photographed meteors in relation to the apparent flux of meteors seen by visual observers of the Dutch Meteor Society in southeastern Spain.

The sample of reduced meteor orbits covers most of the 1999 Leonid storm activity profile, which is shown by a solid line with error bars in Figure 1. This profile represents mean counts from two groups of visual observers of the Dutch Meteor Society at observing sites in southeastern Spain. In the same figure, the Leonids presented in Table I and II are counted in black columns.

The geocentric radiant positions of the meteors are plotted in Figure 2. We find that most meteors are in a very compact cluster, but there is also a significant halo around that cluster. The central position is at $\alpha = 153.67 \pm 0.05$ and $\delta = 21.70 \pm 0.05$ (J2000) for a mean solar longitude of 235.282. The dispersion in $\alpha \cos(\delta)$ and δ is equal within error: 0.046 ± 0.009 and 0.050 ± 0.009 degrees, respectively. The estimated error in the measurement has a median value of 0.06 and 0.04 degrees, respectively, which is in good agreement. Hence, we conclude that the central cluster has not been resolved. The halo, however, is significantly larger than our measurement error.

4. Discussion

In relation to past Leonid showers, we compared radiant positions by correcting for the daily changing direction of motion of Earth itself to that at arbitrary solar longitude 235.0, i.e., $\Delta\alpha = +0.99$ degrees and $\Delta\delta =$

-0.36 degrees per degree solar longitude. In this system, the 1999 Leonid storm radiant was at $\alpha = 153.39 \pm 0.05$ and $\delta = 21.80 \pm 0.05$ (J2000). Compare this to the radiant position of the 1998 outburst on Nov. 17/18, at $\alpha = 153.43 \pm 0.09$ and $\delta = 21.97 \pm 0.14$ (Betlem *et al.*, 1999). We conclude that both are identical, which implies that the second outburst in 1998 was in fact caused by the same dust trailet responsible for the 1999 Leonid storm. Based on the models by McNaught and Asher (1999), both must be due to ejecta from 1899.

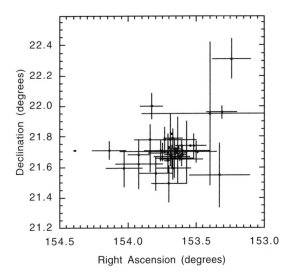

Figure 2. Geocentric radiant position of Leonids at the time of the 1999 Leonid storm.

With at least 13 out of 47 meteors (28 %) part of this halo, it is unlikely that the halo is due to the relatively weak annual Leonid shower or the (almost) absent Leonid Filament (Jenniskens and Betlem, 2000). The halo implies strong wings in the radiant dispersion of the storm, unlike a Gaussian distribution. Instead, the spatial radiant distribution appears similar to the Lorentzian shape of the nodal distribution (Jenniskens *et al.*, 2000). Indeed, the relative flux in the background profile versus the total flux curve of that Lorentzian curve is 28%, which is in agreement with the distribution of radiant positions.

The distribution of orbital elements may serve to improve meteor stream models. It is very significant that all observed orbits are well dated, being ejected during the return of comet 55P/Tempel-Tuttle in

1899. This is a very unusual situation. Further efforts at data reduction will expand the data set and thus improve the statistical accuracy.

Acknowledgements

We thank Jacob Kuiper for timely weather advise, Joseph Trigo and colleagues for their Spanish hospitality, and all Dutch Meteor Society teams that participated in the 1999 Leonid campaign for their achievement and encouragement. The Punto Alto team consisted of Michelle van Rossum, Hans Betlem, and Annemarie Zoete of the Dutch Meteor Society, and Ales Kolar of Ondrejov Observatory. The Casa Nueva team consisted of Pavel Spurný and Pavel Trepka of Ondrejov Observatory, and Guus Docters van Leeuwen of the Dutch Meteor Society. The effort was supported by the Leidse Kerkhoven Bosschafonds, the Dutch Physics Foundation, and by the NASA Planetary Astronomy program in support of the 1999 Leonid Multi-Instrument Aircraft Campaign. In kind support was received from Honda Europe, U-freight Company (Schiphol Airport), and Autocaravan Express (Madrid). The work of Pavel Spurny has been partly supported by a grant of the Grant Agency of the Czech Republic. *Editorial handling:* Mark Fonda.

References

Asher, D.J., Bailey, and M.E., Emelyanenko, V.V.: 1999, *MNRAS* **304**, L53–L56.
Betlem, H., ter Kuile c.R., Van 't Leven, J., de Lignie, M., Ramon-Bellot, L., Koop, M., Angelos, C., Wilson, M., and Jenniskens, P.: 1997, *Planet. Space Sci.* **45**, 853–856.
Betlem, H., ter Kuile C.R., de Lignie, M., Van 't Leven, J., Jobse, K., Miskotte, K., and Jenniskens, P.: 1998, *Astron. Astorphys. Suppl. Ser.* **128**, 179–185.
Betlem, H., Jenniskens, P., Van 't Leven, J., ter Kuile, C., Johannink, C., Zhao, H., Lei, C., Li, G., Zhu, J., Evans, S., and Spurny, P.: 1999, *Meteoritics Planet. Sci.* **34**, 979–986.
Ceplecha, Z., Borovicka, J., Elford, W.G., Revelle, D.O., Hawkes, R.L., Porubcan, V., and Simek, M.: *Space Sci. Rev.* **84**, 327–471.
Jenniskens, P.: 1998, *Earth Planets Space* **50**, 555–567.
Jenniskens, P. and Betlem, H.: 2000, *Astrophys. J.* **531**, 1161–1167.
Jenniskens, P.: 1996, *Meteoritics Planet. Sci.* **31**, 177–184.
Jenniskens, P.: 1999, *Meteoritics Planet. Sci.* **34**, 959–968.
Jenniskens, P., Crawford, C., Butow, S.J., Nugent, D., Koop, M., Holman, D., Houston, J., Jobse, K., Kronk, G., and Beatty, K..: 2000, *Earth Moon and Planets* **82–83**, 191–208.
McNaught, R.H. and Asher, D.J.: 1999, *WGN, Journal of the IMO* **27**, 85–102.

PHOTOGRAPHIC LEONIDS 1998 OBSERVED AT MODRA OBSERVATORY

JURAJ TÓTH AND LEONARD KORNOŠ
*Astronomical Institute, Comenius University, 842 48 Bratislava,
Slovak Republic
E-mail: Juraj.Toth@fmph.uniba.sk*

and

VLADIMIR PORUBČAN
*Astronomical Institute of the Slovak Academy of Sciences,
84228 Bratislava, Slovak Republic
E-mail: astropor@savba.sk*

(Received 1 June 2000; Accepted 18 August 2000)

Abstract. Results of photographic observations of the 1998 Leonids performed at Modra Observatory (Slovakia) are presented and discussed. During an exposure time of 7 hr 14 min on November 16/17 a total of 168 meteors were recorded. Photographic rates of bright Leonid meteors as well as their magnitude distribution are presented and compared with the results obtained by other techniques. The photographic 1998 Leonids exhibit a maximum at the solar longitude 234.52°, Equinox 2000.0 (Nov. 17, 01:40 UT).

Keywords: Fireball, flux, Leonids 1998, meteoroids, meteors, meteor shower

1. Introduction

In 1998 an exceptionally high Leonid activity was anticipated, with rates that might go up to thousands of meteors per hour (Kresák, 1993; Jenniskens,1996). The parent comet 55P/Tempel-Tuttle approached the Earth to 0.36 AU on January 17, 1998, and the perihelion passage took place on February 28 (Yeomans,1996). The Earth intersected the comet orbit on November 17 that year, only 257 days after the comet and outside of the comet orbit. The peak of the shower was anticipated shortly after passage of the comet orbital plane on Nov. 17 19:43 UT (Yeomans *et al.*, 1996), at around solar longitude 234.64 ± 0.05, or 21h UT Nov. 17 (Jenniskens, 1996). Indeed, this was the time when a "second" maximum in the activity curve was observed over eastern Asia (Arlt, 1998). Conditions for

Europe were about equally good for the early mornings of Nov. 17 and 18. However, there was enough uncertainty about the time of the expected shower to keep an eye on both nights. For example, Kresák (1993) predicted that the maximum should appear in the morning hours of November 17 at about 8 UT. As it happened, the "second" maximum was preceded by a shower rich in bright fireballs in the night of November 16/17.

A particularly nice record of this exceptional event was obtained by the photographic cameras at Modra Observatory, Slovakia (48.4° N, 17.3° E). In this paper, the first results are presented of an analysis of these all-sky images from the night November 16/17, 1998.

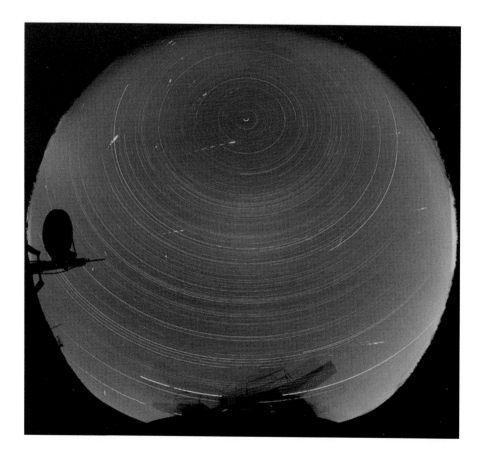

Figure 1. The photo obtained by the fixed camera in the night Nov. 16/17, 1998 at Modra Observatory.

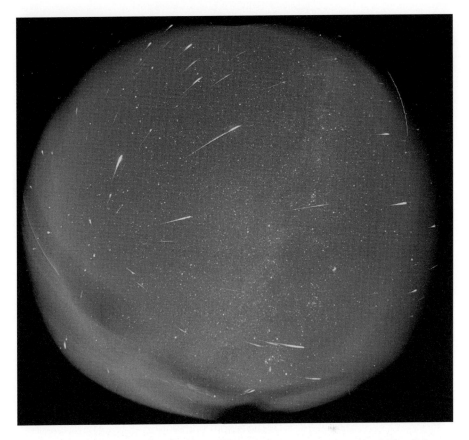

Figure 2. The photo obtained by the guided camera during the peak of the Leonid 1998 activity at Modra Observatory (exposure time: Nov. 16, 23:33:00 - Nov. 17, 03:37:10 UT).

2. Observations

Photographic meteor observations at Modra Observatory are carried out regularly on each clear night as a part of the Slovakian fireball network coordinated by the Astronomical Institute of the Slovak Academy of Sciences and being at the same time a part of the European fireball network coordinated by Ondrejov Observatory. The station consists of two fish-eye cameras equipped with Opton Distagon 3.5/30mm optics. One camera is operated in a fixed mount mode, while the second camera is guided on the stars. This provides the possibility to determine the time of appearance of the photographed meteors. The fixed camera is equipped with a rotating shutter for angular velocity measurements.

The exposures started following the routine program at 16:15 UT on Nov. 16 and were terminated due to a bad weather at 03:37 on Nov. 17. Four plates were exposed during this time and their exposures were as follows. For the fixed camera: Nov. 16, 16:15:00 – Nov. 17, 03:37:10 UT. For the guided camera: Nov. 16, 16:15:00 – 20:17:33 UT; Nov. 16, 20:20:00 – 23:30:00 UT; Nov. 16, 23:33:00 – Nov. 17, 03:37:10 UT.

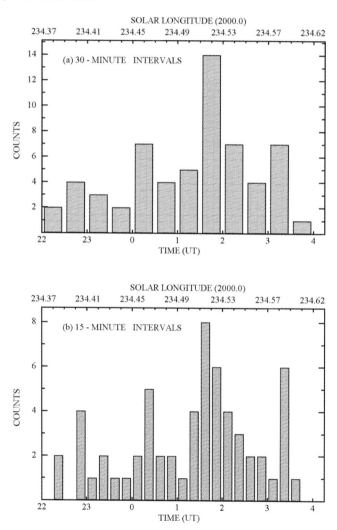

Figure 3. The counts of photographic Leonids 1998 recorded at Modra Observatory (November 16/17) in (top) 30-minute intervals and (bottom) in 15-minute intervals.

3. Results

The 1998 Leonids suprised the observers by a very high number of bright bolides appearing in the night November 16/17. During the second guided exposure that night, already 12 meteors were recorded, while during the final third guided exposure as many as 156 meteors were recorded in a four hour period. Thus, altogether 168 meteors were recorded by the guided camera in 7 hr and 14 min. Only about half of these were also recorded by the fixed camera.

The photos obtained by the fixed camera and by the guided camera in the last observing interval (Nov. 16, 23:33:00 - Nov. 17, 03:37:10 UT) are presented in Figure 1 and 2, respectively.

Unfortunately, none of these meteors were photographed from a second station for stereoscopic measurements. The night of Nov. 16/17, 1998, was cloudy in most parts of Europe, and the Modra station remained the only one from the European Fireball Network where this unique phenomenon was recorded. Although no double or multi-station data from other stations of the network are available, some limited information can be obtained regarding the radiant of the meteors, the brightness distribution of bright Leonid fireballs, and the photographic Leonid rates.

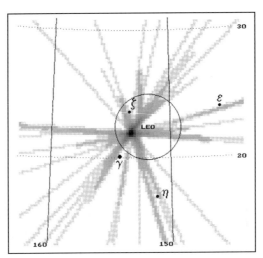

Figure 4. The radiant of the Leonids 1998 derived from 20 best recorded meteors. The big circle is expected radiant area for the annual Leonid shower (5° in diameter – IMO). An equatorial grid is drawn with 10° steps.

We derive the times of appearance for 60 meteors from a comparison of the guided and fixed plates. The identification of another 20 very faint meteors was not certain enough. Therefore, our rates are derived for the brighter meteors only. We did not take into account any correction for the non-overlapping of the fields of view between both cameras. This effect is considered minor. The difference of the sky coverages is a few percent, as demonstrated by the only four additional meteors recorded on the fixed camera. The resulting meteor rates in 30-minute and 15-minute intervals are shown in Figures 3. The highest rate was at about 01:40 UT (Nov. 17, solar longitude 234.52°, J2000) and is consistent with the estimates obtained from visual observations (Arlt, 1998). We note that the efficiency of detection was not different than at other times. The photo obtained by the fixed camera in that time interval had four additional meteors not recorded on the guided camera.

We derived an approximate position of the shower radiant by applying the software RADIANT 1.42 developed by Arlt (1992). To convert the plate coordinates to celestial coordinates, special transformation formulae were applied that were developed for the reduction of all-sky photographs by Borovi_ka, et al. (1995). The result was a good astrometric RMS = 0.017°. The shower radiant presented in Figure 4 is based on the 20 best defined meteors, for which the error of determination of the right ascension and declination of the beginning and ending points were smaller than 0.017°. The radiant position was found to be at $\alpha = 153.29° \pm 0.05°$, $\delta = 22.07° \pm 0.05°$ for the time of the maximum (solar longitude 234.52°, 2000.0). This translates to $\alpha = 153.77 \pm 0.05°$ and $\delta = 22.24 \pm 0.05°$ for solar longitude 235.0, for which time precise multi-station orbits by Betlem et al. (1999) gave a geocentric radiant position at: $\alpha = 153.80 \pm 0.08°$, $\delta = +22.10 \pm 0.03°$ (Jenniskens and Betlem 2000).

The absolute photographic magnitude of the meteors was derived from photometry of the meteor trails and stars on the fixed plate according to a procedure applied before to fireballs by Ceplecha (1987). A precision of $\pm 0.1 - 0.2$ stellar magnitudes was claimed over a wide range of zenith distance, down to 70°. However, for even larger zenith distances the precision is of the order of ± several tenths of magnitudes. In the case of bright fireballs the extrapolation of the characteristic density curve sometimes yields results with standard deviation exceeding one magnitude. We measured the magnitudes for the brightest parts of the fireball trails. The characteristic density curve is based on 11 stars and the magnitudes were derived only for those meteors for which also the time of appearance was deduced (60 meteors). The results are presented in Figure 5, which shows the distribution of absolute photographic magnitudes, that is normalized to 100 km distance at zenith, taking into account extinction, angular velocity and rotating shutter occultation corrections.

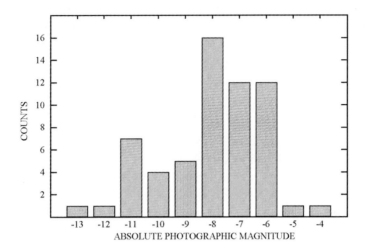

Figure 5. The distribution of the absolute photographic magnitudes of the Leonid meteors recorded on November 16/17, 1998 at Modra (60 Leonids for which the time of appearence could be derived).

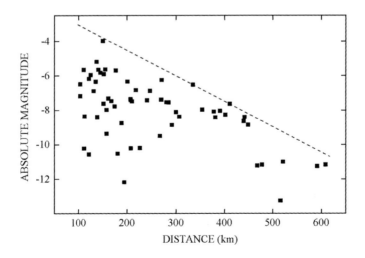

Figure 6. The distribution of the absolute photographic magnitudes of the Leonid meteors recorded on November 16/17, 1998 at Modra vs. the distance from the station.

Further away from Modra Observatory, the detection sensitivity falls off. In order to derive the fireball distance from the station for the absolute magnitude calculation, we had to assume a mean height for all Leonid meteors, which we set to 100 km. Figure 6 shows a distribution of the absolute photographic magnitudes of Leonids vs. the distance from the Observatory. The dashed line approximates the decrease of limiting magnitude for this camera with respect to the distance of the meteor from the station, which changes from −3 up to −10 magnitude from the zenith to horizon, respectively.

The variation of the mean absolute photographic magnitude in one hour intervals around the shower maximum is plotted in Figure 7 (solid line). A dotted line shows those meteors that were at distances less than 300 km. Since the trend of both variations is practically identical, it may reflect a real distribution of magnitudes of the Leonids around their maximum and may show that the brightest population of the swarm preceded the peak by about one hour. Of course, both datasets are not independent. The observed variation of photographic Leonids is consistent with visual results (Arlt, 1998).

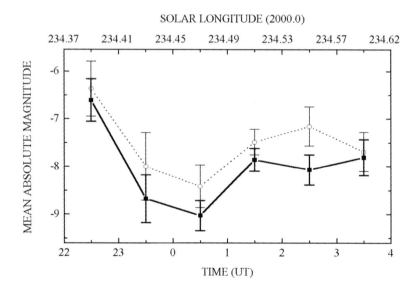

Figure 7. The distribution of the mean absolute photographic magnitudes of the Leonids (in one hour intervals) on Nov. 16/17, 1998 observed at Modra: all meteors (solid line) and those recorded within the distance of 300 km from the station (dotted line). The error bars are standard deviations in corresponding intervals.

4. Radar Observations

At Modra Observatory, the Leonids 1998 were simultaneously monitored by a forward scatter meteor system (Porubcan *et al.* 1998). The system operated along a 600 km baseline with the transmitter at Budrio near Bologna 44.6° N, 11.5° E and the receiver at Modra. The peak activity was observed between 22 h UT, Nov. 16, and 12 h UT, Nov. 17, when the Leonid radiant was above the horizon. The detection was characterized by a very high number of bright bolides. The smoothed curve of Figure 8 represents the observed numbers of echoes ≥8 s in 30–minute intervals corrected for the sporadic background echoes. The forward scatter data exhibit few not significant local maxima for echoes at the solar longitudes e.g. 234.5° and 234.8° (Nov. 17 at about 2.0 and 8.5 UT), respectively. Although the maximum is rather flat, it is of interest that the peak at 8.5 UT is consistent with the prediction of Kresák (1993). It is not clear why this agreement would exist, however. The mass exponent derived for the period of peak activity is very low s = 1.2 (Porubcan *et al.*, 1998).

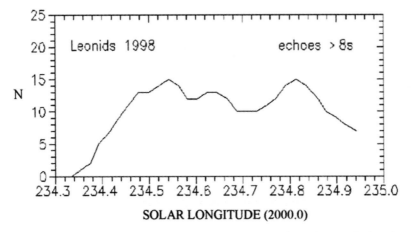

Figure 8. The peak of activity of the Leonids 1998 for echoes of duration ≥ 8 s observed by a forward scatter system at Modra Observatory (N is the number of echoes in 30 minute intervals).

Acknowledgements.

The authors are indebted to Dr. Borovi_ka for providing them the procedure for reducing the all-sky photographs and thank the referees and Dr. Jenniskens for helpful and constructive reviews. This research was supported also by VEGA - the Slovak Grant Agency for Science. *Editorial handling:* Noah Brosch.

References

Arlt, R.: 1992, *WGN, Journal of the IMO* **20**, 62–69.
Arlt, R.: 1998, *WGN, Journal of the IMO* **26**, 239–248.
Betlem, H., Jenniskens, P., Van 't Leven, J., ter Kuile, C., Johannink, C., Zhao, H., Lei, C., Li, G., Zhu, J., Evans, S., and Spurny, P.: 1999, *Meteoritics Planet. Sci.* **34**, 979–986.
Borovicka, J., Spurny, P., and Keclíková J.: 1995, *Astron. Astrophys. Suppl. Ser.* **112**, 173–178.
Ceplecha, Z.: 1987, *Bull. Astron. Inst. Czechosl.* **38**, 222–234.
Kresák, L.: 1993, in J. Stohl and I.P. Williams (eds.), *Meteoroids and Their Parent Bodies*, Slovak Acad. Sci., Bratislava, 147–156.
Jenniskens, P.: 1996, *Meteoritics Planet. Sci.* **31**, 177–184.
Jenniskens, P. and Betlem, H.: 2000, *Astrophys. J.* **531**, 1161–1167.
Porubcan V., Hajduk A., Kalmancok D., Kornos, L., and Tóth J.: 1998, *Meteor Reports SAS* **19**, 19–26.
Yeomans, D.K., Kevin, K.Y., and Weissman, P.R.: 1996, *Icarus* **124**, 407–413.

TEMPORAL VARIATION IN THE ORBITAL ELEMENT DISTRIBUTION OF THE 1998 LEONID OUTBURST

MARC C. DE LIGNIE, MARCO LANGBROEK, AND HANS BETLEM
Dutch Meteor Society, Lederkarper 4, 2318 NB Leiden, The Netherlands
E-mail: M.C.deLignie@planet.nl

and

PAVEL SPURNÝ
Astronomical Institute, Ondrejov Observatory, 251 65 Ondrejov, Czech Republic

(Received 10 July 2000; Accepted 7 September 2000)

Abstract. Double-station video observations of the 1998 Leonid shower from China resulted in 55 trajectories and orbits of meteoroids in the visual magnitude range from +0 to +6 magn. The 1998 Leonid outburst consisted of a relatively long duration shower that was rich in large meteoroids and peaked in the night of November 16/17, and an outburst of shorter duration that was rich in smaller meteoroids and peaked in the next night. The average orbit obtained during the first night agrees well with that from photographic observations. During the second night, the combined set of video and photographic observations shows temporal variation in the radiant distribution. This adds to the earlier discovery of an unusual asymmetric flux profile. In addition, the radiant distribution is shown to be mass dependent. The data suggest the presence of at least two merged dust components or a single dust component perturbed by planetary encounters.

Keywords: Leonids 1998, meteor, meteor shower, meteoroid, orbits, video

1. Introduction

With the increasing understanding of the dynamic evolution of fresh cometary ejecta a direct comparison between model studies and observations of meteor outbursts has become possible (Jenniskens, 1998). Dynamical simulations of the Leonid meteoroid stream (Kondrat'eva *et al.*, 1997; Asher *et al.*, 1999; Asher, 1999; Arlt and Brown, 1999) have aimed at reproducing observed meteor rate profiles. In this paper it is argued that the observed structure of the radiant area provides additional insight in the dust distribution of the Leonid meteoroid stream and can be used to validate model calculations.

In an earlier paper we reported on radiant structure on the basis of precise double-station photographic observations of the 1998 Leonid outbursts (Betlem et al., 1999). In the present paper we extend this study using double-station video observations made of these same outbursts. With these observations we nearly double the sample of orbits of the 1998 outbursts. In addition we extend the magnitude range of the observed meteors to +6, allowing us to search more sensitively for mass dependent structure in the radiant area of the Leonid stream and thus to provide more stringent boundary conditions for model calculations to agree with.

2. Observations

Video observations were made at four locations in the Peoples Republic of China during three nights from November 16-18, 1998. The observations were part of a larger ground-based effort (Betlem et al., 1999; Langbroek and De Lignie, 1999) that supported NASA's 1998 Leonid Multi-instrument Aircraft Campaign (Jenniskens and Butow, 1999). Locations were chosen in the areas of Xing Long (Hebei province) and Delingha (Qinghai province) such that stereoscopic observations could be made in two consecutive time zones. The Xing Long network covered the nights November 16 and 17, while the Delingha network covered the nights November 17 and 18. Exact geographic locations and camera details are listed in Table I. The video cameras typically consist of a second generation image intensifier and a Hi-8 or S-VHS camcorder. The field of view is about 25–40 degrees allowing for an astrometric resolution of 0.02 degree. The limiting magnitude for meteors is about +6 magn.

Deriving atmospheric trajectories and heliocentric orbits from the double-station video observations was done in a standard way described in (De Lignie, 1996). Magnitudes are estimated by visually comparing the surrounding stars on the video frame with the brightest meteor image. Corrections for distance were not made, but are typically smaller than 0.5 magnitude due to the high pointing elevation of the cameras.

Video observations of fast meteors only allow to measure average velocities along the trajectory. Therefore, pre-atmospheric entry velocities were estimated by adding 0.14 km/s to the measured average velocity, the value of 0.14 km/s being the typical difference between measured entry and average velocities for photographed meteors of that speed. This correction mainly influences the semi-major axis and has little effect on other orbital elements.

TABLE I

Location	Northern Latitude	Eastern Longitude	Camera type	Field (°)
Xing Long	40° 23' 48"	117° 34' 28"	2nd generation MCP + SVHS	25
Lin Ting Kou	39° 37' 47"	117° 30' 17"	2nd generation MCP + Hi-8	40
Delingha	37° 22' 42"	97° 43' 44"	3 stage 1st generation + Hi-8	28
Ulan	37° 08' 52"	98° 23' 48"	2nd generation MCP + Hi-8	28

3. Results

The full sample of 55 multi-station orbits is listed in Tables II and III. Seven of the 55 Leonids are from the night November 16/17, when an outburst of bright meteors was ramping up to a peak over Europe (Arlt, 1998; Jenniskens, 1999), while 46 are from the night of November 17/18 during passage of the node of comet Temple-Tuttle. In addition, two are from the following night of November 18/19.

Table II gives the orbital elements (J2000.0) of the 55 Leonid meteoroids. These are: the perihelion distance of orbit = q, semi-major axis = a, eccentricity = e. Node is short for the angle of ascending node and ω indicates the argument of perihelion of the orbit, while pi is Node + ω. Averages are listed separately for November 16 and 17. Meteors observed near Xing Long have video code numbers in the series that start with the numbers 982 and 983, while those observed near Delingha have codes starting with 984.

Table III gives the corresponding trajectory data (J2000.0). Velocity index G = geocentric, H = heliocentric, INF = topocentric before deceleration. Tolerances are given. Individual columns refer to beginning height (HB) and end height (HE) of the meteor, apparent radiant position and Geocentric radiant position. Z is the zenith distance of the radiant and Q_{max} is the angle between the two planes through stations and meteor trails. Again, averages are listed separately for November 16 and 17.

The basic parameters obtained from double-station observations are the geocentric radiant and entry velocity (direction and magnitude of the meteoroid's velocity vector relative to the Earth). The velocity mainly affects the semi- major axis a. For the total sample, a averages 9.5 ± 1.3 AU, in agreement with the current semi-major axis of 10.3 AU for the parent comet 55P/Tempel-Tuttle.

TABLE II

code	day	Mv	q	tol	a	1/a	tol	e	tol	i	tol	w	tol	node	pi	tol
98224	16.7856	3	0.981	0.0012	50.0	0.020	0.07	0.980	0.068	162.8	0.4	169.9	0.9	234.23	44.1	0.9
98238	16.8062	3	0.985	0.0009	16.4	0.061	0.07	0.940	0.067	162.0	0.4	172.4	0.8	234.25	46.6	0.8
98245	16.8218	1	0.984	0.0009	64.2	0.016	0.07	0.985	0.068	161.8	0.4	171.9	0.8	234.27	46.2	0.8
98254	16.8437	0	0.984	0.0013	4.2	0.238	0.19	0.766	0.182	161.2	0.6	170.9	1.6	234.29	45.2	1.6
98259	16.8495	0	0.984	0.0009	19.8	0.050	0.07	0.950	0.067	162.3	0.4	171.7	0.8	234.29	46.0	0.8
98270	16.8711	3	0.983	0.0010	9.4	0.106	0.07	0.896	0.066	161.9	0.4	170.9	0.9	234.32	45.2	0.9
98278	16.8823	-1	0.986	0.0005	-13.0	-0.077	0.07	1.076	0.071	161.2	0.4	174.3	0.6	234.33	48.6	0.6
98297	17.6632	0	0.982	0.0009	11.7	0.085	0.07	0.916	0.067	163.5	0.4	170.3	0.8	235.11	45.4	0.8
98298	17.6643	2	0.984	0.0010	9.3	0.107	0.07	0.894	0.066	162.4	0.4	171.6	0.9	235.12	46.7	0.9
98308	17.7004	1	0.985	0.0007	36.8	0.027	0.07	0.973	0.068	161.5	0.4	173.2	0.8	235.15	48.4	0.8
98309	17.7030	5	0.984	0.0009	13.0	0.077	0.07	0.925	0.067	161.1	0.4	171.7	0.8	235.15	46.9	0.8
98315	17.7140	2	0.985	0.0008	5.9	0.170	0.07	0.833	0.064	161.8	0.4	172.4	0.8	235.17	47.5	0.8
98317	17.7170	2	0.983	0.0010	5.6	0.177	0.07	0.826	0.064	162.3	0.4	171.1	0.9	235.17	46.2	0.9
98319	17.7257	0	0.984	0.0009	39.7	0.025	0.07	0.975	0.068	163.3	0.4	171.9	0.8	235.18	47.1	0.8
98401	17.7275	3	0.980	0.0027	-5.6	-0.177	0.21	1.174	0.203	162.4	0.5	169.8	1.8	235.18	45.0	1.8
98321	17.7312	2	0.984	0.0010	9.1	0.109	0.07	0.892	0.066	161.9	0.4	171.8	0.9	235.18	47.0	0.9
98403	17.7355	4	0.980	0.0018	12.9	0.078	0.07	0.924	0.073	158.6	0.3	168.9	1.2	235.19	44.1	1.2
98333	17.7916	1	0.984	0.0010	5.3	0.190	0.06	0.814	0.064	162.0	0.4	171.6	0.9	235.24	46.8	0.9
98334	17.7927	2	0.983	0.0010	11.6	0.086	0.07	0.915	0.067	162.9	0.4	171.0	0.9	235.24	46.3	0.9
98415	17.8021	5	0.971	0.0060	8.6	0.116	0.19	0.887	0.181	160.7	0.7	164.2	3.2	235.25	39.4	3.2
98340	17.8026	2	0.984	0.0009	9.6	0.104	0.07	0.898	0.066	162.3	0.4	172.1	0.9	235.25	47.3	0.9
98341	17.8048	3	0.984	0.0009	-224.3	-0.004	0.07	1.004	0.069	163.0	0.4	171.8	0.8	235.26	47.1	0.8
98344	17.8072	2	0.984	0.0007	5.0	0.198	0.08	0.805	0.081	162.0	0.5	171.9	0.8	235.26	47.1	0.8
98417	17.8089	5	0.989	0.0002	107.6	0.009	0.07	0.991	0.069	163.8	0.3	179.1	1.2	235.26	54.4	1.2
98418	17.8115	3	0.985	0.0011	15.9	0.063	0.07	0.938	0.071	162.3	0.3	172.4	1.0	235.26	47.7	1.0
98419	17.8145	5	0.984	0.0042	2.2	0.460	0.21	0.548	0.202	162.7	0.8	170.2	4.6	235.27	45.5	4.6
98424	17.8285	1	0.986	0.0009	33.3	0.030	0.07	0.970	0.068	162.3	0.3	173.7	1.0	235.28	48.9	1.0
98426	17.8316	5	0.967	0.0053	4.2	0.238	0.10	0.770	0.099	161.7	0.5	161.9	2.6	235.28	37.2	2.6
98428	17.8342	4	0.984	0.0014	4.0	0.247	0.13	0.757	0.127	162.0	0.4	171.5	1.5	235.29	46.7	1.5
98429	17.8416	2	0.986	0.0009	33.3	0.030	0.07	0.970	0.068	163.0	0.3	173.7	1.0	235.29	48.9	1.0
98357	17.8422	6	0.985	0.0010	5.8	0.174	0.10	0.829	0.099	162.6	0.4	172.6	1.1	235.29	47.9	1.1
98430	17.8447	1	0.986	0.0009	15.7	0.064	0.07	0.937	0.067	162.3	0.3	173.4	1.0	235.30	48.7	1.0
98359	17.8448	1	0.984	0.0010	10.7	0.093	0.12	0.908	0.117	162.3	0.4	172.0	1.0	235.30	47.3	1.0
98431	17.8449	4	0.985	0.0011	9.9	0.101	0.07	0.901	0.066	163.1	0.3	172.6	1.1	235.30	47.9	1.1
98360	17.8450	-1	0.985	0.0008	20.7	0.048	0.07	0.952	0.068	162.8	0.4	172.7	0.8	235.30	47.9	0.8
98362	17.8467	3	0.985	0.0008	-53.2	-0.019	0.07	1.019	0.070	162.7	0.4	173.4	0.8	235.30	48.7	0.8
98369	17.8537	4	0.982	0.0017	12.3	0.081	0.10	0.920	0.099	163.1	0.6	170.6	1.4	235.31	45.9	1.4
98435	17.8537	3	0.983	0.0012	15.4	0.065	0.07	0.936	0.067	161.6	0.3	170.9	1.0	235.31	46.2	1.0
98436	17.8547	-1	0.984	0.0017	2.7	0.370	0.19	0.636	0.191	162.0	0.5	171.2	2.2	235.31	46.5	2.2
98370	17.8556	5	0.980	0.0014	5.6	0.177	0.08	0.826	0.074	163.5	0.4	168.6	1.0	235.31	43.9	1.0
98437	17.8565	2	0.985	0.0011	-70.1	-0.014	0.18	1.014	0.175	162.4	0.4	172.8	1.2	235.31	48.1	1.2
98373	17.8585	6	0.983	0.0014	3.1	0.321	0.10	0.685	0.097	162.7	0.4	170.0	1.4	235.31	45.3	1.4
98374	17.8586	2	0.985	0.0008	-17.4	-0.058	0.07	1.057	0.070	162.4	0.4	173.3	0.8	235.31	48.6	0.8
98380	17.8643	3	0.986	0.0007	-36.6	-0.027	0.07	1.027	0.070	162.5	0.4	173.6	0.8	235.32	48.9	0.8
98442	17.8730	3	0.985	0.0012	7.1	0.140	0.12	0.862	0.120	163.1	0.4	172.8	1.3	235.33	48.1	1.3
98443	17.8735	3	0.984	0.0015	4.2	0.237	0.18	0.767	0.178	162.5	0.5	171.2	1.7	235.33	46.5	1.7
98444	17.8735	3	0.985	0.0015	4.7	0.213	0.24	0.790	0.239	161.9	0.6	172.1	1.9	235.33	47.4	1.9
98390	17.8792	2	0.985	0.0008	11.8	0.085	0.08	0.917	0.077	162.2	0.5	172.5	0.8	235.33	47.9	0.8
98445	17.8817	0	0.984	0.0010	5.5	0.183	0.08	0.820	0.075	162.5	0.4	172.1	1.0	235.33	47.4	1.0
98394	17.8878	2	0.984	0.0009	7.0	0.142	0.07	0.860	0.065	162.5	0.4	171.9	0.9	235.34	47.2	0.9
98395	17.8878	5	0.985	0.0007	26.4	0.038	0.07	0.963	0.068	162.5	0.4	172.9	0.7	235.35	48.2	0.7
98397	17.8892	-1	0.986	0.0006	-7.1	-0.142	0.07	1.140	0.073	162.5	0.4	174.0	0.7	235.34	49.3	0.7
98398	17.9014	0	0.986	0.0008	46.0	0.022	0.07	0.979	0.068	162.8	0.4	173.5	0.8	235.35	48.9	0.8
98457	17.9512	1	0.983	0.0009	3.2	0.317	0.10	0.688	0.099	162.1	0.5	170.7	1.1	235.40	46.1	1.1
98458	17.9522	4	0.981	0.0014	3.1	0.319	0.10	0.687	0.097	162.6	0.5	168.9	1.3	235.41	44.3	1.3
98467	18.8207	3	0.968	0.0116	1.7	0.574	0.29	0.445	0.273	160.2	0.9	158.8	10.5	236.28	35.1	10.5
98472	18.8573	3	0.988	0.0004	-12.5	-0.080	0.15	1.079	0.146	161.7	0.4	177.5	1.0	236.32	53.8	1.0
average	16.84	1.3	0.984		16.9	0.059		0.942		161.9		171.7		234.28	46.0	
st. dev	0.03	1.7	0.002			0.097		0.095		0.6		1.4		0.04	1.4	
average	17.82	2.4	0.983		8.9	0.112		0.890		162.3		171.6		235.27	46.9	
st. dev	0.07	1.8	0.003			0.124		0.122		0.8		2.5		0.07	2.5	

TABLE III

code	VG	VH	VINF	<V>	tol	HB	Hmax	HE	RA	tol	DE	tol	RAG	DEG	cos Z	Qmax
98224	71.4	42.1	72.6	72.4	0.7	128.1	108.1	99.5	153.29	0.24	21.81	0.21	153.41	21.72	0.622	48
98238	71.0	41.7	72.1	71.9	0.7	127.8	113.3	97.4	152.77	0.22	22.46	0.22	152.82	22.38	0.706	54
98245	71.4	42.2	72.5	72.4	0.7	138.3	104.8	94.3	153.10	0.21	22.54	0.23	153.10	22.46	0.754	60
98254	69.0	39.8	70.1	69.9	2.1	125.5	101.8	91.9	153.12	0.19	22.62	0.24	153.05	22.55	0.820	90
98259	71.1	41.8	72.1	72.0	0.7	132.3	104.5	91.9	153.09	0.23	22.21	0.21	153.00	22.14	0.831	83
98270	70.5	41.2	71.5	71.3	0.7	123.0	109.7	99.8	153.38	0.22	22.32	0.22	153.23	22.24	0.879	79
98278	72.3	43.2	73.3	73.2	0.7	153.6	122.3	109.4	152.90	0.19	23.14	0.25	152.72	23.08	0.905	57
98297	70.8	41.5	72.0	71.9	0.7	124.0	117.6	116.7	153.31	0.15	21.36	0.27	153.87	21.06	0.103	35
98298	70.5	41.2	71.7	71.6	0.7	118.9	116.4	113.0	153.14	0.26	22.06	0.19	153.71	21.77	0.116	36
98308	71.3	42.1	72.5	72.3	0.7	122.1	113.3	109.4	153.21	0.21	22.61	0.23	153.63	22.42	0.282	39
98309	70.7	41.6	71.9	71.8	0.7	125.9	118.0	114.2	153.66	0.19	22.69	0.24	154.09	22.49	0.289	38
98315	69.8	40.5	71.0	70.9	0.7	112.9	109.1	98.2	153.17	0.19	22.29	0.24	153.56	22.11	0.343	48
98317	69.7	40.5	71.0	70.8	0.7	112.1	109.1	100.1	153.39	0.22	21.89	0.22	153.77	21.71	0.349	48
98319	71.5	42.1	72.6	72.5	0.7	133.4	-	108.1	153.30	0.20	21.50	0.23	153.64	21.34	0.382	41
98401	73.4	44.2	74.6	74.5	2.1	127.1	123.0	120.6	154.32	0.57	21.85	0.14	154.85	21.62	0.126	18
98321	70.4	41.2	71.6	71.5	0.7	123.8	111.4	100.0	153.51	0.24	22.15	0.21	153.83	21.99	0.410	46
98403	70.4	41.5	71.6	71.5	0.8	119.0	116.4	111.7	154.97	0.36	23.70	0.09	155.52	23.47	0.176	23
98333	69.6	40.3	70.7	70.6	0.7	118.1	110.6	97.6	153.66	0.22	21.99	0.22	153.77	21.89	0.654	61
98334	70.8	41.5	71.9	71.8	0.7	121.5	114.5	102.5	153.82	0.22	21.48	0.22	153.92	21.39	0.649	49
98415	70.2	41.1	71.4	71.2	2.0	114.0	111.1	106.0	155.95	0.78	21.90	0.16	156.24	21.78	0.456	26
98340	70.5	41.3	71.7	71.5	0.7	122.9	114.6	99.1	153.67	0.23	21.87	0.21	153.73	21.78	0.690	54
98341	71.7	42.4	72.8	72.7	0.7	125.4	107.8	101.5	153.81	0.23	21.56	0.22	153.86	21.48	0.693	50
98344	69.5	40.2	70.6	70.5	0.9	120.7	108.6	96.3	153.63	0.13	21.99	0.28	153.68	21.90	0.709	65
98417	71.7	42.3	72.9	72.7	0.7	120.4	115.0	106.5	151.30	0.41	21.86	0.09	151.50	21.77	0.537	26
98418	71.0	41.7	72.2	72.0	0.8	122.5	110.5	93.3	153.51	0.31	21.94	0.09	153.73	21.85	0.523	26
98419	66.6	37.2	67.8	67.7	2.4	118.2	115.2	109.1	153.00	0.85	21.41	0.17	153.25	21.30	0.537	25
98424	71.3	42.0	72.5	72.4	0.7	124.6	110.0	95.4	153.28	0.31	22.07	0.10	153.44	21.99	0.598	39
98426	68.9	39.8	70.1	70.0	1.1	116.1	108.5	105.7	155.97	0.61	21.15	0.14	156.16	21.05	0.573	30
98428	69.0	39.7	70.1	70.0	1.4	122.8	106.5	98.4	153.56	0.32	21.93	0.07	153.71	21.85	0.617	35
98429	71.4	42.0	72.5	72.4	0.7	124.2	110.9	99.9	153.19	0.31	21.68	0.09	153.29	21.61	0.646	34
98357	69.8	40.5	70.9	70.8	1.1	117.3	114.1	105.6	153.50	0.26	21.73	0.17	153.43	21.65	0.814	71
98430	71.0	41.7	72.1	72.0	0.7	132.2	108.8	96.5	153.40	0.31	22.04	0.10	153.50	21.98	0.659	37
98359	70.7	41.4	71.7	71.6	1.3	127.3	110.8	97.2	153.90	0.22	21.86	0.22	153.83	21.79	0.817	63
98431	70.7	41.3	71.8	71.7	0.7	115.7	112.1	102.6	153.36	0.32	21.48	0.10	153.46	21.40	0.657	37
98360	71.2	41.9	72.2	72.1	0.7	138.5	111.9	93.5	153.69	0.23	21.71	0.21	153.62	21.64	0.817	58
98362	71.9	42.6	72.9	72.8	0.7	127.7	115.3	101.5	153.62	0.24	21.86	0.20	153.54	21.79	0.823	59
98369	70.9	41.5	71.9	71.8	1.1	103.6	101.2	98.0	154.14	0.37	21.27	0.32	154.05	21.19	0.835	77
98435	70.9	41.7	72.0	71.9	0.7	114.5	98.1	92.6	154.29	0.29	22.13	0.13	154.37	22.07	0.686	52
98436	67.6	38.3	68.8	68.6	2.2	131.3	99.6	91.6	153.43	0.30	21.81	0.10	153.50	21.74	0.695	46
98370	69.8	40.5	70.9	70.7	0.8	120.3	112.6	103.3	154.37	0.22	20.87	0.22	154.27	20.78	0.833	53
98437	71.8	42.5	72.9	72.8	1.9	129.8	106.8	98.8	153.73	0.30	21.95	0.11	153.79	21.89	0.699	41
98373	68.2	38.9	69.3	69.1	1.1	125.6	106.0	103.4	153.83	0.25	21.35	0.19	153.72	21.26	0.848	70
98374	72.2	43.0	73.3	73.1	0.7	126.0	113.0	97.7	153.81	0.25	22.04	0.19	153.70	21.97	0.852	67
98380	71.9	42.6	73.0	72.8	0.7	125.2	107.4	97.1	153.68	0.24	22.02	0.20	153.54	21.94	0.866	71
98442	70.2	40.9	71.3	71.2	1.3	126.6	110.3	105.7	153.37	0.35	21.49	0.16	153.37	21.43	0.758	43
98443	69.1	39.8	70.2	70.1	2.0	118.4	107.2	98.6	153.73	0.29	21.60	0.13	153.73	21.54	0.757	58
98444	69.3	40.1	70.5	70.3	2.7	123.2	109.7	104.5	153.68	0.30	21.99	0.13	153.69	21.93	0.757	41
98390	70.7	41.5	71.7	71.6	0.8	119.9	110.1	102.3	153.92	0.20	21.98	0.26	153.75	21.90	0.893	63
98445	69.7	40.4	70.8	70.7	0.8	129.9	106.5	93.4	153.65	0.24	21.70	0.20	153.62	21.64	0.783	86
98394	70.2	40.8	71.1	71.0	0.7	124.0	111.7	94.7	153.97	0.23	21.74	0.21	153.77	21.65	0.906	81
98395	71.3	42.0	72.2	72.1	0.7	119.2	104.2	91.9	153.85	0.20	21.85	0.23	153.65	21.77	0.908	75
98397	73.1	43.8	74.1	73.9	0.7	145.4	99.9	98.3	153.78	0.22	22.08	0.22	153.59	22.01	0.910	59
98398	71.5	42.1	72.4	72.3	0.7	133.2	109.1	95.9	153.69	0.25	21.81	0.20	153.47	21.72	0.925	69
98457	68.2	38.9	69.2	69.0	1.1	130.4	102.7	90.1	154.05	0.13	21.70	0.28	153.81	21.63	0.936	62
98458	68.2	38.9	69.2	69.0	1.1	125.3	110.0	101.4	154.35	0.18	21.27	0.25	154.12	21.20	0.933	54
98467	64.9	35.9	66.2	66.0	3.6	114.7	110.3	101.3	156.11	0.33	21.52	0.10	156.37	21.40	0.541	32
98472	72.4	43.2	73.5	73.4	1.5	107.7	96.6	94.4	153.56	0.34	22.53	0.11	153.60	22.48	0.717	52
average	71.0	41.7	72.0	71.9		132.7	109.2	97.7	153.09		22.44		153.05	22.37		
st. dev	1.0	1.0	1.0	1.0		10.5	6.9	6.1	0.21		0.41		0.24	0.41		
average	70.4	41.2	71.6	71.4		123.4	110.2	101.2	153.73		21.84		153.83	21.73		
st. dev	1.3	1.3	1.3	1.3		7.1	5.0	6.8	0.69		0.44		0.71	0.42		

Since the radiant coordinates can be determined with much greater precision than the entry velocity, these parameters are most suitable to look for structure in the distribution of meteoroid orbits. Figure 1 shows the observed radiant points, together with the photographically observed radiant points from Betlem *et al.* (1999). In this diagram the radiant points are corrected for the daily motion of the radiant due to the changing velocity vector of the Earth. This correction amounts to +0.99 and –0.36 degree per degree of solar longitude for the right ascension and declination, respectively. The correction was applied towards the arbitrary value of 235.0 degrees of solar longitude.

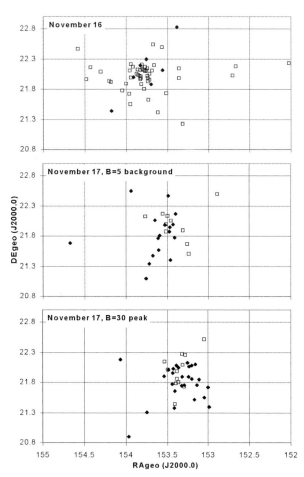

Figure 1. Radiant areas observed during the November 16 and 17 Leonid outbursts with all points moved to solar longitude 235.00 degrees. Closed diamonds refer to video observations, while open squares refer to photographic observations from Betlem *et al.* (1999).

In Figure 1, the radiant points of Nov. 16 are drawn separately, while in addition the sample of Nov. 17 is subdivided in a sample containing the radiant points observed between solar longitudes 235.27 and 235.36 degrees and a sample covering the remaining ranges of solar longitude. The subdivision of the Nov. 17 sample was chosen in line with the earlier discovery of an unusual asymmetric rate profile (Langbroek and De Lignie, 1999; Jenniskens, 1999). The rate profile was shown to be fitted well with a sum of two symmetric exponential distributions with maxima at solar longitudes 235.260 and 235.316 and steepness B = 5 and B = 30, respectively. We now find that the radiant coordinates observed during this proposed B = 30-peak have a systematically smaller right ascension than the radiant coordinates of the B = 5-peak. Note that this difference is probably even larger than visible from the diagram, because during the activity of the B = 30-peak, the B = 5-peak still contributed about 50% of activity.

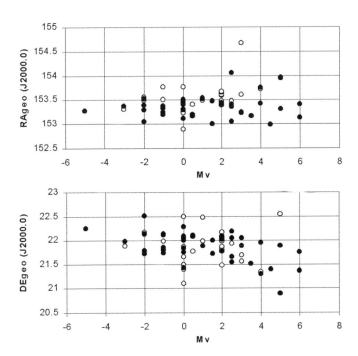

Figure 2. Radiant coordinates as a function of absolute visual magnitude of both the video and photographic observations of November 17 (the data sets only overlap for Mv=0). Open dots: B=5 structure. Filled dots: B=30 narrow peak.

In the B = 30 radiant diagram for November 17, the video and photographic observations do not perfectly coincide, as the video radiants

TABLE IV

Component	Source	N	RA_{geo}	DE_{geo}
Nov 16	Video	7	153.76 ± 0.10	22.11 ± 0.16
Nov 16	Photo	51	153.80 ± 0.06	22.05 ± 0.03
Nov 17, B=5	Video	17	153.57 ± 0.07	21.82 ± 0.09
Nov 17, B=5	Photo	10	153.46 ± 0.08	21.93 ± 0.09
Nov 17, B=30	Video	27	153.35 ± 0.05	21.80 ± 0.06
Nov 17, B=30	Photo	12	153.35 ± 0.04	21.99 ± 0.08

have a slightly smaller declination on average. As indicated in Table 4 the difference is about 0.19 degrees, which is just on the edge of statistical significance. Figure 2 gives a more detailed view of the mass dependence of the radiant coordinates during the nodal outburst activity of November 17. This way of presenting the data visualizes a small but significant mass dependent correlation between the visual magnitude and the declination of the radiant points of the B = 30 peak. The right ascension of the B=30 peak and the radiant coordinates of the B=5 peak might also have a small mass dependence, but if present it is hidden in the dispersion of the radiant distribution.

Finally, one can see that the average position of the radiants obtained from video observations during Nov. 16, coincides with the photographic one (Betlem et al., 1999). As pointed out before, the radiant is different from that of the Nov. 17/18 outburst. Table IV gives average radiants and standard errors, with all radiants moved to solar longitude 235.00 degrees. The outliers with video code numbers 98370, 98401, 98403, 98415, 98417 and 98426 were not included in these values.

4. Discussion

The newly obtained data allow for a review of possible associations between theoretically proposed structures in the Leonid stream and the three dust components observed in 1998. Existing models discriminate between young single ejecta from the parent comet, old single ejecta trapped in an orbital resonance, the so-called Leonid filament consisting of multiple ejecta (Jenniskens and Betlem, 2000) and the annual stream.

According to dynamical simulations of the Leonid meteoroid stream (Kondrat'eva et al., 1997; Asher et al., 1999; Asher, 1999; Arlt and Brown, 1999) the nearest single ejecta dust trail to the Earth's orbit in 1998 was that of dust ejected in 1899. However, it was thought that the Earth had created a gap in the dust distribution during its previous encounter in 1965 and no outburst was expected (McNaught and Asher, 1999). The younger trail ejected in 1932 was significantly further away from Earth.

It is shown in Asher *et al.* (1999) that parts of single ejecta can survive perturbations by the major planets during many revolutions, when trapped in orbital resonances with the major planets. In particular, it was shown by model calculations that a resonance part of the ejecta of 1333 would be visible in 1998.

In Jenniskens and Betlem (2000) the so-called Leonid Filament was proposed, which is visible in the years around perihelion passage of 55P/Tempel-Tutle. The filament is populated by the ejecta of multiple perihelion passages of the comet, but has nevertheless a finite extent due to small ejection velocities in combination with protection against planetary perturbations due to the comet's orbit close to the 5:14 resonance with Jupiter and the 8:9 resonance with Saturn.

The association of the three observed dust components in 1998 with these theoretical structures is not straightforward. In Jenniskens and Betlem (2000), the Nov. 16 fireball outburst was associated with the Leonid filament. In this interpretation the $B = 5$ and $B = 30$ outbursts of Nov. 17 are associated with the recent 1899 or 1932 ejecta. Arguments in favor of this association scheme are the corresponding durations and mass distributions of the 1994–1997 outbursts and the Nov. 16 outburst. However, the deviating node of the Nov. 16 outburst and the occurrence of a rather wide $B = 5$ outburst from a single ejecta are not easily accounted for. This would require explanations in terms of planetary perturbations such as the 1965 encounter of the 1899 ejecta with the Earth.

In an alternative scheme of associations, the Nov. 16 outburst is associated with the 1333 resonant structure (Asher *et al.*, 1999), the $B = 5$ Nov. 17 outburst is associated with the Leonid filament and the $B = 30$ outburst is associated with a recent ejecta. Arguments in favor of this scheme are the similar node and radiant positions of the 1995–1997 outbursts and the $B = 5$ component of Nov. 17. However, the 1994 outburst showed that significant deviations between the filament and the node of the comet are possible. Additionally, the outbursts of 1994–1997 were much wider ($B = 1$) and richer in large particles than the $B = 5$ outburst of 1998. On the other hand, width and mass distribution of the filament might depend on the position relative to the comet.

In either association scheme the observed mass dependence in the radiant distribution of the $B = 30$ peak could be the result of the original mass dependent ejection velocity distribution from the parent comet in combination with the differences in evolution due to radiation pressure and the required intersection with the Earth's orbit.

5. Summary

The double-station video observations of the 1998 Leonid outbursts extend the picture obtained from earlier reported photographic observations. Temporal variations in the radiant distribution were shown to correspond to features in the activity rate curves, which corresonds to the nodal distribution of orbits. Associations between observed dust components and theoretically modelled structures in the Leonid stream cannot be made unambiguously. It is made plausible that in the combined set of video and photographic observations, a small mass-dependence is present in the November 17 B = 30 radiant distribution. It is suggested that this mass dependence derives from the original ejection process at the parent comet.

Acknowledgements

We thank referee David Asher and Peter Jenniskens for their helpful comments in improving the paper. Klaas Jobse, Casper ter Kuile and Romke Schievink made an essential contribution to the achieved results by building and operating the video cameras. Carl Johannink and Koen Miskotte made a large effort in speeding up the data reduction. Dr. Guangyu Li, Dr. Jin Zhu, Dr. Chengming Lei and Dr. Haibin Zhao from Purple Mountain Observatory enabled the expedition by providing overall coordination of transport, lodging facilities and import. Financial funding was made available by the Royal Dutch Academy of Sciences (KNAW), the NASA Planetary Astronomy program, the Dutch Physics Foundation (Stichting Physica), the Leids Kerkhoven-Bosscha Fonds (KBF) and air cargo company Ufreight from Amsterdam Schiphol. Material support was provided by Kodak Netherlands and Canon Benelux. *Editorial handling*: Peter Jenniskens.

References

Asher, D.J.: 1999, *MRNAS* **307**, 919–924.
Asher, D.J., Bailey, M.E., and Emel'yanenko, V.V.: 1999, *MNRAS* **304**, L53–L56.
Arlt, R.: 1998, WGN, *Journal of the IMO*, **26**, 239–248.
Arlt, R. and Brown, P.: 1999, *WGN, Journal of the IMO*, **27**, 267–285.
Betlem, H., Jenniskens, P., Leven, J. van `t, Kuile, C. ter, Johannink, C., Haibin, Zhao, Chengming Lei, Guangyu Li, Jin Zhu, Evans S., and Spurny, P.: 1999, Meteorit. Planet. Sci. **34**, 979–986.
De Lignie, M. and Jobse, K.: 1996, *WGN, Journal of the IMO*, **24**, 20–26.
Jenniskens, P.: 1998, *Earth Planets Space* **50**, 555–567.
Jenniskens, P.: 1999, *Meteorit. Planet. Sci.* **34**, 959–968.
Jenniskens, P. and Butow S.J.: 1999, *Meteorit. Planet. Sci.* **34**, 933–943.
Jenniskens, P. and Betlem, H.: 2000, *Astrophys. J.* **531**, 1161–1167.
Kondrat'eva, E.D., Murav'eva, I.N., and Reznikov, E.A.: 1997, *Sol. Syst. Res.* **31**, 489–492.
Langbroek, M. and De Lignie, M.: 1999, *WGN, Journal of the IMO*, **27**, 30–32.
McNaught R. and Asher, D.J. : 1999, *WGN, Journal of the IMO*, **27**, 85–102.

1997 LEONID SHOWER FROM SPACE

PETER JENNISKENS AND DAVID NUGENT
SETI Institute, NASA ARC, Mail Stop 239-4, Moffett Field, CA 940351-000
E-mail: pjenniskens@mail.arc.nasa.gov

ED TEDESCO
*TerraSystems, Inc., Space Science Research Division, 59 Wednesday Hill Road, Lee,
NH 03824-6537*
E-mail: etedesco@TerraSys.com

and

JAYANT MURTHY
*Dept. of Physics and Astronomy, The Johns Hopkins University, Applied Physics
Laboratory, Baltimore, MD 21218-2686*
Now at: Indian Institute of Astrophyics, Koramangala, Bangalore - 34, India
E-mail: murthy@iiap.ernet.in

(Received 13 July 2000; Accepted 28 July 2000)

Abstract. In November 1997, the Midcourse Space Experiment satellite (MSX) was deployed to observe the Leonid shower from space. The shower lived up to expectations, with abundant bright fireballs. Twenty-nine meteors were detected by a wide-angle, visible wavelength, camera near the limb of the Earth in a 48-minute interval, and three meteors by the narrow field camera. This amounts to a meteoroid influx of $5.5 \pm 0.6 \; 10^{-5}$ km^{-2} hr^{-1} for masses > 0.3 gram. The limiting magnitude for limb observations of Leonid meteors was measured at $M_v = -1.5$ magn. The Leonid shower magnitude population index was 1.6 ± 0.2 down to $M_v = -7$ magn., with no sign of an upper mass cut-off.

Keywords: Flux, Leonids 1999, meteors, meteor shower, MSX, population index, space

1. Introduction

Space based observations of meteors are at a disadvantage in being further away from the meteors than ground-based observers and instrumentation being more expensive to operate, so less observing time is available to catch an elusive phenomenon. However, the larger effective surface area that is covered from such a distant vantage point does make space based observations potentially a suitable technique for measuring the influx and population index of the rare bright fireballs. Occasionally, bright < –17 magn. sporadic meteor fireballs are reported from the routine monitoring of rocket launches by the USA Department of Defense satellites (Reynolds, 1992; Tagliaferri et al., 1994).

Space based observations are also uniquely suited for UV spectroscopy of meteors, at wavelengths that are inaccessible from the ground. The Midcourse Space Experiment (MSX) has unique capability for both imaging of meteors from space and for UV spectroscopy over a wide spectral range (Mill et al., 1994). MSX is a Ballistic Missile Defense Organization project. The satellite was launched on April 24, 1996 by a delta rocket from Vandenberg Air Force Base, CA into a nominal circular orbit with an altitude of 908 km and an inclination of 99.6 degrees. The Johns Hopkins University Applied Physics Laboratory developed, integrated, and operated MSX. The Ultraviolet and Visible Imagers and Spectrographic Imagers (UVISI) instrument houses 11 optical sensors that are precisely aligned so target activity can be viewed simultaneously by multiple sensors covering a wide wavelength range.

The sensors were first deployed to study a meteor shower on November 17, 1997, when the Leonid shower showed a broad maximum in activity centered at 14 ± 2h UT (Arlt and Brown, 1998). The shower is thought to have been a recurrence of the "Leonid Filament", a broad structure of old debris causing abundant bright fireballs and also responsible for the fireball outburst of Nov. 1998 (Asher et al., 1999; Jenniskens and Betlem, 2000). Indeed, the UVISI imagers detected numerous Leonid meteors. These images are unique in being the first record of a meteor shower from space. The results of UV spectroscopy will be discussed elsewhere.

2. Methods.

Two of the UVISI imagers were used during these observations. The UVISI Wide Field Visible Imager is sensitive over the spectral range of 440–695 nm and has a field of view of 13.1 x 10.5 degrees. The Narrow Field UV and Visible Imager covers the spectral range from 300 to 723 nm and has a field of view of 1.6 x 1.3 degrees. The UVISI cameras were run at high gain and gate and provided a white-light (open filters) record of meteors near the slit of the spectrograph. Each image was integrated for 0.5 seconds, with alternating images for each camera every second.

The viewing geometry was chosen to have the spectrographs look to the nighttime limb of the Earth in fixed anti-Sun direction, with the slit parallel to the Earth's surface. Bright Leonid fireballs have the brightest point at about 95 km altitude, while fainter Leonids tend to peak near 100 km altitude. In order to increase our chances of detecting a persistent train and capture different parts of the meteor track, we covered the altitude range between 120 and 80 km in 10 mirror steps perpendicular to the Earth's surface, taking into account the curvature of the Earth. As a result, the cameras are oriented parallel to the Earth's limb and centered in a direction corresponding to 100-km altitude at the limb.

Figure 1 Leonid meteor ablating above the airglow layer in a UVISI Narrow Field UV and Visible Image.

The optimum observing period was chosen to be the 1–2 hour interval centered on 13:34 UT on November 17, when the Earth passed the orbital plane of comet 55P/Tempel-Tuttle (Yeomans et al., 1996). This time was close to the actual peak of the shower (Arlt and Brown, 1998). Specifically, the observations were timed to coincide with the satellite being between +0 and +40 degree northern latitude, while moving in a south to north orbit over the apex (morning) side of the Earth, the part which was exposed to the Leonid meteors. To avoid twilight at the 120 km altitude layer and resulting increase in the airglow emission, the solar zenith angle at the tangent point had to be greater than about 102 degrees. In fact, the higher the better in order to get away from any off axis internal scattering due to the bright limb or sunlit Earth. The full Moon was not expected to interfere with the observations other than causing a bright cloud deck at low viewing elevations.

3. Results

Two target of opportunity events were allotted to this program on 17 November. The first run started with sampling the evening sky at 14:28:16 UT, gradually turning towards the morning sky until about 15:12 UT. Problems with data transmission caused a loss of many video frames and the cameras did not detect meteors during this period. During the second run as many as 29 meteors were observed in the wide field imager and 3 meteors in the narrow field imager. These were all Leonids that appeared between 15:12:16 and 15:58:00 UT.

Figure 1, for example, shows a Leonid meteor detected at 15:31:12 UT in the narrow field imager. The meteor is seen in a direction $\alpha = 36°$, $\delta = 17°$, and is positioned mostly above the airglow layer. The brightness distribution in the airglow layer peaks at about 89 km altitude and much reflects the distribution of ablated meteoric metals from mainly smaller and slower meteoroids that make up the sporadic meteor background. This illustrates that the fast Leonid meteors tend to ablate at higher altitudes than the meteoroids that dominate the mass influx. The meteor light curve is characteristic for many Leonids: an exponential increase, followed by a broad maximum, a rapid decline and a brief end flare.

The meteors move in parallel paths from a direction $\alpha = 153.6°$, $\delta = +22°$ which is the radiant of the shower (Jenniskens and Betlem, 2000).

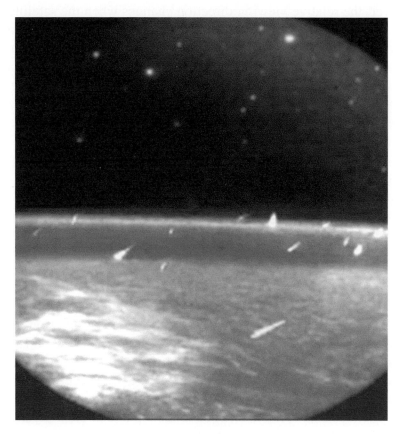

Figure 2a Composite of Leonid meteors in wide-angle camera. We chose the most striking star background and cloud pattern observed between 15:20 and 15:59 UT. Full Moon glare is visible on the clouds in the lower left of the image.

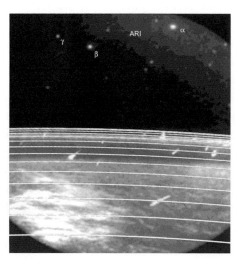

Figure 2b Equidistant lines from the satellite to a layer at altitude 100 km. Stars serve as magnitude calibration. Stars in the constellation of Aries are marked.

However, the relative viewing direction changes when the satellite scans along the limb of the Earth in each revolution. At the time of first detection, the meteors appear to come from the left in projection to the horizon, while later they appear to come from the right. To give a sense of how the Leonid shower is seen from space, a composite image of meteors detected in the wide field imager between 15:20 and 15:59 UT is shown in Figure 2a.

To uniquely determine the meteor flux and size distribution in the Leonid shower, the distance to each meteor is determined. Figure 2b shows equidistant lines from the satellite to a layer at 100 km altitude. Distances range from 1,900 km at the bottom in Figure 2b to 3,300 km at the limb. Beyond the limb, meteors are detected as far away as 4,700 km between airglow layers and the cloud deck. As a result of these large distances, relatively bright meteors appear faint enough to not cause problems with non-linear effects of blooming, and their intensity can be directly compared to that of background stars. Individual magnitudes were determined by comparing the integrated intensity of the meteors with the integrated intensity of the stars. A good correlation between visual magnitude and the log of integrated intensity of comparison stars was found in the range +2 to +7 magnitude, with the expected slope, which implies that the system is linear over this regime. Apparent magnitudes of meteors covered about the same range: +1 to +6 magnitude. Meteors located between cloud top and airglow layers can be at one of two distances, depending on whether they are on the near side or the far side. We find that five out of ten meteors in between the airglow layers and the horizon lack a recognizable exponential increase in brightness and show the bright central part of the light curve more compressed. They are likely on the far side. Thus, the light curve of each meteor can discriminate between near and far meteors, especially in cases close to the Earth's cloud deck where the difference in distance is largest, and offers a unique measure of distance for all meteors.

The resulting magnitude distribution is shown in Figure 3. The wide field visible imager detects nearly all Leonid meteors of magnitude -2 and brighter. The exponential slope implies a magnitude distribution index of $r = 1.6 \pm 0.2$, with a most likely value of $r = 1.7$. The light of a full Moon hampered ground-based observations in 1997. Brown and Arlt (1998) stated "It is almost certain that any attempt to use magnitude data from the peak night will be heavily biased and produce artificially low values for the population index r". Nevertheless, Arlt and Brown (1998) found a low value of $r = 2.0 \pm 0.2$ from apparent visual meteor magnitudes in the range +0 to +5 magn. A general dominance of bright meteors was observed by forward meteor scatter radar (Foschini et al., 1998). From ground-based video observations, Hawkes et al. (1998) found, s = 1.71 ± 0.07, which corresponds to r = 1.92 ± 0.13. Hence, we can confirm that the trend for meteor magnitudes between +0 and +5 continued until at least -7 magnitude, without any sign of an upper mass cut-off.

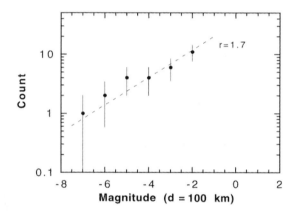

Figure 3 Meteor count in intervals of absolute magnitude distribution.

With this information, it is possible to calculate the influx of meteoroids of magnitudes less than −1.5, or masses larger than about 0.3 gram (Jacchia et al., 1967). For, r = 1.7, we count 40 meteors below the integral dashed line in Figure 1. The observing interval covers a period of 48 minutes. The wide field imagers were recording data during about 40% of that time. The effective surface area perpendicular to the shower is about 1.1×10^6 km^2. The radiant of the shower is about 36 degrees out of the zenith on average over the spatial and temporal interval. Hence, the influx of meteoroids > 0.2 gram (< −1.5 magn.) was $5.5 \pm 0.9 \times 10^{-5}$ km^{-2} hr^{-1} at the peak of the 1997 Leonid shower between 15:12 and 15:59 UT. Arlt and Brown (1998) reported ZHR = 96 ± 13 at the peak at 12:15 UT, which translates to about 1–5 $\times 10^{-2}$ Leonids km^{-2} hr^{-1} of limiting absolute magnitude +6.5 and brighter. Extrapolation of our magnitude distribution, r = 1.7, gives $1.6 \pm 0.3 \times 10^{-2}$ km^{-2} hr^{-1} at 15.6 UT, in general agreement with the results reported by Arlt and Brown (1998).

The good agreement is encouraging for future applications of space based observations in meteor shower research.

Acknowledgements

We are grateful for the constructive reviews by Bob Hawkes and Frans Rietmeijer to improve the presentation of this paper. We thank Steve Price of Hanscomb AFB for support of this project. Rich Dragonette of John Hopkins University handled the UVISI operations (CB11_34 and CB11_35). PJ is supported by NASA's Planetary Atmospheres program. *Editorial handling*: Frans Rietmeijer.

References

Arlt, R. and Brown, P.: 1998, *WGN, Journal of the IMO* **26**, 161–165.
Asher, D.J., Bailey, M.E., and Emelyanenko, V.V., 1999, *MNRAS* **304**, L53–L56.
Brown, P. and Arlt, R.: 1998, *WGN, Journal of the IMO* **26**, 11–12.
Foschini, L., Cevolani, G., and Sbenaglia, E.: 1998, *WGN, Journal of the IMO* **26**, 13–19.
Hawkes, R.L., Babcock, D., and Campbell, M.: 1998, *Analysis Procedures and Final Electro-Optical Results for 1997 Leonids*, Contractor Report under CRESTech Contract #5FUSA-7-J151/001/SV. (57 pages) (Toronto: CRESTech).
Jacchia, L.G., Verniani F., Briggs, R.E.: 1967, *Smithsonian Contr. Astrophys.* **10**, 1–45.
Jenniskens, P. and Betlem, H.: 2000, *Astrophys. J.* **531**, 1161–1167.
Mill, J.D., ONeil, R.R., Price, S., Romick, G.J., Uy, O.M., and Gaposchkin, E.M.: 1994, *Journal of Spacecraft and Rockets* **31**, 900–907.
Reynolds, D.A.: 1992, in G.H. Canavan, J.C. Solem, and J.D.G. Rather (eds.), *Proc. Near-Earth-Object Interception Workshop*, Los Alamos Nat. Lab., Los Alamos, pp. 221–226.
Tagliaferri, E., Spalding, R., Jacobs, C., Worden, S.P., and Erlich, A.: 1994, in T. Gehrels (ed.) *Hazards Due to Comets and Asteroids*, University Arizona Press, Tucson, Arizona, p. 199–220.
Yeomans, D.K., Yau, K.K., and Weissman, P.R.: 1996, *Icarus* **124**, 407–413.

FROM COMETS TO METEORS

J. MAYO GREENBERG

Raymond and Beverly Sackler Laboratory for Astrophysics,
Leiden Observatory, Leiden University, Niels Bohrweg 2,
Leiden 2300 RA, The Netherlands
E-mail: greenber@strw.leidenuniv.nl

(Received 8 May 2000; Accepted 15 July 2000)

Abstract. A summary of comet nucleus and dust properties is used to suggest a basis for predicting the properties of meteor shower particles originating as comet debris.

Keywords: comets, dust, meteors, meteoroids

1. Introduction

The Leonid showers are a classic case of periodic meteors associated with the passage of a comet - Tempel/Tuttle - through the orbit of the earth. The largest particles in the dust distribution - those that would be seen as the antitail - remain in an orbit approximately that of the comet itself and as they return to the region of the Earth are seen as meteor showers. Each shower is associated with a particular Earth orbit crossing of 55P/Tempel-Tuttle. It is the purpose of this paper to provide a general background on the chemical and physical properties of comets and comet dust which may provide a basis for understanding the observed properties of cometary debris and, in particular, of the Leonid meteors.

I find it interesting to quote here from the conclusion of a paper written earlier in which the *comet properties were derived from the character of meteors* "The aggregated dust model makes it possible to derive comet nuclear densities from a comparison of evaporated comet debris with meteor densities. It is shown that a high degree of porosity is to be expected with at least 60% of a comet being vacuum." (Greenberg, 1986a,-b). The large uncertainty in extracting particle density from meteor data could have rendered that result suspect but it turns out to have been confirmed. The situation now is reversed. The post Halley era has led to more direct evidence of the low density nature of comets and we can more usefully invert the meteor–comet connection to make predictions of meteor properties based on comets rather than the other way around. The additional results now available on interplanetary dust particles (IDPs) should help to tie down the connections.

2. Chemical composition of comets

The chemical composition of a comet nucleus can be very strictly constrained by combining the latest results on: the core-mantle interstellar dust model, the solar system abundances of the elements, the space-observed composition of the dust of comet Halley, and the latest data on the volatile molecules of comet comae. A detailed discussion of how interstellar dust comes to be incorporated into comets during the formation of the solar system is beyond the scope of this paper. However, it is certainly recognized that both the comet coma volatiles and the comet dust composition are very closely related to what we infer to be the composition of the primitive solar nebula dust as it existed 4.6 billion years ago. There are striking similarities between the volatile composition of comets and hot cores of regions of star formation. In the very beginning there are the silicate particles blown out of cool evolved stars. These accrete mantles in the denser clouds, which are photoprocessed. The mantles contain molecules created by surface reactions, by gas phase reactions and by photoprocessing. What is followed here are the "large" tenth micron (mean radius) grains which contain much of the mass of the dust - all the volatiles and all the silicates. Other populations of interstellar dust consist of very small carbonaceous particles and even smaller particles, which are presumed to resemble large polycyclic aromatic hydrocarbons. The large grains cycle between low-density (diffuse) clouds and high-density molecular clouds and star forming clouds.

Those that are left over from star formation are shown at the top left of Figure 1. They consist of silicate cores with highly photoprocessed organic mantles. The very small particles/large molecules are also present in the diffuse cloud phase as separate particles. Going back to the molecular cloud phase the process is repeated. Those particles that are confined to the region of star formation in the final collapse phase are presumed to have accreted all the remaining molecules and the small particles as part of the outer mantle. Some of the evidence for these final accreted phases is provided by the observation of gas phase species found in molecular hot cores. They are dense warm clumps located close to the massive young stars where the molecules are presumed to have been evaporated from dust grains that did not aggregate in the stellar nebula. Actually a closer comparison must be found by probing the envelopes of low mass stars more characteristic of our Sun. As stated in Bockelée-Morvan *et al*. (2000) "A quantitative comparison shows that chemical abundances in Hale-Bopp parallel those *inferred* (my italics) in interstellar ices, hot molecular cores and bipolar flows around protostars". The organic mantle beneath the ice had its confirmation with the mass spectra obtained in situ for comet Halley dust which led Kissel and Krueger (1987) to infer a core-mantle structure of the dust particles. Thus according to Jessberger and Kissel (1991) "The existence of the previously postulated (Greenberg, 1982) core-mantle grains seems to be substantiated by data".

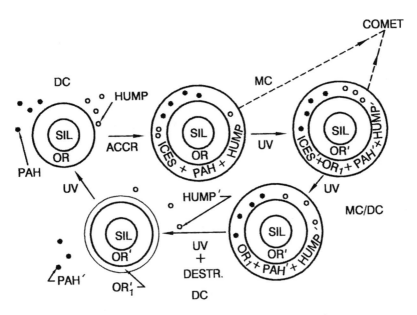

Figure 1. Cyclic evolution of interstellar grains. Upper left is an average tenth micron radius silicate core-organic refractory mantle grain in the diffuse interstellar medium (DC). The mantle is the heavily processed organic material as in Greenberg et al (2000). Schematically illustrated are the hundreds of thousands to millions of very small carbonaceous (hump, denoting their effect on the interstellar extinction) particles and even smaller PAH particles as in Li and Greenberg (1997). Following clockwise, the next phase depicts what happens after entering a molecular cloud (MC) showing the accretion of a complex ice mantle along with the very small particles and, simultaneously with accretion, the ultraviolet photoprocessing of the outer mantle and the organic inner mantle as well as the PAH and hump particles (primes denote modified material). The next phase corresponds to the emergence out of the molecular cloud (MC/DC) after star formation when the ices are evaporated/destroyed leaving first generation organics (OR_1). And finally, the ultraviolet processing and partial destruction of the newly added first generation organic material ($OR_1 \rightarrow OR_1'$) as well as reemergence and reforming of PAH and hump particles leading back to the "original" diffuse cloud (DC) dust. The arrows leading upward depict the kinds of dust, which would make up the protostellar material aggregating to form comets. In this representation it is assumed that this occurs with little or no evaporation and reforming of ices. A single cycle lasts about 10^8 years and as many as 50 may occur before the dust is consumed in star formation.

The basic model of interstellar dust consists of three populations of particles (Li and Greenberg, 1997). The major mass is in tenth micron particles consisting of silicate cores with organic refractory (complex organic molecules) mantles. Additionally there are very small carbonaceous particles/large molecules. In molecular clouds the large particles accrete additional mantles of frozen molecules and in the dense clouds there is also accretion of the very small particles which are imbedded in the "ices". This is schematically shown in Figure 2.

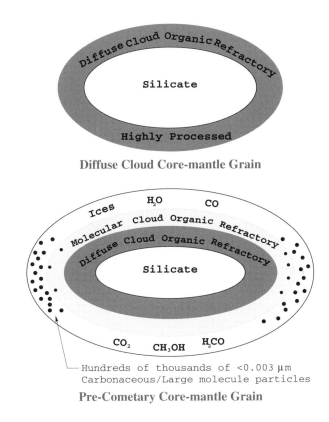

Figure 2: Upper figure depicts a diffuse cloud silicate core-organic refractory mantle particle. It has to be nonspherical (here elongated) in order to provide for interstellar optical polarization. Lower figure depicts a fully accreted grain in the protosolar nebula.

The nature of the organic mantle material varies depending on whether the dust is in a low density diffuse cloud or a molecular cloud (Tielens *et al.*, 1996; Greenberg and Li, 1997). There are significant variations in the relative proportions of C, N, O and H in the complex organics in different regions. In diffuse clouds the organic mantle is strongly depleted in oxygen and hydrogen, whereas in molecular clouds complex organic molecules are present with more abundant fractions of oxygen and hydrogen. Furthermore, the ratio of the mass of organic mantles to the silicate core is highly variable. In the unified model for diffuse cloud dust of Li and Greenberg (1997) this ratio is $V_{OR}/V_{sil} = 0.95$, whereas matching the silicate polarization in the Orion B-N object requires V_{OR}/V_{sil} is about 2 (Greenberg and Li, 1996a). It is of interest to note that the mass spectra of comet Halley dust - as obtained by Kissel and Krueger (1987) and presumably representing the ultimate molecular cloud collapse phase - gave about equal masses of organics and silicates in the dust which implies a volume ratio of about 2. While the presence of organics is observed via a 3.4 μm absorption feature this represents only the aliphatic molecules and much if not most of the diffuse cloud organic mantles consist of aromatics (Greenberg et al., 2000). We shall assume that the organic refractory mantles in the final stages of cloud contraction are most closely represented by the properties obtained for Halley dust; i.e. $M_{OR}/M_{sil} = 1$ and with an atomic distribution as given in Table I for comet dust organics. Table I gives the stoichometric distribution of the elements in laboratory organics (residues of ultraviolet photoprocessed ices) (Greenberg and Li, 1997) compared with the comet Halley mass spectra (Krueger and Kissel, 1987) normalized to carbon.

TABLE I

	Laboratory Organics			Comet Halley		
	Volatile*	Refractory*	Total*	PICCA(gas)[@]	dust[@]	total[@]
C	1.0	1.0	1.0	1.0	1.0	1.0
O	1.2	0.6	0.9	0.8	0.5	0.6
N	0.05	> 0.01	> 0.03	0.04	0.04	0.04
H	1.70	1.3	1.5	1.5	1.0	1.2

* Division between volatile and refractory is here taken at a sublimation temperature less than or greater than ~ 350K respectively.
[@] Assuming equal amounts of dust and gas.

Combining a representative distribution of volatile components from comet comae with the inferred chemical composition of the silicate core-organic refractory component one can arrive at a "canonical" distribution of comet nucleus chemical components as given in Table II. Table II gives the distribution by mass fraction of the major chemical constituents of a comet nucleus: as derived from comet volatiles and dust refractories (Greenberg, 1998). Included in "others" are SO, SO_2, HC_3N, NH_2CHO, HCOOH, $HCOOH_3$, etc. (Bockelée-Morvan et al., 2000). Note that water, while abundant, is certainly not dominant.

TABLE II

Materials	Mass fraction
Silicates	0.26
Carbonaceous (very small)	0.086
Organic Refractory	0.23
H_2O	0.31
CO	0.024
CO_2	0.030
CH_3OH	0.017
H_2CO	0.005
(Others)	0.04

3. Aggregate properties of comet dust

The thermal emission from comet dust, and particularly the presence of a 10 μm excess characteristic of the silicates, have been used to demonstrate the very fluffy character of the particles. I will discuss a limited number of comets for which a fluffy dust model has been applied.

3.1. COMET HALLEY

The uniqueness of comet Halley with regard to the dust was that for the first time three major properties were simultaneously observed: chemical composition, size (mass) distribution, and infrared emission. Greenberg and Hage (1990) showed that, in order to satisfy simultaneously such independent properties of Halley coma dust as (1) 9.7 μm emission (amount and shape), (2) dust mass distribution, and (3) mass spectroscopic composition of both rock and organic elements, one demanded – as the most consistent configuration - that the dust be very fluffy aggregates of sub-micron interstellar dust silicate core-organic refractory mantle particles. The major thrust of this is that comet dust consists of intimately related silicate and carbonaceous materials (core-mantle structure) rather than separate silicate and carbon components. In light of the current understanding that interstellar volatile species are well preserved as seen in comet comae (see-Bockelee-Morvan et al., 2000 and references therein) it seems even more reasonable to expect the refractory silicate core-organic mantle underlying the molecular cloud ices to be well preserved as well. One of the observational supports of the model is that the in situ mass-spectra of Halley dust with high dynamic range show that, except for the very small (attogram) grains (Utterback and Kissel, 1990), neither pure organic (so-called CHON) nor pure silicate particles exist. Instead, they are intimately mixed on a very fine scale in such a manner that they form the sub-units with a core-mantle structure in the aggregates (Lawler and Brownlee, 1992). This is additionally reflected by the fact that the CHON ions have on the average a higher initial energy than the silicate ions in measuring the mass spectra (Krueger and Kissel, 1987).

In summary, the result of the intertwining of the three basic Halley dust observations is: (1) comet dust consists of aggregates of ~0.1 μm silicate core-organic refractory mantle particles; (2) the average porosity of the comet dust is $0.93 < P < 0.975$. The inferred Halley comet dust density is $0.07 < \rho_{CD} < 0.19$ g cm^{-3}; i.e. $\rho_{CD} \sim 0.1$ g cm^{-3} is a suggested canonical value. Note that we have used $\rho_{CD} = \rho_{solid} \times (1-P)$, where ρ_{solid} is the mass density of the compact core mantle particles assumed to be about 2.7 g cm^{-3}. If one reconstitutes the original comet material by adding back the volatiles on the comet dust skeleton, as well as the *very* small interstellar dust particles, and about 1/2 of the original (relatively volatile) organic refractories, which were removed by the solar heating, the reconstituted comet nucleus density may be inferred to be $0.26 < \rho_C < 0.51$ g cm^{-3}. Later works (Greenberg and Li, 1998a; Greenberg, 1998) have not modified these results significantly and representative values are suggested for comet Halley dust density as $\rho_{CD} \sim 0.1$ g cm^{-3} and for its nucleus density as $\rho_C \sim 0.33$ g cm^{-3}. The latter is consistent with the low density suggestion proposed by Rickman (1986) based on the analysis of non-gravitational forces although, using the same kind of data, Sagdeev et al. (1988) arrived at a value 0.61 g cm^{-3}.

3.2. COMET P/BORRELLY (1994*l*): A JUPITER-FAMILY SHORT-PERIOD COMET

The fluffy aggregate comet dust model has also been applied to short-period (SP) comets (see Li and Greenberg, 1998a). As an example, we have calculated the dust thermal emission spectrum of comet P/Borrelly (1994l), with an orbital period P ~ 7 years, from 3 - 14 μm as well as the *weak* 10 μm silicate feature in terms of the comet modeled as a porous aggregate of interstellar dust (Li and Greenberg, 1998a). It seems that, compared to the Halley dust, the dust grains of P/Borrelly appear to be relatively more processed (more carbonized), and less fluffy. A *not so fluffy* aggregate model of silicate core- *amorphous carbon* mantle grains with a porosity P = 0.85 appears to match the observational data obtained by Hanner *et al.* (1996) quite well. This would imply that comet P/Borrelly is substantially denser than Halley. Since P/Borrelly has passed through the inner solar system many more times than P/Halley and therefore been subjected much more to the solar irradiation, could it be that because of thermal processing the outer layers of the nucleus could have been significantly modified? Could a layer of more compacted material have been produced? In particular, could the organic refractory materials have undergone further processing and annealing (Jenniskens *et al.*, 1993)? Observations do show that some Jupiter-family short-period comets are depleted in C_2 and C_3 but are approximately constant in CN (A'Hearn et al., 1995). This is consistent with the idea of carbonization since CN is mostly produced from grains while some C_2 and C_3 come directly from the volatile nuclear ices which are relatively depleted in SP comets (A'Hearn et al., 1995). Perhaps the dust of short-period comets lacks the small particles seen in the Halley size spectra. Are they more strongly bound and less susceptible to fragmentation? These questions require substantial further discussion but clearly while it appears that comets may start out consisting of the same material they can evolve to look rather different (see Greenberg and Li, 1999).

Since up to now only two SP comets were known to have silicate emission and *weak at that* (P/Borrelly and P/Fay; see Hanner et al., 1996), we are not able to generalize the dust properties of short-period comets. Systematic observations of the thermal emission spectra and the silicate features for a large set of samples are needed.

3.3. COMET HALE-BOPP (C/1995 O1): A VERY LARGE LONG-PERIOD COMET

Comet Hale-Bopp (C/1995 O1) is an exceptionally bright long-period comet (P ~2000 years). It was so active and so bright that it became visible even at a heliocentric distance of ~7 AU. Its strong activity and strong thermal emission features provide a rare opportunity to study the origin of comets and to constrain the comet dust morphology, composition and size. Li and Greenberg (1998b) have calculated the dust thermal emission spectrum based on the model of comet dust consisting of very porous aggregates of interstellar dust. Both the continuum emission and the 10 μm silicate feature are well matched (see Li and Greenberg, 1998b for details). The presence of large numbers of very large particles in Hale-Bopp was confirmed by the submillimeter continuum emission observation (Jewitt and Matthews, 1999). It has been argued that these large particles may dominate the total dust mass of the coma (Fulle, 1999; Jewitt and Matthews, 1999). Assuming a spherically symmetric dust coma with uniform radial outflow, adopting the water production rate on Feb.23.9, 1997 (Dello Russo et al., 1997) of 4.3×10^{30} mols/s, and an average dust outflow velocity of 0.12 km s^{-1} (calculated from $v_d \approx 05 \ (r_h/6.82)^{-0.5}$ (where r_h is the heliocentric distance in AU; Sekanina, 1996), the dust-to-water production rate ratio was estimated to be as high as 41 or even higher (see Li and Greenberg, 1998b). If a higher dust outflow velocity of 0.60 km s^{-1}, which may be more realistic, is adopted, the dust-to-water ratio would be about 200! However, one should keep in mind that the IR emission alone can not give a reliable dust production rate since very large particles are too cold to contribute to the limited wavelength range of the infrared radiation considered here (as long as the size distribution for those cold particles is not too flat). Therefore the total mass of the large particles is not well constrained, as was already noted by Crifo (1987). Of equal importance is the fact that the large particles actually can act like mini-comets and preserve their ices until far from the nucleus so that *their* volatiles are not observed and therefore do not contribute to the gas part in the dust to gas ratio.

3.4. COMET 55P/TEMPEL-TUTTLE

For comet 55P/Tempel-Tuttle no silicate emission feature was seen (Lynch *et al.*, 2000). This is along the lines indicated by comet Borrelly, that Jupiter-family short-period comets have weak or no silicate features. A possibility is that this difference from long-period comets can be attributed to the different degree of evolution of the comet nucleus. This could lead to the fact that even though the basic composition of the comet is similar the dust appears different either because the porosity is lower, or that the dust is less fragmented; i.e. contains less small particles in its size distribution. In any case we suggest that the basic composition of the initial *large* cometary dust fragments should be derivable from aggregates of core-mantle interstellar dust particles. However, after lift off perhaps there survives more processed material on the grains so that the volatility of the mantle is less and the particles have a greater strength against fragmentation. Thus even the relatively small ones are initially poor silicate emitters, being filled with more non-absorbing material (ices) and having a lower porosity, both of which lead to lower temperatures. However, ultimately even the very large ones must lose their volatiles so that, as meteors, we shall assume them capable of achieving the ultimate porosity of the skeleton core-organic refractory particles as a reasonable possibility. By ultimate I mean during the time spent near the first perihelion passage. However, the physical evolution of the remaining fluffy structure may subsequently undergo some compaction as could result from the process of differential evaporation considered by Mukai and Fechtig (1983). Solar wind effects are not expected to modify more than the outermost (tenth micron or less) layer of the large comet dust particles.

4. Large comet dust particles

4.1. METEOIDS

Meteors are generally much larger than the comet dust particles observed in the visual or infrared. While these observations may be used to derive the morphological structure they tell us little about the population which shows up as meteors. Submillimeter observations (Jewitt and Matthews, 1999) extend the evidence for larger particles. The existence of even larger particles has been obtained from dust impacts on spacecraft detectors by comets Halley (McDonnell *et al.*, 1991) and Gicobinni-Zinner (McDonnell *et al.*, 1993). Other evidence has been provided by radar echoes from near-Earth comets (Goldstein *et al.*, 1984; Harmon *et al.*, 1989, 1997).

4.2. INTERPLANETARY DUST PARTICLES

Interplanetary dust particles (IDPs) collected in the earth's atmosphere are presumed to be debris from comets and asteroids. Since they were first discovered by Brownlee (1978, and references therein) it was hoped that they would provide the closest available link to the material of the protosolar nebula. The most important point about these particles is that they can be studied in the laboratory (for a recent review see Rietmeijer, 1998). Since some of the IDPs are likely to be of comet origin it would be handy to be able to use them as representative of comet dust. The fact that this is often done leaves open a number of questions. If comets are aggregated interstellar dust, why do we not see the tenth micron core mantle structure? If comet dust is extremely porous (P = 0.95) how have the particles been compacted to, say 0.7–0.8, as in one of the groupings of densities by Rietmeijer (1998)? The mean density of IDP's in the 10 μm to 50 μm size range is about $\rho_{IDP} \approx 0.5$ g cm^{-3} according to Rietmeijer (private communication) although densities on average much higher have been obtained by Love *et al.* (1994). Even the low density implies a porosity of P_{IDP} less than P = 0.8, if the material density of the solid components is = 2.5 g cm^{-3} (a mixture of silicate with a small fraction of organics). The dust of Comet Borrelly comes close to this value. One thing to be considered is that the IDPs have probably been around in the solar system for 100,000 years or more while comet dust, in the form of periodic meteor showers, is only hundreds of years old. Perhaps more fully evolved cometary particles such as meteors resemble the IDPs although the latter correspond to the smaller end of the pre-meteor size spectra. Chemically, it has been shown that H and N isotopic anomalies in the more fragile (porous) cluster IDPs which may be attributed to surviving (but likely altered) organic molecules are closer to those for interstellar molecules than in other IDPs or meteorites (Messenger, 2000). Furthermore cluster IDPs have fine grained structure comparable to the tenth micron characteristic of interstellar dust and are rich in volatile elements and carbon although the core mantle structure is not seen.

5. Concluding remarks

It appears that meteors may be thought of as intermediate between comet dust and IDPs with the further proviso that the comet dust we are thinking of is in the millimeter to meter size range rather than the 10 –100 micron size range of IDPs. What is needed to simulate expected properties of the Leonid meteors is a calculation (or simulation) of the evolution of mg to kg-size cometary dust "grains". These "grains" *initially* consist of aggregates of protosolar dust with mean comet porosity which, including the ices as well as the organics would be 0.5 < P < 0.8. The evolution of mg to kg mass pieces of such mini-comet nucleus material should be further studied particularly with regard to losses in organics and to possibility of compaction. As of now it appears that a working model approximation to meteor properties is to assume something *between* aggregates of silicate core-organic refractory tenth micron particles ($m_{sil}/m_{OR} \geq 2$, mean aggregate density of < 0.5 g cm^{-3}, porosity of at least P = 0.7) and large cluster IDPs with P = 0.7.

Acknowledgments

I like to thank Peter Jenniskens and Noah Brosch for inviting me to present my ideas on comets as related to meteors. I would also like to thank the editor Frans Rietmeijer for his very helpful suggestions and ideas and Don Brownlee for his useful criticisms related to IDPs. An anonymous reviewer probably led to my clarifying some obscure points. *Editorial handling:* Frans Rietmeijer.

References

A'Hearn, M.F., Millis, R.L. Schleicher, D.G., Osip, D.J., and Birch, P.V.: 1995, *Icarus* **118**, 223–270.
Bockelée-Morvan, D., Lis, D.C., Wink, J.E, Despois, D., Crovisier, J., *et al.*: 2000, *Astron. Astrophys.* **353**, 1101–1114.
Brownlee, D.E.: 1978, in J.A.M. McDonnell (ed.) *Cosmic Dust*, J. Wiley, N.Y., pp.295–336.
Crifo, J. F.: 1987, in Z. Ceplecha and P. Pecina (eds.), *Interplanetary Matter, Publ. Astron. Inst. Czech. Acad. Sci.* **67**, 59–66.
Dello Russo, N., DiSanti, M. A., and Mumma, M.J.: 1997, *IAU Circ.* **6604**.
Fulle, M. 1999, *Adv. Sp. Res.* **24**, 1087–1093.
Goldstein, R., Jurgens, R., and Sekanina, Z.: 1984, *Astron. J.* **89**, 1745–1754.
Greenberg, J.M.: 1982, in L.L. Wilkening (ed.), *Comets*, Univ. of Arizona Press, pp. 131–163.
Greenberg, J.M.: 1986a, in C.-I. Lagerkvist and B.A. Lindblad, et al (eds), *Asteroids, comets, meteors II*, Reprocentrum HSC, Uppsala, 1986, pp. 221–223.
Greenberg, J.M.: 1986b, *Nature* **321**, 385.
Greenberg, J.M.: 1998, *Astron. Astrophys.* **330**, 375–380.
Greenberg, J.M. and Hage, J.I.: 1990, *Astrophys. J.* **361**, 260–274.
Greenberg, J.M. and Li, A.: 1997, *Adv. Space Res.* **19**, 981–990.
Greenberg, J.M. and Li, A.: 1998, *Astron. Astrophys.* **332**, 374–384.
Greenberg, J.M. and Li, A.: 1999, *Planet. Space. Sci.* **47**, 787–795.
Greenberg, J.M., Gillette, J.S., Munoz-Caro, G.M., Mahajan, T.B., and Zare, R.N, *et al.*: 2000, *Ap.J. Lett.* **531**, L71–L73.
Hanner, M.S., Lynch, D.K., Russell, R.W., Hackwell, J.A., and Kellogg, R.: 1996, *Icarus* **124**, 344–351.

Harmon, J.K., Campbell, D.B., Hine, A.A., Shapiro, I.I., and Marsden, B.G.: 1989, *Ap.J.* **338**, 1071-1093.

Harmon, J. K. Ostro, S.J., Benner, L.A.M., Rosima, K.D., and Jurgens, R.F.: 1997, *Science* **278**, 1921-1924.

Jenniskens, P., Baratta, G.A., Kouchi, A., de Groot, M.S., Greenberg, J.M., and Strazzulla, G: 1993, *Astron. Astrophys.* **273**, 583-600.

Jessberger, E.K., and Kissel, J.: 1991, in R. Newburn, M. Neugebauer, and J. Rahe (eds.), *Comets in the Post-Halley Era*, Kluwer, Dordrecht, pp. 1075-1092.

Jewitt, D., and Matthews, H.: 1999, *Astron. J.* **117**, 1056-1062.

Kissel, J., and Krueger, F.R.: 1987, *Nature* **326**, 755-760.

Krueger, F.R. and Kissel, J.: 1987, *Naturwissenschaften* **44**, 312-316.

Lawler, M.E., and Brownlee, D.E.: 1992, *Nature* **359**, 810-812.

Li, A., and Greenberg, J.M.: 1997, *Astron. Astrophys.* **323**, 566-584.

Li, A., and Greenberg, J.M.: 1998a, *Astron. Astrophys.* **338**, 364-370.

Li, A., and Greenberg, J.M.: 1998b, *Astrophys. J.* **498**, L83-L87.

Love, S.G., Joswiak, D.J., and Brownlee, D.E.: 1994, *Icarus* **111**, 227-236.

Lynch, D.K., Russell, R.W., and Sitko, M.L.: 2000, *Icarus* **144**, 187-190.

Mcdonnell, J.A.M., Lamy, P., and Pankiewicz, G.: 1991, in R. Newburn, M. Neugebauer and J. Rahe (eds.), *Comets in the Post-Halley Era*, Kluwer, Dordrecht, pp. 1043-1073.

McDonnell, J.A.M., et al.: 1993, *Nature* **362**, 732-734.

Messenger, S.: 2000, *Nature* **404**, 968-971.

Mukai, T. and Fechtig, H.: 1983, *Planet Space Sci.* **31**, 655-658.

Rickman, H.: 1986, in O. Melitta (ed.), *Comet Nucleus Sample Return Mission*, **ESA SP-249**, pp. 195-205.

Rietmeijer, F. J. M.: 1998, in J.J. Papike (ed.) *Planetary Materials, Reviews in Mineralogy* **36**, 2-1 – 2-95, The Mineralogical Society of America, Washington, D.C.

Sagdeev, R.E., Elyasberg, P.E., and Moroz,, V.I.: 1988, *Nature* **331**, 240-242.

Sekanina, Z.: 1996, in K.S. Noll et al. (eds.), *The collision of comet Shoemaker-Levy 9 and Jupiter*, Cambridge Univ. Press, pp. 55-80.

Tielens, A.G.G.M., Wooden, D.H., Allamandola, L.J., Bregman, J., and Witteborn, F.C: 1996, *Astrophys. J.* **461**, 210-222.

Utterback, N.G., and Kissel, J.: 1990, *Astron. J.* **100**, 1315-1322.

COLLECTED EXTRATERRESTRIAL MATERIALS: CONSTRAINTS ON METEOR AND FIREBALL COMPOSITIONS

FRANS J.M. RIETMEIJER
Institute of Meteoritics, Department of Earth and Planetary Sciences,
University of New Mexico, Albuquerque, N.M. 87131, USA.
E–mail: fransjmr@unm.edu

and

JOSEPH A. NUTH III
Laboratory for Extraterrestrial Physics, MS 691,
NASA Goddard Space Flight Center, Greenbelt, MD 20771, USA.
E–mail: uljan@lepvax.gsfc.nasa.gov

(Received 26 May 2000; Accepted 31 July 2000)

Abstract. The bulk density and bulk porosity of IDPs and various meteorite classes show that protoplanet accretion and evolution were arrested at different stages as a function of parent body modification. The collected IDPs, micrometeorites and meteorites are aggregates of different structural entities that were inherited from the earliest times of solar system evolution. These structural entities and the extent of parent body lithification will determine the material strength of the meteoroids entering the Earth's atmosphere. There is a need for measurements of the material strength of collected extraterrestrial materials because they will in part determine the nature of the chemical interactions of descending meteors and fireballs in the atmosphere. High–precision determinations of meteor and fireball compositions are required to search for anhydrous, carbon-rich proto-CI material that has survived in the boulders of comet nuclei.

Key words: Asteroid, bolide, chemistry, comet, cosmic dust, fireball, interplanetary dust particle (IDP), meteorite, meteor, mineralogy

"The search for truth (in Science) is the offspring of silence and unbroken meditation."
– Sir Isaac Newton

1. Introduction

Periodic comets, objects in the asteroid belt and near-Earth asteroids are the sources of the meteoroids that cause meteors and fireballs (or bolides). These meteoroids are the result of impacts, magnetic levitation and possibly volcanism on asteroids and sublimation of ice and tidal disruptions on comet nuclei (Flynn, 1994). The Earth's atmosphere is continuously bombarded by asteroidal and comet debris ranging from dust to 10–meter–sized objects. During deceleration in the atmosphere between ~150 to 80 km altitude incoming meteoroids experience ablation via evaporation and, finally, fragmentation in the stratosphere (Ceplecha *et al.*, 1998, and references therein). The nature and intensity of the interactions with the atmosphere depend on the meteoroid's material strength, its size and density and the atmospheric entry angle and velocity. These interactions vary from texturally-intact survival albeit with some degree of flash heating causing thermal modification, (partial) melting and chemical fractionation to sphere formation, fusion crust formation and fragmentation. The surviving fraction after meteoroid deceleration in the atmosphere can be collected as micrometeorites (MMs), as the fragments of larger bodies at the Earth's surface (and throughout the geological column including land-ice deposits), and as interplanetary dust particles (IDPs) in the lower stratosphere (Rietmeijer, 1998a; Zolensky *et al.*, 1994a). The meteor trail and persistent–train compositions may be affected by rapid, often complex, chemical reactions between meteoric vapors and the middle atmosphere, some of which may be unique to very high-velocity meteors (Zinn *et al.*, 1999). Between meteoroids and collected materials there is a 'black box' that relates these observations to chemical data on meteors, fireballs and laboratory analyses but much work to match them still needs to be done.

Rietmeijer (2000a) looked for quantitative chemical links between the compositions of collected IDPs, MMs, meteorites and the meteor and fireball data and, while there are experimental problems still to be solved, establishing such a connection is both possible and promising. The interpretations of chemical phenomena observed in meteors and fireballs rely on understanding the physical processes involved in the atmospheric interactions. Yet, some observations, such as differential ablation of refractory elements, can find an explanation in the textures, mineralogy and chemical compositions of the meteoroids (Rietmeijer, 2000a). The material strength of IDPs, MMs and meteorites are important in order to assess how meteoroids break up during catastrophic (impact) release from the parent body, erosion during solar system sojourn, and fragmentation on entering and during descent in the Earth's atmosphere.

Some extremes are obvious. For example, the collected cm-sized fragments of type CI carbonaceous meteorites indicate a friable material while large iron-meteorites are a uniformly strong material. Most ordinary chondrites are coherent stones but the friable Bjurböle meteorite falls easily apart into its constituent chondrules and matrix. Such examples show that there are obvious differences in material strength among the unequilibrated and equilibrated ordinary chondrites. Material strength may explain why centimeter-sized types CI and CM carbonaceous meteorites represent only 2% of all meteorite falls while ~95% of collected MMs (50–100 micrometer in size) are type CM spheres, and unmelted type CM IDPs are common non-aggregate chondritic IDPs (Rietmeijer 2000a). Engrand and Maurette (1998) refer to the MMs as new "chondrites-without-chondrules" materials. This designation might reflect more on their physical properties than indicating a new class of objects. Rietmeijer (1996) noted that the stratospheric collection times of type CM IDPs correlate with the recorded dates of CM meteorite falls. A dearth of fragile type CI IDPs suggests that they break-up in the Earth's atmosphere or that they are debris from high-velocity (>20 km s^{-1}) 'cometary meteor' streams and therefore will be more readily evaporated than asteroidal dust of the same mass and cross-section.

We will here consider the bulk density and porosity of collected extraterrestrial materials, as well as their bulk composition and textural heterogeneity as a function of size. This does raise the issue of whether it is reasonable to assume that millimeter to cm-sized asteroidal and cometary meteoroids have a CI or solar composition, or not. What are the chemical compositions and physical properties of the constituents in IDPs, MMs and meteorites? and does the bulk composition vary as a function of nano- and micrometeoroid size?

2. Meteorite and IDP density and porosity

The Fireball classification relies on meteor ablation efficiency and fragmentation behavior to obtain a match with known meteorite types and a putative (mostly asteroidal) origin, or for the lowest-density (< 1 g cm^{-3}) objects, an inferred cometary origin (Ceplecha *et al.*, 1998). The elemental composition of meteoroids is obtained by spectroscopic analyses of meteors and fireballs (Borovicka, 1999; Ceplecha *et al.*, 1998). Their chemistry provides an obvious link between these phenomena and the compositions of the surviving fraction of the meteoroids that are analyzed in the laboratory (Papike, 1998; Jessberger *et al.*, 2000). Consolmagno and Britt (1998), Corrigan *et al.* (1997) and Flynn *et al.* (1999) measured the density and porosity of meteorites from the major classes. The measurements were made in different laboratories and by different techniques.

TABLE I

Samples	Bulk density g cm^{-3}				Porosity %			
	Mean	±σ	Range	N	Mean	±σ	Range	N
Aggregate IDPs	0.1				95			
Aggregate IDPs	0.7				75			
CI	1.58	–	–	1	35	–	–	1
CM	2.23	–	2.08 – 2.34	2	22.7	6.5	16 – 29	3
CV	2.91	0.20	2.72 – 3.18	14	12.5	8.9	2 – 24	7
CO	2.67	–	2.36 – 2.98	2	8	–	4 – 12	2
H	3.35	0.12	3.15 – 3.53	10	8.6	4.0	3 – 18.1	13
H	3.56	–	3.47 – 3.65	2	0	–	–	3
L	3.14	0.2	2.74 – 3.35	11	10.2	5.6	2.9 – 21.5	16
L	3.36	0.07	3.27 – 3.46	8	0.4	0.6	0 – 2	11
LL	3.33	–	–	1	15	–	–	1
LL	3.15	–	–	1	1	–	–	1
Igneous					–	–	<1 – 6	2
Stony irons	4.49	0.365	4.16 – 4.97	6	5	–	0 –13	5
Irons	6.93	1.2	4.01 – 7.59		3.5	–	0 – 13	8

The results from different groups on fragments of the same meteorite show that systematic differences are small and are mostly within the errors of measurement. We calculated the mean and standard deviations of meteorites within each meteorite class using the published data (Table I) for the measured bulk density and porosity of aggregate IDPs, including porous aggregate IDPs (Rietmeijer, 1993; 1998a), and meteorites (Corrigan *et al.*, 1997; Consolmagno and Britt, 1998; Flynn *et al.*, 1999). The 'igneous' group includes one eucrite (Juvinas) and a SNC meteorite (Nakhla). N refers to the number of measurements, which is larger than the number of meteorites analyzed. A remarkable feature of the chemical subtypes, H (high iron), L (low iron) and LL (very low iron), of ordinary chondrites is that some fraction has zero porosity. We treated these meteorites as a separate group (Table I). All ordinary chondrites in this Table belong to the petrographic grades 4, 5 and 6 designating increasingly intense thermal alteration at temperatures below the melting point. Porosity measurements of IDPs were made by Mackinnon et al. (1987), Strait et al. (1996) and Corrigan et al. (1997), while Fraundorf et al. (1982), Flynn and Sutton (1991), Love et al. (1994) and Zolensky et al. (1989) measured densities of IDPs. Porosity and density data for the same IDP are rare; in addition different techniques were used to obtain either IDP porosity or density. The data show a consistent pattern which is fairly well summarized by the data points in Table I (Rietmeijer, 1998a).

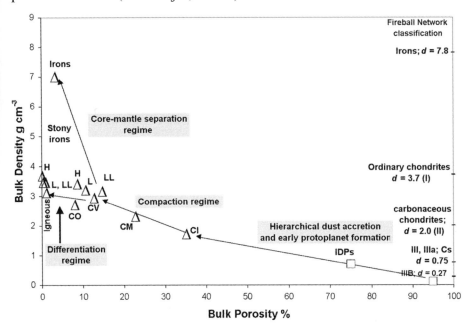

Figure 1. Bulk porosity (%) and bulk density (g cm^{-3}) of IDPs and the major meteorite classes and subtypes, viz., CI, CM, CV and CO carbonaceous chondrites, H, L and LL ordinary chondrites, the 'igneous' group (Table I), and stony-iron and iron meteorites. The Fireball network classification (Ceplecha et al., 1998) is shown at the right-hand side. A recent modeling study found values ranging from 5.8×10^4 dyn cm^{-2} to 2.6×10^6 dyn cm^{-2} for the tensile strength of comet nuclei and an average comet nucleus density of 0.5 g cm^{-3} and up to 1.65 g cm^{-3} when compacted (Sirono and Greenberg, 2000).

The density - porosity distributions (Figure 1) show a trend beginning with hierarchical dust accretion forming IDPs. This stage of solar system evolution merged into a period of early protoplanet evolution that involved the formation, disruption and re-accretion of 'rubble piles' (Rietmeijer, 1998a, 2000a). We have no samples of these early 'rubble piles' but they may have been preserved among the boulders of comet nuclei as "proto-CI" material (Weidenschilling, 1997; Rietmeijer, 2000b). Continued accretion and protoplanet formation of dust, chondrules, calcium-aluminum-rich inclusions (CAIs) and collisional debris of protoplanets (*e.g.* foreign meteorite clasts incorporated in other meteorite classes, see Brearley and Jones, 1998) in the inner solar system yielded larger parent bodies. Compaction of these bodies occurred due to a number of different lithification processes operating with variable relative efficiency and importance among the protoplanets, *viz.:*

(1) thermal metamorphism during prolonged periods of internal heating (at sufficiently high temperatures this will lead to (partial) melting and segregation, *e.g.* core-mantle formation),

(2) aqueous alteration when water was available during thermal alteration at temperatures between ~20°C and 125°C,

(3) transient impact shock heating with the possibility of (localized) parent body melting (shock-melt veins); note that many meteorites are breccias of fractured rocks, and

(4) overburden compaction.

We refer the reader to excellent reviews on each of these topics in the volume "Meteorites and the early solar system" (Kerridge and Matthews, 1988). As presented in Figure 1 it might appear that the ordinary chondrites resulted from continued compaction of C-type meteorites. This is not the case. These meteorites represent different parent bodies with their unique textural and chemical properties (Brearley and Jones, 1998 for a review of meteorites). When temperatures were high enough for melting to occur some parent bodies entered a 'core-mantle separation regime' or a 'differentiation regime' leading to the formation of high-density, low porosity meteorites. The 'igneous meteorites' are a group apart. They include gabbros and basalts from once (partially) molten parent bodies (see Papike, 1998). The simple view using two correlated rock properties can not describe the complexity of parent body modification but it offers a framework to describe properties of materials from different protoplanets.

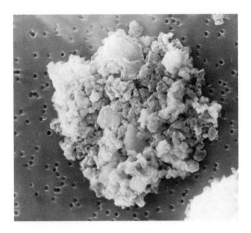

Figure 2. Scanning electron microscope image of a large porous, or fluffy, chondritic aggregate IDP collected in the lower stratosphere showing the matrix with embedded larger platy silicate grains. The particle is placed on a nucleopore-filter (background) during electron microscope imaging. The image is 20 micrometers across. Courtesy of the National Aeronautics Space Administration, Particle W7029B13 (NASA number S-82–27575, Cosmic Dust Catalog 2(1), NASA Johnson Space Center Curatorial Branch Publication 62, 1982)

3. Hierarchical dust accretion

3.1. INTERPLANETARY DUST PARTICLES

The Earth's atmosphere is a gentle decelerator for low-density nanometeoroids. Many of these particles survive 'intact', that is flash heating temperatures (~300 to ~1,000°C) generated during atmospheric entry are below the melting point of chondritic material at ~1,500°C (Brownlee, 1985). The IDPs are collected on silicone-oil coated flat-plate Lexan collectors mounted underneath the wings of high-flying aircraft between 17 – 19 km in the stratosphere (Zolensky *et al.*, 1994a). Detailed reviews of the physical, mineralogical and chemical properties of these particles, including their isotopic compositions, are available [Mackinnon and Rietmeijer (1987), Sandford (1987), Bradley *et al.* (1988), Flynn (1994), Jessberger *et al.* (2000)]. Aggregate IDPs (Figure 2) and cluster IDPs consist of a limited number of minerals that vary primarily in grain size, and to a lesser extent in their mode of occurrence within these particles (Rietmeijer, 1998*a*). Table II gives a classification of IDPs 5–15 µm collected in the Earth's lower stratosphere (based on Rietmeijer, 1998a). The minerals are olivine [$(Mg,Fe)_2SiO_4$], (Ca-free and Ca-bearing) Mg,Fe-pyroxene [$(Mg,Fe)_2Si_2O_6$], Ca-rich pyroxene [$Ca(Mg,Fe)Si_2O_6$], Fe,Ni-sulfide (mostly pyrrhotite, Fe_7S_8) and Fe,Ni-metal (kamacite). None of these minerals has a chondritic (CI or solar) composition for its elements.

TABLE II

Chondritic IDPs	Non-chondritic IDPs
solar abundances within a factor 2 for the major rock-forming elements	*chondritic material adheres to the surface*
(1) Aggregate IDPs: a matrix of principal components with embedded (variable) amounts of Mg,Fe- and Mg,Fe(Ca)-silicates, Ni-free and low-Ni pyrrhotite and iron oxides that are <0.5 to 10 µm	
(2) Non-aggregate IDPs of the CI and CM type	
	(1) Silicate IDPs, mostly Mg,Fe-silicates and Ca,Mg,Fe-silicates
	(2) Sulfide IDPs, mostly Ni-free and low-Ni pyrrhotite
	(3) Refractory, Ca,Ti,Al-rich IDPs
	(4) IDPs that are admixtures of variable amounts of IDPs types 1–3
Cluster IDPs up to 100 µm in size are admixtures of variable amounts of chondritic aggregate IDPs and the non-chondritic IDPs (1) – (4)	

Furthermore, the matrix of aggregate IDPs has been shown to consist of unique smallest entities, known as principal components (PCs), that are 100 – 1,000 nm in diameter and that may represent were the first dust particles to accrete in the solar system (Rietmeijer, 1998a):

1. Carbonaceous PCs of refractory hydrocarbons, amorphous (often-vesicular) carbons, poorly graphitized carbons and pre-graphitic turbostratic carbons. These PCs are often fused into contiguous patches and clumps (Thomas et al., 1993, Figure 2),
2. Carbon bearing ferromagnesiosilica PCs of ultrafine (2 to ~50 nm) platy Fe,Mg-olivines, Fe,Mg-pyroxenes, Fe,Ni-sulfides, Fe-oxides embedded in a carbonaceous matrix of refractory hydrocarbons and amorphous carbons. They contain sulfur and minor Al, Ca, Cr, Mn, Ni, and traces of phosphorous and zinc (Rietmeijer, 1989, Figures 2 and 3), and
3. Ferromagnesiosilica PCs (Rietmeijer et al., 1999, Figure 9; Rietmeijer, 1998a, Figures 32 and 34; Bradley, 1994, Figure 1B):

   (I)   *Coarse-grained PCs* with a $(Mg,Fe)_6Si_8O_{22}$ (smectite dehydroxylate) bulk composition and Fe/(Fe+Mg) *(fe)* = 0 – 0.33. These Mg-rich PCs consist of a coarse-grained (10 – 410 nm) Mg,Fe-olivine and Mg,Fe-pyroxene plus amorphous aluminosilica material (± traces Ca, Mg, Fe),

   (II) *Ultrafine-grained PCs* with a $(Mg,Fe)_3Si_2O_7$ (serpentine dehydroxylate) bulk composition and *fe* = 0.3 – 0.83 (modal *fe* = 0.67). These Fe-rich units consist of an amorphous matrix that contains embedded Mg,Fe-olivines, Mg,Fe-pyroxenes, Fe,Ni-sulfides, magnetite, and metal (kamacite) grains (< 50 nm). Magnetite is presumably a product of terrestrial oxidation; the sulfur content is variable.

3.2. CHEMISTRY

The compositions of coarse grained Mg-rich ferromagnesiosilica PCs are the direct result of kinetically controlled gas-to-solid condensation of an Mg-Fe-SiO-H_2-O_2 vapor in O-rich circumstellar environments and agglomeration of the condensed amorphous magnesiosilica and ferrosilica dusts (Rietmeijer et al., 1999). Each of these dusts has a uniquely defined metasTable eutectic composition (Figure 3). Compositions of ultrafine-grained Fe-rich ferromagnesiosilica PCs (also known as GEMS: glass with embedded metal and sulfides; Bradley, 1994) are the result of mixing of a pure Mg silicate (olivine or pyroxene) with an Fe,Ni-sulfide grain facilitated by irradiation by highly energetic

H and He nuclei in interstellar space (Bradley, 1994) or by mixing a condensed Mg-rich ferromagnesiosilica PC and a condensed Fe-oxide grain with homogenization facilitated by the high "internal free energy" content of the metasTable solids (Rietmeijer, 1999a; Rietmeijer *et al.*, 1999).

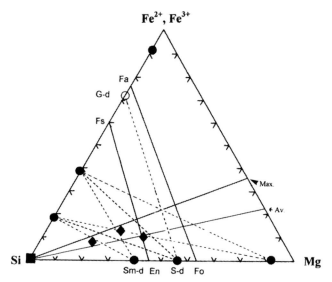

Figure 3. The Mg-Fe-Si (element wt %) diagram is used to compare major element compositions of IDPs, and the matrix of carbonaceous and ordinary chondrites. Cotectic-mixing lines (dashed) connect the composit-ions of the amorphous, condensed solids (dots). These compositions are (1) smectite (Sm-d) and serpentine (S-d) dehydroxylates and low-FeO smectite-like dehydoxylates, (2) Si-bearing Fe- and Mg-oxides, and (3) silica grains (solid square). Fe^{2+},Fe^{3+} symbolizes variations in Fe^{2+}/Fe^{3+} ratios. Average fe (Av.; 0.23) and maximum (max.) *fe* ratio and three bulk compositions of these PCs (black diamonds) are indicated. A metasTable cotectic mixing line (dashed) connects the S-d and greenalite (G-d) (open circle) metasTable eutectics. It is located in between the stoichiometric olivine (Fo-Fa) and pyroxene (En-Fs) lines (from Rietmeijer *et al.*, 1999). © 1999 the American Astronomical Society. Reproduced by courtesy of the American Astronomical Society.

The C-free and C-bearing ferromagnesiosilica PCs, plus the carbonaceous PCs, are found mixed in variable proportions forming the matrix of aggregate IDPs. The above mentioned silicate and sulfide minerals, of sizes <5 to ~15 micrometer (Table II), accreted to form 5 – 15 micrometer-sized aggregate IDPs. A ternary diagram of the three PC types shows how the relative proportions of accreting PCs (Figure 4) define the carbon content of IDPs, comet Halley and in the matrix of carbonaceous and ordinary chondrites. Jessberger *et al.* (1988) found that dust in comet Halley was Si- and Mg-rich compared to the CI or solar abundances. Combining mass and composition of each detected dust particle Fomenkova *et al.* (1992) concluded that the dust in comet Halley consisted of small (1) CHON (carbon-hydrogen-oxygen and nitrogen), (2) mixed carbon-silicate and (3) silicate particles. These particles occur in aggregate IDPs as (1) carbonaceous PCs (2) C-bearing ferromagnesiosilica PCs and (3) ferromagnesiosilica PCs (Rietmeijer, 1998a) (Figure 4).

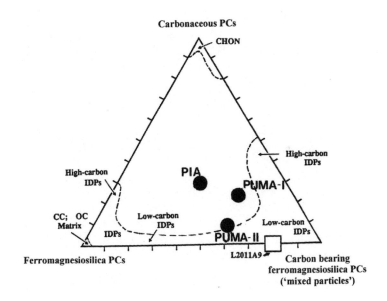

Figure 4. Diagram of the principal components in aggregate IDPs. The dashed line in the lower portion of the Figure denotes the estimated mixing ratios in aggregate IDPs (open square is for IDP L2011A9) and the matrix of carbonaceous (CC) and ordinary chondrites (OC). The dashed line near the apex denotes CHON-like particles as seen in comet Halley but that have not yet been found among the collected stratospheric IDPs. These particles were most likely fused together during flash heating (see text). The black dots are the proportions of the end-members among Halley's dust detected by the PIA and PUMA-1 instruments (Langevin *et al.*, 1987) and both PUMA instruments (Fomenkova *et al.*, 1992). The scatter of these points could be real or may also reflect different instrument detection capabilities and data reduction procedures.

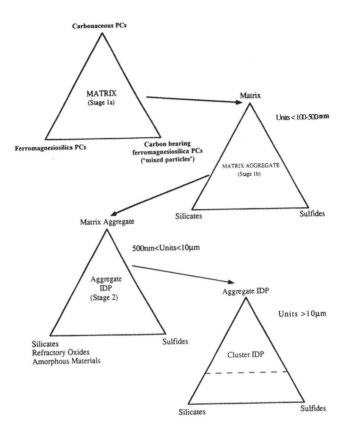

Figure 5. Ternary diagrams depicting the hierarchical accretion of non-chondritic dusts into 'matrix aggregates' (stages 1a and 1b), aggregate IDPs (stage 2) and cluster IDPs (stage 3). The second stage offers an explanation for the interrelationship between refractory IDPs and common chondritic aggregates. The dashed line in the third diagram delineates the relationships among chondritic aggregate IDPs and non-chondritic IDPs with chondritic material attached to the surface. Modified after Rietmeijer (1998a; courtesy of the Mineralogical Society of America).

3.3. ACCRETION

During initial dust accretion the nanometer-sized PCs along with Mg,Fe(Ca)-silicates and Fe,Ni-sulfides < 100 to ~500 nm in size coagulated to form matrix aggregate materials with variable proportions of PCs (Figure 5). Accretion continued with matrix aggregates, < 500 nm to < 10 micrometer in diameter and similarly-sized Mg,Fe(Ca)-silicates,

Fe,Ni-sulfides and (rare) refractory oxides (Table II) to form aggregate IDPs, 10 – 15 μm in size. Olivine and pyroxene crystals of this size were found in impact features on the Solar Maximum satellite (Rietmeijer and Blanford, 1988) and the Long Duration Exposure Facility (Zolensky et al., 1994b). The bulk compositions of these IDPs are characterized by variable proportions of non-chondritic dusts but, nevertheless, are chondritic to within a factor 2–3 for most rock-forming elements. The Ca, Al and Ti abundances are mostly determined by the relative amounts of accreted refractory dust.

TABLE III

Element ratio	Bulk comet P/Halley dust	Coarse-grained PCs	Ultrafine-grained PCs	Aggregate IDPs	Cluster IDP L2008#5	Bulk CI
Sources	(1)	(2)	(2)	(3)	(4)	(3)
C/Si	4.4	Zero	Zero	No data	1.1	0.7
Mg/Si	0.54	0.48	0.64	0.85	0.7	1.06
Fe/Si	0.28	0.18	0.46	0.63	0.7	0.90
S/Si	0.39	0.0	0.2	0.35	0.4	0.46
Al/Si	0.037	0.09	0.09	0.063	0.1	0.085
Ca/Si	0.034	0.04	0.027	0.048	0.07	0.071
Na/Si	0.054	No data	No data	0.049	0.2	0.060
Ni/Si	0.022	0.001	0.032	0.037	0.04	0.051
Cr/Si	0.005	0.0	0.018	0.012	0.01	0.013
Mn/Si	0.003	0.004	0.007	0.015	0.01	0.009
Ti/Si	0.002	No data	No data	0.002	No data	0.002
K/Si	0.001	No data	No data	No data	No data	0.01

Continued accretion produced cluster IDPs, >60 μm in size, made up of variable proportions of the aggregate IDPs and >10 μm-sized Mg,Fe(Ca)-silicate, Fe,Ni-sulfide and Fe-oxide particles (Table II). Cluster IDP bulk compositions resemble the CI composition but with higher carbon abundances (Thomas et al., 1993; 1995). Because carbonaceous PCs are the major carrier of carbon, the carbon content of aggregate and cluster IDPs is fixed early during accretion. There is no reason to believe that aggregate IDPs and cluster IDPs are fundamentally different in any other property than size. A cluster IDP may contain any number of fragments identical to aggregate IDPs and coarse-grained non-chondritic IDPs. We don't know what occurred after cluster IDP formation. There is no reason to believe that hierarchical accretion could not have continued to form ever-larger clusters that eventually became the matrix of undifferentiated meteorites.

The hypothesis of hierarchical accretion predicts that accretion of only a limited number of non-chondritic dust types gradually produced larger, and thus relatively younger, aggregates that ultimately reach a CI bulk composition (Rietmeijer, 1998a). This compositional change with particle size, up to the scale of meteorite matrix is shown by the major element ratios Mg/Si, Fe/Si and S/Si, but is less evident for minor elements. Table III gives the average compositions (atomic ratios) of various primitive solar system materials, including CI meteorites with >>99% matrix taken from the literature: (1) Jessberger *et al.* (1988), (2) Rietmeijer (1998a), (3) Brownlee (1978), and (4) Thomas *et al.* (1995). This hypothesis also predicts that mineralogical heterogeneity may arise during accretion depending on the available dust types in the accretion regions as a function of time and space.

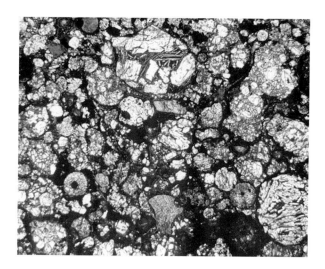

Figure 6. Photomicrograph of the Tieschitz (H3) ordinary chondrite, taken in plane-polarized, transmitted light. This chondrite consists of rounded chondrules surrounded by fine-grained matrix. Individual mineral grains in the chondrules are mostly olivine and pyroxene. Matrix appears black because it is so fine-grained. Metal and sulfide (troilite, FeS) grains that occur both within chondrules and interstitial to the chondrules also appear black. The image is 12 millimeters across (courtesy of R.H. Jones, University of New Mexico).

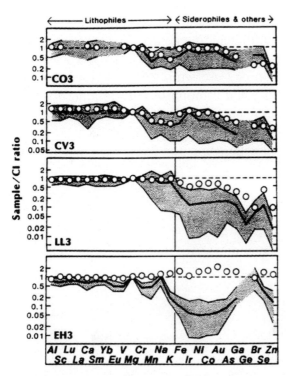

Figure 7. Normalized to Mg and the CI abundances minor and trace elements in chondrules from four major-chondrite (meteorite) groups are arranged in order of their volatility (except iron that occurs in both metals and silicate minerals). The stippled pattern encompasses >90% of the chondrules in each group with the solid line indicating the arithmetic mean. Circles indicate the composition of the host-chondrite. Reproduced from Figure 9.1.5 of Grossman, Rubin, Nagahara and King (1988) in J.F. Kerridge and M.S. Matthews (eds.), Meteorites and the Early Solar System. By courtesy of Jeff Grossman and with the permission of the University of Arizona Press, Tucson (AZ) (copyright © 1988 The Arizona Board of Regents).

4. Carbonaceous, ordinary and enstatite chondrites

The CI, CM, CV and CO chondrites, the ordinary chondrites and enstatite chondrites all contain a fine-grained matrix, chondrules, refractory Ca,Al-rich inclusions (CAIs) and Fe,Ni metal grains. Table IV gives the approximate proportions (vol. %) of components in various chondrite meteorite groups (modified after Brearley and Jones, 1998). The relative proportions of these components vary considerably and range from meteorites that are basically just fine-grained matrix (CI meteorites) to meteorites that are mostly chondrules (H, L and LL ordinary chondrites and EL and EH enstatite chondrites) (Table IV). Chondrules are sub-millimeter quenched-melt spheres of uncertain origin (Brearley and Jones, 1998) (Figure 6).

Chondrule compositions are similar to the composition of the host-chondrite meteorite. Their lithophile (*i.e.* silicate-forming) elements show CI abundances but on average chondrules are depleted in siderophile (*i.e.* metal-forming) and chalcophile (*i.e.* generally sulfide mineral-forming) elements compared to the host meteorite (Grossman *et al.*, 1988) (Figure 7).

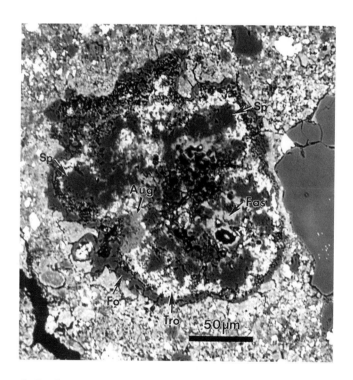

Figure 8. Back-scattered electron microscope image of a CAI from the Bells CM carbonaceous chondrite. It is a nodular spinel-pyroxene CAI that consists of an irregular core of spinel (SP) intergrown with the pyroxene minerals, fassaite (Fas) and augite (Aug), and the sulfide mineral, troilite (FeS). A narrow mantle of forsterite (Fo) rims the inclusion. By courtesy of Adrian Brearley (University of New Mexico). Reprinted from Geochimica et Cosmochimica Acta, 59, A.J. Brearley, Aqueous alteration and brecciation in Bells, an unusual, saponite-bearing, CM chondrite, 2291–2317, copyright 1995, with permission from Elsevier Science.

The CAIs can be spheres or irregular objects with sizes ranging from < 1 mm to > 1 cm that are dominated by minerals with high vaporization temperatures (MacPherson et al., 1988). The largest CAIs, > 1 cm and up to 2.5 centimeter, are found in CV meteorites (Brearley and Jones, 1998) (Figure 8). The CAIs are rich in Al, Ca and Mg (Table V) and titanium is also common in these objects. Iron is generally not associated with CAIs. It occurs in the matrix, in chondrules and as metal inclusions (Table IV). Table V shows the major element abundances in CAIs from CM and CV3 meteorites (modified after MacPherson et al., 1988, Table I0.3.2). The CM data show the range in these meteorites. The CV data are an average for a wide range of CAI types in CV3 meteorites. The renormalised CV bulk composition is from McSween and Richardson (1977). Meteorites are lithified aggregates of components that have their own unique history in the solar system.

TABLE IV

	Matrix	Chondrules [diameter, millimeters]		Refractory inclusions [CAIs]	Fe,Ni-metal
CI	>99; with Fe,Ni-sulfides and Fe-oxides	<<1		<<1	0
CM	70; with Fe,Ni-sulfides and Fe-oxides	20	[0.3]	5	0.1
CR	30–50	50–60	[0.7]	0.5	5–8
CO	34	48	[0.15]	13	1–5
CV	40	45	[1.0]	10	0–5
H	10–15	60–80	[0.3]	0.1–1?	10
L	10–15	60–80	[0.7]	0.1–1?	5
LL	10–15	60–80	[0.9]	0.1–1?	2
EH	<2–15?	60–80	[0.2]	0.1–1?	8
EL	<2–15?	60–80	[0.6]	0.1–1?	15

TABLE V

	CM		CV (CAIs)	CV (bulk)
SiO_2	zero	zero	30.5	
Al_2O_3	91.5	59.9	30.4	
TiO_2	< 0.19	1.9	1.3	
FeO	0.6	0.3	1.4	
MgO	0.5	13.2	9.9	
CaO	6.2	24.7	26.7	
Al/Si (atomic)			1.18	0.115
Ti/Si			0.03	–
Fe/Si			0.04	0.745
Mg/Si			0.48	1.08
Ca/Si			0.94	0.08

Table VI

	Proto–CI	Orgueil
C/Si (*)	0.05	0.08
Mg/Si	0.80	1.00
Al/Si	0.07	0.08
Fe/Si	0.67	0.87
Cr/Si	0.01	0.02
Mn/Si	0.0055	0.01
Ca/Si	0.04	0.06
Na/Si	0.05	0.05
Ni/Si	0.03	0.05

5. Discussion

5.1. COMETESIMALS

The bulk density values used in the Fireball Network classification are generally slightly higher than the measured meteorite values (Figure 1; Table I) but otherwise support the use of bulk density as a useful

classification parameter. There are a few simple, first-order observations that follow from the bulk density - porosity diagram:
(1) meteoroids with a bulk density greater than ~2.5 g cm^{-3} represent parent bodies wherein geological activity had compacted and consolidated the accreted constituents,
(2) meteoroids with bulk densities between ~1 and ~2.5 g cm^{-3} mostly consist of matrix material that is much less consolidated than those of the first group, and
(3) meteoroids with a bulk density <1 g cm^{-3} structurally resemble aggregate IDPs.

The CI meteorites straddle the transition from the 'hierarchical dust accretion and early protoplanet formation' regime and the 'compaction' regime (Figure 1). These meteorites are fully hydrated objects, in fact just "chunks of clay", and are almost free of inclusions of any kind (Table IV). Hydration and the subsequent formation of layer silicates require a parent body wherein temperatures are high enough and are maintained long enough for water to react with the initial aggregate. With our current understanding of small solar system bodies, this environmental constraint uniquely points to an asteroidal source of CI meteorites. The formation of layer silicates will cause the collapse of any porous structure, which will lead to densification of the original material (Rietmeijer and Mackinnon, 1997). There is reason believe that this type of aqueous alteration has occurred in the past and still occurs during perihelion passage in active comet nuclei (Rietmeijer and Mackinnon, 1987). This alteration may be restricted to dirty-ice of the nucleus in a process known as hydrocryogenic alteration at temperatures below the melting point of ice. These reactions occur due to a few nanometer-thick layer present at the interface of dust grains and ice, and which serves as a conduit for dissolved chemical species (Rietmeijer, 1985; 1998b). Comet nuclei are probably primitive 'rubble piles' of boulders that are held together by self-gravitation or by a cement of dirty-ice (so-called Whipple glue') (Weissman, 1986; Gombosi and Houpis, 1986). The boulders might resemble CI meteorites (Gombosi and Houpis, 1986). Such a hypothesis is consistent with fireball observations (Ceplecha *et al.*, 1998). Boulders that evolved in the comet accretion regions may have high carbon contents given the high carbon content observed in comets. Texturally coherent, C-rich boulders with sizes from pebbles to objects many hundreds of meters in diameter may therefore be held together by glue that will be sensitive to radiation exposure and thermal heating. As a result these proto-CI objects may be more susceptible to fragmentation than hydrated CI materials during entry and descent into the Earth's atmosphere.

5.2. PROTO-CI MATERIALS

The existence of anhydrous carbon-rich proto-CI materials can be verified by high precision chemical analyses of meteors associated with known comets such as the Leonid meteors. The bulk compositions of the hydrated CI and CM meteorites include between 2 – 20-wt % H_2O. However, the SiO_2 content of these 'wet' meteorites is between 22 – 28-wt % compared to 33 – 40-wt % SiO_2 in the dry ordinary and enstatite chondrites (Mason, 1971, Table III). Rietmeijer (1995) reported hydration experiments of Mg-SiO materials with 24-wt % MgO that were obtained by gas to solid condensation of a $Mg-SiO-O_2-H_2$ vapor. During initial hydration there was massive leaching of silica from the sample with a resulting change in bulk composition to ~33-wt % MgO. This composition is similar to the smectite dehydroxylate composition of one of the metastable eutectics in this system $MgO-SiO_2$ (Rietmeijer et al., 1992) and is comparable to the CI abundances. The leached silica precipitated as massive opal-like globules (Rietmeijer, 1995). This experiment suggests that H_2O replaced some fraction of the silica in this Mg-SiO material. If similar processes acted on the dry precursors of CI meteorites then their low silica content is due to hydration in the parent bodies and these bodies should have developed silica veins. Replacing (by mass) the amount of water in the Orgueil meteorite (Mason, 1971, Table III) by SiO_2, we calculated the major element composition of proto-CI material (Table VI). This Table gives calculated Si-normalized major element ratios (atomic) of proto-CI material based on the Orgueil meteorite composition after replacing (by mass) water by SiO_2 and re-normalization to 100-wt %. The metal/Si (atomic) ratios for this meteorite are shown for comparison; (*) IDP data suggest that this value is a lower limit. The abundance of sodium, a highly volatile and soluble element, could be as high as Na/Si = 0.11 (Rietmeijer, 1999b). These normalized proto-CI elemental abundances are significantly lower than the CI abundances. These differences may be great enough for detection in carbon-rich proto-CI meteor compositions.

5.3. MATERIAL PROPERTIES

Fragmentation of aggregates on atmospheric entry and during deceleration will occur along the weakest structural bonds of their constituent entities. Table VII gives the potential meteoroid sizes inferred from collected extraterrestrial materials.

Table VII

Size	Types of objects
<5 microns	matrix aggregates of PCs and mineral grain inclusions (Table II)
<40 microns	individual IDPs (chondritic particles and non-chondritic silicate and sulfide grains (Table II) (Rietmeijer, 1999c)
<100 microns	cluster IDPs
20 – 500 microns	micrometeorites; the pre-entry sizes were between ~30 µm – 100 µm and ~0.4 to ~1 millimeters
0.2 – 1 mm	Chondrules
a few mm up to 2.5 cm	CAIs
a few cm and up	Meteorites

Micrometeorites making up the peak in the annual terrestrial mass-accretion rate of extraterrestrial materials (Love and Brownlee, 1993) are ablation products of objects that were initially 1.5–2 times larger than the resulting sphere diameter (Love and Brownlee, 1991). Their typical lack of chondrules might suggest that micrometeorites are matrix fragments of CM-type meteorites that broke up during atmospheric entry (Table VII). In this scenario and depending on the material strength of individual CM, CO, CV, the ordinary and enstatite chondrites, fragmentation might yield matrix fragments, chondrules and CAIs with variable efficiency during descent and fragmentation in the atmosphere (Table VII). Thus, it is a pity that there is very little systematic information on the size distribution of fragments in the strewnfield of a meteorite shower. In the Mbale meteorite (an L-type ordinary chondrite) shower (Figure 9) the largest fragments resulted from the final break-up along zones of structural weakness in the meteorite (shock fractures?) to produce angular fragments while smaller fragments broke-up at higher altitudes (Jenniskens *et al.*, 1994). There is no apparent structural feature in this meteorite that would cause the smaller fragments, but a combination of material strength and the character of the physical interactions at the higher altitudes might be responsible. This example highlights the fact that the dimensions of recovered meteorite fragments from single events, as well as separately, and in combination with structural analyses and determinations of the material strength, will provide information on ablation and fragmentation behavior of meteoroids of various meteorite classes.

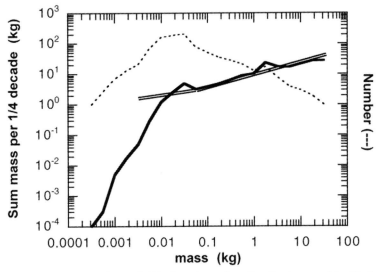

Figure 9. Total mass per bin (solid line) and number distribution per bin (dashed line) of collected Mbale meteorite fragments. The double line shows the expected distribution of fragments for a single catastrophic fragmentation event (Adapted from: Jenniskens *et al.*, 1994).

A statistical analysis of the mass distribution of fragments in sixteen different meteorite showers showed two different types of scaling exponents: a simple power law was attributed to a single fragmentation event while a multiple scaling regime indicated multiple major fragmentation events (Odderschede *et al.*, 1998). This work showed single fragmentation behavior for the Sikhote-Aline and Mbale meteorites, among others (Odderschede *et al.*, 1998) which for the Mbale shower is at odds with the conclusion reached by Jenniskens *et al.*

(1994). Meteorite showers of the types L5, L6 and H4–7 ordinary chondrites showed both single and multiple fragmentation. Stony-iron and iron meteorites predictably showed single-fragmentation, as did the Allende (CV) meteorite. The Murchison (CM) meteorite showed multiple fragmentation, which supports a 'structurally weak' bolide and might contribute to the fact that ~95% of all collected micrometeorites are of this petrologic type. These types of analyses are necessary in order to evaluate the chemical evolution of meteors and fireballs descending in the atmosphere. For example, chondrules and matrix have similar CI-like compositions (within the errors of measurement of meteors and fireballs). Chondrules are massive objects while the latter is a fine-grained compacted material. This structural difference must influence the interactions of these materials with the atmosphere. Similarly, CAIs with refractory elements might break apart from the main decelerating meteoroid and these structurally coherent objects will behave differently from the main body. At a smaller scale, aggregate IDPs show that break-up of cometary dust will also provide structurally different entities with different chemical and textural properties, *e.g.* massive silicate and sulfide grains and matrix aggregates.

6. Conclusions

The bulk density and bulk porosity of IDPs, carbonaceous, ordinary and enstatite chondrites, stony-iron and iron meteorites show protoplanet accretion and evolution that were arrested at different stages as a function of parent body modification. These yielded comet nuclei and a wide range of different asteroids. The collected IDPs, micrometeorites and meteorites represent some fraction of these small solar system bodies. They are in fact aggregates of different structural entities inherited from the earliest times of solar system evolution that underwent various levels of parent body lithification. The bulk material strength of individual meteoroids as well as the material strength of their entities is an important, yet still unknown, property that might have a defining impact on the chemical and physical interactions of meteors and fireballs descending in the atmosphere. This work highlights a strong need for measurements of material strength of collected extraterrestrial materials as well as high-precision determinations of meteor and fireball compositions in a quest for proto-CI materials.

Acknowledgments

We are grateful to the reviewers for the suggestions to improve this paper. We thank Marcy Marcilla for technical assistance at IOM. This work was supported by a NASA grant. *Editorial handling:* Noah Brosch.

References

Borovicka, J.: 1999, in W.J. Baggaley and V. Porubcan (eds.), *Meteoroids 1998, Astron. Inst. Slovak Acad. Sci., Bratislava*, 355–362.
Bradley, J.P.: 1994, *Science* **265**, 925–929.
Bradley, J.P., Sandford, S.A. and Walker, R.M.: 1988, in J.F. Kerridge and M.S. Matthews (eds.), *Meteorites and the Early Solar System*, University Arizona Press, Tucson (AZ), pp. 861–898.
Brearley, A.J.: 1995, *Geochim. Cosmochim Acta* **59**, 2291–2317.
Brearley, A.J. and Jones, R.H.: 1998, in J.J. Papike (ed.), *Planetary Materials, Rev. Mineral* **36**, The Mineralogical Society of America, Washington (DC), pp.1– 398.
Brownlee, D.E.: 1978, in T. Gehrels (ed.), *Protostars & Planets*, University Arizona Press (AZ), pp. 134–150.
Brownlee, D.E.: 1985, *Ann. Rev. Earth Planet. Sci.* **13**, 147–173.
Consolmagno, G.J. and Britt, D.T.: 1998, *Meteoritics Planet. Sci.* **33**, 1231–1241.
Ceplecha, Z., Borovicka, J., Elford, W.G., ReVelle, D.O., Hawkes, R.L., Porubcan, V. and Simek, M.: 1998, *Space Sci. Reviews* **84**, 327–471.
Corrigan, C.M., Zolensky, M.E., Dahl, J., Long, M., Weir, J., Sapp, C., and Burkett, P.J.: 1997, *Meteoritics Planet. Sci.* **32**, 509–515.
Engrand, C. and Maurette, M.: 1998, *Meteoritics Planet. Sci.* **33**, 565–580.
Flynn, G.J.: 1994, *Planet. Space Sci.* **42**, 1151–1161.
Flynn, G.J. and Sutton, S.R.: 1991, *Proc. Lunar Planet. Sci.* **21**, 541–547.
Flynn, G.J., Moore, L.B. and Klöck, W.: 1999, *Icarus* **142**, 97–105.
Fomenkova, M.N., Kerridge, J.F., Marti, K. and McFadden, L.–A.: 1992, *Science* **258**, 266–269.
Fraundorf, P., Hintz, C., Lowry, O., McKeegan, K.D., and Sandford, S.A.: 1982, *Lunar Planet. Sci.* **XIII**, 225–226.
Gombosi, T.I. and Houpis, H.L.F.: 1986, *Nature* **324**, 43–44.
Grossman, J.N., Rubin, A.L., Nagahara, H., and King, E.A.: 1988, in J.F. Kerridge and M.S. Matthews (eds.), *Meteorites and the Early Solar System*, University Arizona Press, Tucson (AZ), pp. 619–659.
Jenniskens, P., Betlem, H, Betlem, J., Barifaijo, E., Schlütter, T., Hampton, G., Laubenstein, M., Kunz, J., and Heusser, G.: 1994, *Meteoritics* **29**, 246–254.
Jessberger, E.K., Christoforidis, A., and Kissel, J.: 1988, *Nature* **332**, 691–695.
Jessberger, E.K, Stephan, T., Rost, D., Arndt, P., Maetz, M., Stadermann, F.J., Brownlee, D.E., Bradley, J., and Kurat, G.: 2000, in E. Grün, H. Fechtig and B. Gustafson (eds.), *Interplanetary Dust*, in press.

Kerridge, J.F. and Matthews, M.S. (eds.): 1988, *Meteorites and the Early Solar System*, University Arizona Press, Tucson (AZ), 1269+ xvii pp.
Langevin, Y., Kissel, J., Bertaux, J–L. and Chassefière, E.: 1987, *Astron. Astrophys.* **187**, 761–766.
Love, S.G. and Brownlee, D.E.: 1991, *Icarus* **89**, 26–43.
Love, S.G. and Brownlee, D.E.: 1993, *Science* **262**, 550–553.
Love, S.G., Joswiak, D. and Brownlee, D.E.: 1994, *Icarus* **111**, 227–236.
Mackinnon, I.D.R. and Rietmeijer, F.J.M.: 1987, *Rev. Geophys.* **25**, 1527–1553.
Mackinnon, I.D.R., Lindsay, C., Bradley, J.P., and Yatchmenoff, B.: 1987, *Meteoritics* **22**, 450–451.
MacPherson, G.J., Wark, D.A. and Armstrong, J.T.: 1988, in J.F. Kerridge and M.S. Matthews (eds.), *Meteorites and the Early Solar System*, University Arizona Press, Tucson (AZ), pp. 746–807.
Mason, B. (ed.): 1971, *Handbook of elemental abundances in meteorites*, Gordon and Breach Science Publishers, New York, Paris, London, 555 pp.
McSween, Jr., H.Y. and Richardson, S.M.: 1977, *Geochim. Cosmochim. Acta* **41**, 1145–1161.
Odderschede, L., Meibom, A., and Bohr, J.: 1998, *Europhys. Lett.* **43**, 598–604.
Papike, J.J. (ed.): 1998, *Planetary Materials, Reviews in Mineralogy, vol. 36*, The Mineralogical Society of America, Washington (DC), 1052pp.
Rietmeijer, F.J.M.: 1985, *Nature* **313**, 293–294.
Rietmeijer, F.J.M.: 1989, *Proc. 19th Lunar Planet. Sci. Conf.*, 513–521.
Rietmeijer, F.J.M.: 1993, *Earth Planet. Sci. Lett.* **117**, 609–617.
Rietmeijer, F.J.M.: 1995, *Lunar Planet. Sci.* **XXVI**, 1163–1164.
Rietmeijer, F.J.M.: 1996, *Meteoritics Planet. Sci.* **31**, 278–288.
Rietmeijer, F.J.M.: 1998a, in J.J. Papike (ed.), *Planetary Materials, Rev. Mineral., vol. 36*, The Mineralogical Society of America, Washington (DC), pp. 1–95.
Rietmeijer, F.J.M.: 1998b, in A.S. Marfunin (ed.), *Advanced Mineralogy, vol. 3*, Springer Verlag Berlin–Heidelberg, pp. 22–28.
Rietmeijer, F.J.M.: 1999a, *Lunar Planet. Sci.* **XXX**, CD ROM #1060.
Rietmeijer, F.J.M.: 1999b, *Astrophys. J.* **514**, L125–L127.
Rietmeijer, F.J.M.: 1999c, *37th Amer. Inst. Aeronautics Astronautics Meeting & Exhibit, paper #99–0502*, 12p.
Rietmeijer, F.J.M.: 2000a, *Meteoritics Planet. Sci.* **35**, 1025–1042.
Rietmeijer, F.J.M.: 2000b, *SPACE 2000 Proc. Conf. Amer. Soc. Civil Engineers*, 695–702.
Rietmeijer, F.J.M. and Blanford, G.E.: 1988, *J. Geophys. Res.* **93(B10)**, 11,943–11,948.
Rietmeijer, F.J.M. and Mackinnon, I.D.R.: 1987, in *Symposium on the Diversity and Similarity of comets, 6–9 April 1987, Brussels, Belgium*, ESA SP–278, pp. 363–367.
Rietmeijer, F.J.M. and Mackinnon, I.D.R.: 1997, in S. Chang (ed.), *Analysis of Returned Comet Nucleus Samples*, NASA Conf. Publ. **10152**, pp. 249–253.
Rietmeijer, F.J.M., Nuth, III, J.A. and Karner, J.M.: 1999, *Astrophys. J.* **527**, 395–404.
Sandford, S.A.: 1987, *Fundamentals of Cosmic Physics* **12**, 1–73.
Sirono, S. and Greenberg, J.M.: 2000, *Icarus* **145**, 230–238
Strait, M.M., Thomas, K.L., and McKay, D.S.: 1996, *Lunar Planet. Sci.* **XXVII**, 1285–1286.
Thomas, K.L., Blanford, G.E., Keller, L.P., Klöck, W., and McKay, D.S.: 1993, *Geochim Cosmochim Acta* **57**, 1551–1566.

Thomas, K.L., Blanford, G.E., Clemett, S.J., Flynn, G.J., Keller, L.P., Klöck, W., Maechling, C.R., McKay, D.S., Messenger, S., Nier, A.O., Schlutter, D.J., Sutton, S.R., Warren, J.L., and Zare, R.N.: 1995, *Geochim Cosmochim Acta* **59**, 2797–2815.
Weidenschilling, S.J: 1997, *Icarus* **127**, 290–306.
Weissman, P.R.: 1986, *Nature* **320**, 242–244.
Zinn, J., Wren, J., Whitaker, R., Szymanski, J., ReVelle, D.O., Priedhorsky, W., Hills, J., Gisler, G., Fletcher, S., Casperson, D., Bloch, J., Balsano, R., Armstrong, W.T., Akerlof, C., Kehoe, R., McKay, T., Lee, B., Kelley, M.C., Spalding, R.E., and Marshall, S.: 1999, *Meteoritics Planet. Sci.* **34**, 1007–1015.
Zolensky, M.E., Lindstrom, D.J., Thomas, K.L., Lindstrom, R.M., and Lindstrom, M.M.: 1989, *Lunar Planet. Sci.* **XX**, 1255–1256.
Zolensky, M.E., Wilson, T.L, Rietmeijer, F.J.M., and Flynn, G.J. (eds.): 1994a, *Analysis of Interplanetary Dust., Amer. Inst. Physics. Conf. Proc.* **310**, Amer. Inst. Physics, New York, (NY), 357 pp.
Zolensky, M.E., Hörz, F., See, T., Bernhard, R., Dardano, C., Barrett, R.A., Mack, K., Warren, J. and Kinard, W.H.: 1994b, in M.E. Zolensky, T.L. Wilson, F.J.M. Rietmeijer, and G.J. Flynn (eds.), *Analysis of Interplanetary Dust, Amer. Inst. Physics. Conf. Proc.* **310**, Amer. Inst. Physics, New York (NY), pp. 291–304.

COMPARISON OF 1998 AND 1999 LEONID LIGHT CURVE MORPHOLOGY AND METEOROID STRUCTURE

IAN S. MURRAY AND MARTIN BEECH
*Department of Physics, University of Regina, Regina, Saskatchewan,
CANADA S4S 0A2
E-mail: murray1i@mail.uregina.ca*

MICHAEL J. TAYLOR
Space Dynamics Laboratory, Utah State University, Logan, Utah 84322-4145, USA

PETER JENNISKENS
SETI Institute, NASA ARC, Mail Stop 239-4, Moffett Field, CA 94035, USA

and

ROBERT L. HAWKES
*Mount Allison University, Physics Department, 67 York St., Sackville, New Brunswick,
CANADA E4L 1E6*

(Received 27 June 2000; Accepted 14 July 2000)

Abstract. Photometric low-light level video observations of 1999 Leonid storm meteors have been obtained from two airborne platforms during the Leonid multi-instrument aircraft campaign (Leonid MAC). The 1999 Leonid light curves tend to be skewed towards the end point of the trajectory, while the 1998 Leonid light curves were not. The variation in the light curves from 1998 and 1999 can be explained as an overall reduction in the mass distribution index, α from ~ 1.95 in 1998 to ~ 1.75 in 1999. We have interpreted this behaviour as being either indicative of a gradual loss of the "glue" that keeps the grains together, or the fact that the meteoroids sampled in 1998 had a different morphological structure to those sampled in 1999. The early fragmentation of a dustball meteoroid results in a light curve that peaks sooner than that predicted by classical single body ablation theory.

Keywords: Comets, dust, Leonids 1999, lightcurves, meteoroids, meteors

Earth, Moon and Planets **82–83**: 351–367, 2000.
©2000 *Kluwer Academic Publishers. Printed in the Netherlands.*

1. Introduction

Until the various *in situ* cometary rendezvous, dust sample and return missions presently underway and planned are completed, we must infer the physical and chemical properties of large (> 50 micron) cometary meteoroids through their interactions with the Earth's atmosphere as meteors. This paper is concerned with the morphology of dust from comet 55P/Tempel-Tuttle, as it was measured from light curve characteristics during the 1998 and 1999 Leonid MAC (Jenniskens, 1999; Jenniskens *et al.*, 2000).

The light that constitutes the passage of a meteor is produced through collisions between ablated meteoric atoms and atmospheric molecules. In the case of small meteoroids the light produced is assumed to be proportional to the rate of change in its kinetic energy. Also, since the velocity remains approximately constant over the luminous trail, the light produced is proportional to the rate of change of meteoroid mass. As a consequence of the rapid atomic excitation and decay process, the light produced gives a direct indication of the instantaneous rate of ablation. The dimensions of the meteoroids studied in this work (typically hundreds of μm) are much less than the mean free path at the heights of atmospheric ablation, and therefore the interaction between the atmosphere and the meteoroid is essentially molecular, with no air cap or shock wave formation.

A solid, compact, non-fragmenting meteoroid will produce a classical light curve with the point of maximum luminosity appearing near the end of the trail (Cook, 1954; Öpik, 1958; McKinley, 1961). The classical light curve reflects both the exponential increase of air density with decreasing altitude and the accelerated decrease in meteoroid surface area at the end of the trajectory. Faint television meteors have been found to have light curves that are on average nearly symmetrical (Hawkes *et al.*, 1998; Fleming *et al.*, 1993; Campbell *et al.*, 1999; Murray *et al.*, 1999). In order to explain such behaviour, we will use the dustball meteoroid ablation model developed by Hawkes and Jones (1975). In this model meteoroids are pictured as a collection of silicate and metallic grains bonded together by a secondary low-boiling point "glue". In this paper we will consistently refer to the sub-units of a meteoroid as grains, and reserve the word meteoroid for the entire solid object. Several authors (Hapgood *et al.*, 1982; Beech, 1984; 1986) have successfully applied the Hawkes and Jones (1975) model to the Perseid,

Draconid, and α-Capricornid meteor showers. Here the dustball model has been expanded to include a description of the mass distribution of grains in a meteoroid. A distribution in the masses of the constituent grains results in an overall light curve that is broader and is earlier skewed than that produced by a classical light curve (Campbell, 1999).

During the 1998 Leonid Multi-Instrument Aircraft Campaign, we measured light curves of a sample of Leonid meteors and described the asymmetry of the light curves, many of which were skewed towards the early part of the trajectory (Murray et al., 1999). We have now measured a sample of light curves from the 1999 Leonid storm under similar conditions and find a quite different behaviour.

2. Experimental Observations

The experimental conditions in this study were similar to those during the 1998 Leonid MAC (Murray et al., 1999). Two co-aligned intensified cameras were pointed at a constant elevation angle of 75° out of a high window port on the FISTA (Flying Infrared Signature Technologies Aircraft). This time, however, both cameras were synchronised using an AC coupling feature of the individual camera systems; this allowed for accurate synchronous frame information to be obtained. Also, one of the two cameras, designated N, was equipped with a narrow band (bandwidth: 50% transmission at 8.96 nm and 10% at 13.03 nm) sodium filter centred at 589.50 nm. With peak transmittance of 64%, the filtered camera reached a limiting apparent stellar magnitude of approximately +4.0.

In parallel with these measurements, narrow field observations were made from the ARIA (Advanced Ranging Instrumentation Aircraft) using two co-aligned Xybion intensified video cameras mounted at an elevation of ~30° on the starboard side of the aircraft. One camera, type RG-350 fitted with a GEN III image intensifier, had a spectral range of ~350-900 nm and was fitted with a variety of filters during the night of the storm including two narrow band interference filters; one centered on the magnesium emission at ~520 nm and the other on the sodium emission at 589 nm. Both filters had a bandwidth of 10 nm (full-width at half maximum) and a peak transmission of ~50%. This imager was fitted with a 75 mm, f/1.4 lens (field of view 8° x 7°) and well over 200 meteors were recorded at various emission wavelengths of which 43 were imaged using the Na and Mg filters. To aid the interpretation of these data simultaneous "white-light" video observations were also made using a wider field of view (23° x 18°) GEN II intensified Xybion camera. However, this system developed an intermittent fault during the mission and the data are only suitable for pointing registration. Additional white-light observations were obtained with wide angle (39° x 28°) for cameras used for flux measurements (Jenniskens et al., 2000). Video imagery from ARIA was recorded onto NTSC standard Hi-8 tapes. A time-date signal was also added to the audio track of each videotape to enhance timing analysis studies.

3. Light curve symmetry

The data were analysed as before (Murray et al., 1999). The meteor images were calibrated to an apparent magnitude scale using a number of background stars. The peak brightness and integrated brightness were determined. The integrated brightness was then converted to photometric mass. The results are presented in Table I. Columns give the camera M photometric results. Maximum luminosity is the brightest recorded point expressed in astronomical magnitudes. The photometric mass is based upon an integration of the observed light curve and is expressed in kilograms. The skew parameter F is an average over the five computed values. The last three columns indicate if the beginning (B), maximum luminosity (M) and end (E) point of the meteor trail occurs in (1) or outside (0) the field of view.

All meteors in Table I are Leonids, determined by using a test for radiant match and angular velocity. Their photometric light curves are shown in Figure 1, and are normalised in intensity and duration to 'draw-out' their overall morphology. The light curves tend to have a fairly sharp rise to a rounded somewhat "flat toped", right-of-centre skewed maximum. This behaviour differs from our Leonid observations of 1998, when the light curves tended to be peak towards the beginning of the trajectory. We observed only a few exceptions to this "flat topped" or symmetrical nature. Two curves, for example, showed a double humped profile (see the 01:38:40 and 01:57:57 light curves in Figure 1). Two other light curves (see the 00:52:16 and 01:53:47 light curves in Figure 1) have characteristics that can be associated with nebulous meteors (LeBlanc et al., 2000).

TABLE I

Time (UT)	Maximum Luminosity (0^M)	Photometric Mass (kg)	$F_{average}$	B	M	E
00:04:07	3.2	1.1×10^{-6}	0.76	0	1	1
00:21:21	4.5	1.6×10^{-7}	0.51	1	1	1
00:47:39	4.4	2.9×10^{-7}	0.74	1	1	1
00:47:47	5.7	3.3×10^{-8}	0.74	0	1	1
00:52:16	3.5	1.1×10^{-6}	0.79	1	1	1
00:56:21	3.6	6.5×10^{-7}	0.37	1	1	1
01:07:36	3.9	3.8×10^{-7}	0.65	1	1	1
01:10:57	4.4	1.9×10^{-7}	0.57	1	1	1
01:17:28	5.8	2.9×10^{-8}	0.79	1	1	1
01:22:00	3.5	4.3×10^{-7}	0.58	1	1	0
01:29:00	3.6	4.8×10^{-7}	0.56	1	1	1
01:32:13	4.2	2.6×10^{-7}	0.57	1	1	1
01:37:01	4.4	1.5×10^{-7}	0.49	1	1	1
01:37:33	3.8	3.2×10^{-7}	0.52	0	1	1
01:38:40	5.7	6.1×10^{-8}	0.54	1	1	1
01:38:48	3.8	3.8×10^{-7}	0.63	0	1	1
01:41:38	3.8	4.0×10^{-7}	0.81	1	1	0
01:43:14	5.0	1.0×10^{-7}	0.68	1	1	1
01:45:57	3.9	2.8×10^{-7}	0.63	0	1	1
01:49:48	6.0	2.6×10^{-8}	0.76	0	1	1
01:51:03	4.2	1.5×10^{-7}	0.65	1	1	1
01:51:05	4.5	1.9×10^{-7}	0.84	1	1	1
01:53:47	4.1	5.2×10^{-7}	N.V.	1	1	0
01:54:36	2.7	1.4×10^{-6}	0.56	1	1	1
01:57:57	4.0	3.1×10^{-7}	0.71	1	1	0
01:58:59	4.0	2.8×10^{-7}	0.46	1	1	1
02:01:45	3.6	6.7×10^{-7}	0.72	1	1	1
02:10:38	2.9	1.1×10^{-6}	0.54	1	1	1
02:11:52	2.8	1.3×10^{-6}	0.69	1	1	1
02:12:49	3.0	9.5×10^{-7}	0.69	0	1	1
02:13:41	3.2	1.0×10^{-6}	0.59	1	1	1
02:15:46	4.2	1.5×10^{-7}	0.56	1	1	1
02:16:22	3.8	3.9×10^{-7}	0.58	1	1	1
02:18:20	6.1	2.3×10^{-8}	0.14	1	1	1
02:19:27	5.0	9.2×10^{-8}	0.51	1	1	1
02:21:29	4.8	7.9×10^{-8}	0.30	0	1	1
02:22:42	3.5	5.4×10^{-7}	0.57	1	1	0
02:25:50	3.7	4.9×10^{-7}	N.V.	1	1	1
02:32:22	3.6	4.1×10^{-7}	0.40	1	1	1
02:56:12	2.6	1.5×10^{-6}	0.67	1	1	1
02:56:27	3.5	6.5×10^{-7}	0.82	1	1	1
03:01:42	3.8	3.0×10^{-7}	0.73	1	1	1
03:09:59	4.3	1.8×10^{-7}	0.63	1	1	1
03:31:26	3.6	4.8×10^{-7}	0.42	0	1	1
04:07:10	5.5	4.8×10^{-8}	0.54	1	1	1
04:12:45	3.7	3.5×10^{-7}	0.71	1	1	1
04:21:59	3.6	5.4×10^{-7}	0.72	1	1	1
04:52:00	3.7	3.6×10^{-7}	0.55	1	1	1

356 MURRAY ET AL.

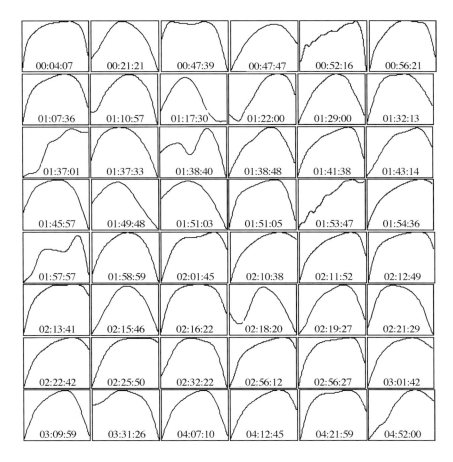

Figure 1. Photometric light curves for Leonid meteors observed with camera M. The vertical axis is a relative scale proportional to the peak magnitude and the horizontal axis is a relative time scale. The curves shown are smooth fits to the data points collected at time steps of 1/30th of a second (the standard video frame rate).

As before, we performed a statistical analysis of the light curve profiles. The F parameter listed in Table I essentially quantifies the nature of a light curve's skew (the relative position of maximum). The F-parameter is defined as the ratio of the distance to the point of maximum brightness to the entire length of the curve (see e.g., Fleming *et al.*, 1993; Murray *et al.*, 1999):

$$F_{\Delta m} = \frac{t_{B\Delta m} - t_{M\Delta m}}{t_{B\Delta m} - t_{E\Delta n}} \quad (1)$$

where t_{max} is the time of light curve maximum and $t_{B\Delta m}$ and $t_{E\Delta m}$ are the beginning and end times at which the brightness is Δm magnitudes fainter than the maximum. The F-values were calculated at magnitude intervals of Δm equal to 0.25, 0.50, 0.75, 1.0 and 1.25 fainter than maximum brightness. The F-values where then averaged to give the values listed in Table I.

TABLE II

	Camera System	$F^*_{0.25}$	$F_{0.50}$	$F_{0.75}$	$F_{1.00}$	$F_{1.25}$
Mean	50mm 1999	0.58	0.61	0.62	0.63	0.61
Std. Dev.		0.17	0.15	0.15	0.14	0.15
Mean	50 mm 1998	0.46	0.45	0.48	0.47	0.49
Std. Dev.		0.20	0.18	0.14	0.15	0.14

*) Mean F-values for the Leonid light curves sampled in 1998 (penultimate row) and 1999 (first row).

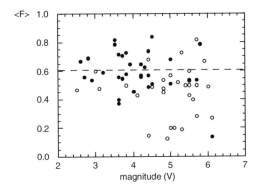

Figure 2. The F-parameter as a function of peak brightness of the meteors. Dashed line shows the mean F value for a classical light curve while (o) and (•) designate 1998 and 1999 data points.

The result of averaging the F-values for all of the sampled light curves (Table II) is a mean of 0.61, essentially the value expected for a classical light curve with a peak occurring towards the end of the trajectory. A perfectly symmetric light curve will have an F-value of 0.5 for all Δm; a light curve with an early maximum will have F < 0.5; a late maximum will have F > 0.5. However, this result is somewhat misleading since the

F-parameter fails to account for the 'flat-topped' nature of the light curves and in general the light curves of Figure 1 are not well represented by the classical curve. It is important to note, however, that the 1998 Leonid meteors showed a skew towards the early part of the trajectory, with a mean F of 0.47 (Murray *et al.*, 1999).

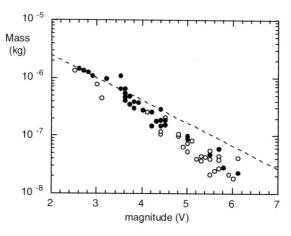

Figure 3. The integrated light curve as reflected in the photometric mass calculation versus the peak brightness. Here the dashed line is the expected relationship, while (o) and (•) designate 1998 and 1999 data points respectively.

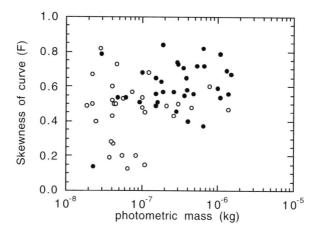

Figure 4. The F-parameter as a function of photometric mass. Where (o) and (•) designate 1998 and 1999 data points.

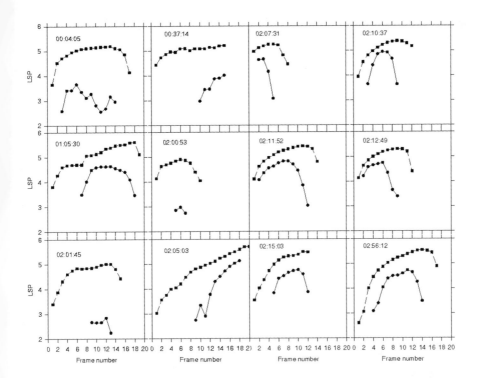

Figure 5. Comparison of the sodium filtered and unfiltered light curves for 12 Leonid meteors. In each diagram the Na light curve is always the lower of the two displayed. In this diagram the vertical axis corresponds to the log sum pixel (LSP) count and the horizontal axis is a relative time scale showing individual sequential video frames. Each frame is separated by a time interval of $1/30^{th}$ of a second. The filled circles and squares are individual data points.

Figure 2 shows the skew (averaged F-values) of both 1998 and 1999 light curves as a function of the peak brightness. A trend towards lower F values for the 1998 light curves is apparent in Figure 2. Seven light curves from 1998 have $F < 0.4$, while only 2 from 1999 are smaller than 0.4. Four light curves from 1998 have $F > 0.6$ while 14 of the 1999 light curves have averaged F-values larger than 0.6. The difference in the light curves also shows up in the calculated photometric masses. For a given peak brightness, the 1998 meteors have a smaller integrated intensity (photometric mass) as shown in Figure 3. The dashed line in Figure 3 shows the expected relationship if all the meteor light curves are identical and a constant fraction of the kinetic energy is transformed into light (this assumption implies $\log M \sim 0.4 m_v$). When plotting the F-parameter as a function of photometric mass (Figure 4), we notice that neither data set shows a mass-dependence. However, the 1999 data are mostly displaced to higher F-values and higher photometric masses.

4. Sodium Filtered Light Curves

An important clue to the fragmentation and ablation properties of the Leonid meteoroids is differential ablation, where one mineral evaporates earlier than another. Borovicka *et al.* (1999) have argued that several of the fireballs observed during the November 1998 Leonid display exhibited evidence for the early depletion of sodium. Building upon this result, we have extended the analysis of sodium depletion to fainter Leonid meteors.

Our procedure has been to compare the light curves of simultaneously observed meteors. That is, we have obtained the simultaneous light curves of Leonid meteors at visual wavelengths and through a narrow band sodium filter. A total of 12 Leonid meteors imaged from FISTA proved to be bright enough (brighter than the approximately +4 limiting magnitude of the sodium filter camera) for a comparative analysis. Accurate timing information and the synchronised nature of the two camera systems allowed for frame-by-frame comparisons of the unfiltered (visual) and filtered images. This capability allows for a straightforward log sum pixel comparison of the observed light curves.
The visual and Na light curves are shown in Figure 5. Note that the 00:37:14 and 02:05:03 events show only the beginning part of the meteor. Much variability is observed between individual Na light curves.

We seemingly confirm the discovery by Borovicka *et al.* (1999) that sodium is depleted before all ablation has stopped. Typically the sodium light curve falls off sooner than its visual counterpart. Also, we notice that the sodium curves have a later onset than the visual light curves. This suggests the Na-rich phases are not part of the meteoroid "glue" but belong to the grain component.

The relative intensity of sodium and white light emission varies considerably from one light curve to the next. From Figure 5 it appears that the stronger sodium signal is observed for light curves with the steepest increasing slope, or the highest F value. This result, however, will require further study and modelling.

An initial investigation of the sodium and magnesium light curves recorded with the ARIA filtered camera was performed to determine their overall morphology. As first indicated by Borovicka *et al.* (1999) one might expect to find systematic differences between the light curves observed through sodium and magnesium filters. This situation can arise because sodium and magnesium reside in different host phases in IDPs (Rietmeijer, 1998; 1999; 2000) and meteorites (Papike, 1998). To date 14 complete (5 magnesium filtered and 9 sodium filtered) Leonid light curves have been examined (see the scaled sequence of light curves in Figure 6).

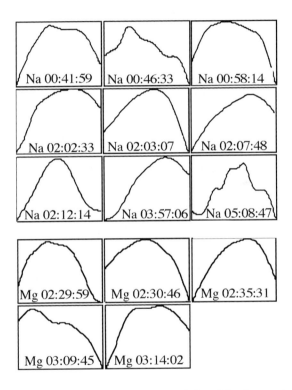

Figure 6. Nine calibrated filtered Na (top) and five Mg (bottom) light curves from the ARIA observations. As with the earlier presented light curves, they are plotted with relative scales to show overall morphologies.

We find that, in general, the morphology of the filtered light curves is similar to those shown by the non-filtered observations (see Figures 5). Some sodium light curves show significant intensity variations along the meteor trajectory. Although low number statistics prevail, F-values were computed for the ARIA filtered light curves at log sum pixel intervals equivalent to apparent magnitude intervals of Δm = 0.2, 0.4, 0.6, 0.8 and 1.0. Mean values for the five intervals considered result in an average F-value of 0.55 for the magnesium filtered light curves and an average F-value of 0.50 for the sodium filtered light curves.

5. Dustball model calculations

Since we have at present no clear knowledge of how Leonid meteoroids are 'constructed', we shall assume for modelling purposes a mass distribution for which the number of constituent grains, with masses between m and m + dm, is proportional to $m^{-\alpha}$, where α is the mass distribution index. In this fashion, by fixing upper and lower mass limits on the constituent grains present, we may 'build-up' any given initial mass Leonid meteoroid by adding together the appropriate numbers of component grains. The modelling discussed here considers twelve fundamental grain masses; the largest grains have masses of 10^{-7} kg, while the smallest grains have masses of 5×10^{-13} kg. The mass range adopted for the fundamental grains is somewhat arbitrary, but reflects the range of grain masses deduced from observed flare events in large, bright meteors (Smith, 1954; Simonenko, 1969; Campbell *et al.*, 1999). In general we note, that the smaller the value of the mass distribution index the greater the relative number of 'large' grains; the larger the mass distribution index the greater the relative number of 'small' mass grains.

The dustball model that we have constructed follows the classical, single-body ablation of individual grains and determines the variation in light intensity as a function of initial mass and atmospheric height. We assume that all of the grains have the same 'generic', that is stone-like composition as specified by Fyfe and Hawkes (1986) and used by Campbell *et al.*, (1999). The single-body ablation equations for mass loss and deceleration are solved numerically with atmospheric density being interpolated from the MSIS-E-90 Earth atmosphere model. The instantaneous luminous intensity, *I*, produced by a small single-body meteoroid is assumed to be proportional to the rate of change of its kinetic energy. The overall light curve, however, is synthesised by combining the number-weighted intensities of the constituent grains as a function of atmospheric height. Figure 7 shows a series of synthesised light curves for a 10^{-6} kg Leonid meteoroid (V = 71 km/s and a zenith angle of 45 degrees). The curves are labelled according to the assumed mass distribution index (α). It can be seen that 'flat topped' light curves require a mass distribution index $\alpha \sim 1.8$.

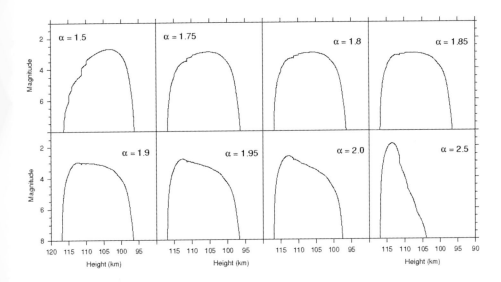

Figure 7: Synthesised light curves for a 10^{-6} kg Leonid meteoroids for a range of grain size distributions.

For small values of $\alpha < 1.5$ the synthesised light curve approaches the classical light curve of the largest mass grain. Likewise the synthesised light curves constructed with $\alpha > 2.5$ would approach that of the classical light curve of the lowest mass grain. For $1.5 < \alpha < 2.5$, a range of variable light curve morphologies are realised. We see, for example, that when $\alpha \sim 1.5$ the light curve has a near linear increase to its maximum and is late skewed. For $\alpha \sim 1.85$ the light curve is very nearly 'flat topped' and symmetrical about the maximum. For $\alpha > 1.9$ the light curve has an early peak and displays a near linear decrease of magnitude with time after the maximum. We see also from Figure 7 that there is a systematic shift in the height of the light curve maximum as α increases from 1.5 to 2.5. For $\alpha = 1.5$ the maximum is at 103.4 km altitude, at $\alpha = 2.5$ the maximum is at 113.4 km altitude (assuming a zenith angle of 45 degrees). The change in height of the light curve maximum, as a function of the mass distribution index, is at its most dramatic for $1.8 < \alpha < 1.9$. In this range the height of maximum increases by some 7 kilometres. In principle it appears that one may be able to construct a monotonic relationship between the mass distribution index and the height of maximum for a given initial meteoroid mass – we intend to pursue the development of this idea in a subsequent publication.

Table III compares the classical single-body meteoroid of mass 10^{-6} kg, with those for several dustball meteoroids constructed with various mass-distribution indices. The F-values for a classical, single-body ablation light curve are given for comparison in the last row of the Table. We acknowledge that the F-parameter does characterise the basic change in symmetry of a light curve but note that it offers little physical information about the meteoroid. In future it will be desirable to phase out the use of the F-parameter and substitute a more physically meaningful morphological parameter (e.g. Campbell et al., 1999). Still, the observed variation in the light curves from 1998 and 1999 Leonid meteor storms can be explained as an overall reduction in α from ~ 1.95 in 1998 to ~ 1.75 in 1999.

TABLE III

α	$\Delta m = 0.25$	$\Delta m = 0.50$	$\Delta m = 0.75$	$\Delta m = 1.00$
1.50	0.56	0.58	0.59	0.59
1.85	0.53	0.50	0.50	0.48
2.00	0.55	0.37	0.30	0.26
2.50	0.52	0.53	0.53	0.54
Classical	0.58	0.61	0.63	0.65

A few of the Leonid light curves observed in 1999 appear to be 'odd'. For example the double peaked, or 'humped' light curves (e.g. the 01:38:40 and 01:57:57 light curves in Figure 1) do not fit the single mass distribution index scheme described above. We do not believe, however, that the second, brighter maximum shown by these meteors is due to flaring or fragmentation, but find that the light curves may be readily described as a dustball with an additional 'massive' grain. Figure 8 illustrates, by way of example, a synthesised light curve for the 01:57:57 light curve shown in Figure 1. We find the light curve is reasonably well described by a 2.4×10^{-7} kg dust ball constructed with $\alpha = 1.85$, and an additional single grain of mass between 2 to 2.5×10^{-7} kg.

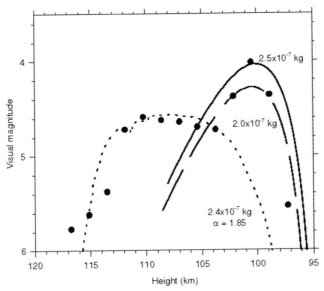

Figure 8. Composite light curve model for the 'humped' Leonid meteor observed at 01:57:54 UT, 1999 November 17. The dots correspond to the derived magnitudes. The dashed line shows the synthesised light curve for a 2.4 × 10^{-7} kg meteoroid with $\alpha = 1.85$. The solid and broken-solid lines correspond to the classical light curves for 2.0 and 2.5 × 10^{-7} kg meteoroids.

6. Discussion

The apparently higher mass distribution index α for the constituent grains of the 1998 Leonid meteoroids is highly interesting. One reason for this apparent enhancement could be that the 1998 meteoroids have a significantly different ejection age than the meteoroids sampled in other years. While the 1999 Leonid shower was predominantly composed of debris ejected in 1899 (McNaught and Asher, 1999), the material sampled in 1998 was ejected from comet 55P/Tempel-Tuttle several perihelion passages before 1899 (Asher *et al.*, 1999; Jenniskens and Betlem, 2000). If older grains dominate the 1998 Leonid sample set it would imply that meteoroids become progressively more fragmented over time during their exposure to the interplanetary environment. This effect may be mitigated through repeated heating and cooling episodes associated with returns to perihelion, and possibly the "glue" that holds the grains together may be gradually lost or made less sticky by UV photoprocessing. Likewise, the "glue" could be modified through irradiation by energetic solar wind and solar flare nuclei, or both. It is highly probable that the 'glue' component responsible for bonding a dustball meteoroids grains together is organic in nature and compositionally it maybe similar to the CHON particles detected during the GIOTTO spacecraft encounter with comet 1P/Halley (see e.g., Jessberger *et al.*, 1988). The organic component probably produces very little luminosity in the visual and near-visual regions of the electromagnetic spectrum, and consequently it might well remain undetected with the typical equipment used to record 'visual' light curves. The height distributions of meteors observed with radar, however, have recently been interpreted by Steel (1998) and Elford *et al.* (1997) as being indicative of the fact that most meteoroids must have a high, heavy organic make-up. The low-boiling point of the constituent organic compounds makes the meteoroids susceptible to on-going fragmentation as they descend through the Earth's upper atmosphere.

Acknowledgements

We extend our many thanks to Joe Kristl who has un-begrudgingly helped with many tape-editing and copying tasks. The help and guidance of Mr. D. Prohor of the Faculty of Film and Video at the University of Regina is gratefully acknowledged. Many thanks to Mr. D. Kolybaba and Mr. K. Wolbaum of the Faculty of Science at the University of Regina for their technical assistance. Special thanks go to Andrew LeBlanc for his helpful suggestions and comments. We also gratefully acknowledge and thank the 452nd Flight Test Squadron at Edwards Airforce Base for their logistical support during the 1999 Leo MAC and the many people who have made this mission possible. This work was partially supported with funds provided by the National Science and Engineering Research Council of Canada, the Space Dynamics Laboratory at Utah State University and the US National Science Foundation grant No ATM-9612810. Leonid MAC is supported by the NASA Planetary Astronomy and Exobiology programs, the NASA Advanced Missions and Technologies Program for Astrobiology (AAMAT), and USAF/XOR. *Editorial handling*: Frans Rietmeijer.

References

Asher, D.J., Bailey, M.E., and Emelyanenko, V.V.: 1999, *MNRAS* **304**, L53–L56.
Beech, M.: 1984, *MNRAS* **211**, 617–620.
Beech, M.: 1986, *Astron. J.* **91**, 159–162.
Borovicka, J., Stork, R., and Bocek, J.: 1999, *Meteoritics Planet. Sci.* **34**, 987–994.
Campbell, M.D., Hawkes, R.L., and Babcock, D.D.: 1999, in V. Porubcan and W.J. Baggaley (eds.), *Meteoroids 1998*, Slovak Academy of Sciences, Bratislava, Slovakia, 363–366.
Cook, A.F: 1954, *Ap. J.* **120**, 572–577.
Elford, W.G., Steel, D.I. and Taylor, A.D.: 1997, *Adv. Space. Res.* **20**, 1501–1504.
Fleming, D.E.B., Hawkes, R.L. and Jones, J.: 1993, in J. Stohl and I.P. Williams (eds.), *Meteoroids and Their Parent Bodies,* Slovak Academy of Sciences, Bratislava, Slovakia, 261–264.
Fyfe, J.D.D. and Hawkes, R.L: 1986, *Planet. Space. Sci.* **34**, 1201–1212.
Hapgood, M., Rothwell, P., and Royvrik, O.: 1982, *MNRAS* **201**, 569–577.
Hawkes, R.L. and Jones, J.: 1975, *MNRAS* **173**, 569-577.
Hawkes, R.L., Campbell, M.D., Babcock, D., and Brown, P.: 1998, D. Lynch, et al. (eds.), *Proc. 1998 AIAA Leonid Threat Conf.* 11 pp.
Jenniskens, P.: 1999 *Meteoritics Planet. Sci.* **34**, 959–968.
Jenniskens, P. and Betlem, H.: 2000, *Ap. J.* **531**, 1161–1167.
Jenniskens, P., Butow, S. and Fonda, M.: 2000, *Earth, Moon and Planets* **82-83**, 1–26.
Jessberger, E. K., Christorforidis, A., and Kissel, J.: 1988,*Nature* **322**, 691–695.
LeBlanc, A.G., Murray, I.S., Hawkes, R.L., Worden, P., Campbell, M.D., Brown, P., Jenniskens, P., Correll, R.R., Montague, T. and Babcock, D.D: 2000,*MNRAS* **313**, L9–L13.
McNaugth R.H. and Asher D.J.: 1999, *WGN, Journal of the IMO* **27**, 85–102.
Murray, I.S., Hawkes, R.L., and Jenniskens, P.: 1999, *Meteoritics Planet. Sci.* **34**, 949–958.
McKinley, D.W.R.: 1961, *Meteor Science and Engineering.* McGraw-Hill, New York, 308 pp.
Opik, E.J.: 1958, *Physics of meteor flight in the Atmosphere.* Interscience, New York., p. 63–68.
Papike J.J (ed.), 1998; In *Planetary Materials, Reviews in Mineralogy* **36**, 1039 pp., The Mineralogical Society of America, Washington, DC, 1052 pp.
Rietmeijer, F.J.M.: 1998. in A.S. Marfunin (ed.) *Advanced Mineralogy,* **3**, Springer Verlag Berlin-Heidelberg, 22p.
Rietmeijer, F.J.M.: 1999, *Ap. J.* **514**, L125–L127.
Rietmeijer, F.J.M.: 2000, *Planetary Space Science,* in press.
Simonenko, A.N.: 1969, *Astron. Vest.* **3**, 26–35.
Smith, H.J.:1954, *Ap. J.* **119**, 438–442.
Steel, D.I.: 1998, *Astron. Geophys.* **39**, 24–26.

FIRST RESULTS OF HIGH-DEFINITION TV SPECTROSCOPIC OBSERVATIONS OF THE 1999 LEONID METEOR SHOWER

SHINSUKE ABE
The Graduate University for Advanced Studies,
National Astronomical Observatory,
2-21-1 Oosawa, Mitaka, Tokyo 181–8588, Japan
E-mail: avell@pub.mtk.nao.ac.jp

HAJIME YANO
Planetary Science Division,
Institute of Space and Astronautical Science, Kanagawa 229–8510, Japan
E-mail: yano@planeta.sci.isas.ac.jp

NOBORU EBIZUKA
The Institute of Physical and Chemical Research, Saitama 351–0198, Japan
E-mail: ebizuka@atlas.riken.go.jp

JUN-ICHI WATANABE
Public Relations Center,
National Astronomical Observatory, Tokyo 181–8588, Japan
E-mail: jun.watanabe@nao.ac.jp

(Received 13 July 2000; Accepted 2 September 2000)

Abstract. During the 1999 Leonids, an intensified HDTV camera was used for slit-less meteor spectroscopy at visible and near-UV wavelengths in three night flights of the Leonid Multi-instrument Aircraft Campaign. The HDTV system provided a high dynamic range (digital 10-bit) and a wide field of view of 37 x 21 degrees. The maximum spectral sensitivity is at 405 nm, while the resolving power of the spectrograph was $\lambda/\Delta\lambda \sim 250$. Here, we report on the results for one particularly nice spectrum from a Leonid meteor at 03:24:40 UT. Numerous atomic emission lines of magnesium and iron were detected, with an excitation temperature of T = 5,000 ± 1,000. It is confirmed that sodium is released from the meteoroid earlier than iron and magnesium, just as in meteors of the 1998 Leonid shower. Rovibrational bands of N_2 dominate the spectrum in the visible and near-IR. We searched for CN band emission at 389 nm, but could not detect it due to the overlap of numerous iron lines.

Keywords: Comet 55P/Tempel-Tuttle, high definition TV, interplanetary dust, Leonids 1999, meteor spectroscopy, meteors

1. Introduction

Meteors are the luminous ablation of dust particles from parent comets that enter the Earth's atmosphere at a few tens of kilometers per second. They are heated, vaporised, and even partially ionized in the upper atmosphere. Meteor emission originates from a mixture of atoms and molecules ablated from the meteoroid itself as well as from the surrounding air. The Leonid meteor shower is one of the most interesting meteor showers and have occurred roughly every 33 years at least in the last 100 years recorded in the history. This corresponds to the orbital period of the comet, 55P/Tempel–Tuttle. Comets are the surviving bodies since the genesis of our solar system. They are thought to be remnants of planetesimals at the edge of the protoplanetary disk that could not grow into planets. Through detailed meteor observations and analysis of their interaction with the Earth's atmosphere, physical and chemical properties of cometary meteoroids can be studied. Spectroscopic observations of the flash heating and evaporation reveal not only the chemical composition of the interplanetary dust but also emission processes of hypervelocity impacts in the atmosphere, which are difficult to reproduce in laboratory experiments at present.

Millman et al. (1971) first observed meteor spectra by television techniques. TV observations can record fainter meteors with higher temporal resolution than photographic spectroscopic techniques. Most early papers were concerned with line identification. In recent years, J. Borovicka et al. (Borovicka, 1995; Borovicka and Bocek, 1993; Borovicka et al., 1999) have applied this technique to quantitative analysis of spectroscopy of meteors and their persistent trains. Typically, the intensified video data are recorded in an 8-bit($2^8 = 256$) analogue video system in PAL or NTSC format.

Here, we report on the first such spectroscopic observations using Intensified High-Definition TV. Earlier, HDTV imaging was used to monitor the prominent activity of the 1998 Giacobinid meteor shower, associated with comet 21P/Giacobini-Zinner (Watanabe et al., 1999). This technique increases the number of TV lines from about 576(NTSC) to 1150, and has higher 10-bit($2^{10} = 1024$) dynamic range, 4 times higher than previous video systems. For a given field of view, the system is more sensitive than conventional intensified CCD cameras. Meteors as faint as 8th magnitude and stars of 10th magnitude can routinely be observed even with a wide $37° \times 21°$ field of view.

Unlike the 1998 Leonid MAC, when we operated both cameras in imaging mode only for flux measurements and stereoscopic observations (Jenniskens and Butow, 1999), this year we operated one of the cameras part time as a visible - near-UV spectrometer. At the time of writing, only one Leonid spectrum could be digitized from the original HDTV format due to a lack of access to convergence systems. In this report, we will describe briefly the analysis of this spectrum and explore the performance of the system.

Figure 1. Composite of meteor images recorded from ARIA at the peak of the 1999 Leonid storm.

2. Observations

Our spectroscopic observations were performed with the "blue-sensitive" intensified HDTV camera equipped with a transmission grating with 600 grooves per mm, blazed at 550 nm, made by Jobin Yvon. We operated two HDTV cameras provided by the Japanese Broadcasting Company (NHK) that were mounted on both the "FISTA" and "ARIA" aircraft in NASA's 1999 Leonid Multi-instrument Aircraft Campaign (Leonid MAC) (Jenniskens et al., 2000a). Each HDTV camera was composed of a large diameter image intensifier along with a blue sensitive 1-inch 2M-pixel FIT CCD, which has a resolution of 1150 TV lines (Sunasaki et al., 1997; Yamazaki et al., 1998). Figure 1 shows a composite of images from the ARIA camera around the time of the peak.

During flight, the grating was rotated to set the dispersion direction perpendicular to the projected direction of Leonid meteor trajectories on the sky. The camera was mounted in one of the high $\sim 61°$ ports of the FISTA aircraft. This system not only recorded intrinsically faint meteors, but also had a wide field of view of 37 x 21 degrees by using a f50mm F1.0 photographic lens. The resulting dispersion was 0.59 nm/pixel in the first order. That is a resolving power of 250 at 590 nm. The images were recorded in 10-bit monochrome frames at a rate of 33 ms, i.e. 30 frames per second. The limiting magnitude for meteor spectra corresponded to the first order meteor of about +5th apparent visual magnitude.

The spectroscopic observations were carried out intermittently for a total of 297 minutes in the nights of Nov. 16–19, 1999, consisting of 87,

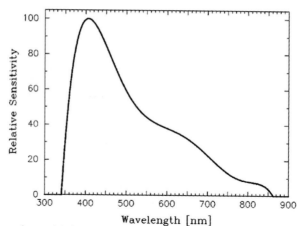

Figure 2. Spectral sensitivity curve of the intensified HDTV camera operated on FISTA.

46 and 164 minutes respectively for the 1st, 2nd and 3rd nights. In total, 105 meteor spectra were recorded, of which 76 were Leonids, 11 Taurids and 18 sporadics. Most of the Leonid spectra recorded in the night of Nov. 17/18 are meteors belong to the 1999 Leonid meteor storm. The meteor studied here appeared at 03:24:40 UT, in the Lorentzian tail of the meteor storm (Jenniskens *et al.*, 2000b).

3. Data archives and Analysis

Selected segments of the HD-CAM tapes were digitized at the studio facilities of Keisoku Giken Co., Ltd., in Yokohama, Japan. The converted frames are 1920 x 1035 pixels 10-bit monochrome files. After the conversion, the image reductions were conducted by using the astronomical image processing software named IRAF(released by National Optical Astronomy Observatories). Both a sky frame subtraction and flat-fielding are applied to the raw data. Flat frames were taken from inside of the aircraft by filming a diffuse white board outside the aircraft before each flight. Also, the same frames obtained before and after the meteor spectrum were used for sky frames.

The effective spectral sensitivity curve of the instrument was constructed by measuring a number of spectra of bright main-sequence stars in the observing field. The sensitivity curve is shown in Figure 2. The effective spectral sensitivity covers the range of 360 - 850 nm, with the maximum sensitivity at 405 nm.

The two dimensional images (Figure 3) were reduced to one dimensional spectra by a shift-and-add task, which was repeated to integrate

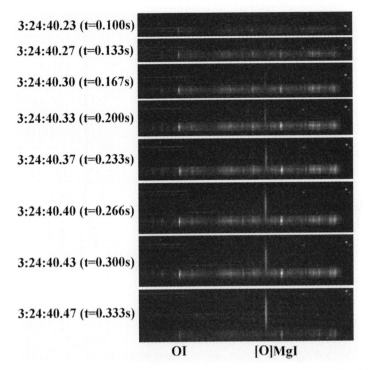

Figure 3. Time sequence (top to bottom) of spectra recorded from Leonid meteor 03:24:40 UT at a rate of 30 frames/s.

the emission signals along the meteor trajectory and to achieve the highest signal-to-noise ratio (S/N). Typically, the spectrum in each frame conisisted of about 10 TV lines. Once the lines were aligned, the combined spectrum was extracted along a trace.

A deconvolution procedure was applied for improving the spectral resolution. The convolution function was based on the point spread function of standard stars in the same frame. The resulting spectral resolution was obtained 2.2 nm in FWHM(Full Width at Half Maximum), which improves over typical video observations by a factor of 4 (Borovicka et al., 1999).

4. Results and Discussion

Figure 3 shows a a set of time series spectra between 3:24:40.17 and 3:24:40.53 UT on November 18, 1999. The time resolution is 0.033 seconds. The zeroth order meteor image disappears from the field of view. The dispersing direction is from right(short-wavelength) to left(long-wavelength). The forbidden OI line at 557.7 nm is delayed be-

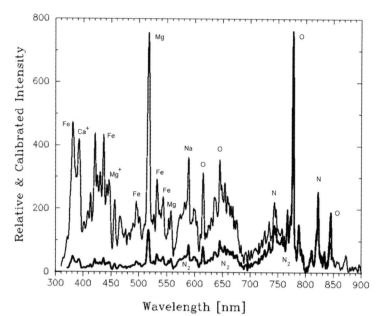

Figure 4. The identification of spectral lines in the 03:24:40 UT Leonid at t = 0.266s. Thin line shows the observational spectrum after a deconvolution procedure. The thick line shows the spectrum after calibrated by the spectral sensitivity of the observed system.

hind the meteor image (Halliday, 1958, Millman, 1962). The spectrum covers the first order image up to 720 nm and an overlap of 1st and 2nd order images above that.

The spectra consist of numerous overlapping emission lines. Figure 4 shows the spectrum taken at 3:24:40.30 UT, at time t = 0.167s along the meteor trajectory (Figure 3). The thin line is the spectrum as it was observed. The thick line shows the spectrum after correction for the spectral sensitivity of the system (Figure 2). About 60 emission lines are seen in the 360 - 850 nm wavelength range.

Once the wavelength and FWHM of each line are determined, the temperature and abundances of each atom can be calculated by assuming We adopted the simplex method for separation of blended emission lines (Nelder and Mead, 1965). The simplex method makes it possible to determine the wavelength and FWHM of each emission line based on an assumed spectral profile. We adopted a Gaussian curve rather than Lorentzian for the fitting function, with free parameters as intensity, position and width of each line.

Once the wavelength and FWHM of each line are determined, the temperature and abundances of each atom can be calculated by assum-

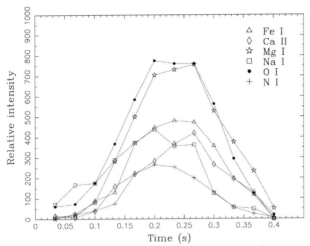

Figure 5. Relative intensity curves of the spectral emissions from six atomic species in the 03:24:40 UT Leonid.

ing local thermodynamic equilibrium(LTE). It is an assumption that all energy levels are populated by collisional and radiative processes which were characterized by a Maxwellian velocity distribution for all radiating particles. Relative intensities obtained by calibrated spectrum allow us to estimate the excitation temperature. The identified lines imply an excitation temperature of about $T = 5,000 \pm 1,000$. This excited temperature is calculated mainly from intensities ratio of the Mg and Fe lines and applies to the mean intensities along the meteor trajectory shown in Figure 3.

Figure 5 shows the relative intensities of the most dominant emissions. Iron and magnesium follow a similar ablation profile than atmospheric oxygen emission. However, the atomic sodium emission is depleted significantly earlier than the other metal atom emissions. This effect was earlier observed for the 1998 Leonid meteors, and is now confirmed for the 1999 Leonids. This confirmation is significant in light of results by Murray *et al.* (2000), that the light curves for 1998 and 1999 Leonids differed significantly, with the 1999 Leonids being less finely fragmented. Further study should reveil whether the sodium release from the 1999 Leonids is significantly different from those of 1998.

There are many broadband emissions from molecular excitations. The near-IR and visible region is dominated by molecular emission from first positive bands of N_2. Other broad features are simply blends of atomic lines.

Figure 6 shows the magnified spectrum (uncorrected data) of the Leonid meteor in the 360–410 nm wavelength region. This part of the

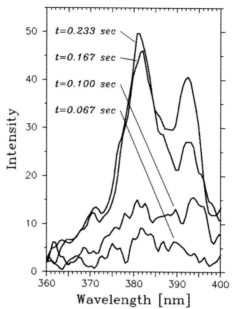

Figure 6. The 03:24:40 UT Leonid spectrum in near-ultraviolet region. These four spectra show a time profile of one Leonid meteor. The two strong features are caused by neutral FeI, MgI at 382, 383 nm and ionised CaII at 392 and 395 nm. The lower spectra were observed at the beginning of meteor ablation, with relatively weak CaII lines. The upper spectra show meteor spectra at the lower altitudes.

spectrum is dominated by lines of iron, causing the broad features at 382 and CaII, causing the band at 392–395 nm. Our HDTV camera has a high sensitivity in the near-UV region and thus can be used to search for the band head of CN at 388 nm, of particular interest to astrobiology (Jenniskens and Butow, 1999). CN emission was not detected significantly above the blends of metal atom lines, not even at the higher altitudes where metal atom lines are less dominant. A comparisson with theoretical line spectra is needed to bring out any emission hidden in the blended atomic lines.

5. Conclusions

The large pixel format of the HDTV camera makes it a very suitable instrument for time-resolved slit-less spectroscopy of meteors. The large number of resolution elements helps to resolve lines in the meteor spectrum and still be able to compare spectral line intensities at wavelengths far apart. In addition, the large dynamic range improves the measurement of the line intensity ratios for lines that are much different in intensity. In the near future, we expect to have our own system for digitization of the HDTV images, which will make a more comprehensive analysis of the present data set possible.

Acknowledgements

The 1999 Leonid Multi-Instrument Aircraft Campaign was made possible by grants from the NASA Exobiology, Planetary Astronomy, and Astrobiology Advanced Missions and Technology Programs, NASA Ames Research Center, and the U.S. Air Force. The authors thank NHK for kindly providing the HDTV-II system to make this observation possible. We appreciate cooperation with Keisoku Giken Co., Ltd. for digitization of the HDTV images. We also thank Dr. Y. Hirahara and Dr. J. Takahashi for their reviews of this paper. SA was supported by a travel grant of the National Astronomical Observatory of Japan. *Editorial handling:* Peter Jenniskens.

References

Borovička, J.: 1995, *Earth, Moon, and Planets* **71**, 237–244.
Borovička, J., and Bocek, J.: 1993, *Astron. Astrophys.* **279**, 627–645.
Borovička, J., Stork, R., and Bocek, J.: 1999, *Meteoritics Planet. Sci.* **34**, 987–994.
Halliday, I.: 1958, *ApJ* **128**, 441–443.
Jenniskens, P. and Butow, S. J.: 1999, *Meteoritics Planet. Sci.* **34**, 933–943.
Jenniskens, P., Butow, S. J., and Fonda, M.: 2000a, *Earth, Moon and Planets* **82–83**, 1–26.
Jenniskens, P., Crawford, C., Butow, S.J., Nugent, D., Koop, M., Holman, D., Houston, J., Jobse, K., Kronk, G., and Beatty, K.: 2000b, Earth, Moon and Planets **82–83**, 191–208.
Millman, P. M.: 1962, *J. RAS. Canada* **56**, 263–267.
Millman, P. E., Cook A. F., and Hemenway C. L.: 1971, *Canadian J. Phys.* **49**, 1365–1373.
Murray, I.S., Beech, M., Taylor, M.J., Jenniskens, P., and Hawkes, R.L.: 2000, Earth, Moon and Planets **82–83**, 351–368.
Nelder, J. A. and Mead, R.: 1965, *The Computer J.* **6**, 163–168.
Sunasaki, S., Yamazaki, J., Okuura, T., Takenouchi, K., Suzuki, H., and Nakamura, H.: 1997, 'A Large Diameter I.I. CCD Camera for HD TV Use', *ITE Technical Report* **21**, 7–12 (in Japanese).
Yamazaki, J., Majima, K., Yamaguchi, S., and Shirotake, K.: 1998, 'Development of an Ultra-high-sensitivity HD TV I.I. Color Camera and Shooting Heavenly Bodies', *ITE Technical Report* **22**, 35–40 (in Japanese).
Watanabe, J., Abe, S., Takanashi, M., Hashimoto, T., Iiyama, O., Ishibashi, Y., Morishige, K., and Yokogaw, S.: 1999, *Geophysical Research Letters* **26**, 1117–1120.

JET-LIKE STRUCTURES AND WAKE IN Mg I (518 nm) IMAGES OF 1999 LEONID STORM METEORS

MICHAEL J. TAYLOR AND LARRY C. GARDNER
Space Dynamics Laboratory and Physics Department, Utah State University, Logan, UT 84322-4145
E-mail: mtaylor@cc.usu.edu

IAN S. MURRAY
Department of Physics, University of Regina, Regina, Saskatchewan, Canada S4S 0A2
E-mail: murray1i@mail.uregina.ca

and

PETER JENNISKENS
SETI Institute, NASA Ames Research Center, Mail Stop 239-4, Moffett Field, CA, 94035-1000
E-mail:pjenniskens@mail.arc.nasa.gov

(Received 13 July 2000; Accepted 15 August 2000)

Abstract. Small meteoric fragments are ejected at significant transverse velocities from some (up to ~8%) fast Leonid meteors. We reach this conclusion using low light intensified image measurements obtained during the 1999 Leonid Multi-Instrument Aircraft Campaign. High spatial resolution, narrow band image measurements of the Mg I emission at 518 nm have been used to clearly identify jet-like features in the meteor head that are the same as first observed in white light by LeBlanc et al. (1999). We postulate that these unusual structures are caused by tiny meteoroid fragments (containing metallic grains) being rapidly ejected away from the core meteoroid as the constituent glue evaporates. Marked curvature observed in the jet-like filaments suggest that the parent meteoroids are spinning and as the whirling fragments are knocked away by the impinging air molecules, or by grain-grain collisions in the fragment ensemble, they ablate quickly generating an extended area of structured luminosity up to about 1-2 km from the meteoroid center. Fragments with smaller transverse velocity components are thought to be responsible for the associated beading evident in the wake of these unusual Leonid meteors.

Keywords: Fragmentation, jet-like, Leonids 1999, meteoroids, meteors, structures, wake

1. Introduction

Classical meteor ablation theories consider meteoroids to be tiny droplets or single bodies that sometimes include a "fudge factor" to account for fragmentation (e.g., Ceplecha *et al.*, 1998). Fragmentation is an elusive feature of meteor ablation that is responsible for long wakes observed behind some meteors (e.g., Jacchia *et al.*, 1950) that can dominate the shape of the light curves (Hawkes and Jones, 1975; Campbell *et al.*, 1999). Recently, LeBlanc *et al.* (2000) have reported observations of short-lived jet-like features emanating from a single bright meteor imaged during the 1998 Leonid meteor shower. The measurements were made in Mongolia on November 16, 1998 using an unfiltered, intensified CCD imager that had a field of view of ~33° horizontal by 25° vertical and an angular resolution of ~3 arc min. Six jets were apparent in their processed data (five video frames), the length and orientation of which varied from frame to frame. Although several possible explanations for this unusual observation were discussed no firm conclusions could be drawn at that time. In the same paper LeBlanc *et al.*, also present white light images showing a diffuse, nebulous glow associated with one high altitude (~138 km) Leonid meteor imaged during the 1998 Leonid Multi-Instrument Aircraft Campaign. In this case the camera had a smaller field of view of 9.5° by 7.3° (angular resolution 0.9 arc min) and clearly showed a diffuse glow surrounding the meteor and extending out to a range of ~600 m (Murray *et al.*, 1999; Jenniskens and Butow, 1999).

In this paper we present new high resolution (~0.85 arc min) image data recorded during the '99 Leonid Multi-Instrument Aircraft Campaign. The measurements were obtained using a narrow field, intensified imager that was filtered to observe light from neutral magnesium (Mg) emission at 518 nm. The high spatial resolution and narrow bandwidth (~10 nm) of these data were expected to enhance significantly the optical measurements of structure and nebulosity of the meteors during ablation and to help confirm or deny the early reports of jet-like structures.

Initial analysis of the raw images clearly shows well-defined, jet-like features, associated with several of the meteors imaged during the Leonid storm night that evolved systematically from video frame to frame as the meteors ablated. These observations favour one particular explanation, namely the ablation of tiny meteoric fragments. Further signs of fragmentation are the detection of beading in the wake of these (and other) meteors.

2. Instrumentation and Observations

The 1999 Leonid MAC mission consisted of two instrumented B707-type aircraft: the FISTA (Flying Infrared Signature Technologies Aircraft) and the ARIA (Advanced Ranging and Instrumentation Aircraft). Each aircraft was fitted with a diverse array of optical instrumentation designed to investigate the Leonids shower characteristics in exceptional detail (Jenniskens et al., 2000). An extensive mission was flown from the USA to the Middle East (Israel) and back during the period November 13–21, 1999. Both the FISTA and the ARIA flew along parallel paths at an altitude of about 11 km over the Mediterranean Sea from Israel to the Azores during the night of the Leonid shower maximum.

The Utah State University instrumentation consisted of four low light TV cameras (two mounted on each aircraft), designed to study the dynamics of meteor ablation, primarily at two metal atom (magnesium and sodium) emission wavelengths, and to perform a novel investigation of longitudinal variability in the near infra-red (NIR) hydroxyl nightglow emission. The nightglow measurements were made mainly from the FISTA aircraft using two co-aligned imagers: a Gen III Xybion camera and an InGaAs camera (spectral ranges 710-850 nm and 1,100-1,600 nm respectively). However, for this study the primary meteor observations were made from the ARIA aircraft, where two Xybion intensified cameras were mounted together at an ~30° elevation, starboard window, and co-aligned to measure the meteor emission morphology and ablation signatures at selected wavelengths in the visible and NIR spectrum.

Previous spectral studies during the 1998 Leonids shower indicated strong magnesium and sodium emission from meteors as well as from their persistent trains (Borovicka et al., 1999; Abe et al., 2000). Hence, one CCD camera, type RG-350 (756 x 484 pixel array) and equipped with a Gen III image intensifier (spectral range ~350-900 nm), was fitted with a range of filters during the storm night including two narrow band interference filters: one centered on the magnesium emission at ~520 nm and the other on the sodium emission at ~589 nm. Both interference filters had a bandwidth of ~10 nm (full width at half maximum) and a peak transmission of ~50%. This imager was fitted with a 74 mm, f/1.4 lens resulting in a field of view of approximately 8° horizontal by 6° vertical. Video data were recorded onto NTSC standard Hi-8 tapes and the overall system resolution was estimated to be ~560 x 410 lines yielding an angular resolution of 0.85 arc min. This corresponds to a spatial resolution of ~50 meters at the distance of the meteors (~210 km), assuming a mean Leonid height of 105 km altitude.

To aid the interpretation of these data simultaneous "white-light" video observations were also made using a wider field (23° horizontal by 18° vertical) Gen II intensified Xybion camera fitted with a 25 mm, f/1.4 lens. Unfortunately, this system developed an intermittent fault during the mission night and the data are currently only suitable for pointing registration. A time-date signal (accurate to one ms) was also added to each video frame for the narrow field imager to enable easy identification of individual meteor signatures and for detailed comparative studies with other instruments onboard, in particular the Japanese high-definition TV measurements. In due course, it is hoped that these data will provide important additional information on the height distribution of the nebulous, and jet-like meteors discussed here, and to help model their characteristics.

3. Results

Over a hundred meteors were recorded by the narrow field camera at Mg and Na emission wavelengths during the storm night. Estimates of the magnitudes of these events have yet to be made but analysis of white-light imagery from the FISTA aircraft suggest that many of the meteors exhibited peak magnitudes in the range +4 to +5 (Murray et al., 2000). Visual inspection of the raw images has revealed unusual features in several of the meteors recorded in both the Mg and Na filtered data. In particular non-uniformity or "beading" of the light emanating from the meteor trails, and nebulosity with associated transverse "jet-like" features in the meteor head emission appears to be present in up to ~8% of the filtered data. However, further detailed image processing may well increase this estimate.

3.1 STRUCTURE IN METEOR WAKE

Figure 1 shows an example of bead-like structure in the wake of a Leonid meteor imaged in Mg emission at 00:37:15.893 UT. The meteor entered the field of view from the upper-right of the image (~0.4 sec earlier). For clarity the figure shows an enlarged region (~5.6° x 3.4°) of the camera field of view. The meteor trail is well developed and exhibits several persistent luminous "beads" of light separated by depleted regions of luminosity extending along its entire length. Analysis of the video data indicates that these features can endure for several frames. Similar structure in the meteor wake is visually evident in several, but certainly not all, of our filtered meteor image data.

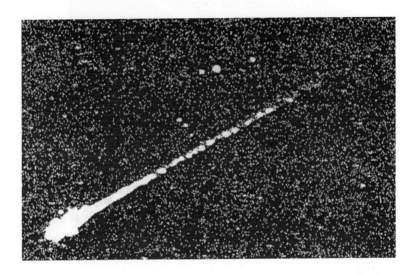

Figure 1. Wake in the path of a Leonid meteor imaged in the MgI emission at 00:37:15 UT showing "beading".

3.2 JET-LIKE FEATURES

Our analysis, to date, has revealed several Leonid meteor events exhibiting "jet-like" features as described by LeBlanc *et al.* (1999). The meteor image of Figure 1 is one such example. This event is shown again in Figure 2 as four consecutive images (each separated by ~33 ms) as the meteor transited across our field of view. The meteor was first detected at 00:37:15.508 UT (as it entered the upper-right of the camera field) and already exhibited a significant transverse spread or glow. This nebulosity was observed to increase in size around the meteor "head" until 00:37:15.659 UT when it reached its maximum extent (~ 1.8 km wide) and jet-like features became evident. The four images of Figure 2 show the evolution of these jets within the luminous region after the meteor had attained maximum brightness.

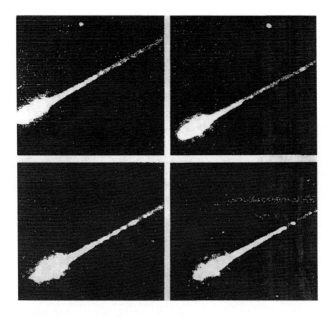

Figure 2. Sequence of images showing meteor "jets" in the Mg I emission at 00:37:15 UT. Note the non-uniformity of the emission and the large jet on the lower side of the meteor image.

About 10 jet-like features can be seen in the Mg emission surrounding the meteor. The jets are not aligned in any one direction and appear to tilt both forward and backwards, as reported by LeBlanc *et al.* (2000). Some jets also appear to have detached from the main transverse spread region and appear as tiny diffuse patches. In particular, one jet (present on the lower side of the meteor images) is most pronounced and extended over several pixels transverse to the meteor motion. This jet was present in several frames suggesting it was a relatively long-lived feature (>0.2 sec) and was still evident as the meteor passed out of our field of view. In comparison, the jets observed by LeBlanc *et al.* were evident only in a single frame of video (i.e. <33 ms in duration).

Figure 3 shows a second example of jet-like features. The meteor was first imaged at 03:08:48.722 UT as it entered the top-right of our camera field. At this time it appeared as a faint dot of light which proceeded to develop into a well-formed nebulous meteor as it grew in luminosity. The two images show a ~50% enlargement of the meteor "head" (265 ms later) as it transited the lower half of our field of view. The top image shows several very well developed jets originating within the luminous region and extending well outside the nebulous area of transverse spread. Unfortunately the image reproduction does not show the details (or contrasts) evident in the raw data. Nevertheless, the jets are prominent and striking in their appearance and over 10 filaments are seen curving both forwards and backwards in an apparently systematic fashion suggesting that the meteor is rotating. The lower figure (33 ms later) shows that the luminous region has decreased significantly in area but the same jet-like features are still prominent. This meteor was somewhat fainter than that of Figure 2 (magnitude not yet estimated), but it still exhibited prominent jets.

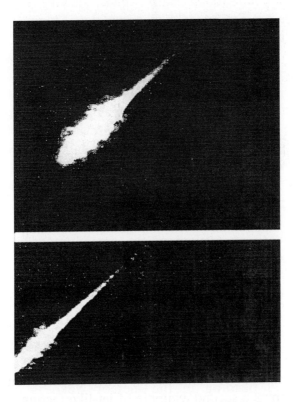

Figure 3. As Figure 2 but for Leonid meteor imaged at 03:08:48 UT. Note, the individual jets are seen to surround the meteor and are initially aligned almost perpendicular to the meteor track but then appear to curve both forwards and backwards suggesting that the meteor is spinning.

Estimates of the scale size of the jets are in good agreement with those determined by LeBlanc et al. who found lengths up to ~ 2km. For example, from the tip of the jets to the center of the meteor trajectory, we measure lengths of typically 0.5–1.0 km. This size scale may be relatively independent of meteor size as the jet-like event reported by LeBlanc et al. was one of the brightest they recorded during the 1998 Leonids shower, much brighter than the meteors observed here.

One obvious question concerning these measurements is whether the observed jet-like features are real and not simply an artifact of registering a high-speed meteor using an intensified camera system. This question has been addressed by LeBlanc et al. (albeit for a different camera system) who found that by rapidly slewing their camera across a bright object, e.g. Saturn (+0.8 magnitude) and Jupiter (–2.3 magnitude), to simulate a meteor signal, they could under certain conditions induce some streaking into the video data. However, they concluded that these features were only evident in the Jupiter data (which is much brighter than the Leonid meteor signals) and that the induced streaks were short and straight and were all directed away from the brightest point. In contrast, the jet-like features reported here and by LeBlanc et al. (1999) were more numerous and exhibited clear transverse type structuring at a variety of orientations. Furthermore, our data show marked curvature in some of the jet filaments that appears to be regular indicating that the meteor was probably spinning rapidly. Although we have yet to perform similar tests it is difficult to imagine how such complex structuring could be attributed to an artifact of the camera/recording system alone. Finally, as jets have been found only in a fraction of the Leonid meteors imaged (~8 %) and do not appear to be associated only with bright events there can be little doubt of their authenticity.

4. Discussion

The observed beading in the meteor wake is a strong sign of fragmentation of the meteoroid. Such fragmentation is consistent with the "dustball" model developed by Hawkes and Jones (1975) which pictures the meteoroid as a collection of tiny silicate and metallic grains bonded together by a secondary component of low boiling point "glue", possibly of organic origin (see also Murray et al., 2000). Smaller fragments released as the glue evaporates are quickly slowed down by the impinging air molecules and will lag the main mass thereby creating the observed beaded trail.

The presence of marked jet-like features in some Mg-filtered images that move with the ablating meteoroid indicates that they are not due to excited atmospheric emissions such as the N_2 first positive band emission (which has a spectral signature within the pass band of the Mg filter). Rather, the data show that it is the ablating material itself that is responsible for these optical structures. A fundamental question concerning the detection of jet-like features associated with the Leonids meteors is that the meteoroids are much smaller than the mean free path (~1 m) at the heights at which the majority of the Leonids ablate in the atmosphere (around 105 km). The interaction between the atmosphere and the meteoroid is therefore expected to be essentially molecular, with no air cap or shock waves generated. Thus, the size of the luminous region should be quite small (a few meters in diameter) (LeBlanc et al., 1999; Boyd, 2000). However, the jet-like structures can clearly surpass this region by a factor of 100.

These data strongly support the concept that the jets may be the signature of plasma effects caused by small fragments containing metallic grains explosively spinning away from the central (rotating) body at speeds as large as 15–30 km/s (perpendicular to the meteor trajectory), as the glue binding them evaporates rapidly. Once the fragments are free from the parent meteoroid, they ablate rapidly in the ambient air causing the whirl-like distributions evident in the filtered emissions. The net effect is deposition of ablated material over a much wider region surrounding the meteor and possibly a higher efficiency of aerothermochemistry than implied by single-body models.

Recent studies indicate that the smallest silicate sub-units in the Leonid meteoroids are about a micron (Rietmeijer and Nuth, 2000) or sub-micron (Greenberg, 2000) in size. However, the fragments responsible for the jets must be significantly larger in order to account for their observed luminosity. Campbell et al. (1999) and Murray et al. (2000) have considered meteoroid breakup into sub-units to explain the unusual light curve of cometary meteors. Campbell et al. *(1999)* find grain sizes from 10^{-6} to 10^{-12} kg necessary to account for the observed light curves of Leonid and Perseid meteors. In comparison, Murray, et al., have measured light curves for Leonid meteors, similar to the ones studied here, indicating sub-units covering the mass range 5×10^{-13} to 10^{-7} kg (which is equivalent to grain sizes in the range 1–60 microns assuming a mean density of 1 g/cm^3). If the mass-luminosity equation of Jacchia et *al.* (1967) applies, i.e. log M (kg) = –4.16 – 0.44 mv, then these sub-units would be expected to produce white light in the range +6.5 to +18.5 magnitude. As mentioned earlier we have not yet estimated the magnitude of the parent meteors. However, the contrast between the jets and the peak intensity of the meteors is clearly large. Assuming a limiting value for the contrast of about 100 (i.e. 5 magnitudes), for a peak white light meteor brightness of +1, the fragment would be about +6 magnitudes which corresponds to ~50 microns in size. This estimate, however, does not take into account the explosive nature of the ablation that appears to be typical for these small fragments. Indeed, their sudden release from the meteoroid in high-densities, often with relative velocities > 72 km/s, may cause significantly higher peak luminosities. Hence, the mass estimate would tend towards an upper limit. This said, these data are consistent with the range of Leonid particle sizes that have been modeled using light curve analysis, and provide independent support for one aspect of the proposed mechanism.

In summary, there can be little doubt that the unusual nebulosity and jet-like features that we have imaged during the 1999 Leonids meteor storm are identical in morphology to those recently reported by LeBlanc *et al.* (2000). These new data confirm and extend on their measurements and suggest that this unexpected structuring, which is present in a significant fraction (up to ~8%) of Leonid meteors observed, may in part, be due to explosive ejection of meteoroid fragments possibly in association with rapid rotation. Further, more detailed studies of the jet-like features will be used to provide independent information on the typical grain sizes of meteoroid fragments and the kinetics of fragmentation. It is hoped that these observations will help stimulate new theoretical and modelling studies on the high altitude (typically >100 km) ablation characteristics of fast Leonid meteors.

Acknowledgements

We gratefully acknowledge the constructive reviews by Bob Hawkes and an anonymous referee which have significantly enhanced the presentation of this paper. We are most indebted to R. Sturz of Xybion Electronic Systems Corp. for detailed technical assistance and the generous loan of two Gen III intensified cameras used for these measurements, and to J. Kristl and colleagues at Stewart Radiance Laboratory for their considerable help with systems integration and support throughout the mission. We thank the 452nd Flight Test Squadron at Edwards Air Force Base, the FISTA and ARIA aircrews, and the NASA logistic support staff for their dedication towards the mission and tireless help. Support for the USU image measurements was provided, in part, by the National Science Foundation Grant # ATM-9612810 and by the Space Dynamics Laboratory, Utah State University. The Leonid MAC was supported by NASA's Exobiology, Planetary Astronomy and Suborbital MITM programs, by NASA's Advanced Missions and Technologies program for Astrobiology and by the NASA Ames Research Center. *Editorial handling:* Mark Fonda.

References

Abe, S., Ebizuka, N., and Watanabe, J.: 2000, *Meteoritics Planet. Sci.*, in press.
Borovicka, J., Stork, R., and Bocek, J.: 1999, *Meteoritics Planet. Sci.* **34**, 987–994.
Boyd, I.D.: 2000, *Earth, Moon and Planets* **82–83**, 93–108.
Campbell, M., Hawkes, R.L., and Babcock, D.: 1999, in W.J. Baggaley and V. Porubcan (eds.), *Meteoroids 1998,.*, Astronomical Institute, Slovak Academy of Sciences, Bratislava, 363–366.
Ceplecha, Z., Borovicka, J., Elford, G. W., Revelle, D., Hawkes, R., Porubcan, V., and Simek, M.: 1998, *Space Science Reviews* **84**, 327–471.
Greenberg, J.M.: 2000, *Earth, Moon and Planets* **82–83**, 313–324.
Hawkes, R. L. and Jones, J.: 1975, *MNRAS* **173**, 339–356.
Jacchia, L.G., Kopal, Z., and Millman, P.M.: 1950, *Astrophys. J.* **111**, 104–133.
Jacchia, L.G., Verniani, F., and Briggs, R.E.: 1967, *Smits. Contr. to Astrophys.* **10**, 1–139.
Jenniskens, P. and Butow, S.J.: 1999, Meteoritics Planet. Sci. **34**, 933–943.
Jenniskens P, Butow, S., and Fonda, M.: 2000, *Earth, Moon and Planets* **82–83**, 1–26.
LeBlanc, A. G., Murray, I. S., Hawkes, R. L., Worden, P., Campbell, M. D., Brown, P., Jenniskens, P., Correll, R. R., Montague, T., and Babcock, D. D.: 2000, *MNRAS* **313**, L9–L13.
Murray, I.S., Hawkes, R.L., and Jenniskens, P., 1999: *Meteoritics Planet. Sci.* **34**, 945–947.
Murray, I.S., Beech, M., Taylor, M.J., Jenniskens, P., and Hawkes, R.L.: 2000, *Earth, Moon and Planets* **82–83**, 351–368.
Rietmeijer, F.J.M. and Nuth, J.A.: 2000, *Earth, Moon and Planets* **82–83**, 325–350.

GROUND-BASED LEONID IMAGING IN THE UV

ELCHANAN ALMOZNINO
Wise Observatory & School of Physics and Astronomy,
Raymond and Beverly Sackler Faculty of Exact Sciences,
Tel-Aviv University
E-mail: nan@wise.tau.ac.il

JEREMY M. TOPAZ
El-Op Ltd., Remote Sensing Operation, POB 1165, Rehovot, Israel 76111.
E-mail: jeremsis@netvision.net.il

(Received 29 May 2000; Accepted 07 Aug 2000)

Abstract. During the 1999 Leonid meteor storm, a camera with a UV sensitive image intensifier and CCD readout (ICCD), was operated at the Wise Observatory in Mitzpe Ramon, Israel. The photocathode spectral response, together with the transmittance of the atmosphere, limited the sensitivity to a narrow band around 320 nm. The aim was to obtain quantitative information on emitted radiance in this band, for comparison with recordings of the same tracks in other wavelengths. The field of view of $7°.2 \times 5°.7$ was pointed to the East at about $60°$ elevation and the video recorded from 21 until 03 UT, Nov. 18, 1999. At least five clear images of meteor tracks were recorded, and compared with the signals from known stars. It was concluded that the brightest track gave an estimated monochromatic magnitude at 320 nm peaking at ~ -4. Some of the peculiarities of astronomical observation with a photon-counting imager are discussed.

Keywords: Leonids 1999, meteor, meteor physics, photon counting imager, ultraviolet

1. Introduction

Since meteor trails are in essence an atmospheric phenomenon, it is impossible, by definition, to detect them at wavebands in which the atmosphere is opaque. However, in order to expand our knowledge we can try and look in the near UV regime, longer than ~ 300 nm where the atmosphere is barely transmitting.

The spectrum of meteor trails consists mainly of atomic spectral lines of various elements, either neutral or ionized, including heavy metals. Although information about this spectrum was gathered in the past in the visible range (e.g. Borovicka 1994, 1996) very little is known about wavelengths shorter than 350 nm. In normal cases the spectral lines are expected to dominate the radiation in the UV. In case of large particles, however, a hot plasma in the meteoroid head heats the surrounding air to a temperature of up to $\sim 10,000$ degrees, which may

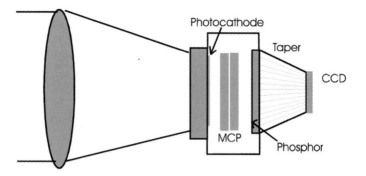

Figure 1. Schematic View of the Intensified CCD Solar Blind UV Camera.

increase the UV continuum considerably. It is interesting, therefore, to have UV observations of meteor trails.

Atmospheric and instrumental constraints make it practically impossible to perform any spectroscopic measurements in the UV waveband due to a very weak signal. UV imaging, on the other hand, is more practical, and this is what was attempted in this work.

2. Observations

For the observations we used an ICCD with a UV so-called 'solar blind' photocathode manufactured by Delft Electronic Products (DEP) of Roden, The Netherlands, that was on loan to Ofil Ltd., which is developing a UV camera for more earthbound pursuits. The schematic view of this camera is shown in Figure 1.

In order to find the effective spectral response of the camera operating on the ground, we took into account the DEP Rubidium Telluride 'Solar Blind' photo-cathode quantum efficiency, together with the atmospheric transmittance. The DEP detector has a tail of sensitivity out to about 360 nm, while for the atmospheric calculation we used the MODTRAN atmospheric transmittance program with the conditions of Mitzpe Ramon. It transpired that the transmittance in this wavelength region was not negligible, so the effective efficiency peaks at 320 nm. This can be seen in Figure 2.

The camera was set up at the Wise observatory in the night of Nov. 17/18, 1999. It recorded from 23:10 UT continuously until about 3 UT, except for a short break to change the tape at 02:10 UT. The field of view was $7°.2 \times 5°.7$. The MCP gain of this camera (which is controlled by varying the voltage on the MCP) can be set to adjust the visibility of the 'blobs' appearing on the screen. The setting was a compromise

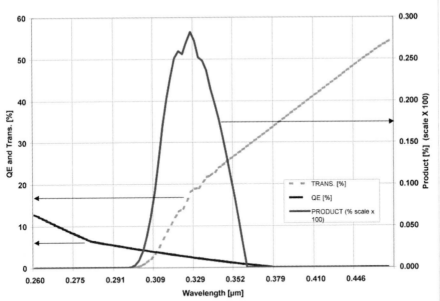

Figure 2. Atmospheric transmittance in Mitzpe-Ramon, the quantum efficiency of the camera, and their product yielding the effective efficiency of the system. The scale of the efficiency is enlarged by 100 and is marked on the right side axis.

between having enough signal to be clear above the dark noise of the CCD and not being so high as to saturate the detector and prevent quantitative measurement.

In spite of the difficulties of detecting meteor tracks, five of these have been located, mostly around the peak of the Leonid activity. Unfortunately, no optical images are available. The camera was pointed in the same general direction (Eastwards at approx. 60 deg. elevation) as one of the CCD's of the Canadian team (Brown et al., 2000), but the detected meteors were not recorded by the Canadian cameras because of small differences in pointing (Hawkes, priv. correspondence).

3. Data Reduction and Analysis

The difference between imaging with a photon counting image intensified CCD camera and ordinary imaging with a CCD camera is very significant.

A UV ICCD can be, and in our case was, operated in the photon counting mode. This means that for each photon which is collected by the photocathode we obtain an 'event' which is then amplified to a visible 'blob'. A strong star may give one or a few blobs every frame of

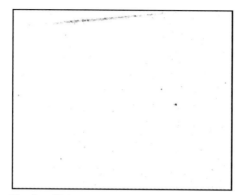

Figure 3. Complete Track of Meteor #3. Combination of 11 Video Frames.

the TV (1/25 or 1/30 of a second, depending on which TV standard is used). However, in a given frame, a weak star, or a single photon of scattered light from ambient sources, can produce a blob which is equally visible. Since there are many stars and many scattering particles, there are random blobs in the frame, additional to those from stars or meteors. The meteors which were found were quite visible because of the track of their blobs.

Figure 3 shows the track of the strongest meteor we recorded, which we have arbitrarily named #3. The signals of the track in individual frames were reduced in the following manner: First, the noise was removed by substracting a certain threshold from the images and considering only positive pixels. Then, every frame was divided to its two fields – odd lines and even lines. Every field has an exposure time of 40 ms but is only 20 ms apart from the next field taken, so the fields are interlaced in time with each field overlapping with half of the previous field and half of the next one. This was seen in the images where the meteor spot is shifted between the two fields of the same frame. We have then cut the meteor image of each field in two halves representing the beginning and end of the exposure time of that field. The end-half of one field was merged with the beginning-half of the next field to give a full image (odd + even) of a time interval of 20 ms. This way better time resolution and signal is obtained.

A strip containing the meteor was cut out of each such processed frame and put together one below the other to illustrate the full movement of the meteor in time. This is shown in Figure 4. One can see there also noise blobs in the images.

As explained earlier, each of these meteor images is a stack of blobs piled up during the exposure of the frame. It is worthwhile, therefore, to calibrate the intensity of the meteor in relation to the number of

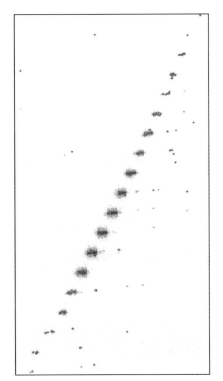

Figure 4. Composite of slices from video frames of meteor #3.

blobs it contains. For this, all the random blobs in the images were measured and found to have a mean of 2680 ± 890 counts, where 890 is the standard deviation from the mean. The counts in each meteor image were summed up and divided by this number to yield the UV intensity of the meteor in each frame in units of 'blobs per frame'.

In order to obtain absolute calibration one needs to measure known stars in the images. This was difficult because of the low throughput of the system in this waveband. In order to obtain significant signal from stars a total of 1845 frames were reduced. One field in each frame was automatically searched for existing blobs and a list of blobs (x and y positions) was prepared for each field. These were done in several groups of about 240 frames each. When one sums up all the blobs in each group and plots them as points according to their locations, one or two stars are visible as concentration of points. These stars change their position from one group to the other since the camera was fixed and did not track the sky. According to this movement one can correct the blob location in all the lists to match that of the first group. Then, all the corrected blob locations are considered together to construct a

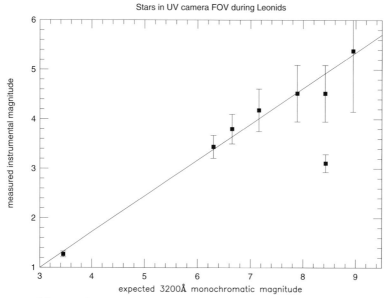

Figure 5. Measured instrumental magnitude vs. the predicted UV magnitude of the stars on the images. The linear best-fit used for calibration is shown.

synthetic image, each blob giving one 'count' in its position. We have detected eight stars in this image, and by comparing them to a sky map which fits the exact time, date and location of the observation, we could identify them.

The next step was measuring these stars in the image as a normal aperture photometry of a CCD frame, taking into account the 'sky' background. This gives for each star the number of blobs it produces per 1845 frames. The error is calculated as a normal photon-statistics error with a gain of 1 photon per count. The results were divided by 1845 to convert to the standard unit of blobs per frame, in which the meteor was measured. The counts of the brightest star (Tania Borealis), however, were divided by 1400 instead of 1895 since it was not present on all the frames as it moved out of the field in the last two groups of frames.

All these stars appear in the Hipparcos input catalog, and have spectral classification as well as V and B magnitudes. In order to find their UV magnitude we have used an algorithm developed at the Wise Observatory (Shemi *et al.* 1993). This algorithm predicts the UV magnitude of stars based on their optical parameters, together with statistical UV properties of stars measured by the IUE satellite. It was checked to have an uncertainty from 0.3 mag. for bright blue stars and up to 4 mags. for very faint red stars. We have run this program for the

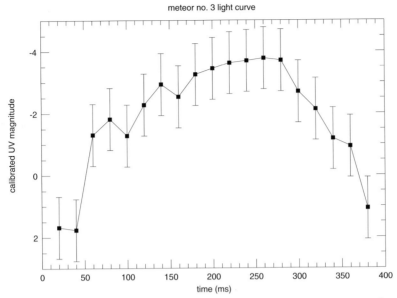

Figure 6. Calibrated UV monochromatic magnitude of meteor #3 as a function of time.

band in question of 320 nm and obtained the predicted UV magnitudes of the observed stars.

The measured intensity of the stars was converted into an instrumental magnitude as $M_{inst} = -2.5 log(I)$ where the intensity (I) is in blobs per frame. In Figure 4 we show the instrumental magnitude vs. the predicted UV magnuitude of the stars. It is seen there is a very good agreement between these values if we disregard the one point that deviates considerably from the rest. This point turns out to be a bright red giant of type M0, the spectrum of which drops considerably in the UV, so the UV prediction algorithm may be very inaccurate due to a large color term across the bandpass.

A linear best-fit was derived for these magnitudes (Figure 5). It can be seen that the slope of this fit is not unity. This is because of the noise removal in the images, which reduces the intensity of fainter objects more significantly than bright ones. Using this fit the UV monochromatic magnitudes of the meteor blobs were calculated. The error of magnitudes obtained from the fit is about 0.3 mag. while the error of the measured stars gives an additional 0.3 mag. The error of measurement of the Leonid blobs is around 0.4 mag., taking into account both the non-uniformity of the blobs and the photon-statistics of the meteor counts. We can therefore realistically estimate the errors of UV monochromatic magnitude of the meteor to be of ± 1 mag.

The result of calibrated UV monochromatic magnitudes of the meteor is shown in Figure 6, where the time difference between each field is 20ms. It can be seen that the meteor trail peaks at about -4 mag. indicating this is a very bright meteor or fireball.

This is the first time a meteor trail has been measured in the UV 300–350 nm band. Unfortunately, no correlation with broadband visible emission could be made. However, the technique has been demonstrated and can be improved for future meteor observations.

Acknowledgements

We would like to thank Dr. Malka Lindner of Ofil Ltd. for permission to borrow the ICCD camera despite a tight schedule. We thank Dr. Noah Brosch, TAU for encouraging the use of the UV camera in Mitzpe-Ramon. We also thank Galore Ltd. and El-Op Ltd. for image processing which enabled us to locate the stars in the images. UV research at Tel-Aviv University is supported by grants from the Ministry of Science through the Israel Space Agency to develop TAUVEX, a UV space telescope. *Editorial handling:* Peter Jenniskens.

References

Borovicka, J.: 1994, *Astron. Astrophys. Suppl.* **103**, 83–96.
Borovicka, J., Zimnikoval, P., Skvarka, J., Rajchl, J., and Spurny, P.: 1996, *Astron. Astrophys.* **306**, 995–998.
Brown, P., Campbell, M.D., Ellis, K.J., Hawkes, R.L., *et al.*: 2000, *Earth, Moon and Planets* **82–83**, 149–166.
Shemi, A., Mersov, G., Brosch, N., and Almoznino, E.: 1993, in D.R. Crabtree, R.J. Hanisch, J. Barnes (eds.), *Astronomical Data Analysis Software and Systems III*, Astron. Soc. Pacific Conference Series, Vol. 61, Conference on Astronomical Data Analysis Software and Systems, San Francisco, CA: 1994, 541 pp.

TIME RESOLVED SPECTROSCOPY OF A LEONID FIREBALL AFTERGLOW

JIŘÍ BOROVIČKA

Astronomical Institute, Ondřejov Observatory, 251 65 Ondřejov, Czech Republic
E-mail: borovic@asu.cas.cz

PETER JENNISKENS

SETI Institute, NASA ARC, Mail Stop 239-4, Moffet Field, CA 94035-1000, USA
E-mail: pjenniskens@mail.arc.nasa.gov

(Received 7 July 2000; Accepted 31 July 2000)

Abstract. Two video spectra of a meteoric afterglow were obtained for the first time during the 1999 Leonid aircraft campaign. The train was produced by a -13 magnitude Leonid fireball at a relatively low height between 91–75 km. The meteor spectrum has a strong hydrogen emission, proportional to 10-20 H atoms per one Fe atom The train spectrum consisted of a red continuum, yellow continuum, and about 50 atomic lines between 3700–9000 Å. The yellow continuum, possibly due to NO_2, was also detected in the persistent train. The red continuum is interpreted as a thermal radiation of dust from meteoric debris at about 1400 K. Evidence for secondary ablation is found in the afterglow. The atomic lines decayed within seconds of the meteor. The lines of Fe I, Mg I, Na I, Ca I, Ca II, Cr I, Mn I, K I, and possibly Al I were present in the glow together with the 5577 Å forbidden O I line. The gas temperature in the train was close to 5000 K at the beginning and decayed to 1200 K within two seconds. However, thermal equilibrium was not satisfied for all populated levels.

Keywords: Afterglow, debris, dust, H, meteor, persistent train, spectroscopy

1. Introduction

On November 18, 1999 at $04^h00^m29^s$ UT, about two hours after the maximum of the 1999 Leonid meteor storm, a very bright Leonid meteor appeared over the island of Corse, France, and produced an afterglow evolving into a persistent train that was visible to the naked eye for several minutes. The meteor and the glow were observed by several instruments on-board two aircrafts flying above the Mediterranean during the 1999 Leonid Multi-Instrument Aircraft Campaign (Jenniskens et al., 2000). For the first time, spectra of the afterglow were captured by video spectrographs. These data provide information about the physical conditions in the path of the meteor just after the train formation. A low resolution spectrum of the persistent train at later stages was obtained with the same instrument as well.

The spectra were obtained by two spectrographs of similar construction, the Ondřejov spectrograph assembled at the Ondřejov Observat-

Table I.

	Ondřejov spectrograph	BETSY spectrograph
image intensifier	Dedal 41	Mullard XX1332
camcorder	Panasonic NV-S88E	Sony CCD-TRV69E
video format	S-VHS, PAL	Hi-8, PAL
frames per second	25	25
lens diameter	36 mm	140 mm
focal length	50 mm	105 mm
grating diameter	52 mm	140 mm
grating grooves density	600 mm^{-1}	230 mm^{-1}
first order dispersion	11 Å pixel^{-1}	16 Å pixel^{-1}
field of view	25°	20°
aircraft	FISTA	FISTA
window elevation	12°	62°
camera elevation	35°	62°

ory and the BETSY spectrograph assembled at the SETI Institute. Both instruments consist basically of an optical sequence of transmission grating, lens, image intensifier, and video camcorder. The actual system parameters are given in Table I. The spectrographs are slitless, so the given dispersion is valid when imaging a narrow object. For objects with significant angular width, the spectral resolution will be worse.

The spectrographs were placed on-board the FISTA aircraft and directed to different positions on the sky. The fireball occurred outside the fields of view, but, fortunately, the gratings were in such positions that high order spectra of the terminal parts of the fireball and of the afterglow fell within the fields of view of both instruments. As the train weakened, its spectrum ceased to be visible in the high orders after about five seconds. After that, the Ondřejov spectrograph was moved manually with the aim to capture the first order spectrum of the persistent train. Due to a difficult orientation from the airplane, this succeeded for only a short period of time 37 seconds after the train formation and, finally, from 1 minute 23 seconds after the train formation.

The fireball itself was imaged by a direct non-spectral image intensified camera on-board the other aircraft, ARIA, located 160 km south of FISTA. The combination of observations from both aircrafts enabled us to determine the fireball trajectory.

Table II.

	beginning	maximum light	end
relative time [s]	0.00	1.77	2.17
longitude [East]	9°.57	8°.93	8°.78
latitude [North]	41°.51	41°.83	41°.90
altitude [km]	191	82	56
absolute magnitude	+2	< −10	0
distance to ARIA [km]	354	303	298
distance to FISTA [km]	—	203	191
elevation at ARIA	29°	12°	7°
elevation at FISTA	—	19°	13°
velocity [km/s]	72	72	(60)
geocentric radiant	$\alpha = 153°.3$, $\delta = +21°.2$, $\pm 2°$		
	$Az = 303°$, $h = 59°$		

In this paper we present all available data and their basic interpretation. The trajectory, light curve and spectrum of the mother fireball are given first. Then the structure of the afterglow and its spectrum are described, including detailed identification of spectral lines. The intensities and intensity decay rates of important lines are given and a very simple model is used to derive the temperature decay in the train during the first seconds. Then the spectrum of the persistent train at later times is discussed. Finally, our data are compared to the spectra of meteor trains obtained by other authors.

2. Fireball trajectory and light curve

At 04:00:29 UT, when the fireball occurred, the position of the aircrafts, as measured by the GPS system, was 7°11'15" E, 40°45'05" N, 11 230 m altitude for FISTA and 7°03'44" E, 39°36'19" N, 11 253 m for ARIA. Since no direct image of the fireball from FISTA exists, we used the spectrum from the Ondřejov spectrograph to derive the position of the fireball. The extrapolation introduces some uncertainty into the fireball trajectory computation, nevertheless, we were able to obtain quite consistent results. The consistency of the geometry was confirmed by the fact that the computed fireball velocity was close to 72 km s^{-1} when using measurements from either aircraft, and that the altitudes of the fireball terminal point and of the lower end of the train agreed

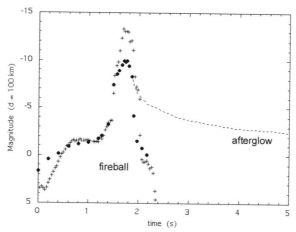

Figure 1. Lightcurve of the fireball and the train measured on the ARIA camera data (•). Since the camera was saturated, the real maximum brightness was certainly higher. An independent assessment (Jenniskens & Rairden 2000) yielded a value of −13 mag (+).

to within several hundreds of meters. We, therefore, conclude that the precision of the derived altitudes is better than 1 km. The final data on fireball trajectory are given in Table II. Indeed, there is good agreement with the independent trajectory calculations by Jenniskens & Rairden (2000), who derived the FISTA fireball position from back tracking the persistent train.

The fact that the fireball lasted for 2.17 sec when the Leonid radiant was 60 degrees above horizon means that the altitude span must have been about 135 km. The beginning altitude was found to be 191 km with a hint that the meteor was visible even a few kilometers higher. This high beginning is consistent with the observations of bright Leonids by Spurný et al. (2000a). Above the altitude of 140 km, the meteor had the same diffuse appearance as described by Spurný et al. (2000b). Between 90–75 km the fireball exhibited bright flare, evidently caused by a disintegration of the meteoroid. Nevertheless, part of the meteoroid survived and disappeared only at the altitude of 56 km, which is probably a record low altitude for a Leonid. A −13 mag Leonid 98023 of Spurný et al. (2000a,b) had a similar behavior with bright outburst and a surviving remnant continuing down, but the outburst terminated at 82 km and the remnant disappeared at 73 km. Another −13 mag Leonid 98041 terminated the burst at 73 km similarly to the present fireball but there was no surviving fragment.

The approximate meteor light curve is given in Figure 1. where we included the intensity decay of the afterglow. The magnitude is

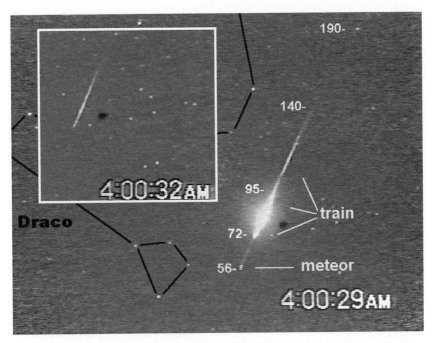

Figure 2. A single frame, including the original time stamp, taken by the direct image camera from the ARIA aircraft. The meteor just before its disappearance and the bright train above it can be seen. The numbers mark altitudes in km. The position of the constellation of Draco is indicated. The inset figure shows the train three seconds later. The inset is an average of nine frames.

given in absolute scale, i.e. recomputed to 100 km distance. The actual distance from ARIA was nearly 300 km (see Table II). Due to camera saturation, the meteor peak brightness could not be well determined and the measured value of −13 magn. is somewhat uncertain. The photometric mass of the meteoroid is about 1 kg.

In Figure 2, the meteor just before its disappearance and the bright afterglow left behind the meteor are shown in a single video frame from the ARIA camera. The bright afterglow lies at the position where the meteor flared a fraction of second earlier. Fainter afterglow continue up to the altitude of 140 km. The initial total brightness of the train was about −8 mag (see Figure 1).

3. Fireball spectrum

The information on the fireball spectrum is limited. During the flare, the spectrum was mostly saturated, although BETSY has useful records of the reflected spectrum from the aircraft window (see Figure

Table III.

Line		eV	S.	Line		eV	S.
3829.4–38.3	Mg I 3	5.9	1	5371.5	Fe I 15	3.3	2
3859.9	Fe I 4	3.2	1	5397.1–05.8	Fe I 15	3.2	2
3933.7	Ca II 1	3.2	6	5429.7–34.5	Fe I 15	3.3	2
3968.5	Ca II 1	3.1	6	5446.9–55.6	Fe I 15	3.3	2
4226.7	Ca I 2	2.9	2	5497.5–06.8	Fe I 15	3.2	1
4375.9)	Fe I 2	2.8	2	5528.4	Mg I 9	6.7	1
4383.6)	Fe I 41	4.3		5577.3	O I 3F	4.2	2
4404.8	Fe I 41	4.4	1	5615.6)	Fe I 686	5.5	1
4427.3	Fe I 2	2.9	1	5616.5)	N I 24	14.0	
4481.2	Mg II 4	11.6	4	5666.6–79.6	N II 3	20.6	2
4558.7	Cr II 44	6.8	1	5890.0–95.9	Na I 1	2.1	5
4583.8	Fe II 38	5.5	1	6156.0–58.2	O I 10	12.8	3
4630.5	N II 5	21.2	2	6347.1	Si II 2	10.1	5
4703.0	Mg I 11	7.0	1	6371.4	Si II 2	10.1	4
4861.3	H β	12.7	3	6562.8	H α	12.1	5
4871.3–72.1	Fe I 318	5.4	1	7423.6	N I 3	12.0	4
4890.8–91.5	Fe I 318	5.4	1	7442.3	N I 3	12.0	5
4919.0–20.5	Fe I 318	5.4	2	7468.3	N I 3	12.0	5
4923.9	Fe II 42	5.4	3	7771.9–75.4	O I 1	10.7	6
4957.3–57.6	Fe I 318	5.3	2	8184.8–88.0	N I 2	11.8	3
5005.2	N II 19	23.1	3	8216.3–23.1	N I 2	11.8	5
5018.4	Fe II 42	5.4	3	8242.5	N I 2	11.8	2
5056.0	Si II 5	12.5	2	8446.3–46.8	O I 4	11.0	4
5110.4	Fe I 1	2.4	2	8498.0	Ca II 2	3.2	2
5167.3–83.6)	Mg I 2	5.1	4	8542.1	Ca II 2	3.2	6
5169.0)	Fe II 42	5.3		8629.2	N I 8	12.1	2
5204.5–08.4	Cr I 7	3.3	1	8662.1	Ca II 2	3.1	5
5227.2	Fe I 37	3.9	2	8680.2–83.4	N I 1	11.8	4
5269.5)	Fe I 15	3.2	3	8711.7	N I 1	11.8	2
5270.4)	Fe I 37	4.0					
5328.0)	Fe I 15	3.2	2				
5329.1–30.7)	O I 12	13.1					

3). Before the flare, part of the spectrum is present on the BETSY spectrograph, while the spectrum was out of field of view on the Ondřejov spectrograph. The surviving remnant was well placed for both spectrographs but it was not very bright.

The combination of the data yielded the list of atomic lines present in the spectrum of the fireball (Table III). The excitation potential of the upper level is given for each transition. Unresolved lines are joined by parentheses. The relative strength of the lines is given on a scale from 1 to 6, the lines marked by 6 being the strongest. We refrain from a more meaningful intensity measure at this point because the spectrum changed along the trajectory.

All atomic lines in Table III have previously been found in photographic spectra of high velocity fireballs – see the detailed list of Halliday (1961). This is the reason why we do not give more details about the identification procedure for the fireball spectrum, in contrast to the afterglow spectrum in the next sections. In fact, a photographic spectrum of a fireball of this brightness would be superior to our video spectra and would undoubtedly yield much more line identifications.

From Table III it can be seen, that lines of very different excitation+ionization energy are present: from 2 eV for Na I lines to about 35 eV needed to ionize and excite nitrogen atoms to produce the N II lines. In addition, lines of different origin are present: nitrogen and a substantial part of oxygen come from the atmosphere, while other elements come from the meteoroid. This assignment follows from comparing relative abundances of elements in the radiating gas and also from the behavior of individual emissions along the trajectory.

We can therefore divide the lines in several groups or components according to their origin and energy, following partly Borovička (1994) and Borovička and Boček (1995):

		origin	
		atmospheric	meteoric
	low	[O I] 5577 Å	Na I, Mg I, Fe I, Ca I, Cr I, Ca II
energy	high	O I, N I	Ca II, Mg II, Si II, H I, Fe II, Cr II
	very high	N II	*none*

The low energy meteoric group is identical to Borovička's (1994) main component of temperature of about 5000 K and the high energy meteoric group corresponds to the second component of about 10,000 K. The temperatures of the atmospheric lines are difficult to establish.

The relative strength of individual groups of lines changed along the meteor trajectory. We have no information about the meteor spectrum at high altitudes above 110 km. Before the flare, at about 100 km, the spectrum was dominated by the atmospheric lines of O I and N I. As the brightness increased, the low energy meteoric lines become strong. At the peak brightness, in the flare, the high energy meteoric lines dominated the spectrum and the N II lines were also present. In the spectrum of the surviving remnant, the O I and N I lines were the brightest again, but the low and high energy meteoric lines were also present.

The changes of the spectrum are intriguing. For example, the 8500 Å region is dominated by the N I and O I lines before and after the flare and the Ca II lines are hardly visible. In the flare, on the other hand, the Ca II lines are extremely bright, in fact the brightest lines in the whole spectrum.

A detailed explanation of the above facts is not the scope of the present paper. The general picture seems, however, to be clear. The flare was caused by dramatic fragmentation and violent evaporation of the meteoroid. The amount of the meteoric vapors increased rapidly and the intensity of meteoric lines therefore increased much more than the intensity of the atmospheric lines. The lines of the second meteoric component brightened more than the lines of the main component because the second component is optically thin (Borovička and Majden 1998). It seems that the development of the second component was accompanied with the atmospheric lines of N II. These emissions are probably connected with the development of meteor shock wave, while the main component and the O I and N I lines are present also in the absence of shock wave at higher altitudes. At low altitudes, the shock wave is present also in the absence of a meteor flare. It is interesting that the atmospheric lines in both cases fall to higher energy groups than the related meteoric lines. The correlation of the brightenings of Si II and N II lines was found also in the spectrum of the Benešov fireball (Borovička and Spurný 1996).

The presence of N II lines in meteor spectra was first found by Halliday (1961). Their excitation energy by far exceeds other lines in the spectrum, nevertheless, it is still only a fraction of the kinetic energy of N_2 molecule at 72 km s^{-1} (750 eV). O II lines of atmospheric origin can be expected in the spectrum as well but O II lines are fainter than N II and their presence could not be firmly confirmed.

Of special interest are the hydrogen lines since they may suggest the presence of organic matter in the meteoroid. Hα and Hβ are well seen in the flares and in the surviving remnant spectrum. Their presence is not surprising, H lines have been seen e.g. in Perseid spectra (first found by Millman 1953). However, they are brighter in the present spectrum. Assuming the temperature of 10,000 K for the second component the ratio of number of atoms H/Fe $\approx$ 10–20 was found, while the values for Perseids are 2–6 (Borovička and Betlem 1997).

4. The spatial structure of the afterglow

The image of the afterglow in Figure 2 has a low resolution. The structure of the bright part of the glow can be better seen on the spectrograph records. In Figure 3 it can be seen that the glow after its formation consisted of four segments, which we numbered 1 to 4. The segments are separated by gaps with almost no emission (except for the 5577 Å line). The uppermost segment 1 is the brightest, longest and has a sharp leading edge. The lower segments are fainter, shorter and more symmetrical. The reality of the structure is confirmed on the image from the Ondřejov spectrograph in Figure 4. Here the altitude scale is also drawn. Segment 1 goes from about 85 km to 80 km, the center of segment 4 lies at 73 km. We believe that individual segments were formed at the positions of individual consecutive meteoroid break-ups and dust releases.

When playing the video, it was noted that the leading edge of the glow segment 1 moves down. The motion relative to a background star was evident as well as the fact that the gap between segments 1 and 2 became narrower. The movement was measured on the Ondřejov spectrograph data and the result is displayed in Figure 5. The train travelled 1.5 km within one second, with the initial velocity of about 2 km s^{-1}. This observation confirms that the train was formed by ablated material, dust and gas, which has its own inertia and continued the motion through (probably rarified) air in the track of the main body.

5. Afterglow spectrum, line identifications

The spectrum of the afterglow consists of numerous lines and a continuous radiation. Figures 3 and 4 show that the spectra of all glow segments are nearly the same and contain different lines than the meteor spectrum. The BETSY spectrum (Figure 3) is well placed on the diagonal of the frame, however, it is an overlap of five different spectral orders (3rd to 7th) which makes the line identification difficult. The Ondřejov spectrum consists of the red part of the unblazed first order spectrum and a part of the second order spectrum. The wavelength scale is shown in Figure 4. The system is sensitive in the range 3800–9000 Å but there is no information on the interval 6000–7000 Å. For the brightest segment 1, an even smaller part of the spectrum lies in the field of view. About 2.5 seconds after the fireball the camera was slightly moved, so the segment 1 moved nearly to the position of the meteor in Figure 4 and the whole wavelength range was accessible. At that time, however, only four lines remained visible. Later, the grating

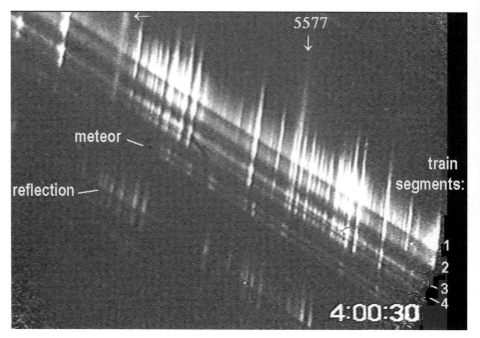

Figure 3. A single frame taken by the BETSY spectrograph. High order spectra of the meteor and the afterglow can be seen. Four separate train segments are numbered. An internal reflection of the bright part of the glow spectrum is also present. The position of the green oxygen line at 5577 Å (in the 4rd and 5th order) is marked by arrows. The camera time was not quite synchronized with the ARIA camera (Figure 2).

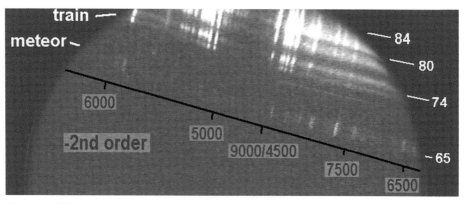

Figure 4. Upper part of a single frame taken by the Ondřejov spectrograph with circular field of view. Meteor and afterglow spectrum can be seen. The white numbers mark altitudes in km, the dark numbers show approximate wavelengths in Å. The end of the −1st order and the beginning of the −2nd order overlap.

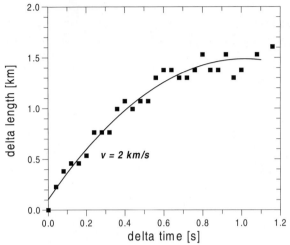

Figure 5. The displacement of the leading edge of glow segment 1 as a function of time after the segment formation. A parabolic fit is drawn through the measured points.

was rotated by 180 degrees and the blazed first order spectrum was searched for persistent train emissions. This succeeded only 1 minute 23 seconds after the train formation. The inspection of the tape revealed, that the first order spectrum appeared in the field of view for a short period of time also at 37 s. At both these times the train spectrum consisted only of a continuos radiation as shown in Figure 6.

Most of this paper is devoted to an analysis of the first 2–3 seconds of the afterglow, when we have a plenty of information from the atomic lines. The persistent train at later times is discussed in Section 8.

As an example, the plot of the spectrum of glow segment 3 is presented in Figure 7. Besides the lines, strong continuum is also present in the glow spectrum. We divided the continuum into two parts which have probably different origins, the red continuum observed in the first order and a yellow continuum seen in the second order (see Figure 7). The red continuum could be fitted relatively well with a blackbody of the temperature of about 1400 K and we attribute it tentatively to a thermal radiation of dust. The yellow continuum is probably due to molecular emissions. Possible identifications will be discussed in Sect. 8.

The identification of the atomic lines was done in several steps. First, an an approximate wavelength scale was determined by using the lines in the meteor spectrum. The identification of the brightest lines of Na and Fe then became immediately clear. Other lines were identified after revising the wavelength scale and some faint lines could be identified only after the physical conditions in the train were considered (Sect. 7).

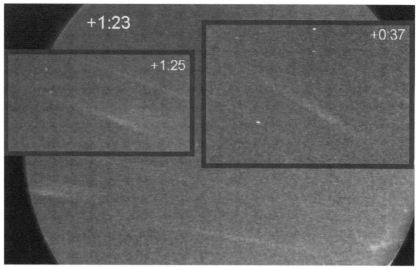

Figure 6. First order spectra of the persistent train. The background image shows the train (zero order) at lower right and a part of its spectrum on the left 1 minute and 23 seconds after the train formation. The insets show whole train spectra at different times.

In all cases the presence of the line in the spectra of both spectrographs and in different spectral orders was checked. The final identifications for the Ondřejov spectrum are given in Table IV and for the BETSY spectrum in Tables V–VIII.

The arrangement of both tables is similar. The visually estimated intensity of the line is given on scale 1–5. The observed intensity is affected by the wavelength dependent sensitivity in different spectral orders. The observed wavelengths are ambiguous in most cases and are given for all possible orders. The identification contains the laboratory wavelength, atom, ionization degree, and multiplet number according to multiplet table (Moore 1945).

With the exception of the oxygen green line 5577 Å, the afterglow spectrum contains only lines of ablated metals. Lines of neutral Fe, Mg, Na, Ca, Cr, Mn, and K and ionized Ca are present. The lines of Fe are the most numerous. The presence of Al is not certain. The identification of the K I line at 7699 Å seemed doubtful, because there is an alternative explanation of this line by the Fe I 3859 Å line in the second order and another K line at 7664 Å, which should be brighter, is very faint. However, the 7664 Å line is absorbed by the underlying atmospheric O_2 and we concluded that the 7699 Å line really is the main contributor to the observed feature. The only ionic lines present belong to Ca II. Calcium is the only abundant element with relatively

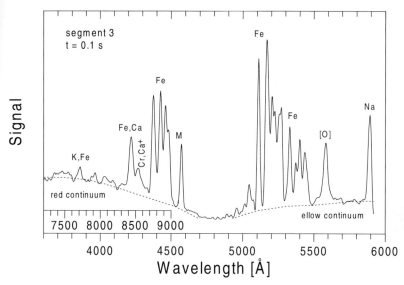

Figure 7. The wavelength calibrated spectrum of the afterglow segment 3 0.1 s after its formation extracted from the Ondřejov spectrograph. The lower wavelength scale corresponds to the second order, the upper scale is for the first order. In the overlapping region the wavelengths are ambiguous. Important lines are identified. The brightness of the Na line is underestimated because this line lies at the edge of the field of view.

low ionization potential (6.1 eV) and strong low excitation (3.1 eV) ionic lines. The excitation potential of all lines in the train is up to 7 eV. The high excitation potential of the allowed transitions of O and N is probably the reason why these lines are not present.

The strongest lines are the multiplets 1 and 2 of Fe I and multiplets 1 of Na I and Ca I. All have low excitation potential but they differ in transition probability. The Na line is a strong allowed transition, while the other lines are semi-forbidden intercombination transitions which are normally relatively faint. The explanation of their strength in the train is given in Sect. 7.

The forbidden green oxygen line has an evidently different origin than other lines. This line forms a short-living train in all Leonids and other fast meteors, even in faint ones. In our case the green line does not follow the structure of the glow segments, but dominates the upper part of the train up to the altitude of 140 km. The intensity of the green line is therefore not related to the ablation rate, which is consistent with the atmospheric origin of the line.

Table IV.

No	Int.	Wavelength −1	−2	Identifications	No	Int.	W.l. −2	Identifications
1	1	7430	3715	3719.9 Fe I 5	22	2	4892	4891.5 Fe I 318
2	1	7670	3835	7664.9 K I 1				4890.8 Fe I 318
3	3	7710	3855	3859.9 Fe I 4	23	2	4918	4920.5 Fe I 318
				7699.0 K I 1				4919.0 Fe I 318
4	2	7880	3940	3944.0 Al I 1	24	2	4956	4957.6 Fe I 318
				3930.3 Fe I 4				4957.3 Fe I 318
				3927.9 Fe I 4	25	2	5012	5012.1 Fe I 16
				3933.7 Ca II 1	26	3	5042	5041.8 Fe I 36
5	3	7930	3965	3961.5 Al I 1	27	1	5072	5072.9 Cr I 8
				7912.9 Fe I 12	28	5	5110	5110.4 Fe I 1
6	3	8060	4030	4030.8 Mn I 2	29	5	5168	5166.3 Fe I 1
				4033.1 Mn I 2				5168.9 Fe I 1
				4034.5 Mn I 2				5167.5 Fe I 37
				4045.8 Fe I 43				5171.6 Fe I 36
				8047.6 Fe I 12	30	4	5204	5204.6 Fe I 1
				8075.1 Fe I 12				5208.4 Cr I 7
7	2	8200	4100	8194.8 Na I 4				5206.0 Cr I 7
				8183.5 Na I 4				5204.5 Cr I 7
8	2	8290	4145	4134.3 Fe I 3	31	4	5227	5225.5 Fe I 1
				4143.9 Fe I 43				5227.2 Fe I 37
				4152.2 Fe I 18	32	4	5255	5255.0 Fe I 1
9	1	8360	4180	4177.6 Fe I 18				5250.2 Fe I 1
				4174.9 Fe I 19				5247.0 Fe I 1
				4172.7 Fe I 19	33	4	5269	5269.5 Fe I 15
10	4	8440	4220	4216.2 Fe I 3				5270.3 Fe I 37
				4226.7 Ca I 2	34	4	5330	5328.0 Fe I 15
				4206.7 Fe I 3	35	3	5370	5371.5 Fe I 15
11	3	8540	4270	4274.8 Cr I 1	36	3	5400	5405.8 Fe I 15
				8542.0 Ca II 2				5397.1 Fe I 15
12	2	8580	4290	4291.5 Fe I 3				5394.7 Mn I 1
				4289.7 Cr I 1	37	3	5435	5434.5 Fe I 15
13	2	8660	4330	8662.1 Ca II 2				5429.7 Fe I 15
14	5		4377	4375.9 Fe I 2				5432.5 Mn I 1
15	5		4428	4427.3 Fe I 2				5446.9 Fe I 15
16	4		4463	4461.7 Fe I 2	38	2	5502	5501.5 Fe I 15
17	4		4483	4482.2 Fe I 2				5506.8 Fe I 15
				4489.7 Fe I 2				5497.5 Fe I 15
18	4		4570	4571.1 Mg I 1	39	2	5528	5528.4 Mg I 9
19	2		4651	4646.2 Cr I 21	40	4	5581	5577.3 O I 3F
				4652.2 Cr I 21	41	5	5892	5890.0 Na I 1
20	2		4706	4703.0 Mg I 11				5895.9 Na I 1
21	2		4870	4871.3 Fe I 318	42	(5)	6570	6572.8 Ca I 1

Table V.

Line No.	Int.	Observed wavelength					Identifications
		−3	−4	−5	−6	−7	
1	4	6100	4575				4571.1 Mg I 1
2	2	6160	4620				6162.2 Ca I 3
3	2	6280	4710	3768			6280.6 Fe I 13
							4703.0 Mg I 11
4	3	6361	4771	3817			6358.7 Fe I 13
5	2	6400	4800	3840			6400.3 Fe I 13
6	1	6440	4830	3864			3859.9 Fe I 4
							6439.1 Ca I 18
7	2	6492	4869	3895			4871.3 Fe I 318
							6498.9 Fe I 13
8	5	6565	4924	3939			6572.8 Ca I 1
9	3	6607	4955	3964			4957.6 Fe I 318
							4957.3 Fe I 318
							3961.5 Al I 1
10	2	6679	5009	4007			5012.1 Fe I 16
11	2	6723	5042	4034			5041.8 Fe I 36
							4030.8 Mn I 2
							4033.1 Mn I 2
							4034.5 Mn I 2
12	2	6773	5080	4064			5072.9 Cr I 8
							4063.6 Fe I 43
13	5	6812	5109	4087			5110.4 Fe I 1
14	5	6889	5167	4134			5166.3 Fe I 1
							5168.9 Fe I 1
							5167.5 Fe I 37
							5171.6 Fe I 36
15	4	6941	5206	4165			5204.6 Fe I 1
							5208.4 Cr I 7
							5206.0 Cr I 7
							5204.5 Cr I 7
16	4	6973	5230	4184			5225.5 Fe I 1
							5227.2 Fe I 37
17	4	7013	5260	4208			5255.0 Fe I 1
							5250.2 Fe I 1
							5247.0 Fe I 1
18	4	7033	5275	4220			5269.5 Fe I 15
							5270.3 Fe I 37
							4216.2 Fe I 3
							4226.7 Ca I 2
19	4	7111	5333	4266			5328.0 Fe I 15

Table VI.

Line No.	Int.	Observed wavelength					Identifications
		−3	−4	−5	−6	−7	
20	3	7163	5372	4298			5371.5 Fe I 15
21	3	7203	5402	4322			5405.8 Fe I 15
							5397.1 Fe I 15
							5394.7 Mn I 1
22	3	7247	5435	4348			5434.5 Fe I 15
							5429.7 Fe I 15
							5432.5 Mn I 1
							5446.9 Fe I 15
23	4	7300	5475	4380			4375.9 Fe I 2
24	2	7340	5505	4404			5501.5 Fe I 15
							5506.8 Fe I 15
							5497.5 Fe I 15
25	4	7383	5537	4430			4427.3 Fe I 2
26	4	7437	5578	4462	3719		4461.7 Fe I 2
							5577.3 O I 3F
27	4	7477	5608	4486	3739		4482.2 Fe I 2
							4489.7 Fe I 2
28	4	7620	5715	4572	3810		4571.1 Mg I 1
29	1	7664	5748	4598	3832		7664.9 K I 1
30	2	7700	5775	4620	3850		3859.9 Fe I 4
							7699.0 K I 1
31	1	7763	5822	4658	3881		4646.2 Cr I 21
							4652.2 Cr I 21
							3886.3 Fe I 4
							3878.6 Fe I 4
32	5	7855	5891	4713	3927		5890.0 Na I 1
							5895.9 Na I 1
33	3	7933	5950	4760	3967		3961.5 Al I 1
							7912.9 Fe I 12
							5956.7 Fe I 14
							5949.3 Fe I 14
							5958.2 Fe I 14
34	2	8047	6035	4828	4023		8047.6 Fe I 12
							4030.8 Mn I 2
35	1	8083	6062	4850	4041		8075.1 Fe I 12
36	1	8140	6105	4884	4070		6102.7 Ca I 3
							6122.2 Ca I 3
							4891.5 Fe I 318
							4890.8 Fe I 318

Table VII.

Line No.	Int.	Observed wavelength					Identifications
		−3	−4	−5	−6	−7	
37	2	8204	6153	4922	4102		4920.5 Fe I 318
							4919.0 Fe I 318
							6162.2 Ca I 3
							8194.8 Na I 4
							8183.5 Na I 4
38	1	8261	6196	4957	4131		4957.6 Fe I 318
							4957.3 Fe I 318
39	1	8347	6260	5008	4173		6280.6 Fe I 13
							5012.1 Fe I 16
40	2	8400	6300	5040	4200		5041.8 Fe I 36
41	3	8427	6320	5056	4213		4216.2 Fe I 3
							4206.7 Fe I 3
							4226.7 Ca I 2
42	5	8523	6392	5114	4261		5110.4 Fe I 1
43	5	8613	6460	5168	4307		5166.3 Fe I 1
							5168.9 Fe I 1
							5167.5 Fe I 37
							5171.6 Fe I 36
44	4	8667	6500	5200	4333	3714	5204.6 Fe I 1
							5208.4 Cr I 7
							5206.0 Cr I 7
							5204.5 Cr I 7
45	3	8707	6530	5224	4353	3731	5225.5 Fe I 1
							5227.2 Fe I 37
46	5	8760	6570	5256	4380	3754	4375.9 Fe I 2
							6572.8 Ca I 1
47	4	8849	6637	5310	4425	3793	4427.3 Fe I 2
48	2	8880	6660	5328	4440	3806	5328.0 Fe I 15
49	4	8920	6690	5352	4460	3823	4461.7 Fe I 2
50	4	8969	6727	5382	4485	3844	4482.2 Fe I 2
							4489.7 Fe I 2
							5371.5 Fe I 15
51	3	9000	6750	5400	4500	3857	5405.8 Fe I 15
							5397.1 Fe I 15
							5394.7 Mn I 1
							3859.9 Fe I 4
52	3		6795	5436	4530	3883	5434.5 Fe I 15
							5429.7 Fe I 15
							5432.5 Mn I 1
53	2		6810	5448	4540	3891	5446.9 Fe I 15
							5455.6 Fe I 15

Table VIII.

Line No.	Strength	Observed wavelength				Identifications
		−3	−4	−5	−6	−7	
54	4		6855	5484	4570	3917	4571.1 Mg I 1
55	2		6882	5506	4588	3933	5501.5 Fe I 15
							5506.8 Fe I 15
							5497.5 Fe I 15
							3933.7 Ca II 1
56	2		6945	5556	4630	3969	3961.5 Al I 1
							3968.5 Ca II 1
57	2		6978	5582	4652	3987	5577.3 O I 3F
58	5		7370	5896	4913	4211	5890.0 Na I 1
							5895.9 Na I 1
59	3		7395	5916	4930	4226	4226.7 Ca I 2
60	2		7445	5956	4963	4254	5956.7 Fe I 14
							5949.3 Fe I 14
							5958.2 Fe I 14
61	2		7570	6056	5047	4326	5041.8 Fe I 36
62	5		7665	6132	5110	4380	4375.9 Fe I 2
							5110.4 Fe I 1

6. Temporal evolution of the afterglow spectrum

The decay of the individual spectral lines and continua was measured on the Ondřejov spectrum. The automatic gain control of the image intensifier of the BETSY spectrograph decreased the gain after the fireball brightening and the gain was slowly recovering during the afterglow radiation. The BETSY spectrum would therefore be difficult to use for photometric purposes. Moreover, the spectrograph has not been calibrated for spectral sensitivity in the high spectral orders.

The evolution of the intensity of the continuous radiation is shown in Figure 8. At the beginning, the intensity decreased rapidly. The decrease was followed by a standstill, pronounced especially in segment 1. Finally, a gradual exponential decrease occurred. The red and the yellow continuum behaved similarly. No big change of the shape of the continua during the decay was evident.

The measurement of the decay of the atomic lines was complicated by the fact that the intensity range of the video system is only 8 bits. Bright lines are saturated at the beginning and could be measured only after some time. Faint lines, on the other hand, are visible only at the beginning. Nevertheless, when in the measurable range, the decay of

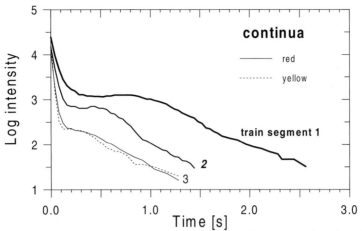

Figure 8. The decay of intensity of the continuous radiation in the afterglow spectrum during the first seconds. The intensity is given in relative units. The red continuum was measured at three train segments, the yellow continuum (dashed line) could be measured only on segment 3.

all lines could be fitted by a simple exponential function:

$$\log I = \log I_0 + Bt, \qquad (1)$$

where I is the line intensity, t is time, I_0 is the intensity at time zero and B is the decay coefficient. The values of $\log I_0$ and B for all measured lines at three train segments are given in Table IX. The values of I_0 have been corrected for spectral sensitivity of the instrument. For each line, the time interval t_1, t_2 covered by the measurements is also given. Time zero corresponds to the train formation just after meteor passage at a given place. Note that lines 29, 41, and 42 could be measured at segment 1 only after they appeared in the field of view following the shift of the camera.

For most of lines in Table IX the data of the dominant atomic transition are given: the element, the excitation potential of the upper level, E_k, and $\log gf$, the oscillator strength multiplied by statistical weight of the lower level. These data are not given if there is no one dominant contributor to the observed line, or the line is too faint, or it does not belong to a neutral atom of meteoric origin (i.e. O I and Ca II excluded). Atomic data given in Table IX were used for further consideration of physical conditions in the train. Detailed identification of all lines is given in Table IV.

Lines no. 3, 5, 28, 29, 41, and 42 exhibit the slowest decay. All have low excitation potential not exceeding 2.5 eV. Figure 9 shows that there is a definite dependency of the decay rate on excitation potential.

Table IX.

No	El	E_k eV	log gf	segment 1 I_0 B t_1 t_2	segment 2	segment 3
1						4.03 -4.42 .00 .50
3	K	1.72	-0.17	5.07 -1.04 0.70 2.00	4.47 -1.79 .00 1.10	4.00 -2.44 .00 .80
4				4.93 -2.78 0.08 0.30		
5	Fe	2.43	-4.85	5.02 -1.45 0.08 0.50	4.72 -3.51 .00 0.30	4.05 -3.34 .00 .60
6	Mn	3.08	-0.13	6.66 -2.02 0.65 1.25	5.71 -3.52 .00 0.20	4.96 -4.45 .00 .45
7	Na	3.62	0.74	5.38 -3.25 0.08 0.30	4.95 -5.84 .00 0.20	
8				5.63 -5.37 0.08 0.30	5.07 -8.94 .00 0.16	
9				5.40 -2.67 0.08 0.50		
10	Fe	2.94	-3.35	6.62 -2.28 0.40 1.25	5.78 -3.03 .05 0.70	5.21 -4.11 .00 .70
11				6.48 -3.77 0.20 0.80	5.68 -5.09 .00 0.40	5.28 -10.4 .00 .20
12				6.06 -3.54 0.10 0.70		
13				6.11 -5.16 0.10 0.60		
14	Fe	2.83	-3.03	7.62 -2.68 0.80 1.80	6.90 -4.14 .40 1.00	5.63 -4.70 .00 .80
15	Fe	2.85	-2.92	7.44 -2.66 0.80 1.80	6.70 -4.11 .40 0.80	5.60 -4.60 .00 .65
16	Fe	2.87	-3.21	7.22 -2.63 0.70 1.70	6.57 -4.10 .40 0.80	5.43 -4.63 .00 .65
17	Fe	2.88	-3.50	6.94 -2.69 0.70 1.60	6.12 -4.00 .30 0.80	5.15 -4.70 .00 .65
18	Mg	2.71	-5.40	7.33 -2.61 0.80 1.80	6.41 -4.11 .40 0.90	4.96 -4.44 .00 .60
19	Cr	3.68	-0.50	4.69 -5.63 0.00 0.35		
20	Mg	6.98	-0.37	4.67 -11.1 0.00 0.18		
21	Fe	5.41	-0.41	4.71 -6.46 0.00 0.30	4.46 -13.9 .00 0.16	
22	Fe	5.39	0.04	4.65 -6.20 0.00 0.30	4.68 -15.6 .00 0.16	
23	Fe	5.35	0.20	4.75 -6.44 0.04 0.30	4.82 -14.1 .00 0.16	
24	Fe	5.31	0.32	4.86 -6.56 0.04 0.35	4.99 -14.2 .00 0.16	
25	Fe	3.33	-2.64	4.16 -1.72 0.15 0.80	4.57 -5.79 .00 0.36	
26	Fe	3.94	-2.20	4.63 -2.53 0.24 0.45	5.11 -5.56 .00 0.40	4.04 -5.44 .00 .25
28	Fe	2.43	-3.76	7.49 -1.84 1.20 2.15	7.26 -3.96 .60 1.20	5.57 -3.92 .00 .80
29	Fe	2.42	-3.77	6.78 -1.50 2.50 3.00	7.27 -3.95 .60 1.20	5.70 -3.99 .00 .80
30	Fe	2.47	-4.33		6.38 -4.09 .30 0.80	5.29 -4.92 .00 .50
31	Fe	2.48	-4.79		5.67 -3.84 .10 0.60	5.18 -6.02 .00 .40
32	Fe	2.47	-4.40		5.83 -3.72 .24 0.70	5.03 -5.10 .00 .40
33	Fe	3.21	-1.32		6.02 -4.41 .24 0.70	5.16 -5.52 .00 .40
34	Fe	3.24	-1.47		5.56 -4.66 .10 0.60	5.06 -6.67 .00 .40
35	Fe	3.27	-1.64			4.71 -6.97 .00 .40
36	Fe	3.28	-1.61			4.88 -6.25 .00 .40
37	Fe	3.24	-1.68			4.82 -6.60 .00 .40
40						5.20 -10.2 .00 .20
41	Na	2.10	0.29	5.56 -0.81 2.50 3.50		5.50 -3.45 .04 .90
42	Ca	1.89	-4.30	4.23 -0.63 2.50 3.50		

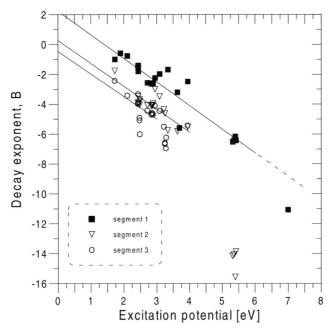

Figure 9. The dependency of the decay rate of lines of neutral atoms on their excitation potential.

Although the spread is relatively large, the decay rate of most lines is nearly proportional to the excitation potential:

$$B = B_0 + DE \qquad (2)$$

The proportion constant is nearly the same for all segments: $D = -1.5 \text{ s}^{-1}\text{eV}^{-1}$. The constant B_0 (a decay rate of a fictive zero eV of excitation line) is, however, different. B_0 may indicate a change in abundance of atomic species, for example. For segments 2 and 3 B_0 is, within the limits of errors, nearly zero but for segment 1 it is about 2 s^{-1}. Lines of the highest excitation do not follow Eq. (2) at all. They decay more quickly. Also the lines of Ca II and O I (not shown in Figure 9) decay more quickly.

The decay rate of the continuous radiation is slower than of most of lines. The asymptotic decays in Figure 8 correspond to the values of B from -1.0 to -1.2. The continuum decay is slightly slower at segment 1 and for the yellow continuum than for the red continuum.

7. Physical conditions in the afterglow

7.1. Initial conditions

In this section we will consider physical conditions during the first seconds of the formation of a persistent train at the time of the afterglow, in particular the gas temperature. We have at our disposal the data from Table IX which enable us to determine the intensities of all lines at one time instant. In the following simple approach, we assume that the train is optically thin. Then, the intensity of a line, I_{ki}, is directly proportional to the population of the corresponding upper level, N_k, (i.e. total number of atoms in the upper state in the whole train segment):

$$I_{ki} \sim A_{ki} N_k, \qquad (3)$$

where A_{ki} is the radiative transition probability, which is directly related to the oscillator strength.

We assume statistical equilibrium, i.e. the number of downward and upward transitions at a given time is the same:

$$N_i(A_i + C_{i0}) = N_0 C_{0i}, \qquad (4)$$

where $A_i = \sum_{j<i} A_{ij}$ is the total radiative deexcitation rate of the level i and C_{ij} are the collisional transitional probabilities. Index 0 designs the ground level. For simplicity, we consider only collisional transitions involving the ground level. The following relations for collisional transitions can be written (Mihalas 1978):

$$C_{0i} = \frac{g_i}{g_0} e^{-E_i/kT} C_{i0} = b(T) C_{i0}, \qquad (5)$$

$$C_{0i} = n_e e^{-E_i/kT} Q_{0i}(T) \qquad (6)$$

where we designed the Boltzmann factor as $b(T)$. Here T is temperature, k is the Boltzmann constant, g are statistical weights of the levels, and n_e is the density of free electrons. Q_{0i} is a complicated function slightly dependent on temperature.

Combining the above equations and ignoring for simplicity the statistical weights, we arrive at the following order-of-magnitude relation for the level population:

$$\frac{N_i}{N_0} = \frac{b(T)}{\frac{A_i}{n_e Q_i(T)} + 1} \qquad (7)$$

The ratio $A_i/n_e Q_i$ is important for a deviation of level population from the value in thermal equilibrium (TE). TE is satisfied as long as

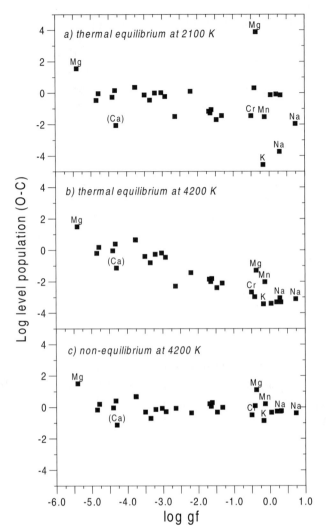

Figure 10. The difference $(O - C)$ of the logarithm of level population computed from the observed line intensity (Eq. 3) and for a given temperature and density (Eq. 7). Each point represents one line. Unmarked points belong to iron lines. The $(O-C)$ is given as a function of log gf of the line. Plot a) $T = 2100$ K and $n_e Q \gg A$ for all lines.; b) the same for $T = 4200$ K; c) $T = 4200$ K and $n_e Q = 10^5$ s^{-1}.

$A_i \ll n_e Q_i$. If electron density is low and collisions are infrequent, levels are depopulated by radiative transitions. The levels where a strong allowed transition originates are the most affected.

Using the data from Table IX, line intensities 0.1 s after the train formation (0.05 s for segment 3) were determined and the level populations were computed using Eq. (3). The results were then compared with predictions from Eq. (7). Two parameters were varied, temperature T, which affects the Boltzmann factor $b(T)$, and the value of $n_e Q_i$, which was, for simplicity, taken the same for all lines and was assumed to be independent of temperature. To use all available data, all three train segments and all elements were combined. The level populations derived from segments 2 and 3 were risen by empirically determined factors of 0.5 and 1.4 in logarithmic scale, respectively, to be on the same scale with segment 1. Chondritic relative abundances of elements were assumed.

Figure 10a shows the observed minus predicted level populations as a function of the $\log gf$ of the corresponding line for a temperature of 2100 K and thermal equilibrium. This temperature fits many iron lines relatively well but some iron lines are underpopulated by and order of magnitude and, moreover, lines of other elements deviate by up to four orders of magnitude to both sides. When assuming a temperature of 4200 K (Figure 10b), the scatter of the $(O - C)$ values is substantially reduced. The $(O - C)$'s, however, do not lie near zero but form a clear dependence on $\log gf$. This suggests that the temperature is correct but the assumption of TE is not correct. Indeed, the assumption of $T = 4200$ K and $n_e Q = 10^5$ s^{-1} (Figure 10c) explains the population of most levels relatively well, considering all the simplifications we made. Only the two lines of Mg deviate. Magnesium seems to be overabundant in the train, for which we have no explanation. In the meteor itself the Mg abundance was normal. The line of Ca deviates a little bit as well, but this line was out of the field of view most of the time and was seen only after the camera movement, when it was already faint.

The average value of $n_e Q = 10^5$ means that radiative depopulation was really important for many levels. For the Na line, for example, the radiative deexcitation rate is $A = 6 \times 10^7$ s^{-1}. For the intercombination lines of Fe, Ca, and Mg, on the other hand, the A value is in the range 10^2 to 10^4 and the radiative deexcitation is not so important. These lines are therefore much brighter when compared to the allowed transitions than in laboratory conditions.

The relative enhancement of the intercombination lines has been observed earlier in the spectra of wakes of bright fireballs (Halliday 1968, Borovička & Spurný 1996). The mechanism of a low collisional rate is clearly the same. In this sense, the afterglow was like a "prolongated wake" The wake normally does not separate from the mother meteor and the radiation as a given point lasts for less than 0.1 s. Now it was observed for several seconds.

7.2. Temporal evolution

As mentioned earlier, we do not have enough data to derive the temperature independently at various time instants. Even the temperature derived in the previous section was obtained by extrapolation of line intensities and combining three train segments. We will continue in this approach and combine several simple equations to obtain the temperature trend.

It was found that the decay of line intensities depends on excitation potential, not on transition probability. The decay is therefore due primarily to the decrease of temperature, not density. The level populations N_i, total abundance of an atom N, and temperature T are functions of time and are connected by the Boltzmann equation (modified by a radiative deexcitation factor in some cases, which, however, is nearly constant and is not important here):

$$\log(N_i(t)) = \log(N(t)) - \frac{5040}{T(t)} E_i. \qquad (8)$$

T is given in Kelvins, E in Electron Volts. The decay of line intensities and thus level populations could be fitted with an exponential function:

$$\log(N_i(t)) = \log(N_i(0)) + B_i t. \qquad (9)$$

The decay rate was found to depend linearly on excitation potential for most lines:

$$B_i = B_0 + D E_i. \qquad (10)$$

Substituting to Eq. (9) from (8) and (10) and comparing the coefficients with (8), we find the solution:

$$\log N(t) = \log N(0) + B_0 t \qquad (11)$$

and

$$\frac{5040}{T(t)} = \frac{5040}{T(0)} - Dt \qquad (12)$$

Equation (11) shows that the total number of atoms in segments 2 and 3, where $B_0 \approx 0$ (Figure 9), was nearly constant, while in segment 1, where $B_0 \approx 2$, the number of atoms increased with time. This is probably connected with a secondary ablation that may be induced by the movement of the whole segment (see Sect. 4). Both effects are further evidence that meteoric debris was left behind. Of course, these conclusions are valid only for the first second or few seconds, when the line decay was measured.

The value of the constant D in Eq. (12) was found to be nearly the same in all segments ($D \approx -1.5$), so the temperature decay was nearly

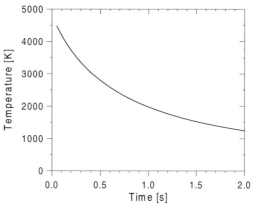

Figure 11. The temperature of the train during the first two seconds as derived from the atomic line intensities and their decay rates.

the same in the whole train. The function (12) is displayed in Figure 11. At the beginning, the train temperature nearly corresponded to the temperature of the main spectral component in the fireball spectrum, i.e. 4500–5000 K. Within one second, the temperature decreased to 2000 K and during the next second to 1200 K. We do not claim that the temperature strictly followed function (12) but Figure 11 gives a good idea on the temperature decay.

8. Persistent train emission

As explained in Sect. 5, the spectrum of the persistent train was captured in the first order at 37 seconds and from 1 minute and 23 seconds after the train formation. The spectrum at these times consisted only from continuous radiation as shown in Figures 6 and 12. The spectrum was virtually identical at both times and no spectral change occurred in the following minutes, either. The continuum only weakened.

The continuum extends from 5000 Å to at least 7500 Å with the maximum near 6200 Å and shows no significant structure. The spectral resolution is worse than for the afterglow because of the physical width of the train, nevertheless, it is still reasonable. The physical width of the train can be seen on the zero order image in the upper panel of Figure 12. Any atomic lines or molecular bands separated more than this width should be resolved, if they were present in the spectrum.

The continuum is probably identical with the yellow continuum observed in the initial stages of the train. In addition, the red continuum may also be present near 8000 Å but the sensitivity in this region is low and the signal is very noisy. The yellow continuum is very prob-

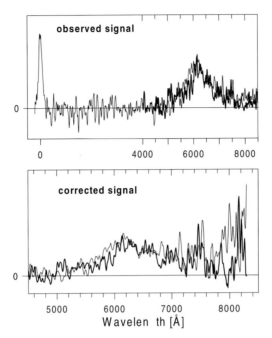

Figure 12. The spectrum of the train 37 s (bold line) and 85 s (thin line) after the train formation. The upper panel shows the uncorrected spectrum and the zero order. The lower panel shows the signal after correction to spectral sensitivity of the instrument.

ably due to molecular emissions but the source was not unambiguously identified. Various molecules like FeO, CaO, TiO, and CN have bands in this region but all should show structure under the given resolution. The most probable source seems to be the NO_2 molecule which has an extended featureless continuum and has been proposed as the possible source of the luminosity of persistent meteor trains (Rajchl 1975, Murad 2000). If NO_2 is really the source, then the energy for the radiation comes from the chemical reaction as proposed by Rajchl (1975) and Murad (2000):

$$NO + O \rightarrow NO_2^* \rightarrow NO_2 + h\nu.$$

According to Murad (2000), the reaction proceeds on the surface of metallic dust grains.

9. Comparison with other data

There are not many meteor train spectra in the literature. Nasirova and Nasirov (1966) presented a photographic spectrum of a Leonid meteor train taken in 1965. Most of the radiation was due to a continuous radiation between 5200–6400 Å, though some lines were present as well. This spectrum is the most similar to our spectrum at later times. Other good photographic spectra were taken by Abe et al. (2000) and Borovička et al. (1996). Both spectra show the Na I line but also a number of other atomic lines which differ from our spectrum. The brightest line of the Leonid train spectrum taken by Abe et al. (2000) 12–22 s after the train formation is the Mg I line at 5180 Å, which in our spectrum is totally overwhelmed by the Fe I 1 lines. The Fe I multiplets 1, 2, and 3 were not found in Abe's spectrum though the intercombination line Mg I 1 at 4571 Å is quite bright. The Perseid train spectra taken in Japan (Rajchl et al. 1995) also contain the Mg I 5180 Å line and not the Fe I 5110 Å line.

Borovička et al. (1996) identified the forbidden nebular lines of O I, O II, O III and S II, in the spectra of Perseid trains taken 5–25 s after the train formation. Since a prism spectrograph was used, the identifications may be questionable. Nevertheless, it seems that the differences between different authors are not largely due to inaccurate interpretation. Train spectra may really differ according to the altitude, density, age and other parameters of the train.

Zinn et al. (1999) reported that a Leonid meteor train was visible for over an hour on an all-sky imager with a Na 5890 Å filter. This, however, does not mean that sodium was the emitter. The signal could well be caused by the yellow continuum. Zinn et al. (1999) also presented a model and concluded that the persistent train emission should be done by the O I 5577 Å airglow with additional contributions from Na I and from NO_2 chemiluminescence.

Numerous color pictures of meteor trains were presented on the internet and in astronomical journals following the 1998 and 1999 Leonids. Many trains show a dominant yellow-orange color which may be due to the yellow continuum. There are, however, also fainter sections which show a distinct blue or red color, suggesting that train spectra may vary also within one persistent train.

10. Summary

In conclusion, we presented here for the first time resolved spectroscopy of a persistent meteor train. The number of observed spectral lines was larger than ever detected in a meteor train. The temperature of the train in the first seconds of its existence was derived. For the first time, the spectrum of the persistent train was also recorded tens of seconds and minutes after the train formation. At that time, the luminosity was done by a yellow continuum.

The facts from the previous sections can be summarized as follows: The fireball reached about -13 mag at a relatively low altitude between about 85 and 80 km. The whole trajectory extended from as high as ≥ 190 km to a very low end for a Leonid, 56 km. At its peak, the fireball spectrum contained relatively bright H lines and the inferred hydrogen abundance is about three times higher than in Perseid fireballs. The lines in the fireball spectrum can be divided into several groups according to their origin (atmospheric or meteoric) and the degree of excitation. The relative brightness of the groups of lines changed with time, in dependence on the ablation rate.

A bright afterglow was formed at the position of bright fireball flares at a relatively low altitude, 85–75 km. The afterglow decayed quickly over a period of a few seconds and a persistent train remained visible for at least eleven minutes. The afterglow emission originates from atomic compounds ablated by the meteor as well as from excited atmospheric gas. We also find evidence for continuum emission of dust at temperatures around the evaporation temperature of silicates (≈ 1400 K), which is likely meteoric debris that has not fully ablated. Part of the train had its own inertia and moved downward for a second, providing additional energy for radiation and ablation.

The upper part of the train was stretching up to the altitude of 140 km and was caused by the forbidden 5577 Å oxygen line which has a long radiative lifetime. This emission is not related to that caused by the ablation matter.

After the formation of the train, the temperature of the gas was the same as the temperature of the main part of the ablated gas in the fireball itself, i.e. 4500–5000 K. The temperature then decreased rapidly, causing a brief initial decrease of the train luminosity. The (electron) density in the train was low and thermal equilibrium was not satisfied in the train. Radiative deexcitation of atomic levels was important and semi-forbidden lines were enhanced relatively to the allowed transitions. The apparently high Mg/Fe ratio in the train was not explained. In the fireball itself the Mg/Fe ratio was normal.

After about 30 s or less, the glow ceased and a persistent train was observed having a featureless continuum, possibly due to NO_2. The prevalence of this emission may be associated with the presence of dust. The comparison with other data suggest that other trains may show different emissions and different structure. There may be various kinds of meteor trains which differ by their altitude, density, dust to gas ratio, and ongoing chemistry. More observations are needed to reveal all possible mechanisms responsible for meteor train phenomenon. Better resolution is needed to reveal the true nature of continuous and quasi-continuous emissions.

Acknowledgements

The 1999 Leonid Multi-Instrument Aircraft Campaign was made possible by grants from the NASA Exobiology, Planetary Astronomy, and Astrobiology Advanced Missions and Technology Programs, NASA Ames Research Center, and the U.S. Air Force. The data analysis was supported by grant 205/99/0146 from the Grant Agency of the Czech Republic. *Editorial handling:* Frans Rietmeijer.

References

Abe, S., Ebizuka, N., Watanabe, J., Murayama, H., and Ohtsuka, K.: 2000, *Meteorit. Planet. Sci.*, in press.
Borovička, J.: 1994, *Planet. Sp. Sci.* **42**, 145–150.
Borovička, J. and Betlem, H.: 1997, *Planet. Sp. Sci.* **45**, 563–575.
Borovička, J. and Boček, J.: 1995, *Earth, Moon, Planets* **71**, 237–244.
Borovicka, J. and Majden, E.P.: 1998, *J. R. Astron. Soc. Canada* **92**, 153–156.
Borovicka, J. and Spurný, P.: 1996, *Icarus* **121**, 484–510.
Borovička, J., Zimnikoval, P., Škvarka, J., Rajchl, J., and Spurný, P.: 1996, *Astron. Astrophys.* **306**, 995–998.
Halliday, I.: 1961, *Publ. Dominion Obs. Ottawa* **25**, 3–16.
Halliday, I.: 1968, in Ĺ. Kresák and P.M. Millman (eds.), *Physics and Dynamics of Meteors*, IAU Symp. 33, D. Reidel Publ. Co., Dordrecht, pp. 91–104.
Jenniskens, P. Butow, S., and M. Fonda: 2000, *Earth, Moon, Planets* **82–83**, 1-26.
Jenniskens, P. and Rairden, R.: 2000, *Earth, Moon and Planets* **82-83**, 459–472.
Mihalas, D.: 1978, *Stellar Atmospheres*, W.H. Freeman and Co, San Francisco, 650 pp.
Millman, P.M.: 1953, *Nature* **172**, 853.
Moore, C.E.: 1945, *A multiplet table of astrophysical interest*, Contr. Princeton Obs. **20**.
Murad, E.: 2000, *Heterogeneous chemical processes as source of persistent meteor trains*, in preparation.
Nasirova, L.I. and Nasirov, G.A.: 1966, *Astron. Tsirk.* **370**, 1–2.
Rajchl, J.: 1975, *Bull. Astron. Insts. Czech.* **26**, 282–288.
Rajchl, J., Boček, J., Očenáš, D., Škvarka, J., Zimnikoval, P., Murayama, H., and Ohtsuka, K.: 1995, *Earth, Moon, Planets* **68**, 479–486.
Spurný, P., Betlem, H., van't Leven, J., and Jenniskens, P.: 2000a, *Meteorit. Planet. Sci.* **35**, 243–249.
Spurný, P., Betlem, H., Jobse, K., Koten, P., and van't Leven, J.: 2000b, *Meteorit. Planet. Sci.* **35**, in press.
Zinn, J., Wren, J., Whitaker, R., Szymanski, J., ReVelle, D.O., *et al.*: 1999, *Meteorit. Planet. Sci.* **34**, 1007–1015.

FeO "ORANGE ARC" EMISSION DETECTED IN OPTICAL SPECTRUM OF LEONID PERSISTENT TRAIN

PETER JENNISKENS AND MATT LACEY
SETI Institute, NASA ARC, Mail Stop 239-4, Moffett Field, CA 94035
E-mail: pjenniskens@mail.arc.nasa.gov

and

BEVERLEY J. ALLAN, DANIEL E. SELF, AND JOHN M.C. PLANE
School of Environmental Sciences, University of East Anglia, Norwich NR4 7TJ, UK
E-mail: J.Plane@uea.ac.uk

(Received 13 July 2000; Accepted 31 July 2000)

Abstract. We report the detection of a broad continuum emission dominating the visual spectrum of a Leonid persistent train. A comparison with laboratory spectra of FeO "orange arc" emission at 1 mbar shows a general agreement of the band position and shape. The detection of FeO confirms the classical mechanism of metal atom catalyzed recombination of ozone and oxygen atoms as the driving force behind optical emission from persistent trains. Sodium and iron atoms are now confirmed catalysts.

Keywords: Airglow, FeO, Leonids 1999, mesosphere, meteors, persistent train

1. Introduction

Bright fireballs of fast meteors leave persistent trains that are visible for many minutes because of a luminous mechanism that is poorly understood. The persistent trains are significant, because they display the wind direction and velocity at altitude and probe the interaction of meteoroids with the atmosphere. A better understanding of the luminous mechanism is needed to make full use of the unique opportunity of probing the physical conditions in the meteor's path many minutes after the meteor (Jenniskens *et al.*, 2000). The first optical spectra of long lasting persistent trains were obtained by eye and with slit-less spectrographs and, consequently, were of low resolution. Visual inspection of persistent trains show two bright lines in the green and yellow, tentatively identified with MgI at 517.3 and 518.4 nm and NaI at 589.0 and 589.6 nm (Trowbridge, 1907). Rajchl *et al.* (1995), Borovicka *et al.* (1996), and Abe *et al.* (2000) have published photographically recorded slit less spectra. Borovicka *et al.* (1996) studied two Perseid spectra (370-640 nm) taken between 5 and 25 seconds after the meteor. They reassigned the green line to OIII rather than to MgI. In contrast, they also assigned a near-UV line at 372 nm to OII and a red line at 630 nm to OI, without explaining the range of excitation conditions. The line identifications are somewhat uncertain because of poor 25-nm resolution and no zero order detection.

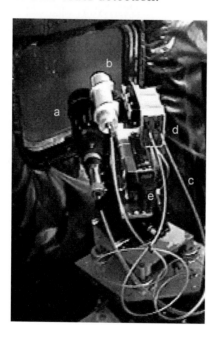

Figure 1. Instrument on optical bench behind aircraft window. (a) F5 / 400 mm telescope for visual and near-IR emission, (b) F3 / 150 mm telescope for near-UV and visual emission, (c) Optical fiber, (d) Miniature spectrometer, (e) Intensified camera for pointing.

To better understand the luminous mechanism of persistent trains at times when the trains become diffuse and faint, modern slit spectroscopy is called for. Slit spectroscopy can shed light on the assignment of the green line, while CCD detectors can extend the spectral range beyond the typical photographic range.

In 1999, fiber-optic coupled slit spectrographs were deployed at Weybourne Atmospheric Observatory in Norfolk (UK) and from aircraft during the Leonid MAC mission. While clouds prevented all but a single 10-second observation from Weybourne, several trains were observed during the airborne Leonid MAC campaign. We report here the first slit-spectra of a persistent train. All spectra were taken minutes after the meteor appeared and should characterize the mechanism of the persistent glow.

2. Instrumental techniques

The airborne instrument consisted of a compact mount of two optical telescopes: a Celestron Firstscope f5.0/400 mm for low 2.1 nm resolution at 200 - 850 nm wavelengths (Figure 1- "a") and an OptoSigma UV achromat lens 47.0 mm, f = 149.4 mm ("b") for high 0.3 nm resolution at 300-440 nm wavelengths. The telescopes are connected with a 2-meter 600-micron fused-silica patch fiber with SMA905 connectors and 74-UV collimating lenses ("c") to a dual channel Ocean Optics miniature fiber optics spectrograph SD2000 ("d"). The near-UV telescope is connected to a 2400 l/mm holographic grating with UV Detector Upgrade and detector collection lens and a fixed 25-micron slit installed (master). The VIS-NIR telescope was connected to a 600 l/mm blazed grating (400 nm) with a fixed 50 micron slit (slave). The whole assembly can be rotated and pointed to a persistent train. A co-aligned f2.8/100mm Mullard XX1332 intensified camera (Figure 1e) is used for training the telescopes at the persistent train. Its field of view is about 19 x 15 degrees and star limiting magnitude about +8.2. The camera is connected to a video headset display (I-goggles) that is worn by the operator, who also handles the data storage on a Sony Notebook laptop computer and carries an external trigger to start the exposure. Several persistent trains were observed with this instrument at its lowest resolution.

3. Results

The train left by the 3:30:33 UT meteor on November 18, 1999 (Figure 2), provided both successful pointing and data gathering in a timely manner. We obtained six 30s exposures for the train starting at 4m59s, 6m00s, 6m52s, 13m29s, 14m8s, and 15m03s after the first appearance of this meteor. The train was observed low in a southern direction towards the coast of Tunisia. Individual spectra are too noisy for analysis. The combined spectrum is reproduced in Figure 3, yet without correction for spectral response of the instrument. The spectra show the forbidden green line at 557.7 nm of OI, Na emission at 589.5 nm, and a broad continuum emission that stretches from about 500 to 700 nm. There is also a spectral feature at about 633 nm. The continuum emission is not observed in the background airglow spectrum taken earlier that night (lower part of Figure 3). The 589.5 Na emission is weaker in the background airglow also, but the 557.7-nm OI line is not.

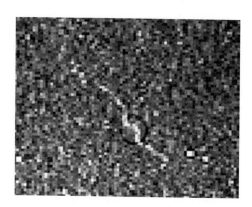

Figure 2. Persistent train of 3:30:33 UT meteor targeted in this work. The dark circle indicates the field of view (0.1° diameter) and the approximate position of measurement.

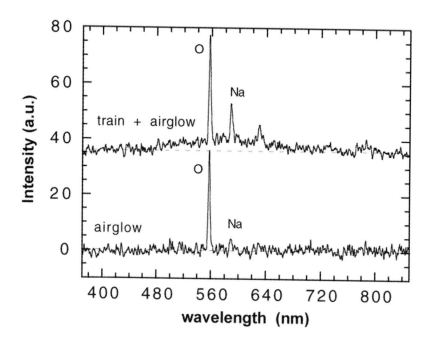

Figure 3. Persistent train spectrum of the 3:30:33 UT meteor and background airglow emission observed with the same instrument.

Figure 4 shows the result after subtraction of the airglow spectrum and correction for the wavelength dependent sensitivity of the system. Note that the sensitivity of the visible channel falls off gradually below 450 and above 700 nm. After background subtraction, a residual 557.7-nm OI emission feature remains plus residual sodium and the broad continuum emission. OH Meinel bands may be present at 610-640 nm (6,1) and 680-700 nm (7,2).

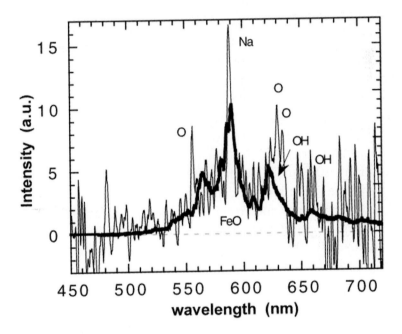

Figure 4. Persistent train spectrum of the 3:30:33 UT meteor after subtraction of airglow background and normalization for the instrument's spectral response. Also shown is an overlay of a laboratory spectrum of the FeO "orange arc" emission bands (thick line).

The spectroscopic observations confirm that the most intense emission arises from the Na D-line, almost certainly through the Chapman airglow mechanism:

$$Na + O_3 \rightarrow NaO + O_2 \qquad (1)$$

$$NaO + O \rightarrow Na(3^2P, 3^2S) + O_2 \qquad (2)$$

where the branching ratio of reaction 2 to produce the Na(3^2P) state, that then emits an orange photon at 589 nm, is ~10% (Clemesha *et al.*, 1995).

The broad continuum points to a molecular emission band, which probably arises from:

$$Fe + O_3 \rightarrow FeO(^5\Delta \text{ etc.}) + O_2 \qquad (3)$$

$$FeO + O \rightarrow Fe + O_2 \qquad (4)$$

where the exothermic reaction 3 produces FeO in excited electronic states, leading to emission in the "orange arc" bands between 570 and 630 nm with more than 2 % efficiency (West and Broida, 1975; Helmer and Plane, 1994). Figure 4 compares the observations with our laboratory spectrum of FeO emission. This spectrum was obtained in a fast flow tube operating at 1 mbarr of N_2 and 298K. Fe atoms were generated by the pulsed laser ablation of a pure iron rod using a Nd:YAG laser at 532 nm (energy ≈ 20 mJ pulse^{-1}), and mixed downstream with O_3 (concentration ≈10^{12} cm^{-3}). Further downstream, corresponding to a flow time of 8 ms, the chemiluminescence spectrum was recorded using a 0.5 m f/6.9 Czerny-Turner spectrometer connected to a 1024 x 256 pixel charge coupled device (CCD) detector. The resolution of the instrument with the slits set to 200 μm was approximately 2.2 nm, corresponding to a sampling ratio of 9 pixels per full width at half maximum (FWHM).

The "orange arc" bands of FeO between 500 and 700 nm match the width and shape of the continuum emission in the meteor train very well, particularly the peak around 590 nm and a significant contribution to the 633 nm feature. However, the feature at 553 nm does not appear to be well matched within the noise of the observations. One should bear in mind that our laboratory spectrum was taken at a much higher pressure than encountered in the upper mesosphere and lower thermosphere. Thus, the relative heights and shapes of the three dominant FeO peaks may be somewhat different at low pressure when quenching is absent.

5. Discussion

It is clear from Figure 2 that significant background emission may have been picked up by our spectrometer. The natural airglow emissions are much like the train emissions. Could the broad band be due to artificial light from coastal villages of Tunisia? Such contamination was not present earlier in the night, when ARIA was farther to the East. Indeed, the difference-spectrum (Figure 4) is not unlike that of airglow emission spectra monitored at Kitt Peak and Mount Hopkins Observatories, for example, which also show a broad emission feature centered at 590 nm (e.g., Massey and Foltz, 2000). This broad emission feature was assigned to high-pressure sodium (HPS) lamps of nearby cities. Upon further inspection, we find that the feature does not increase in intensity with other artificial emissions. Also the shape of the airglow band is somewhat broader and slightly shifted from the HPS emission in the light polluted skies over Silicon Valley (Figure 5).

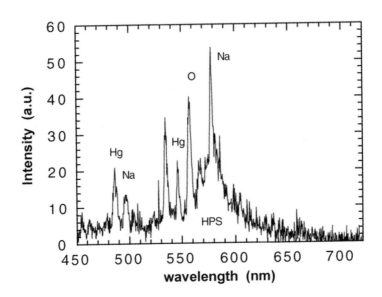

Figure 5. High-pressure sodium (HPS) and other artificial emissions measured with the same instrument in the light-polluted Silicon Valley, California.

Support for the assignment of the key chemical mechanisms of Equations 1-4 comes from the low-resolution slit less spectrum of a bright persistent train reported by Borovicka and Jenniskens (2000). The train itself is visible as excess emission in zero order, while the first order spectrum clearly identifies the train as the source of the spectral feature. This spectrum has a nearly identically shaped broad band as in Figure 4. The NaI emission associated with this train was not observed, possibly because of the lower spectral resolution. The broad band is slightly shifted to higher wavelengths, but the slit less technique can cause wavelength errors of at least ± 20 nm. Borovicka and Jenniskens found the band to be centered on 610 nm, which led to a tentatively assigned to NO_2. Indeed, a complex series of visible NO_2 bands has been observed due to chemi-luminescence from the radiative recombination reaction of NO and O. In the past, a pseudo continuum of the sum of many overlapping bands of this radiative reaction was proposed as the source for the continuum in the airglow (Hertzberg, 1966, p. 507). However, the peak in the NO_2 pseudo continuum is at about 400 nm, with a secondary maximum at, amongst others, 660 nm. Also, it is not clear how to sustain the NO_2 emission in the persistent trains. We conclude that the assignment to FeO emission is the more likely.

The presence of the two red 630.0 and 636.4 nm "auroral" lines of atomic O in the spectrum is surprising. These lines are produced from $O(^1D)$ with a radiative lifetime of over 100s. However, the excited state is rapidly quenched by N_2 and O_2 collisional de-excitation in the lower thermosphere and upper mesosphere. Typically, $O(^1D)$ emission is observed by satellites only in the altitude range 210-270 km where the pressure, and hence the quenching rate, is very low. Unless the highly unlikely scenario that an aurora was in progress simultaneously that could be seen from the Mediterranean looking in a south/south-western direction, OI line emission cannot be entirely responsible for the large 630 nm peak. FeO emission can significantly contribute to this peak, as well as the OH Meinel bands. Indeed, the peak is much broader than due to an atomic emission (see, the width of the 558 nm green line).

Unlike the spectra obtained by Abe *et al.* (2000), there is no strong MgI emission in our spectra. The possible feature around 520 nm, if real, is probably due to $N(^2D-^4S)$ emission at 519.9 nm. A $Mg(^3S-^3P)$ line at 520 nm would be hard to explain because $Mg(^3S)$ is excited by over 5 eV. The only obvious mechanism for exciting Mg to this level is via dissociative recombination of molecular ions with electrons:

$$Mg^+ + O_3 \rightarrow MgO^+ + O_3 \qquad (5)$$

$$Mg^+ + O_2 (+ M, \text{third body}) \rightarrow MgO_2^+ \qquad (6)$$

$$Mg^+ + N_2 (+ M) \rightarrow MgN_2^+ \qquad (7)$$

$$MgX^+ + e^- \rightarrow Mg^* + X \tag{8}$$

where MgX^+ represents the molecular ions formed in reactions 5–7 (Rowe et al., 1981). Abe et al. reported MgI in the spectrum from a bright Leonid fireball at about 10 nm resolution (380–600 nm) wherein this assignment was confirmed by the presence of several other MgI lines at 457, 470, and 553 nm. However, they also identified atomic lines of CaI (443 nm) and FeI (418, 486 and 537 nm), although with less good agreement between theory and observation. Possibly, these emissions are of relatively short duration. The train spectra were obtained shortly after the meteor had extinguished, at a time when the train was still spatially confined on the sky and measured gas temperatures were relatively high. It is possible that the MgI emission is part of a phenomenon called the meteoric afterglow (Borovicka and Jenniskens, 2000), but less important when the train has had time to cool.

In summary, the optical spectrum discussed here confirms the classical mechanism for train luminosity. We find that FeO rather than FeI contributes to the train luminosity. Future work under better observing conditions and with improved instruments are expected to increase the number of metal atom catalysts beyond Na and Fe, and may reveal other chemical processes in the meteor path.

Acknowledgements

We are grateful for the constructive reviews by Frans Rietmeijer and an anonymous referee to improve the presentation of this paper. The NASA Leonid MAC is supported by the NASA Exobiology and Suborbital MITM programs, the NASA Astrobiology Advanced Missions and Technology Program, NASA Ames Research Center and the U.S. Air Force. Matt Lacey held an Astrobiology Academy position at NASA Ames Research Center for the summer of 1999. The laboratory work at the University of East Anglia was supported by the Natural Environment Research Council. *Editorial handling:* Frans Rietmeijer.

References

Abe, S., Ebizuka, N., Watanabe, J.-I., Murayama, H, and Ohtsuka, K.: 2000, *Meteoritics Planet. Sci.* **35**, in press.

Borovicka, J. and Jenniskens, P.: 2000, *Earth, Moon and Planets* **82–83**, 399–428.
Borovicka, J., Zimnikoval, P., Skvarka, J., Rajchl, J., and Spurny, P.: 1996, *Astron. Astrophys.* **306**, 995–998.
Clemesha, B.R., Simonich, D.M., Takahashi, H., Melo, S.M.L., and Plane, J.M.C.: 1995, *J. Geophys. Res.* **100**, 18909–18916.
Helmer, M. and Plane, J.M.C.: 1994, *J. Chem. Soc., Faraday Trans.* **90**, 31–37.
Hertzberg, G.: 1966, Molecular Spectra and Molecular Structure. III. Electronic Spectra and Electronic Structure of Polyatomic Molecules. D. Van Nostrand Company Inc., Princeton, New Jersey, 745 pp.
Jenniskens, P., Nugent, D., and Plane, J.M.C.: 2000, *Earth, Moon and Planets* **82–83**, 471–488.
Massey, P., and Foltz, C.B.: 2000, *Publ. Astron. Soc. Pacific* **112**, 566–573.
Rajchl, J., Bocek, J., Ocenas, D., Skvarka, J., Zimnikoval, P., Murayama, H., and Ohtsuka, K.: 1995, *Earth, Moon and Planets* **68**, 479–486.
Rowe, B. R., Fahey, D. W., Ferguson, E.E., and Fehsenfeld, F.C.: 1981, *J. Chem. Phys.* **75**, 3325–3328.
Trowbridge, C.C.: 1907, *Ap. J.* **26**, 95–116.
West, J. B. and Broida, H.P.: 1975, *J. Chem. Phys.* **62**, 2566–2574.

MID-INFRARED SPECTROSCOPY OF PERSISTENT LEONID TRAINS

RAY W. RUSSELL, GEORGE S. ROSSANO,
MARK A. CHATELAIN, DAVID K. LYNCH,
TED K. TESSENSOHN, ERIC ABENDROTH, AND DARYL KIM

*Space Science Applications Laboratory,
The Aerospace Corporation, M2-266, P.O. Box 92957, Los Angeles, CA 90009-2957
E-mail: Ray.W.Russell@aero.org*

and

PETER JENNISKENS

SETI Institute, NASA ARC, Mail Stop 239-4, Moffett Field, CA 94035

(Received 16 August 2000; Accepted 1 September 2000)

Abstract. The first infrared spectroscopy in the 3–13 micron region has been obtained of several persistent Leonid meteor trains with two different instrument types, one at a desert ground-based site and the other on-board a high-flying aircraft. The spectra exhibit common structures assigned to enhanced emissions of warm CH_4, CO_2, CO and H_2O, which may originate from heated trace air compounds or materials created in the wake of the meteor. This is the first time that any of these molecules has been observed in the spectra of persistent trains. Hence, the mid-IR observations offer a new perspective on the physical processes that occur in the path of the meteor at some time after the meteor itself has passed by. Continuum emission is observed also, but its origin has not yet been established. No 10 micron dust emission feature has been observed.

Keywords: Meteors, meteoroids, mid-IR spectroscopy, persistent trains

1. Introduction

Spectroscopy of meteors and persistent trains in the infrared (IR) part of the spectrum from about 3 to 13 microns has long been expected to be a useful tool in our efforts at understanding the composition of meteoroids and how they interact with the Earth's atmosphere (Jenniskens and Butow, 1999).

Shower meteoroids originate in the dust grains thrown off from parent comets, 55P/Tempel-Tuttle in the case of the Leonids, and follow similar orbital paths for long periods of time. Our group at The Aerospace Corporation has been studying the thermal spectra of comets for many years (e.g., Hanner *et al.*, 1994) in order to determine the temperature, composition, and morphology of cometary dust grains. In the case of comet 55P/Tempel-Tuttle, the thermal IR emission of dust in the comet coma showed a gray body behavior that is typical of large grains or organic materials (Lynch *et al.*, 2000), but not very diagnostic of the detailed nature of the grains.

During the heating of the grains as they interact with the atmosphere as meteors, meteoroid fragments, molecules, or even atoms are separated from the grain in a process of ablation (Boyd, 2000; Popova, 2000), and are heated sufficiently to exhibit spectral signatures that could shed additional light on at least the underlying composition of these grains. The thermal emission from the heated grains striking the atmosphere will appear first in the long wave IR, and shift to shorter wavelengths as the body gets hotter. The meteor's kinetic energy is sufficiently large to completely vaporize the meteoroid.

Moreover, the kinetic energy is sufficiently large to heat a significant volume of air in the mesosphere and lower thermosphere. Long persisting luminous glows are seen at visual wavelengths in the path of bright Leonid fireballs, called "persistent trains," which allow pointing at and tracking of this heated air.

The first published mid-IR broadband detections of meteors, but no spectra, were obtained during the 1998 Leonid Multi-instrument Aircraft Campaign (Jenniskens and Butow, 1999) and are presented in Rossano, *et al.* (2000). The problem with attempts at IR spectroscopy of transient phenomena such as meteors has been the difficulty in capturing the signals themselves with IR sensors with sufficient sensitivity, a large enough spatial field of view and yet fine enough spatial resolution, and spectral resolution to permit the analysis of these phenomena. Major advances in recent years in IR sensor design, detector arrays, and data acquisition systems have created the capability to acquire, track, and measure these phenomena with sufficient sensitivity to provide meaningful datasets with which to investigate meteor-related phenomena.

This paper reports the first mid-IR spectra of Leonid persistent trains obtained with an imaging spectrograph, taking full spectra with each scan, situated onboard the 1999 Leonid Multi-instrument Airborne Campaign (Leonid MAC) on a mission to the Mediterranean, and with a wavelength-scanning spectrometer on the ground at the Starfire Optical Range at Kirtland AFB in New Mexico.

2. The Observations

The data were acquired with the Aerospace Mid-wave InfraRed Imaging Spectrograph (MIRIS) and the Circular Variable Filter wheel spectrometer (CVF). The MIRIS uses a liquid nitrogen-cooled 2D 256 x 256 InSb array and a grism (combination grating and prism, Rossano *et al.* 2000) to obtain long-slit spectra and zeroth order broadband images in the 3-5.5 micron spectral region (ibid). The slit was constructed to permit slitless spectroscopy and imaging of meteors in the center 128 rows of the array, and also to permit narrow slit (5 pixels wide in the dispersion direction) spectroscopy of meteor persistent trains in the 64 rows near the top of the array and the 64 rows near the bottom of the array. MIRIS was deployed aboard the USAF Flying Infrared Signature Technology Aircraft (FISTA) as part of the 1999 Leonid MAC effort (Jenniskens *et al.*, 2000). The spectral resolving power is about 50 (due to the extended nature of the source and the slit width), and spatial pixel size was about 0.8 mrad (0.046 °). The sensor viewed the sky through a ZnSe window while the FISTA was flying at about 10–12 km, where the sensor was above the majority of the Earth's atmosphere and more than 99% of the water vapor. Complete spectra were obtained at 24 frames per second with an observing efficiency of about 40%, generating gigabytes of data during the MAC.

The CVF uses three multi-layer interference wedges and an Si:As back-illuminated blocked impurity band (BIBIB) detector element cooled to liquid helium temperatures to obtain spectroscopy from 2.5 to 14.5 microns with a spectral resolving power of about 60, with an off-axis parabola as the collecting optic in a custom setup created for this event. It was mounted on a steerable alt-azimuth telescope mount at the Starfire Optical Range (SOR), Kirtland AFB (New Mexico), and the field of view was approximately 0.25 degree. Data were obtained in a step-and-integrate mode as DC voltages with the sensor viewing the sky or sky plus meteoroid train. Each spectrum required about 5–10 minutes to acquire. Spectra on the meteor trains were compared to spectra of the sky at the same elevation angle but at a slightly different (delta of about 15°) azimuth. Sky spectra were very consistent over the course of the night, facilitating the accurate subtraction of atmospheric and sensor emissions contributing to the spectral signals measured while observing the persistent meteor trains.

The wavelength calibration of the MIRIS was achieved through the use of known absorption features in calibrated pieces of plastic and was good to about 0.03 microns. The plastic materials, none of which was a commercial calibration product, were taken from a variety of sources and their absorption wavelengths calibrated on a Fourier Transform spectrometer. The CVF wavelength calibration was performed with an Oriel single-pass grating monochromator to better than 0.3 of a spectral resolution element, where a spectral resolution element is about 0.017 times the wavelength. The radiometric calibrations of both sensors were accomplished through observations of extended blackbodies designed and built for this purpose at The Aerospace Corporation. These sources have been compared to NIST-traceable cavity blackbodies and shown to have an emissivity of about 95–98% over the entire spectral range of interest here except in the 8–10 micron region, where a small (~10%) dip is observed due to glass beads in the 3M Black Velvet paint used to coat the blackbodies. The blackbodies are heated by Kapton-coated extended heater elements and controlled through either an automatic servo-controlled heater or by manual control of a Variac transformer, in both instances using thermocouples on the blackbodies to determine the temperature of the sources. There is an absolute temperature uncertainty in the surfaces of the blackbodies of about 1–2 K due to the uncertainty from the thermocouple and the paint thickness and non-uniformities over the surface of the blackbodies. The temperature stability was typically 0.2 K.

Both IR sensors utilized bore-sighted optical sensors (CCDs) to obtain starfield information for pointing determination and for direct simultaneous viewing of the meteor trains that were being measured in the IR. Pointing alignment between the IR and visible scenes was verified using bright sources such as the moon and man-made sources (landing strip lights). At SOR, initial visual sightings of the trains led to verbal direction to steer the sensors toward a particular part of the sky, and then visible CCD images were used to guide the commanding of the telescope mount to perform fine positioning of the IR beam onto the train. For the FISTA observations, initial detections in the visible caused a command to turn the plane to a proper heading, and then the eyeball mount was steered by hand while monitoring the CCD image to position the slit of the IR sensor on the train.

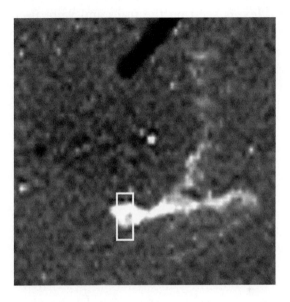

Figure 1. Approximate position of the MIRIS slit on the "Y2K" persistent train at 04:02:19 UT on Nov. 18, 1999. The white box indicates the location of the MIRIS slit and the dark marker in the top center of the image was used for positioning the camcorder relative to the image intensifier.

3. Results

3.1. MIRIS RESULTS

The airborne MIRIS data were obtained the night of the storm, 18 Nov. 1999 UT, with the best spectrum recorded being that of the so-called "Y2K" train (Figure 1), caused by a –13 magnitude Leonid fireball at 04:00:29 UT (Jenniskens and Rairden, 2000). The train was observed from about 04:02 to 04:09 UT. During our observations, the train was at an altitude of 83.2 ± 1.0 km, which corresponds to positions 14–18 in Jenniskens and Rairden (2000), at a distance of 205 km from the FISTA aircraft and 19° above the eastern horizon at the airplane's altitude.

The train was observed by MIRIS in both zeroth and first orders. It was first observed as a zero order image in the slitless region of the spectrograph, and then MIRIS was repositioned to place the brightest part of the train onto the narrow slit which spanned the upper 64 rows of the array. The train spectrum was observed over approximately 8 rows up and down the narrow slit. Adjacent rows on the array do not show the enhanced signal levels seen on the train and were used to subtract the signals due to sky emission and instrumental background from the train spectra. An average scan was made of about 20 frames, representing 10 msec exposure each at 1m50s after the fireball, at a time when the brightest part of the train was first positioned in the slit of the spectrograph (Figure 1).

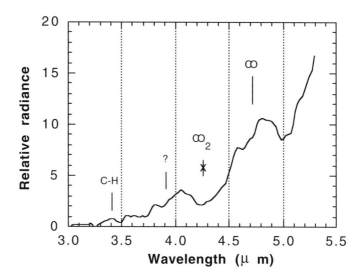

Figure 2a. Linear plot of the intensity from the "Y2K" train at 04:02:19 UT reported by MIRIS.

Figure 2a shows the MIRIS data on a linear plot. Superposed on a continuum that rises towards longer wavelengths are seen several broad emission features. A small feature at 3.4 micron is readily identified as the C-H stretch vibration band, possibly of the molecule CH_4 or some C-H bearing complex organic molecules. The strong emission feature between 4.4 and 5.0 microns is attributed to emission from CO (see Section 4). Surprisingly enough, no CO_2 emission is detected around 4.3 microns. Strong emission bands in the 3.5–4.2 micron region of the train spectrum remain unexplained at this time.

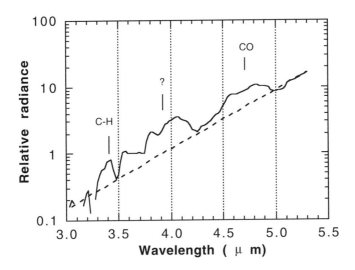

Figure 2b. Semi-log plot of the intensity from the "Y2K" train at 04:02:19 UT reported by MIRIS.

The same data are shown in a semi-log plot to bring out the molecular bands superposed on a continuum rising towards five microns (Figure 2b). Note that the alternative placement of the continuum, with a CO_2 band in absorption, is a less likely choice because it would imply a very high abundance of absorbing molecules and / or solid materials in atmospheric windows from 3.4–4 µm and around 5 µm, for example.

The nature of this continuum is not understood at this time. We are trying to assess the relative likelihood of thermal emission by small solid particles versus a molecular origin such as blended water vapor lines typically found in this part of the spectrum. No optically thick (blackbody) or thin (graybody) thermal emission at $T > 1000$ K, such as would be seen from hot dust grains, has been detected at this time after the passage of the meteor. Such a hot thermal emission would have resulted in a continuum rising towards shorter wavelength in this regime.

3.2. CVF RESULTS

Ground-based CVF data were obtained on five trains that occurred during the nights of Nov. 17 and 19, 1999 UT. Due to cloud cover on the night of the peak of the shower (Nov. 17/18), no spectra were collected that night. The IR signatures were observed to last at least as long as the visible signatures, in some cases more than 20 minutes. On the first night at SOR, some of the trains that were seen in the visible had dimmed by the time the platform was pointed in their direction and the IR data were taken even though the visible evidence of the trains on the CCD images was gone. However, IR spectra similar in nature to those shown here were still obtained, suggesting that the IR signatures were longer-lived than the visible emissions.

We present spectra and discuss two trains. The "Puff Daddy" train was caused by a bright Leonid meteor at 10:05 UT on Nov. 19, Az = 34, El = 35. A second train was observed following a bright Leonid at 12:26 UT on Nov. 19, Az = 0, El = 29 to 34 (the trail started at an elevation of 29 degrees, and drifted up to an elevation of 34 degrees, while the instrument tracked the trail as it drifted). Most data pertain to the "Puff Daddy" train, which was about 150–200 km away from the observers at an elevation angle of about 30 degrees.

Three narrow wavelength regimes were studied in coarse spectral steps first. Those wavelength regimes were chosen with the expectation that nitrogen-bearing molecular and CO_2 molecular emissions similar to those seen in auroral displays might be emitting in the meteor train. These narrow wavelength regions are the small pieces shown in Figures 3, 5, and 6. However, because the signal appeared to peak outside the chosen narrow wavelength pieces and we really did not know what to expect, it was decided to map the full wavelength regime (2.5–14.5 microns), which resulted in the full scans shown in Figures 3, 5, and 6. While each wavelength position was integrated and stored, the train gradually decreased in intensity and the spectral shape may also have changed during the scan. It took about 5 minutes to obtain the three narrow regions, and approximately an additional 20 minutes to obtain the three complete spectra. At this point, the train was about 26 minutes old.

3.2.1. The 2.50–4.35 micron wavelength region (wedge 1)

Figure 3 shows the data obtained with the CVF over the short wavelength filter wedge with the emission due to the C-H stretch vibration band prominent at 3.3–3.4 microns. The unidentified emission band at 3.7–4.2 microns is present in the CVF data as well as in the MIRIS data at about the same relative intensity compared to the C-H stretch vibration. Unlike for the MIRIS data, a very rapid increase coinciding with an onset of the CO emission band occurs above 4.0 microns. The data do not appear to have saturated above 4.4 microns and the rapid increase should represent the spectral band shape.

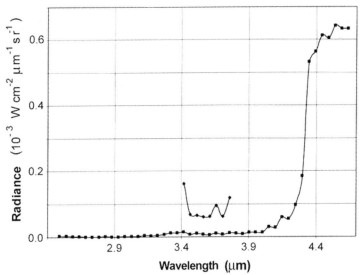

Figure 3. CVF spectra of the train of the meteor that occurred shortly before 10:06 UT, Nov. 19, 1999. The measurements were started at 10:06 UT (short spectrum) and at 10:11 UT (full scan). The effect of the decreasing train intensity between the acquisition of the narrow spectral piece and the full coverage can be seen.

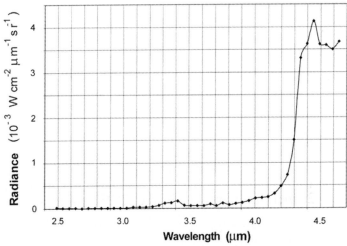

Figure 4. Short wavelength wedge CVF scan of the train of a second meteor at 12:26 UT. These data were taken from 12:27 to 12:32 UT Nov. 19, 1999. Compare to Figure 3.

Given the manner in which the data were taken and processed, that is, by subtracting the spectra obtained while observing the sky at the same elevation as the train but at a different azimuth, one might question the validity of the spectral shapes seen here. The sky signal might not have been constant over the time between meteor data acquisition and sky acquisition, or the elevation angle might not have been accurately set or maintained, or the sky signal shape or amplitude might have been dependent on azimuth. In fact, the shape and amplitude of the sky spectrum repeated well over the entire night for the same elevation angles. However, if spectra of two different meteor trains were to exhibit the same or similar spectral structures while the sky spectra over the night showed differences only at a much smaller signal level, it would lend credence to the data acquisition and processing methods. Thus, for comparison, the spectrum obtained on another meteor in this wavelength region of the spectrum is shown in Figure 4. The spectra of Figures 3 and 4 are very similar. In fact, similarly shaped spectra were obtained on three other Leonid meteor trains during this campaign.

3.2.2. The 4.3–7.8 micron wavelength range (wedge 2)

The detection of the CO_2 band with the CVF occurs only during the first quick scan of one of the three wavelength regions (Figure 5). The band is detected between 4 and 5 microns. In the subsequent full-range scan, the CO_2 emission has decayed considerably.

In the full range spectrum, strong emission attributed to "warm" H_2O is observed. In this context, warm means higher than the typical 200–300 K temperature of water vapor in the lower atmosphere. The CO band emission is not detected, possibly masked by variations in the emission from the ever changing appearance of the train. For the same reason, variations on the gradual rise may also not be spectral features.

The strong rise in the signal above 5 microns was unexpected, considering the normally high opacity of water vapor in the air along the line of sight from the ground to the meteor train. Consequently, the CVF was operated in high gain mode over the 5-8 micron region of the spectrum when observing the trains as well as the (cold) sky. In this mode, the unexpectedly high train signal drove the amplifier into saturation at wavelengths from 6.7 to 7.8 microns.

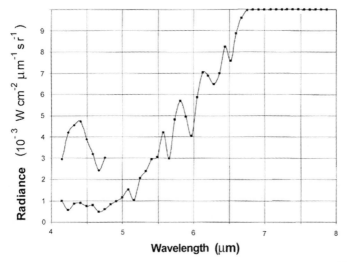

Figure 5. CVF spectrum of the train of meteor 10:05 UT Nov. 19, 1999, taken at 10:16 UT, with a strong H_2O emission peak. Inset shows the earlier narrow scan at time 10:07 UT, with strong CO_2 emission at 4.3 microns (same units as Figure 4).

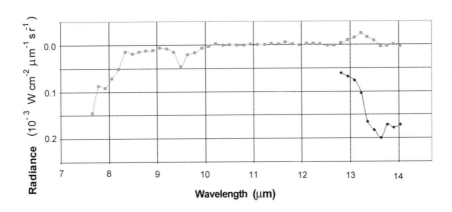

Figure 6. CVF spectrum from 7.5 to 14.5 microns of the same train observed to obtain the data shown in Figures 3 and 5. The data acquisition consisted of taking the short pieces in all three figures first, followed by the acquisition of the entire 2.5 to 14 micron spectrum. The strong absorption seen near 13 – 14.5 microns is unexplained, as is the dip from 7.7 to 8.3 microns (same units as Figures 4 and 5).

3.2.2. The 8–14 micron wavelength range (wedge 3)

The data acquisition in wedge 2 is stopped after 7.8 microns, the gain changed from high to low, the filter wheel advanced to the third filter wedge, and the user prompted for an input that it is OK to continue the data acquisition. This procedure takes more than enough time for the saturation effect of the instrument to go away in the next (8–14 micron) range, and the change to low gain also speeds up the electrical recovery from saturation.

Figure 6 shows the CVF data of the long wavelength regime from the same train that was observed to obtain the data shown in Figures 3 and 5. The scale is significantly expanded. This spectrum exhibits a puzzling downturn at wavelengths greater than 13 microns, which was more prominent in the earlier narrow spectral scan than in the later full scan. This occurs in the part of the spectrum dominated by the edge of the 16-micron atmospheric CO_2 band. The decreased emission longward of 13 microns was seen in all three trains observed with the CVF at these wavelengths.

The full scan taken later in time also exhibits an unexplained dip in the 7.7 to 8.3 micron region. (The small negative signal from ~8.3 to 10 microns is consistent with the uncertainty in the sky subtraction process, and the 9.6 micron dip may well be due to the known azimuth-dependence of ozone emission from the sky.) There is not enough additional data around 8 microns on other trains to address the reality of this dip in the train spectrum, and it seems inconsistent with the large excess observed with wedge 2 (Figure 5) just a short time before. We believe the multiple train observations of the enhanced 5–8 micron emission and the 13–14 micron dips are evidence for those features being real, but we do not have enough data for the 7.7–8.3 micron region to make a similar claim. Additional data will have to be acquired on more trains to address this issue.

4. Discussion

The emissions here reported do not originate from airglow-type chemistry that was proposed as an explanation for the optical luminosity features observed (Kelley *et al.*, 2000; Jenniskens *et al.*, 2000b). The mid-IR spectral features are assigned to enhanced emissions of CO, CO_2, CH_4 and H_2O, which may originate from heated trace air compounds or materials created in the wake of the meteor. None of the molecules detected here has been observed in the visible range. As such, the mid-IR observations offer a whole new perspective on the physical properties of meteor trains.

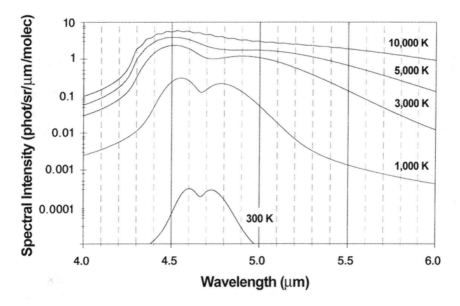

Figure 7. Model calculations for hot, optically-thin CO molecules. These spectra go from cool on the bottom to hot on the top. They have been smoothed to a resolving power of 100 to better compare the shapes with the train data.

4.1. CO

The shape of thermal emission bands can be used to determine the excitation temperature of the responsible molecules. Figure 7 shows calculations for the CO molecule that is believed to be responsible for the band between 4.4 and 5.0 microns. The shape is significantly broader at the high temperatures of T ~ 5,000 K and T ~10,000 K that are reported for the hot and warm visible emissions from meteors. The train emission is more typical for a gas at T ~ 300 K, which is consistent with the observation by Borovicka and Jenniskens (2000) that the Y2K train temperature decreased from ~4,500 K to ~1,200 K in a mere two seconds. It is also consistent with the results from Chu *et al.* (2000), which determined the temperature of sodium emission from persistent trains using LIDAR measurements of the Doppler broadening of the resonant scattering profile. Chu *et al.* (2000) found elevated gas temperatures of about 230 K at 92.2 km altitude and 260 K at 92.35 km altitude, which is ~50 K above the background ambient Na temperature of 210 K, observed 2.9 minutes after the meteor's first appearance. This time lapse is similar to that present for our observations.

4.2. CO_2

Warm CO_2 is observed by the CVF instrument in the persistent train spectra immediately after train formation. The initial scan of Figure 5 about 2 minutes after the meteor shows the emission at the expected position. It is not observed in the MIRIS data 1m52s after train formation and has declined to background levels in a subsequent CVF scan taken 9 minutes later. A rapid decline of the CO_2 emission is implied, perhaps responsible for the lack of CO_2 emission in the MIRIS data.

4.3. C-H

The C-H stretch vibration is observed in all MIRIS and CVF data. The band is not well resolved. It has a gradual rise above 3.27 microns, with possibly two maxima at 3.35 and 3.42 microns and a sharp downturn above 3.46 microns (see first scan Figure 3).

The 3.4 microns feature in room temperature cometary dust typically has a similar asymmetric peak with a broad maximum at 3.38 microns and a steeper downturn at the long wavelength edge (e.g. Encrenaz and Knacke, 1991), and is characteristic of complex organic matter rich in CH_2 and CH_3 groups. Additional modeling will be required for the detailed interpretation of the origin of the emission structure, but the Occam's Razor assignment would be to the simple, easy to form, prevalent molecule CH_4. However, we can't rule out more complex molecules or even organic solids at this time.

4.4. H_2O

As we consider wavelengths long-ward of 5 microns in Figure 5 (and the implication of the trend in the MIRIS data in Figure 2a), we see a gradual rise towards 6.7 microns. It is likely that the gradual rise is followed by a decline of emission at wavelengths above 6.7 microns as suggested by subsequent observations in the 8–14 micron range.

It is not certain that the rise itself must be due to warm water vapor band emission centered at 6.24 microns, or whether there may be thermal dust continuum emission analogous to that seen in cometary spectra. However, the absence of strong continuum dust emission in the 8–13 micron region (wedge 3, Figure 6) argues against a broad thermal dust continuum with or without silicate emission, and is consistent with the rapid decline expected in warm water vapor emission beyond 7 microns.

One difficulty with the warm water vapor interpretation is that the warm gas must have a temperature above 200–300 K relatively long after the meteor, to avoid being completely absorbed by warm lower atmospheric water vapor between the ground-based observation station and the meteor train.

4.5. OTHER FEATURES

Given the detection of CO_2 band emission at 4.2 microns (Figure 5, narrow scan), it was expected that strong CO_2 band emission would arise above 13 microns in the first narrow scan of Figure 6. However, it is unknown how a lack of emission in the long wavelength 16 micron CO_2 band can be explained in relation to the distant train. The constancy of the atmospheric emission at these wavelengths over the course of the night implies much less variation over the relatively short time between the observations of the train and of the sky (at the same elevation angle, but slightly different azimuth, a few minutes later) than the difference between the train and the sky. As long as the atmospheric emission is constant, this strongly suggests that the observed dip is a real phenomenon, and not an artifact of the sky subtraction process. The fact that the dip was seen in all the train spectra at these wavelengths (at least 3) is viewed as evidence for some real effect other than inaccurate sky subtraction. As we expected most of the observed CO_2 emission to originate close to the sensor along the line of sight from the ground to the train, and not to originate in the train itself, additional modeling of the amount of emission and absorption as a function of position along the line of sight will be undertaken in an effort to understand the data.

4.6. SYNTHESIS

Given our understanding of meteor grains based on spectroscopy of parent cometary dust, it is unlikely that the molecules that we believe are responsible for the emission reported here existed in molecular form in the meteor. It is more likely that the observed emissions are due to heating of ambient air molecules (atmospheric molecules that existed prior to the passage of the meteor), all of which are present (albeit at very low densities) at altitudes of ~85 km in the quiet atmosphere before passage of the meteor.

It is also possible that the interaction of the meteor with the Earth's atmosphere at 72 km/sec caused material to ablate and vaporize, forming a hot atomic gas. The laboratory analyses of chondritic interplanetary dust particles that include dust of cometary origin show the presence of silicate minerals and organic matter (Rietmeijer and Nuth, 2000). Thus, the dust grains that made up the meteor could be sources for the atoms in the gas species that we observed. The evolved atomic gas from the ablation process cooled and formed the excited molecules whose IR emissions are reported here.

Future work will use the Modtran software package to ascertain whether the water vapor could be ambient atmospheric molecules excited by the passage of this meteor, or whether the very dry conditions at the meteor altitude of ~85 km combined with the strength of the emissions suggests or requires that the hydrogen and oxygen came from organic and mineral components in the meteor, respectively. Note that this does not require that the meteor body contained water (such as in layer silicate minerals) or water ice. Vaporization of the meteor body may have broken down organic and inorganic materials into atomic species, thus freeing the hydrogen and oxygen necessary to produce the water seen in emission in these data.

Further evidence may also be present in the totality of the current set of data. The MIRIS data from multiple pointings at the train (as opposed to the single spectrum shown here), and over a seven minute time span, will provide a series of spectra that follows the CO and C-H emission evolution over time. These data are expected to provide detailed information about the temperature evolution in the train and the physical conditions in the meteor path for the long time scales at which organic chemistry and metal atom chemistry between meteor material and ambient air molecules occurs.

5. Summary

IR spectroscopy of persistent Leonid meteor trains shows prominent molecular band emissions at 3.4, 4.0, 4.3, 4.7, and 6–7 microns. The commonality of the spectra obtained from two platforms, one on the ground and one airborne, and with two dramatically different sensors, lends strong credence to the validity of the spectral structure seen emanating from these long-lived trains. Not all features have been identified with certainty. Obvious candidates are CH_4, CO, CO_2, and H_2O. No optically thick or thin thermal emission at > 1000 K has been seen to date, but continuum emission from cold ~ 300 K sources may be present. However, we have seen no evidence for expected emission in the wavelength region of the Si-O stretch vibration at 10 micron. The exact emission/excitation mechanisms for the long-lived (in some cases more than 20 minutes) IR signatures are still not understood. Future work will include modeling of the passage of the meteor through the atmosphere to investigate the heating and cooling of meteoric and atmospheric materials, and to model the molecular emissions at the various wavelengths to discriminate between atmospheric and meteoroid sources for the atoms.

Acknowledgments

We are grateful to the staff and crew of the FISTA and SOR, without whose strong support this work could not have been accomplished. This research is supported at The Aerospace Corporation by the Internal Research and Development program. The Leonid MAC was supported by NASA's Exobiology, Planetary Astronomy and Suborbital MITM programs, as well as by NASA's Advanced Missions and Technologies program for Astrobiology and NASA Ames Research Center. We thank Joe Kristl and his group at Space Dynamics Laboratory, Utah State University, for their assistance in the FISTA installation and operation, and for the loan of an eyeball mount and ZnSe window for use with MIRIS. Paul Zittel, The Aerospace Corporation, is acknowledged for providing the molecular emission spectra of CO. We thank referee Diane Wooden, whoes comments helped improve the paper. *Editorial handling*: Frans Rietmeijer.

References

Borovicka, J. and Jenniskens, P.: 2000, *Earth Moon and Planets* **82–83**, 399–428.
Boyd, I.D.: 2000, *Earth Moon and Planets* **82–83**, 93–108.

Chu, X., Liu, A.Z., Papen, G., Gardner, C.S., Kelley, M., Drummond, J., and Fugate, R.: 2000, *Geophys. Res. Lett.* **27**, 1815–1818.

Encrenaz, T. and Knacke, R.: 1991 in R.L. Newburn, Jr., M. Neugebauer, J. Rahe (eds.), *Comets in the Post-Halley Era*, Vol. 1, 107–137.

Hanner, M. S., Lynch, D. K., and Russell, R. W.: 1994, *Astrophys. J.*, **425**, 274–285.

Jenniskens, P. and Butow, S.J.: 1999, *Meteoritics Planet. Sci.* **34**, 933–943.

Jenniskens, P. and Rairden, R.: 2000, *Earth, Moon and Planets* **82–83**, 457–470.

Jenniskens, P., Butow, S.J., and Fonda, M.: 2000a, *Earth Moon and Planets* **82–83**, 1–26.

Jenniskens, P., Nugent, D., and Plane, J.M.C.: 2000b, *Earth, Moon and Planets* **82–83**, 471–488.

Kelley, M.C., Gardner, C., Drummond, J., Armstrong, T., Liu, A., Chu, X., Papen, G., Kruschwitz, C., Loughmiller, P., Grime, B., and Engelman, J.: 2000, *Geophys. Research Letters* **27**, 1811–1814.

Lynch, D.K., Russell, Ray W., and Sitko, M. L.: 2000, *Icarus*, **144**, 187–190

Popova, O.P., Sidneva, S.N., Shuvalov, V.V., and Strelkov, A.S.: 2000, *Earth Moon and Planets* **82–83**, 109–128.

Rossano, G.S., Russell, R.W., Lynch, D.K., Tessensohn, T.K., Warren, D., and Jenniskens, P.: 2000, *Earth Moon and Planets* **82–83**, 81–92.

BUOYANCY OF THE "Y2K" PERSISTENT TRAIN AND THE TRAJECTORY OF THE 04:00:29 UT LEONID FIREBALL

PETER JENNISKENS
SETI Institute, NASA ARC, Mail Stop 239-4, Moffett Field, California 94035
E-mail: pjenniskens@mail.arc.nasa.gov

and

RICK L. RAIRDEN
Lockheed Martin Space Sciences LaboratoryDept L9-42, Bldg 255, 3251 Hanover
Street, Palo Alto, California 94304
E-mail: rairden@spasci.com

(Received 23 June 2000; Accepted 29 July 2000)

Abstract. The atmospheric trajectory is calculated of a particularly well studied fireball and train during the 1999 Leonid Multi-Instrument Aircraft Campaign. Less than a minute after the meteor's first appearance, the train curves into a "2"-shape, which persisted until at least 13 minutes after the fireball. We conclude that the shape results because of horizontal winds from gravity waves with a scale height of 8.3 km at 79–91 km altitude, as well as a westerly wind gradient with altitude. In addition, there is downward drift that affects the formation of loops in the train early on.

Keywords: Fireball, leonids 1999, lower thermosphere, mesosphere, meteor, persistent train, winds

1. Introduction

A bright fireball of absolute magnitude about –13 appeared over the isle of Corsica at 04:00:29 UT in the night of November 18, 1999. The fireball registered on three slit-less spectrographs onboard the Leonid Multi-Instrument Aircraft Campaign, probing various wavelength ranges in the near-UV, visual and optical near-IR. The fireball provided the first spectrum of a meteor's afterglow, which made it possible to study the cooling rate of the emitting gas in the first seconds after the meteor (Borovicka and Jenniskens, 2000). Once the afterglow had subsided, a luminous glow persisted for more than 13 minutes. Such persistent trains have eluded a better understanding for over a century (Lockeyer, 1869).

The train provided the first mid-IR spectroscopy (Russell *et al.*, 2000), and was observed by other techniques as well.

All observed physical conditions are a function of altitude. The two participating aircraft in Leonid MAC, the "Advanced Ranging and Instrumentation Aircraft" [ARIA] and the "Flying Infrared Signatures Technology Aircraft" [FISTA], offered two perspectives that made triangulation possible (Jenniskens and Butow, 1999). In this paper, we calculate the trajectory of the fireball from the available optical records, providing height and range information for each part of the train. The meteor was only observed from ARIA but the persistent train was observed from both aircraft. The train record from FISTA is of high spatial resolution, which enables us to trace the train back to its point of origin. This is made easy because the train has 'billowing' structures that stand out for the entire period of observation.

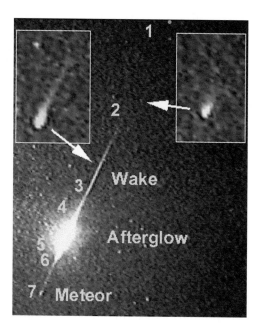

Figure 1. The 04:00:29 UT meteor and its afterglow. Inserts show the onset of persistent emission.

Horizontal winds and vertical drifts are responsible for the observed shape changes of the persistent train over time. Strong horizontal winds are caused by gravity waves in the ambient atmosphere. Vertical drifts may result from the buoyancy of the heated air in the path of the meteor.

Especially the latter aspect is of interest in understanding the physical conditions in the meteor path. We have examined the observations for evidence of such vertical motions

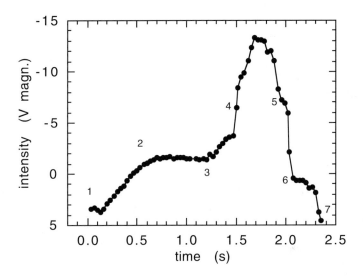

Figure 2 The visual light-curve magnitude of the meteor (normalized at 100 km distance) versus time. Markers as shown in Figure 1.

2. The observations

2.1. THE FIREBALL

The fireball was observed by a wide angle Mullard XX1332-intensified Hi-8 camera (Jenniskens, 1999) onboard ARIA (Figure 1). The video record was digitized with 640 x 480 resolution elements. Figure 1 is taken from a single frame shortly after the bright flare. The full frame is shown in Borovicka and Jenniskens (2000). Note that a small fragment of the meteor survives to the low altitude marker [7]. The field of view of the image is about 20 x 25 degrees. The meteor entered the frame at the top of the field as a faint point source (marker [1]). It soon spreads into the characteristic V-shaped structure first recognized by Spurny *et al.* (2000a). Shortly after, it turns into a droplet shape and at the same time starts to create a wake that persists for two seconds (Figure 1; [2]). At lower altitude, it rapidly brightens [3], followed by a sudden increase in brightness [4]. The persistent train originates from the part of the meteor's trajectory between [4] and [5]. The meteor then fades into a point source [6], which persists for a short period until time marker [7].

The light-curve shown in Figure 2 was reconstructed from the video record, after calibration of the blooming pattern in the intensified camera, and with help of a reflected image in an adjacent ARIA camera. The absolute peak-brightness of the meteor is about −13 magn., which would correspond to a mass of about 1 kg for this meteoroid (Spurny *et al.*, 2000b). At the peak of the light curve, a series of abrupt brightness changes occurred within a single frame that were only recognized clearly in the spatial pattern of the afterglow (Borovicka & Jenniskens, 2000).

2.2. THE AFTERGLOW

A bright afterglow is seen at the position of the intense flare in Figure 1, which persisted for about 5 seconds. This emission is due to low-excitation neutral atom emission lines that are discussed in Borovicka and Jenniskens (2000).

2.3. THE PERSISTENT TRAIN

Once the afterglow has faded, persistent emission remains that was nearly constant in intensity along the path of the meteor (Figure 3). This persistent train is initially a straight line. It stretches only from about point 4 to 5, where the meteor is brighter than the typical train-forming threshold of about −4 magnitude (Jenniskens *et al.*, 1998). Subsequently, "S"-shaped structures form at three different positions in the train, making a corkscrew pattern with large separation between the tightly wounded curls. First, the lower half of the train quickly fades in intensity, wrinkles, and forms a curl (Figure 3 at time 0:38; see also Figure 4). In the middle of the train, a second curl formed that gradually brightens and lengthens in the horizontal direction, whereby its upper part appears to move downward (Figure 3). Finally, at the top of the train another curl forms more gradually, by what initially appears to be a settling motion to establish its "foot". The lower part of this curl subsequently stretches out and retains a relatively high intensity, while the upper part is soon undetectable.

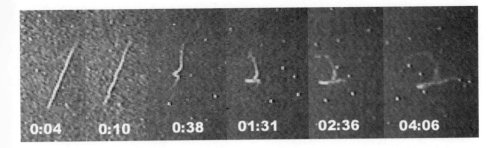

Figure 3 The persistent train as seen from ARIA at different times following the meteor (in minutes and seconds).

After about 1.5 minutes from the time of the meteor (1:31 in Figure 3) the basic shape of a "2" is formed as a composite of the middle curl and the foot of the top curl. Because of its striking shape and occurrence, this particular nature's-own end-of-the-millennium fireworks was soon named the "Y2K train". The "2"-shaped feature, including its many 'billowing' features, does not significantly change over the next 12 minutes (Figure 5). Aircraft motion causes a gradual, but not substantial, drift in azimuth as seen by the projection of the train against the star background (Figure 3). The train was observed until 04:13:29 UT, when it drifted outside the field of view of the ARIA camera.

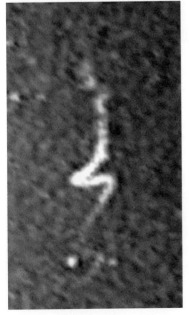

Figure 4. First view from FISTA 50 seconds after fireball (camera: FH50R). Notice the corkscrew pattern.

In response to the unusual sighting, the FISTA aircraft quickly changed its direction in order to bring this train into view of its onboard cameras and spectrometers. The earliest record is taken with the upward looking camera FH50R, only 48 seconds after the meteor appeared (Figure 4). The corkscrew pattern is clearly visible with most of the light intensity now being in the middle and upper parts of the train.

A particularly nice image was recorded with an UV sensitive Xybion intensified CCD camera (Figure 5). With a field of view of only 8.6 x 11.5 degrees that offered the best spatial resolution, while having most of the train in the field of view for the entire event. It was filmed from 3m38s until 8m20s after the meteor's first appearance. The train stays together, with no noticeable expansion or diffusion over the course of the experiment but its internal structures become gradually diffuse. Yet, the 'billowing' features remain visible over the entire period.

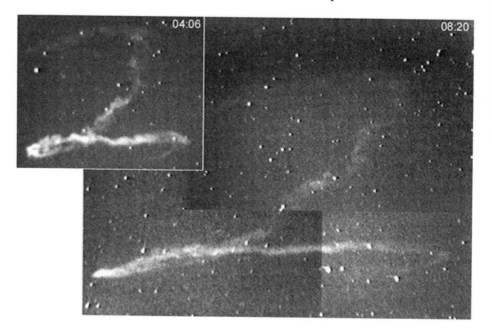

Figure 5. Persistent train as seen from FISTA at times 04:06 and 08:20 (Xybion camera).

3. TRIANGULATION OF THE TRAIN AT 04:04:35 UT

We identified 29 features in the train that can be used to trace changes in the horizontal winds. First, we calculated the position of each feature by triangulation of these individual features. The result is shown in Figure 6 wherein the position of each feature is indicated by a black dot. The numerical results are summarized in Table I, where each feature is numbered from the top down. The upper part of the "2" is situated at an altitude of 91 km, while the curl making its "foot" is at around 84 km.

The lowest visible part of the train (at time 04:06) is at about 79 km altitude.

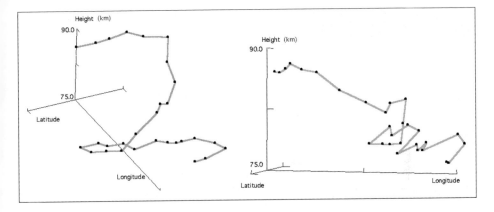

Figure 6. The 3-dimensional structure of the train at 04:04:35 UT, at time 04:06 after the meteor from two perspectives. The positions of train features are indicated.

TABLE I

#	H (km)	R (ARIA) (km)	R (FISTA) (km)	#	H (km)	R (ARIA) (km)	R (FISTA) (km)
1	91.6 ± 3	308 ± 6	191 ± 4	16	82.6	320	203
2	90.9	304	189	17	84.2	320	205
3	91.0	303	188	18	83.2	319	204
4	91.0	300	187	19	83.8	318	205
5	90.2	302	191	20	83.2	318	205
6	89.7	296	186	21	80.3	309	197
7	88.4	300	189	22	81.3	310	199
8	86.7	302	191	23	80.5	309	198
9	85.5	312	195	24	81.4	309	199
10	86.3	309	197	25	80.5	306	196
11	86.6	312	201	26	81.7	308	200
12	85.9	316	202	27	80.5	307	200
13	82.1	318	203	28	79.2	308	199
14	82.1	318	203	29	79.3	310	200
15	82.4	331	216				

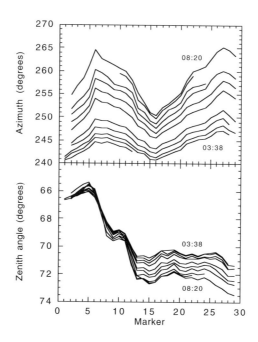

Figure 7. Train drift in azimuth (top) and zenith distance angle (bottom) as seen from the FISTA aircraft. Markers are positions defined in Figure 6.

3.1. TRAIN DRIFT AFTER 01:30

The observed train drift after time 01:30 can be traced back to the time of the fireball, thus providing a record of the position of the meteor as seen from FISTA. For that purpose, we use the images in Figure 5 that were taken between times 3:38 and 8:20. Figure 7 shows the azimuth and zenith distance angle for the 29 positions along the train, as defined in Figure 6, at ten different times during this interval. In this paper azimuth is expressed as an angle from South over West. The particular choice of ten times is dictated by the video record obtained by a hand-pointed camera that only occasionally provided images that were suitable for analysis.

The change in azimuth is unremarkable (closed points in Figure 8). There is a linear component that is caused by the near-linear aircraft motion (Figure 9) and an altitude-dependent variation that is a signature of gravity waves. The change in zenith distance also varies with altitude (open points in Figure 8), but is not easily understood. The lower parts of the train gradually decline in elevation, which is mostly an effect of the aircraft moving away from the train. The expected effect is an increase in zenith distance by about 0.012 deg/s. The upper part, above marker point 14 (above the foot of the "2") stays at nearly constant elevation. This implies that the train drifts either towards FISTA (in west/southwestward direction) at a rate of about 70 m/s, or this part of the train moves upward at a rate of about 25 m/s, relative to the train below marker 14. A relative 75 m/s west/northwestward drift is implied by the shape of the train in Figure 6, suggesting that most of the effect is in fact due to a horizontal wind gradient with altitude.

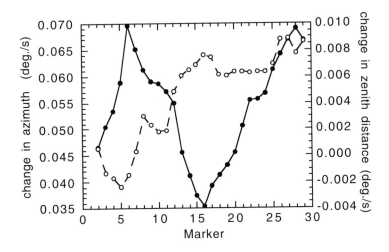

Figure 8. Mean rate of change in azimuth and zenith distance between times 03:38 and 08:20.

3.2. TRIANGULATION OF THE METEOR

By extrapolating the train motion backward in time, we can reconstruct the position of the meteor as seen from FISTA and calculate the trajectory by comparison with the meteor record from ARIA (Figure 1). We use the fact that the FISTA aircraft motion was almost linear during the period of interest (03:38–08:20), as shown in Figure 9. Figure 10 shows the resulting position of each marker point, in azimuth and zenith distance as seen from FISTA. We show the results for both linear and a parabolic extrapolations of the observed trend of position versus time for each feature in the train.

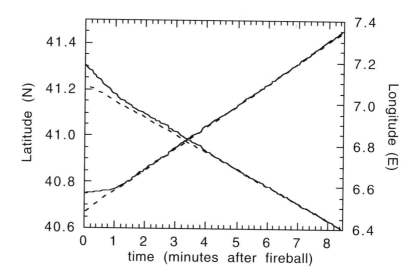

Figure 9 Change of the geographic position of the FISTA aircraft during the time of observation of the "Y2K" persistent train.

The convergence angle between the two planes that are defined by the position of the aircraft and the projection of the meteor against the sky is only 12°. In order to arrive at the correct trajectory, we made the assumptions that the radiant of the meteor is that of the Leonid shower (RA = 153.1°, DEC = 22.7° J2000) and the average meteor speed is that of the Leonids: 71.5 km/s (Jenniskens *et al.*, 1998). The radiant direction was towards Azimuth = 309° (SE) and Zenith Distance = 31°. These are strong constraints that leave little room for a wrong fit. The change in azimuth and zenith distance for different positions of the meteor trajectory as seen from FISTA is shown in Figure 10. The direction of the Leonid radiant determines the slope of the dashed line. It is clear that the "foot" of the "2"-shaped train feature at about 81 km altitude stands out as having significant motion in azimuth other than caused by the aircraft's motion. The motion in azimuth causes the initial stretching of the middle curl (Figure 3).

For any reasonable fit of the meteor's azimuth direction, we can only find a computational solution matching the altitudes of position marker points if the zenith distance of the meteor is about two degrees higher than was back-calculated from the extrapolation of the marker points (lower part Figure 10). The speed of the meteor is also an important additional constraint, because it defines the length of the observed trajectory and thus the position of the meteor's path in distance from the aircraft. No reasonable solution is obtained for a fit differing more than ±0.4 degree in zenith distance from that shown as a dashed line in Figure 10. If the zenith distance is as low as implied by the data in Figure 10 the computed solution for the speed of the meteor will be too high and hence the trajectory too far from both aircraft.

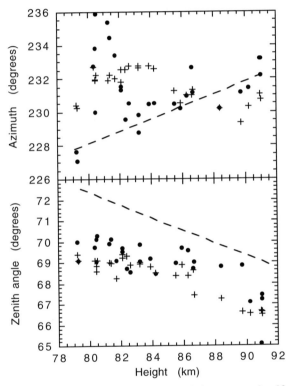

Figure 10. Azimuth and zenith distance angle of the meteor itself as seen from FISTA. The position of the meteor is derived from assumed linear (+) or parabolic (•) extrapolation of train drift. Dashed lines show the trajectory that provides the best triangulation results with ARIA.

The reason for this discrepancy remains unknown. However, the best fitted meteor trajectory (dashed line) is in good agreement with other Leonid fireball trajectories, with key features at similar altitudes. The meteor discussed here was first detected at about 195 km altitude and ended the so-called 'diffuse phase' (Spurny *et al.*, 2000a) at about 136 km altitude. This compares to 195 km and 134–131 km for a −12.5 magnitude Leonid fireball DMS98023 studied by Spurny *et al.* (2000b). The rapid increase in intensity associated with the ablation of silicates (Borovicka *et al.*, 1999) occurred at about 100 km altitude. The bright burst in the train discussed here started at 89-km altitude and lasted until 74-km altitude. The decay in brightness was complete only at 71 km altitude. This compares to a photographic end height of the train at 73.2 km for DMS98023 (Spurny *et al.*, 2000b). A small but structurally coherent fragment of the meteor continued to an unusually low altitude of 55 km.

4. Discussion

We submit that vertical motions are important in the early dynamic evolution of the persistent train of the meteor described here. The formation of "S"-shaped features, and significant stretching of the lower part of those structures shortly after train formation appears to be the result of settling of air in the path of the meteor. Judging from the differences in the initially-formed shape and those that formed later, this phenomenon is not a mere reflection of periodic winds in the ambient air. The settling motion is surprising because heated air will flow upward. On the other hand, the heated gas is expected to expand rapidly after the meteor until pressure equilibrium is established, probably in a manner of seconds. Subsequent cooling of the buoyant air can lead to flow downward until pressure equilibrium is established. The observed 'billowing' features are remarkably constant in time. They must have formed early on during formation, possibly as a result of the intensity variations across the meteor's path that are so conspicuous in the afterglow images.

At later times, ambient winds dominate the change in shape and train expansion. When the 3-dimensional solution of the train geometry is viewed at an angle about 90 degrees from the lines of sight from ARIA and FISTA (Figure 6), a more or less linear appearance of the train results. We considered that this might be an artifact of triangulation because both trains are relatively distant from the observing aircraft. By imposing a periodic wind variation on the solution, we attempted to recreate the "2" shape in a model. The best solution reflects a periodic wind variations between 79 and 91 km with an amplitude of 57 ± 8 m/s and a vertical scale length of 8.3 ± 0.5 km (Figure 11). The shape of the "2" is only well reproduced if a vertical wind gradient is included.

THE "Y2K" PERSISTENT TRAIN AND FIREBALL 469

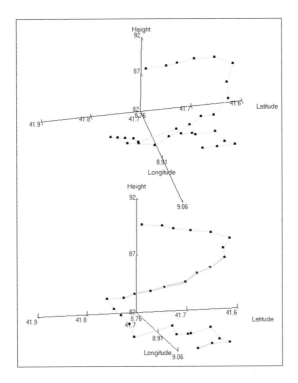

Figure 11. Model of the Y2K train at 04:04:35 UT by imposing a periodic wind variation with altitude but without a wind gradient.

Radar wind measurements in middle-Europe (Singer *et al.*, 2000) showed tidal wind oscillations with an amplitude of about 40 m/s and a scale height of about 8.5 km between 85 and 105 km altitude. At 04 UT, the zonal winds were changing in direction from east to west, more quickly at higher altitude. These observations are consistent with the observed east-west gradient in Figure 6.

Acknowledgements

We are grateful for the constructive reviews by Jack Drummond and Frans Rietmeijer to improve the presentation of this paper. Klaas Jobse operated the Mullard XX1332-intensified Hi-8 camera onboard ARIA. Mike Koop operated the FH50R camera on FISTA. The 1999 Leonid MAC was sponsored by the NASA Exobiology, Planetary Astronomy, and Suborbital MITM programs, by the Advanced Missions and Technologies for Astrobiology Program, and by NASA Ames Research Center. ARIA flight time was supported by the USAF/XOR. *Editorial handling:* Frans Rietmeijer.

References

Borovicka, J. and Jenniskens, P.: 2000, *Earth, Moon and Planets* **82–83**, 399–428.
Borovicka, J., Stork, R., and Bocek, J., 1999: *Meteoritics Planet. Sci.* **34**, 987–994.
Jenniskens, P.: 1999, *Meteoritics Planet. Sci.* **34**, 959–968.
Jenniskens, P. and Butow, S.J.: 1999, *Meteoritics Planet. Sci.* **34**, 933–943.
Jenniskens, P., de Lignie, M., Betlem, H., Borovicka, J., Laux, C.O., Packan, D., and Kruger, C.H.: 1998, *Earth, Moon and Planets* **80**, 311–341.
Lockeyer, N.: 1869, *Nature* **1**, 58.
Russell, R., Rossano, G., Catelain, M.A., Lynch, D., Tessensohn, T., Abendroth, E., and Jenniskens, P.: 2000, *Earth, Moon and Planets* **82–83**, 439–456.
Singer, W., Hoffmann, P., Mitchell, N.H., Jacobi, Ch.: 2000, Earth, Moon and Planets **82–83**, 565–575.
Spurny, P., Betlem, H., Jobse, K., Koten, P., and Van 't Leven, J.: 2000a, *Meteoritics Planet. Sci.* **35(5)**, in press.
Spurny, P., Betlem, H., van 't Leven, J., and Jenniskens, P.: 2000b, *Meteoritics Planet. Sci.* **35**, 243–249.

THE DYNAMICAL EVOLUTION OF
A TUBULAR LEONID PERSISTENT TRAIN

PETER JENNISKENS AND DAVID NUGENT
SETI Institute, NASA Ames Research Center, Mail Stop 239-4, Moffett Field, CA 94035
E-mail: pjenniskens@mail.arc.nasa.gov

and

JOHN M.C. PLANE
School of Environmental Sciences, University of East Anglia, Norwich NR4 7TJ, UK
E-mail: J.Plane@uea.ac.uk

(Received 26 May 2000; Accepted 16 August 2000)

Abstract. The dynamical evolution of the persistent train of a bright Leonid meteor was examined for evidence of the source of the luminosity and the physical conditions in the meteor path. The train consisted of two parallel somewhat diffuse luminous tracks, interpreted as the walls of a tube. A general lack of wind shear along the trail allowed these structures to remain intact for nearly 200 s, from which it was possible to determine that the tubular structure expanded at a near constant 10.5 ms^{-1}, independent of altitude between 86 and 97 km. An initial fast decrease of train intensity below 90 km was followed by an increase in intensity and then a gradual decrease at longer times, whereas at high altitudes the integrated intensity was nearly constant with time. These results are compared to a model that describes the dynamical evolution of the train by diffusion, following an initial rapid expansion of the hot gaseous trail behind the meteoroid. The train luminosity is produced by O (^{1}S) emission at 557 nm, driven by elevated atomic O levels produced by the meteor impact, as well as chemiluminescent reactions of the ablated metals Na and Fe with O_3. Ozone is rapidly removed within the train, both by thermal decomposition and catalytic destruction by the metallic species. Hence, the brightest emission occurs at the edge of the train between outwardly diffusing metallic species and inwardly diffusing O_3. Although the model is able to account plausibly for a number of characteristic features of the train evolution, significant discrepancies remain that cannot easily be resolved.

Keywords: Airglow, chemistry, dynamics, Leonids 1998, lower thermosphere, mesosphere, meteor, persistent train

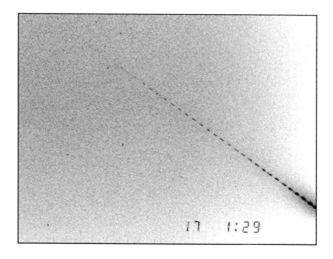

Figure 1. The 01:31:16 fireball photographed by Steve Evans of the British Astronomical Association – Meteor Section.

1. Introduction

Bright Leonid meteors are known for a characteristic long-lasting persistent glow that is called a *persistent train*. The luminous source of persistent trains has not been established, although it is generally believed that the reaction between ozone and atomic oxygen, efficiently catalysed by meteoric metals in the train itself, is the dominant mechanism (Kolb and Elgin, 1976; Poole, 1979; Baggaley, 1980; Hapgood, 1980).

Persistent trains probe upper atmosphere chemistry. Moreover, they enable probes of meteoric aerothermochemistry by providing direction to telescopes many minutes after the meteor has disappeared (Jenniskens *et al.*, 1998). Several such experiments during the Leonid Multi-Instrument Aircraft Campaign (Jenniskens and Butow, 1999a) have provided the first visual, near-IR and mid-IR spectroscopy of trains. In order to better interpret this spectroscopic information, we study here a rather striking example of a persistent train observed over the United Kingdom in 1998, in order to examine the evolution of trail width and intensity with time. Unlike many other trains, this train exhibited little distortion from wind shear along the trajectory and remained fairly linear during the observation period.

2. Observations

At 01:31:16 UT on November 17, 1998, a bright Leonid meteor (Figure 1) erupted over the southern U.K., moving from Southend-on-Sea to just beyond Reading. Amateur astronomer Sandy Osborough, from Chippenham, Wiltshire (51°28'N, 02°07'W), was located near the end point of the trajectory. The meteor was outside the field of view of his intensified video camera, but the scattered light in the atmosphere left a flash in the video record. Osborough adjusted the viewing direction of the camera and obtained a particularly nice record of the "Chippenham" persistent train between 15 seconds and 2m54s after the flash (Figure 2, right part of each frame). The train persisted longer than that, but the camera was pointed elsewhere. He used a 45 lp/mm ITT Night Vision gogles attached to a 3CCD Panasonic digital video camera. The close range of the train (116–80 km) produced a spatial resolution of 0.2 km/pixel.

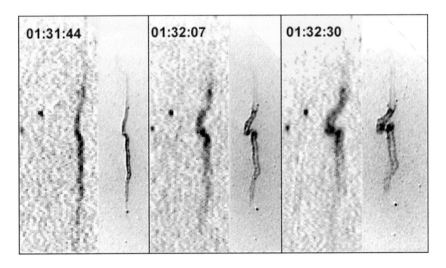

Figure 2. The train as seen by Tim Haymes from the perspective of Knowl Hill (left) and by Sandy Osborough from Chippenham (right).

The train was also filmed by Tim Haymes from Knowl Hill, Berks. (00°48' 51.3" W, 51° 30' 22.1" N), again from relatively short range (~102 km). Haymes used a 28 mm f/2 lens imaging onto an 18 mm 2nd generation MCP image intensifier (30 lpi). The image resolution was also 0.2 km/pixel, but the noisier tube created a less exceptional image.

From his perspective, the train was less foreshortened (Figure 2, left panels). Haymes also captured the diffuse beginning of the meteor, first detected at about 183 km altitude at 01:31:13 UT, until it left the field of view at 145 km. A bright flash was timed at 01:31:16 UT. After that, the camera was hand held and pointed at the train from 01:31:30 until 01:32:33 UT, and again from 01:34:58 until 01:35:17 UT when only the bright loop in the center of the train was visible as a diffuse blob. A faint sonic boom was heard around 01:36:40 UT, consistent with the distance from the meteor train. This is the second sonic boom reported for a bright Leonid fireball (ReVelle and Whitaker, 1999). Unfortunately, the convergence angle between the planes emanating from the two observing sites is only Q = 3°, too small for stereoscopic measurements.

Fortunately, the meteor was photographed by Steve Evans from Thurlow, near Newmarket in Suffolk (52° 7' 58".1 N, 0° 26' 49".1 E, Alt: 83m), in a 5m59s exposure on Ektapress 1600 commencing at 01:29:00 UT (Figure 1). The perspective was good, with a convergence angle Q = 29.8° with Chippenham and Q = 27.1° with Knowl Hill. Triangulation shows that the meteor entered the atmosphere at an angle of 29° with the horizon and came from an eastern azimuth of 87° from North from a direction Right Ascension 150.0 ± 0.3, Decl. 23.8 ± 0.3. The meteor was first detected at an altitude of 116 km and left the photograph at 86 km. The two linear parts of the train span the range 98 until 85 km, while the end point of the Chippenham train is at about 80 km altitude. Photometry of the meteor and stars gave a peak brightness of $M_v = -9 \pm 1$ magn. at the edge of the photograph. Beginning height and end height suggest a peak brightness of -10 ± 1 magn. in comparison to the trajectories calculated for other bright Leonid fireballs (Spurny *et al.*, 2000). The mass of the meteoroid was about 0.1 kg, within a factor of two.

3. Results

3.1. Train Morphology

The train consists of two parallel somewhat diffuse luminous trails, which are visible along the full length of the recorded trajectory. The double structure has been noted previously and is a characteristic feature of long lasting persistent trains (Jenniskens *et al.*, 1998). The diffuse trails show some amount of puffy billowing, which implies some amount of turbulence.

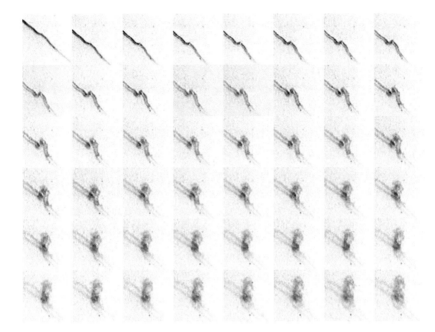

Figure 3. Sequence of images as seen from Chippenham between 0m15s and 02m52s after the meteor, showing the formation of the loop and the evolution of the linear parts.

This train is very unusual because the two straight sections of the trajectory remain almost straight with respect to the star background (Figure 3). Note that the lower part gradually gains upon the higher part and may be rising in altitude. Only the bright middle section forms a loop, which eventually overlaps in the line of sight. At the position of the loop, the wind direction changes dramatically with altitude. It is from the North in the straight sections, while from the East in the distorted middle part and end sections, with strong wind shears in the transition regions. The magnitude of the wind vector changes only by a factor of 2–3. Most of the wind shear is laminar, preserving the tubular structure of the train during distortion.

The morphology of the train is either that of a tube, where the two trails represent the longest line of sight along the walls of the tube, or they represent the turbulent top and bottom of a ribbon-like structure. The most likely morphology is that of a tube, for two reasons. First of all, from the somewhat different perspective of Knowl Hill, the train diameter (in km) is the same as that measured from Chippenham (Figure

2). Secondly, one can see that the distance between the two walls does not change where the train distorts in a knot and the line of sight cuts at a different angle through the train (Figure 3).

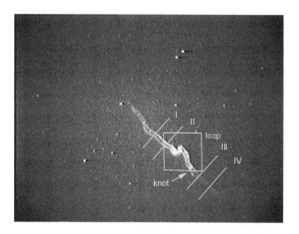

Figure 4. Full frame of the persistent train as seen from Chippenham at a time 45 seconds after the fireball. Four positions are indicated that were studied in detail, at altitudes 86 (I), 89 (II), 95 (III), and 97 (IV) km.

3.2. DYNAMIC EVOLUTION

Over time, one can see the walls of the tube separate and slightly thicken. We analyzed this behavior by fitting a set of two Gaussian curves to the variation of intensity in a slice perpendicular to the train at four positions indicated in Figure 4. The positions I–IV correspond to altitudes of about 86, 89, 95, and 97 km, respectively.

Figure 5a (top) shows the separation (in km) as a function of time. We find a constant expansion velocity of 10.5 ± 0.5 ms^{-1} for all positions. There is no sign of a slowing down over this time interval, except perhaps for the highest position at 97 km. The fact that the expansion is practically altitude independent between 85 and 98 km is surprising. In addition, the least-squares fit through the data does not extrapolate to zero separation at zero time, but yields a positive intercept of 0.13 ± 0.03 km. This may signify an initial rapid expansion of the train, but the small value does not exclude a more mundane artifact of the measurement procedure.

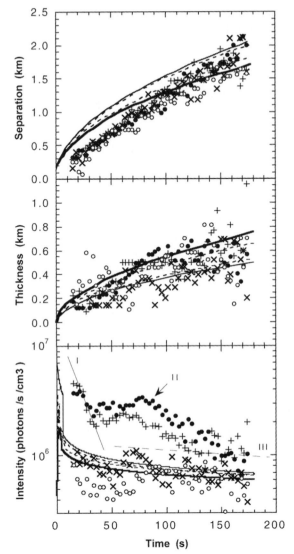

Figure 5. (a) The separation between the peak intensity of the tubular walls; (b) the thickness of the tubular wall; and (c) the integrated intensity of the tubular walls as a function of time. Symbols are: Pos. I (•), Pos. II (+), Pos. III (x), Pos. IV (o). Model results shown are for 86 (solid line), 89 (dashed), 95 (dashed) and 97 km altitude (thick solid line).

The tubular walls are resolved at the end of the exposure. We measured the thickness at half peak intensity as a function of time and deconvolved with a 2-pixel wide Gaussian response curve. We find that the turbulent walls tend to show slightly more billowing over time, gradually thickening, but the expansion slows down after about 100 s (Figure 5b).

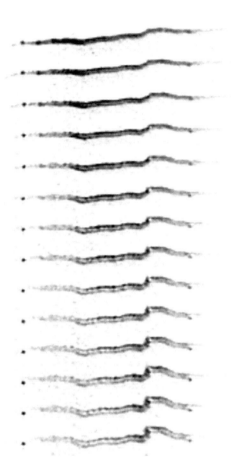

Figure 6. Altitude dependent decay of emission during the first period from 15.0 to 34.5 seconds after the meteor (in intervals of 1.5 seconds). A star (left) is shown for brightness reference.

The two sides of the tube do not have the same intensity at positions I and II (Figure 6). One side is brighter by up to 50% over the other. The integrated intensity of both tubular walls shows three distinct periods of decay, particularly below 90 km (Figure 5c, bottom graph). An initial fast decay (I) is followed by an increase in intensity, which after some time decreases again (II). At the end of the observing time the train decay rate slows significantly (III).

Phase (I) has a decay time of about 15 seconds, which is slightly altitude dependent. This altitude dependence is apparent when viewing the train intensity in the meteor video during the first 30 seconds (Figure

6). The brightness decays fastest at the lower altitude end of the train (left in Figure 6), while the higher altitudes follow in succession.

Phase (II) is characterized by the initial brightening of the train. The intensity peaks earliest towards the middle part of the trajectory, with mechanisms delaying the increase at very high and very low altitudes. The subsequent decay has a time constant of about 63 ± 2 s at 86 km, 70 ± 2 s at 89 km, 150 ± 20 s at 95 km, and about 190 s at 97 km altitude.

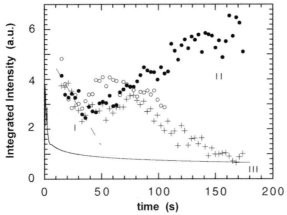

Figure 7. Integrated intensity variation of the knot (•) and the total intensity of the loop (o), scaled to that of the "quiet" train at position I (+). Solid line shows a model fit for the height of 86 km (in units of 10^6 photons cm^{-3} s^{-1}).

This intensity increase during Phase (II) may be related to the brightening phenomenon that enhances the brightness of the loop between 90 and 94 km. The integrated intensity of the loop (box in Figure 4) is shown in Figure 7, in relation to the brightness of the linear part of the train at position II. At the end of the observation, the loop is the only part that remains clearly visible. A remarkable feature is the "knot" (marked in Figure 4), where wind shear appears to be particularly high at the beginning of the observations. We see that the tubular structure breaks down during bending and a wall of billowing emission is observed. Here, we find the quickest brightness increase, which is followed by a decrease that mimics the linear part of the train (open circles in Figure 7).

Phase (III) is most apparent in the high altitude part of the train, but also visible in position II. This phase represents the late stages of train evolution. It is well represented by our model calculations below.

4. Discussion

4.1. A MODEL OF A PERSISTENT METEOR TRAIN

The persistent train luminosity is thought to be caused by the catalytic recombination of ozone and oxygen atoms by meteoric metal atoms (Chapman, 1956; Kolb and Elgin, 1976; Poole, 1979; Baggaley, 1980). Recent spectroscopic observations of persistent trains during the Leonid MAC mission and at the Weybourne Atmospheric Observatory in Norfolk (Jenniskens et al., 2000b) have confirmed that the most intense emission arises from the Na D-line, almost certainly through the Chapman airglow mechanism:

$$Na + O_3 \rightarrow NaO + O_2 \qquad (1)$$

$$NaO + O \rightarrow Na(3^2P, 3^2S) + O_2 \qquad (2)$$

where the branching ratio of reaction 2 to produce the Na (3^2P) state (which then emits an orange photon at 589 nm) is about 10% (Clemesha et al., 1995). Molecular emission bands also probably arise from:

$$Fe + O_3 \rightarrow FeO(^5\Delta \text{ etc.}) + O_2 \qquad (3)$$
$$FeO + O \rightarrow Fe + O_2 \qquad (4)$$

where reaction 3 is sufficiently exothermic to produce FeO in excited electronic states, leading to emission in the "orange arc" bands between 570 and 630 nm with about a 2 % efficiency (Helmer and Plane, 1994). Other metals such as Ca and K will also contribute to the overall emission intensity, but the ablated concentrations of these metals are much lower (Plane, 1991).

If we now assume that the twin tracks of the observed train are due to a luminous tube with a dark center when viewed from below (e.g. Figure 8), then the explanation for the dark center must be that O_3, which "fuels" these chemiluminescent reactions, has been consumed. The same hypothesis was made independently in a recent paper by Kelley et al. (2000). Of course, these emissions also depend on atomic O to complete the catalytic cycles. However, O is in excess over O_3 by about 3 orders of magnitude in the ambient nighttime upper mesosphere, and the meteoric impact on the atmosphere causes the dissociation of O_2 to produce additional O (the green tail of Leonid meteors, in particular, is caused by emission from O (^{1}S) which is highly dependent on the atomic O concentration, as discussed below). The depletion of O_3 could arise both from thermal dissociation in the initially very hot train, and metal-catalysed destruction. Effective catalytic removal places a lower limit on the concentration of metallic species produced by ablation. Note that all neutral metal atoms, as well as metallic ions with the exception of the alkali metal ions, participate in catalytic O_3 destruction (Plane and Helmer, 1994).

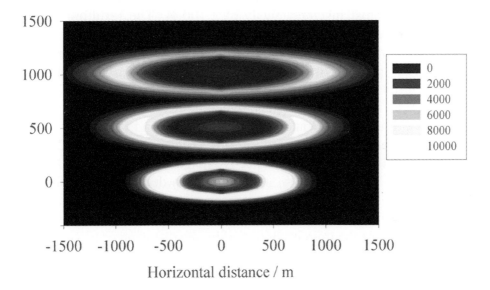

Figure 8. Three cross-sections through the persistent train at an altitude of 86 km, showing the modeled emission intensity at times 50s (bottom), 100s and 150s after the meteor. The 100 and 150 s sections have been displaced upward by 500 and 1000 m, respectively, for the purpose of presentation. The central emission patch visible at 50 s and 100 s is due to [OI] emission, the outer ring is due to chemiluminescence from metal atom reactions with ambient ozone.

We have therefore constructed a model of a Leonid meteor train in order to simulate the train expansion, the increase in wall thickness and the observed brightness variation as a function of position and time. The model makes the following assumptions:

1. The meteoroid ablates according to the deceleration equation (particle density = 3,200 kg m^{-3}, drag coefficient = 0.75 and shape factor = 1.2 for a sphere) and heat transfer equation (heat of sublimation = 2 x 10^6 J kg^{-1} and heat transfer coefficient = 0.5) given

by Hughes (1992). Fragmentation is not considered in this simple model. In order to provide sufficient metallic species to cause effectively total removal of O_3 in the centre of the train after 50 s, the initial meteoroid mass has to be in excess of 0.1 kg. The simulations shown here employed an initial mass of 0.2 kg, at the upper end of the estimated mass of the Chippenham meteoroid (see above).

2. Even at the very low pressures of the upper mesosphere/lower thermosphere ($< 10^{-5}$ bar), the size and velocity of this meteoroid would create a turbulent wake (Reynolds number $> 2,000$). Thus we assume that the train radius is initially 30 m, in which the air is then heated almost instantaneously to 2,100 K. The resulting pressure increase by more than a factor of 10 creates a shock wave which expands radially. Assuming that this expansion occurs adiabatically, then the pressure will equilibrate with the background atmosphere when the radius is about 70 m, leaving a train temperature of about 1,100 K. This train radius and temperature are predicted to be nearly constant between 97 and 86 km (heights IV and I in Figure 4), with the concentration of ablated metallic species ranging from 2.5×10^9 to 9.9×10^9 cm^{-3} at these respective heights. The ambient O_3 within this initial train volume would be thermally decomposed.

3. Following this very rapid expansion on a time-scale of less than a second (the speed of sound is 270 ms^{-1} in this region), the subsequent expansion of the train is controlled by the diffusion of mass and heat. For the modelling exercise presented here, this was allowed to vary as a function of height and be different in the horizontal and vertical, with the lower limit being the atomic diffusion coefficient of Na (Helmer and Plane, 1993).

4. The reaction rate coefficients for 1–4 were taken from Plane and Helmer (1994). The relative concentrations of Na and Fe were assumed to be in their meteoritic ratio, about 1:8 (Plane, 1991).

5. On the relatively short time scale of the train (minutes rather than hours), and particularly in the presence of elevated concentrations of atomic O, it is very unlikely that the metallic species would be able to form more stable reservoir compounds such as $NaHCO_3$ or $Fe(OH)_2$. Indeed, between 85 and 100 km in the background atmosphere the meteoric metals are overwhelmingly in the atomic form (Plane et al., 1999).

6. The "green line" emission at 557 nm from the O ($^1S - ^1D$) transition (termed [OI]) was assumed to be produced by the Barth mechanism with the absolute intensity calculated using the parameterisation of

Murtagh et al. (1990). Assuming that this emission dominates the train emission immediately after the meteor, then in order to simulate the observed decrease in intensity of the Chippenham train at longer times the model requires that about 15% of the O_2 in the initial train was dissociated.

The train model was then run with a spatial resolution of 25 meter and integration time-step of 0.2 seconds. Figure 8 shows cross sections through the modelled train at an altitude of 86 km at 50s, 100s and 150s after the meteor. Initially, strong [OI] emission is observed at the center of the trail because of the enhanced O atom concentration produced from dissocation of atmospheric O_2 by the meteor. The [OI] intensity is dependent on $[O]^3$, so that the intensity falls very quickly as the atomic O diffuses outwards from the centre of the train.

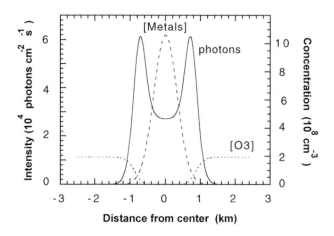

Figure 9. Variation in the metal atom and O_3 densities across the center of the trail at 98 km, 100 s after the meteor. The vertically integrated emission intensity which would be observed from the ground, is shown for comparison.

As shown in Figure 9, after 100 s the O_3 concentration within the train has been reduced by orders of magnitude by the combined effects of thermal decomposition and catalytic destruction. The concentration profile of the metallic species is approximately Gaussian, as expected for diffusion-controlled transport. The metallic emission is strongest at the edge of the train, where fresh ambient O_3 is diffusing inwards.

For reasonable diffusion rates, the model does account for the general widening of the train, the tubular structure, and the increase of the train width. The model also explains phase III in the brightness decay.

The model simulations of the wall separation, the wall thickness, and the integrated intensity are plotted as a function of time in Figure 5 for comparison with the observations at positions I–IV. In the case of the wall separation (Figure 5a), the model is able to simulate the "average" observed rate of separation over the first 180 s by using a horizontal diffusion coefficient ranging from $(5–7) \times 10^6$ cm^2 s^{-1} between 86 and 97 km, and a vertical diffusion coefficient set to the larger of either the vertical eddy diffusion coefficient K_{zz} employed in 2D atmospheric models [Garcia and Solomon 1994], or the molecular diffusion coefficient with a temperature $T^{1.8}$ dependence (Helmer and Plane, 1993). The reason for choosing different horizontal and vertical coefficients is discussed below.

Inspection of Figure 5a shows that the model, being based on diffusive transport necessarily produces a separation that varies as time $t^{1/2}$, whereas the observed separation increases linearly with time at all four heights. This striking observation remains unexplained. Free expansion is ruled out, because the passing meteor cannot have affected the air density over such large volume. Interestingly, the thickness of the walls does increase much more like $t^{1/2}$ (Figure 5b), in accord with a diffusion-controlled process. Note that the rapid initial expansion of the train to a wall separation of about 150 m, driven by a shock-heated pressure wave (see above), is in good accord with the extrapolated intercepts of the observed separations at the four heights (Figure 5a).

Although there is reasonable overall agreement, the model is unable to match the contrast between the dark center and the walls that is apparent in the images. Figure 10 illustrates the vertically integrated emission intensity across the train when viewed from below, comparing model and observation at 89 km, 160 s after the meteor. If the trail is indeed a cylinder with a luminous wall and dark center, then the model predicts that the contrast between the center and the brightest part of the walls, as seen from the ground, is about 0.5, remaining roughly constant with time. This factor arises simply from the fact that when viewed from the ground, the center is seen against the top and bottom of the cylinder and therefore cannot be very dark. By contrast, we observe an exponential decrease of contrast to about 0.1 after 120 s. The symbols in Figure 10 show the train cross-section between 153 s and 165s at positions I and III. The data have been scaled to the model curve to compare the contrast between the central minimum and the wall maxima. In fact, the lack of contrast in the model is all the more striking because we have maximised the contrast by using a smaller vertical diffusion coefficient to reduce the vertical transport of metallic species and fresh O_3, thereby minimising the wall brightness in the top and bottom parts of the tube (Figure 8). The lack of contrast is not explained by decreasing the initial ozone concentration in the center faster, for example as a result of photodissociation of ozone by the meteoric UV light (Zinn et al., 1999).

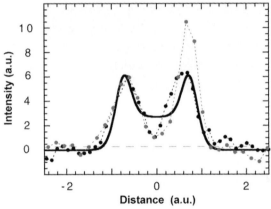

Figure 10. Contrast in brightness of center and walls in model (solid line) and observations. Model in units of 10^4 photons cm^{-3} s^{-1}.

Also, there are significant difficulties with modeling the brightness behavior (Figure 5c). The phase I decay is thought to be due to emission of the forbidden 557 nm O ($^1S - ^1D$) transition. Indeed, meteors of lesser brightness are known to have persistent emission on a time scale of about 10 seconds, sometimes referred to as the meteor "wake" (Halliday, 1958). This is a direct result of molecular oxygen dissociation by the meteor and has been well recorded in photographic and TV video spectra (e.g. Borovicka et al., 1996). Figure 11 reproduces one of our own measurements, where it can be seen that the [OI] emission starts shortly after the meteor itself, peaks, and then rapidly decays. The model predicts that the [OI] emission in the Chippenham train would have been brighter at and below 90 km (positions I and II), since this is where more atomic O is produced in the meteor. The emission is predicted to decay on a time scale of only about 10 s (due to its $[O]^3$ dependence and the rapid outward diffusion of atomic O), rather than the observed 30 - 40 s in phase I. Note also that the [OI] line intensity should be concentrated in the center of the train (Figure 8), rather than the train walls, whereas from Figure 6 it is clear that at least part of the early decay is the result of emission from the train walls.

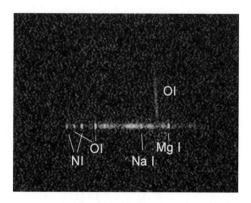

Figure 11. Forbidden line emission of OI in the wake of a −1 Leonid meteor (Nov. 17, 1998, 19:31:11 UT). This first order spectrum was taken with a low-resolution visible spectrometer onboard FISTA during the 1998 Leonid MAC (Jenniskens and Butow, 1999). The meteor moved from top to bottom. Short wavelengths are to the right.

The models predict very little change in intensity over the 174 s of observations that comprise Phase II in Figure 5c. This is because as the peak intensity of the walls decreases so the thickness of the walls increases with time, and hence the integrated intensity hardly changes. This behavior does reproduce the phase III behavior as signified by the decay of intensity observed at altitudes above 90 km, and also correctly predicts that the intensity at lower altitudes of 86 and 89 km will eventually decrease to similar levels. However, the complex time evolution of the intensity at these lower altitudes, particularly the intensity increase of Phase II, remains unexplained

4.2. ALTERNATIVE TRAIN MODELS

The present model does not include a detailed treatment of small-scale turbulent mixing at the boundary of train and ambient air. The observations seem to suggest that wind shear enhances the observed luminosity, and the walls show clear signs of billowing. This could increase the interfacial area between the train and surrounding air, enhancing the rate of chemiluminescent reactions between metallic species and O_3. If such turbulence spread horizontally rather than vertically, this would help to explain the high contrast between train center and walls.

The model assumes that the longer-lived emission is due to metallic atoms reacting with O_3. A rapid decline in train intensity, such as observed in phase II at 86 and 89 km, could be because of depletion of these metals. However, there are no reactions with background atmospheric constituents such as H_2O, CO_2, O_2 etc. that will convert these species to stable forms on the time scale observed. The only other possibility is that these metals are reacting with the high concentration of silicates and other debris in the trail, although again the time scale of 200 seconds is very short.

In summary, the present model coupling meteor ablation with simple diffusive transport of the resulting train is able to account satisfactorily for some of the significant features of this unusual event. These include the appearance of two luminous tracks, the average rate of increase of their separation and thickness, and some aspects of the luminous emission decay. However, the model fails to explain the strikingly constant rate of separation increase, which cannot be diffusive in nature, and the complex variation of the emission with time at some altitudes, amongst others. Clearly, there is still much to be understood about the nature of persistent trains.

Acknowledgments

Amateur observers Sandy Osborough, Tim Hayes and Steve Evans are congratulated with their fine records of train and meteor. We are thankful for their kindness in making these records available for analysis. We also thank John Green and David Stephens at the University of East Anglia for helpful discussions on turbulent transport. This work forwards the goals of the Pro-Amat working group of IAU Commission 22. The work is supported by grants from NASA's Suborbital MITM and Exobiology programs, and by the NASA Advanced Missions and Technology program for Astrobiology.
Editorial handling: Noah Brosch.

References

Baggaley, W.J.: 1980, in Halliday, I., and McIntosh, B.A., (eds.), *Solid Particles in the Solar System*. IAU Symp. 90, pp. 85–100.

Borovicka, J., Zimnikoval, P., Skvarka, J., Rajchl, J., and Spurny, P.: 1996, Astron. Astrophys. **306**, 995–998.

Borovicka, J., Stork, R., and Bocek, J.: 1999, *Meteoritics Planet. Sci.* **34**, 987–994.

Chapman, S.: 1956, in A. Dalgorno, E.B. Armstrong (eds), *The Airglow and the Aurorae.*, Permagon Press, New York, pp. 204–.205.

Clemesha, B.R., Simonich, D.M. Takahashi, H. Melo, S.M.L, and Plane, J.M.C.: 1995, *J. Geophys. Res.* **100**, 18909–18916.

Garcia, R.R., and Solomon, S.: 1994, *J. Geophys. Res.* **99**, 12937–12951.

Halliday, I.: 1958, *Astrophys. J.* **128**, 441–443.

Hapgood, M.A.: 1980, *Nature* **286**, 582–583.

Helmer, M., and Plane, J.M.C.: 1993, *J. Geophysical Res.* **98**, 23207–23222.

Helmer, M., and Plane, J.M.C.: 1994, *Journal of the Chemical Society, Faraday Transactions* **90**, 31–37.

Hughes, D.W.: 1992, *Space Science Reviews* **61**, 275–299.

Jenniskens, P.: 1998, *Meteoritics Planet. Sci.* **33**, 955–957.

Jenniskens, P., de Lignie, M., Betlem, H., Borovicka, J., Laux, C.O., Packan, D., and Kruger, C.H.: 1998, Earth, *Moon and Planets* **80**, 311–341.

Jenniskens P., and Butow S.J., 1999, *Meteoritics Planet. Sci.* **34**, 933–943.

Jenniskens P., Butow, S.J., and Fonda, M.: 2000, *Earth, Moon and Planets* **82–83**, 1–26.

Jenniskens, P., Lacey, M., Allan, B.J., Self, D.E., and Plane, J.M.C.: 2000b, *Earth, Moon and Planets* **82–83**, 433–442.

Kelley, M.C., Gardner, C., Drummond, J., Armstrong, W.T., Liu, A., Chu, X., Papen, G., Kruschwitz, C., Loughmiller, P., and Engleman, J.: 2000, *Geophys. Res. Lett.* **27**, 1811–1814.

Kolb, C. E., and Elgin, J. R. (1976), *Nature* **263**, 488–490.

Murtagh, D.P., Witt, Stegman, J., McDade, I.C., Llewwellyn, E.J., Harris, F., and Greer, R.G.H.: 1990, *Planetary and Space Science* **38**, 43–53.

Plane, J.M.C.: 1991, *International Reviews of Physical Chemistry* **10**, 55–106.

Plane, J.M.C., and Helmer M.: 1994, In: *Research in Chemical Kinetics,* G. Hancock and R.G. Compton, eds, Elsevier, Amsterdam, 313–367.

Plane, J.M.C., Cox, R.M., and Rollason, R.J.: 1999, *Advances in Space Research* **24**, 1559–1570.

Poole, L.M.G.: 1979, *J. of Atmosph. Terr. Phys.* **41**, 53–64.

ReVelle, R.O., and Whitaker, R.W.: 1999, *Meteoritics Planet. Sci.* **34**, 995–1005.

Spurny, P., Betlem, H., Van 't Leven, J., and Jenniskens, P.: 2000, *Meteoritics Planet. Sci.* **35**, 243–249.

Zinn, J., Wren, J., Whitaker, R., Szymanski, J., ReVelle, D.O., Priedhorsky, W., Hills, J., Gisler, G., Fletcher, S., Casperson, D., Bloch, J., Balsano, R., Armstrong, W.T., Akerlof, C., Kehoe, R., McKay, T., Lee, B., Kelley, M.C., Spalding, R.E., and Marshall, S.: 2000, *Meteoritics Planet. Sci* **34**, 1007–1015.

DUST PARTICLES IN THE ATMOSPHERE DURING THE LEONID METEOR SHOWERS OF 1998 AND 1999

NINA MATESHVILI, IURI MATESHVILI, GIULI MATESHVILI, LEV GHEONDJIAN, AND ZURAB KAPANADZE

Abastumani Astrophysical Observatory, Alexander Kazbegi av. 2a, Tbilisi 380060, Republic of Georgia
E-mail: matesh@yahoo.com

(Received 1 June 2000; Accepted 14 July 2000)

Abstract. November twilight sounding experiments carried out in the Abastumani Astrophysical Observatory including the Leonid showers of 1998 and 1999 provided an opportunity to obtain height distributions of dust of meteoric origin between 20 km to 140 km altitudes. The formation of several dust layers and their subsequent descent and dissipation were observed during these periods. The layers at 117 km and 54 km were estimated to consist of particles of 0.01 μm and 21–23 μm radii respectively, according to settling velocities of these layers. The particle number density at 117 km altitude was estimated to be ~30 p cm^{-3}.

Key words: Aerosols, Leonid meteor shower, meteors, terrestrial atmosphere, twilight sounding method.

1. Introduction

We are presenting in this article some observational results from a twilight sounding method carried out at the Abastumani Astrophysical Observatory, Georgia, during the 1998 and 1999 Leonid meteor showers. The ability of twilight events to serve as a sounding mechanism for the vertical distributions of fine dust particles suspended in the atmosphere between 20 km to 150 km relies on the fact that, at every given moment during twilight, the bulk of the scattered sunlight comes to an observer from a distinct and rather narrow atmospheric layer. Its altitude depends on time and may be calculated. The lower cut-off of this scattering layer is determined by the shadow of solid Earth and of a dense tropospheric air above the terminator region. Its upper boundary is a surface above which the total scattered light is negligible in comparison with the light coming from the lowest sunlit layer at the moment of measurement due to an exponential decrease of the air density. An extra amount of a scattering matter at any atmospheric level causes additional twilight sky brightness in a corresponding time interval. Fesenkov (1924) and Bauer *et al.* (1924) originally formulated the theory of the twilight phenomenon. It was later improved by (Brunner, 1935; Divari, 1967; Rozenberg, 1966). Some of the earliest twilight observations had led to a conjecture of the presence of dust particles in the upper atmosphere. The observations by Link (1975) had shown that there was a certain correlation between enhanced twilight scattering above 70 km and the occurrences of Orionid, Geminid and Librid meteor showers. He also registered a significant rise in time variation of the twilight sky brightness on 3 January 1971 during the active period of the Quadrantids (Link, 1971). A strong variation of the twilight light polarization was recorded in Davos during the 1968 Orionids (Steinhorst, 1971). It was also noted that this variation affected the stratosphere for a short time. The twilight observations carried out in the Engelgardt Observatory nearby Kazan, Russia (Gusakovskaya, 1973) during different periods of time had revealed a correlation between meteor activity and enhanced twilight light scattering, especially for the Orionids. Balloon-borne twilight measurement performed in October 1970 during the Orionids also showed this correlation (Fehrenbach and Link, 1974).

Our twilight observations had revealed enhanced abundances of meteoric dust in the upper and middle atmosphere during the meteor showers of the Quadrantids, η-Aquarids, Perseids, Orionids and Geminids (Mikirtumova, 1974; Mateshvili and Mateshvili, 1986a, 1986b, 1988; Mateshvili *et al.*, 1986, 1997). Owing to extremely favorable weather conditions, we succeeded in detailed monitoring of η-Aquarid meteoric particles intruding into the atmosphere and the formation of multiple dust layers in 1987, when we could carry out 18 twilight observations between April 27 and May 17 (Mateshvili *et al.*, 1997). It was possible to derive mean sizes of the particles from the observed settling velocities of the dust layers. A few simultaneous twilight observations carried out in different locations during the 1969 Orionid (Link and Gilbert, 1970; Divari *et al.*, 1973; Mikirtumova, 1974) and the 1969 Giacobinid meteor showers (Divari and Mateshvili, 1973) showed that the observed dust layers were large-scale, possibly global, features.

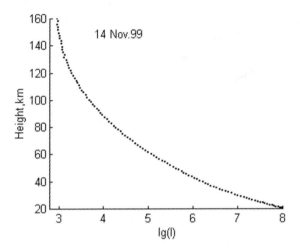

Figure 1. A typical twilight curve.

2. Observation results

Our Leonid 1998 and 1999 observations were performed with the aid of an electrophotometer equipped with an interference filter centered at the wavelength $\lambda = 610$ nm. A more detailed description of the twilight phenomenon and our calculation techniques can be found in (Mateshvili *et al.*, 1998).

We carried out twilight photometry during the period of Leonid shower activity on 1998 November 17, 20, 21 and 22 and on November 17, and on 22, 1999. Unfortunately, bad weather prevented us from observing during the Leonid maximum in 1999. We review here the preliminary results of our observations of 1998 (Mateshvili *et al.*, 1999) and add more precise calculations of particle sizes based on results of the Leonid Multi-Instrument Aircraft Campaigns obtained by Borovicka *et al.* (1999), Murray *et al.* (1999) and Spurny *et al.* (2000). We also calculate meteoric dust concentrations in the atmosphere.

The evaluation of sizes of particles by their settling velocities is based on a few assumptions on their physical and mineralogical properties. In this article we make an attempt to bring these assumptions as close as possible to the currently available data obtained from measurements and analyses of interplanetary dust particles (IDPs) collected in the lower stratosphere (Rietmeijer, 1998).

A typical photometric curve of logarithm of the twilight light intensity I(h), as a function of the height (h) of the Earth's shadow is shown in Figure 1. A simple way to distinguish a rapidly evolving component of atmospheric aerosol brought in by meteor showers is to normalize a current set of data, $I_{cur}(h)$, by a reference twilight set, $I_{ref}(h)$. The reference twilight measurement should be obtained as close as possible to the active period but still free of a significant extra influx of cosmic dust.

Figure 2 shows the I_{cur}/I_{ref} ratio curves of evening twilight measurements on 12, 20, 21 November and on the mornings of 21 and 22 November 1998 (Mateshvili et al., 1999). The evening of 11 November was selected as a reference for calculation of these ratios. I_{cur}/I_{ref} ratio curve of 12 November is given in Figure 2a. The curve shows significantly lower ratios then at the times of the Leonid shower but there are still a few peaks. This is understandable because the atmosphere is hardly ever at conditions for pure molecular scattering to occur due to constant influx of sporadic meteors. Also, this observation was carried out during the time the Taurid meteor shower was active.

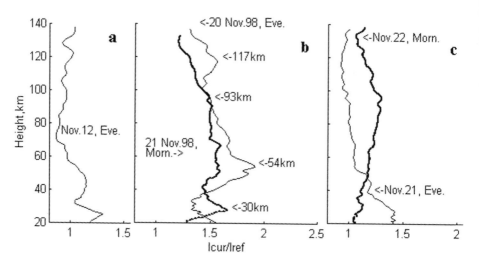

Figure 2. The ratios of current twilight intensities to reference ones versus height for the evenings of November 12 (a), 20 (b) and 21 (c), and on the mornings of November 21 (b) and 22 (c), 1998 (Mateshvili et al., 1999)

Figure 2b shows variations in the observed current to reference twilight ratios obtained on the evening of November 20 and the morning of November 21, 1998. It can be seen that two dust layers had formed by the evening of November 20 at 117 km and 54 km. On the morning of the 21st these layers had descended to 93 km and 30 km, respectively. The atmosphere was gradually restoring itself to pre-shower conditions during the following days as can be seen in the curve for the evening of November 21 (Figure 2c). An unexpected growth of I_{cur}/I_{ref} curve was observed in the morning twilight of November 22 (Figure 2c), which is suggestive of an additional influx of dust.

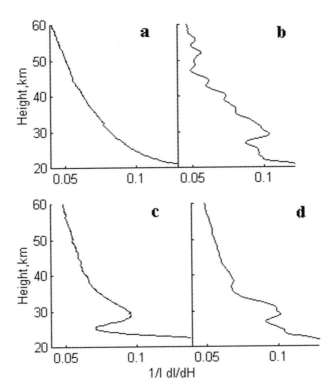

Figure 3. The logarithmic gradients of twilight intensities obtained from the observations carried out during (a) a non-shower period (10 Nov. 1999, evening), and during activity of (b) Leonids (21 Nov. 1998, morning), (c) Orionids (25 Oct. 1999, morning) and (d) Geminids (13 Dec..1999, morning).

It should be noticed that a layer at ~30 km is a rather frequent feature during meteor showers. The four curves in Figure 3 are so called logarithmic gradients that are derivatives of logarithm of the twilight light intensity (1/I dI/dh). Using this parameter, we are able to reveal the stratified structure of atmospheric dust in those cases when a suitable reference twilight curve is not available. The curve of Figure 3a was obtained on a non-shower day. The curves of Figure 3b, c and d were obtained during the active periods of the Leonid, Orionid and Geminid meteor showers.

We can see that the 30-km layer is a prominent feature in these curves. This layer seems to be a short-lived feature. That is, in the events shown by the curves of Figure 3b and 3d, this layer had disappeared by the next day.

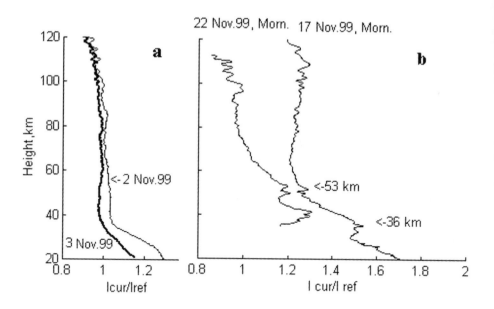

Figure 4. The ratios of current twilight intensities to the reference ones versus height for the evenings of November 2 and 3 (a) and on mornings of November 17 and 22, 1999 (b).

Figure 4a illustrates the atmospheric conditions well before the Leonid shower. Figure 4b shows the current to reference twilight intensities for 17 November and 22 November 1999. The conditions during the evening of 10 November were selected as a reference. According to data published by International Meteor Organisation (Arlt *et al.*, 1999), the observed Leonid activity on November 17, 1999 was about 16 to 20 meteors per hour. The corresponding twilight ratio curve demonstrates that the upper layers of the atmosphere were heavily dusted during the morning of November 17. A broad dust layer peaking at 36 km is also seen in Figure 4b. It should be noticed that the maximum at 53 km is very weak unlike on the morning of 20 November 1998 (Fig 2b). A weak maximum at 50 km and a better-developed one at 40 km were seen on November 22 (Figure 4b).

3. Discussion

The twilight measurements show the general picture of stratification of dust in the atmosphere and the changes in altitude of these layers formed by incursions of meteoric dust. These measurements provide an opportunity to estimate mean values of some physical characteristics of the dust particles. In particular, the mean radius of the dust particles forming distinguishable layers may be estimated from the settling velocity, which is observed in a series of consecutive twilight measurements. We use for this calculation the formula by Stokes-Davis; eddy diffusion is also taken into account (see, Mateshvili *et al.*, 1999).

Let us examine from this point of view some results obtained from our observations during the 1998 Leonids. The data obtained from twilight measurements are insufficient by themselves to estimate particle radii. Certain assumptions should be made on physical properties of the particles, in particular their volumetric density. These values can be adopted from the results of optical observations of Leonid showers and laboratory analyses of IDPs.

The analysis of the meteor light curves obtained during NASA's Leonid Multi-Instrument Aircraft Campaigns had led to a conclusion that the corresponding meteors should be metal or silicate grains bounded by a low boiling point, possibly an organic substance, rather than compact bodies (Murray *et al.*, 1999). Spurny *et al.* (2000) derived from a study of one bright Leonid meteor a density of 0.7 g/cm^3. A similar conclusion on the fragile structure for Leonid meteors was reached by Borovicka *et al.* (1999) based on spectroscopy of meteor trails.

A fraction of IDPs were characterized with more then 70% porosity and their mean density was estimated at 0.7 g/cm^3 (Rietmeijer, 1998). Because Leonid meteors were found to be porous bodies, a value of 0.7

g/cm^3 is a reasonable assumption. We do not imply, of course, that all Leonid particles in the atmosphere had a uniform density of 0.7 g/cm^3. Porous particles are aggregates consisting of units of different density (Rietmeijer, 1998; Rietmeijer et al., 1999). They may undergo fragmentation as they enter the atmosphere. The resulting particles of the individual constituent units are denser than the original aggregates. Ablation is another process in which denser particles may be created. Sufficiently small meteoroids may melt and become dense spheres (Love and Brownlee, 1991). And, finally, very small dense particles may be formed as condensates of meteor vapor (Rosinski and Snow, 1961; Hunten et al., 1980).

Condensation might be responsible for the formation of dust particles of the layer at 117 km that emerged by the evening of November 20, 1998 (Figure 2b). The layer had descended to 93 km by the morning of 21 November. The dust layer at 93-km altitude appears rather weak in comparison with other layers discussed here. We will consider the possibility that this layer was an accumulation of particles descending from the layer at 117 km. It is obvious that the particles in the 117-km layer should be small in order to have a long enough residence time to create a visible layer at this altitude. As an alternative a weak layer at 67 km could have consisted of dust that had settled from the 117-km layer. In this case we would obtain particles as large as 0.5 µm in radius. These particles would fall from 120 km to 110 km, i.e. from the top to the bottom of the layer peaking at 117 km, in about 20 s. Therefore, their accumulation at 117 km would be highly unlikely. It should be noted that a given amount of dust manifests itself as a decreasingly prominent feature in a twilight I_{cur}/I_{ref} ratio curve as its height decreases. That is because its light scattering becomes relatively smaller in comparison with an increasingly stronger molecular light scattering. Confirmation for likelihood of the presence of a dust layer at about 93 km height was provided by a rocket-born probe (Gelinas et al., 1998) that detected a 5-km thick layer of electrically-charged dust at about 90 km in February 1998. A meteoric origin of this dust layer was suggested. Thus, we submit that the dust particles had settled from 117 km to a height of 93 km on the morning of November 21.

The radii of these particles, as derived from the settling velocity at any reasonable assumption of the particle density, are much smaller than the characteristic sizes of principal building blocks (> 0.1 µm) of aggregate IDPs (Rietmeijer et al., 1999). The Fe-Mg-SiO-H$_2$-O$_2$ composition of vapors used in laboratory vapor phase condensation experiments (Rietmeijer et al., 1999) are probably similar to vapor mixtures that are produced by evaporation of incoming meteoroids. The condensates in these experiments were amorphous "FeSiO" and "MgSiO" particles plus some amounts of Mg-, Fe- and Si-oxides. The mean density of these particles of different composition is assumed to be 2.5 g/cm^3. Assuming this density for the particles in the 117-km layer and their settling velocity value, we calculate an average particle radius of 0.01 µm (Table I). This value is in good agreement with the size of the condensed "FeSiO" grains, 0.01 to 0.03 µm in diameter (Rietmeijer et al., 1999).

It is of interest to trace the fate of these small particles. Their settling velocity should decrease as they enter increasingly denser air layers. Thus, they should concentrate at certain altitudes. But, as particle concentrations become high enough, the coagulation process should be intensified. One of the results of the condensation experiments by Rietmeijer et al. (1999) is that the condensed grains tend to form clusters and strands but other factors may decrease the efficiency of this process under the atmospheric conditions. One possible factor is an electric charge of particles (Reid, 1997). The resulting larger particles will accelerate their downward movement. Confirmation of this scenario is that somehow persistent enhanced turbidity features are never found in post-shower twilight curves. If not for this rapid sink mechanism, the smallest meteoric particles would accumulate in the mesosphere for much longer times.

When we know the particle radius, we can also make an attempt to derive their concentration from the additional light scattering they cause in the atmosphere. A circumstance that facilitates such estimates for small particles at high atmospheric levels is that larger particles descend rapidly to lower altitudes causing size separation of the particles. A subsequent injection of larger particles adds to differently-sized particle mixture in the middle atmospheric region. It leaves the rather homogenous fraction of small particles at high levels uncontaminated.

Some assumptions should be made to perform the evaluation of the particle concentration at 117 km on the morning of 20 November 1998. We should assume that the light scattering at this height was purely molecular in nature on the reference day and that the scattering on 20 November at the same height was a sum of molecular scattering and scattering on the meteoric dust particles. We assume also that all of these particles were of the same radius, r = 0.01 µm, as it was estimated by their settling velocity.

A value of the refractive index is also to be assumed. Proceeding from the assumed composition and density of the particles, we postulate that they are pure dielectric with the refractive index about 1.5.

A concentration of scattering particles N_p may be determined as a ratio of the particle volume scattering coefficient β_p and the scattering cross section of a single particle σ_p:

$$N_p = \frac{\beta_p}{\sigma_p} \tag{1}$$

Assuming the radius r of a single particle to be the value determined by the sedimentation velocity of the dust layer, we obtain the scattering cross section (McCartney, 1977):

$$\sigma_p = \frac{128\pi^5 r^6}{3\lambda^4}\left(\frac{n_p^2 - 1}{n_p^2 + 2}\right)^2 \tag{2}$$

where n_p is the particle refractive index; λ is the wavelength of the scattered light. The scattered light intensity I may be expressed as:

$$I = \beta f(\theta) I_0 \tag{3}$$

where β is the volume scattering coefficient of aerosol, $f(\theta)$ is scattering phase function, and I_0 is the intensity of the impacting sun light beam.

With the above assumption on the nature of the light scattering in reference and current twilight, we obtain:

$$\frac{I_{cur}}{I_{ref}} = \frac{(\beta_m + \beta_p) f(\theta) I_0}{\beta_m f(\theta) I_0} \tag{4}$$

where β_m is the Rayleigh volume scattering coefficient for air molecules and β_p - particle volume scattering coefficient. As the particles under consideration are very small, the space distribution of their scattering may be described by an almost molecular phase function. Then:

$$\beta_p = (I_{cur}/I_{ref} - 1)\beta_m . \tag{5}$$

Volume scattering coefficient for air molecules β_m may be expressed as (McCartney, 1977; Penndorf, 1957):

$$\beta_m = \frac{8\pi^3}{3} \frac{(n_S^2 - 1)^2 N}{\lambda^4 N_S^2} \left(\frac{6 + 3\alpha_n}{6 - 7\alpha_n} \right) \qquad (6)$$

where n_s is the refractive index and N_s is the number density of standard air. α_n is the depolarisation factor, N is the number density at any atmospheric temperature and pressure.

Rayleigh volume scattering coefficient for air molecules at any height is:

$$\beta_m(h) = \beta_{m0} \frac{\rho(h)}{\rho_0} \qquad (7)$$

where β_{m0} is the Rayleigh volume scattering coefficient and ρ_0 is the air density at standard conditions; $\rho(h)$ is the air density at a height h.

With the above-mentioned assumptions we obtain that the concentration of scattering particles N_p causing the intensity of the scattered light observed on November 20 1998, which was brighter than that of the reference day by the factor of 1.5, corresponds to a particle concentration 30 p cm^{-3} (Table I). Uncertainties in these particle radii and their concentration result from uncertainties inherent to the twilight sounding method and our assumptions on the particle properties. It is known that the uncertainty of altitude determination by this method increases with altitude to ± 3 km at 100 km (Rozenberg, 1966). Also, not considering finite widths for the dust layers, the resulting uncertainty in the particle radii becomes ± 35%. Similarly, our estimates of particle concentrations will range from 2–200 p cm^{-3}. Assumptions on the particle properties, particularly the refractivity index, will also affect our calculated particle concentrations, as well as ignoring the effects caused by particles with a particular size distribution. Other sources of errors in the particle concentrations could be our assumptions of the pure molecular scattering of the atmosphere during the reference twilight curve acquisition. We are presently unable to estimate the various uncertainties and our results on particle radii and particle concentrations only serve as a cautious approximation.

TABLE I

Height (km)	Density (g/cm³)	Radius (μm)	Mass (g)	Refractive index	Concentration (p cm⁻³)
117	2.5	0.01	$1.0\ 10^{-7}$	1.5	30
54	0.7	58	$5.7\ 10^{-7}$		
54	3.0	23	$1.5\ 10^{-7}$		

It would be interesting to compare our particle concentrations with direct measurements but unfortunately we could not find any results for similar atmospheric situations. Still, the values listed in Table II that were obtained at different heights and under different circumstances show that our estimates of the Leonid dust concentration at 117 km are reasonable.

TABLE II

Height (km)	Radius (μm)	Concentr. (p cm⁻³)	Method and circumstances	Sources
20	0.01	10	Balloon-borne sampling before the St. Helens eruption	Hofmann and Rosen, 1982
20	0.01	550	Balloon-borne sampling after the St. Helens eruption	Hofmann and Rosen, 1982
70-80	< 0.25	1	Photoelectric counter	Bradin and Orishich, 1989
80	0.01< r <0.1	4.2	Satellite horizon scanning	Gray et al, 1971, 1973
100	0.01< r <0.1	$5.5\ 10^{-2}$	Satellite horizon scanning	Gray et al, 1971, 1973
80-90	0.01, 0.05	100	Rocket-borne photometry in a noctilucent cloud	Gumbel et al, 2000

Let us consider the particles that formed a layer at 54 km on the evening of 20 Nov. 1998 that had descended to 30 km on the morning of 21 Nov. 1998 (Figure 2b). As was noted above, the assumed density of Leonid meteors is 0.7 g/cm3. In this case, the particles should be 58 μm in radius based on their settling velocity. But what types of collected particles correspond to the optically detected particles with these sizes and density? Aggregate IDPs are 5 to 25 μm in size and cluster IDPs composed of chondritic aggregate and non-chondritic IDPs may be up to 60 μm in size (Rietmeijer, 1998, 1999). According to Love and Brownlee (1991) particles of this size entering the atmosphere at cometary velocities should melt and settle as dense spheres. The mean density of spherules collected from ocean floor sediments is 3 g/cm^3 (Murrell et al., 1980). Using this value, we obtain that the particles descending from 54 to 30 km are 23 μm in radius. Love et al. (1994) estimated the mean density of the spheres at 3.4 g/cm^3. With this density, the particles we observed at 54 km altitude on 20 November 1998 should have a 21 μm radius. The uncertainty of ± 20 % in this value is the result from our neglect of the peak width of the dust layers. The silicate spherules found among the collected stratospheric IDPs are 1.5 to 15 μm in radius with a mean value of 5.4 μm (Rietmeijer, pers. comm.), which is significantly smaller than our estimates of the particle size. We suggest that this could be because the larger particles rapidly descended, thus lowering probability of being collected in the stratosphere. Therefore, spherules of the sizes we have obtained should be found at the Earth's surface. Brownlee et al. (1997) had studied spherules 50 to 1000 μm in size that were collected in deep-sea sediments and the Antarctic ice sheet. According to their calculations, spherules larger than 50-70 μm should be of asteroidal origin because large particles entering the atmosphere with cometary velocities will be evaporated. Thus, the large particles (r = 21–23 μm) we detected could be only a small fraction that remained after evaporation in the atmosphere. They form an optically detectable layer only because of the atmospheric property that accumulates dust particles at certain levels depending on the particle sizes. It is not possible to calculate their concentration because of strong admixture of particles of other sizes in this atmospheric region.

We like to emphasize that we are unable to draw firm conclusions on a nature of the particles from the twilight data alone but we offer probable interpretations. Other explanations for the presence of large particles at 54-km altitude are possible. For example, they could be produced in situ through coagulation of smaller particles. The presence of smaller particles at this altitude is indicated by the persistence of this layer during the morning (Figure 2b) and as a weakened feature during the evening (Figure 2c) twilights of 21 November 1998. Definitive conclusions have to await particle collections following meteor showers. Let us examine our results of 1999 Leonid shower (Figure 4a, 4b). Because we could not observe just prior to November 17, we were unable to trace the dynamics of the layer at 36 km. Thus, we could not derive the sizes of these particles. A high degree of turbidity of the upper atmosphere can be seen in the twilight ratio curve for the Leonid activity on November 17, 1999 (Figure 4b), in correlation with visual observations by Arlt et al. (1999). The lower part of twilight ratio curve of November 22 resembles the previous curve of November 17 (Figure 4b). There is a weak maximum at 53 km and a stronger one at 40 km. But the upper part of this curve is quite different. The I_{cur}/I_{ref} ratio became close to unity above 70-km altitude. It can be seen that the peak in the Leonid activity is already over and almost all of the fine particles had settled.

4. Conclusions

Twilight measurements carried out during the active periods of the Leonid 1998 and 1999 meteor showers showed that large amounts of dust particles were injected into the atmosphere. The formation of dust layers was observed in addition to an overall enhancement of atmospheric turbidity. Particle sizes were derived from settling velocities of these layers. A layer at 117-km altitude of probably secondary particles, 0.01 µm in radius, that could be condensates of meteor vapors, had particle concentrations of about 30 p cm^{-3}. A significant part of the particles in a dust layer at 54 km, which was simultaneously observed with the 117-km layer, was 21–23 µm in radius. These particles had descended to 30-km altitude by next day. They might be melted spherules. Our estimates of particle sizes derived from twilight measurements have a considerable error but they are comparable to atmospheric particle sizes obtained by different methods and under different circumstances. The twilight observations during the 1999 Leonid shower showed that a dust layer had formed at 36–40 km on 17 November and 22 November. Separation of meteoric particles of different sizes takes place in the atmosphere due to different settling rates. Even the largest particles could have descended to altitudes below 20 km in a few hours. Another feature of an intra-atmospheric particle dynamics is the formation of transient dust layers. The smallest particles accumulate at high atmospheric levels but particles >10 µm in radius settle rapidly to about 50 km. The additional dust layers formed during the Leonid showers had disappeared within a few days. One could expect a more persistent presence of the finest particles between 50 and 80 km altitudes. Probably, simple sedimentation alone can not account for the observed rapid clearing of the upper atmosphere, which might require some other processes, such as coagulation or electrostatic charging of the particles. A comprehensive model based on observations and describing the fate of injected meteoric dust, would be of interest for other applications, such as predicting climatic and other consequences of much more powerful and hazardous invasions of cosmic dust.

Acknowledgments

We thank Frans Rietmeijer for information on the properties of collected IDPs. Reviews by George Flynn and an anonymous referee significantly improved our presentation. *Editorial handling:* Frans Rietmeijer.

References

Arlt, R., Rubio, L.B., Brown, P. and Gyssens, M.: 1999, *WGN, Journal of the IMO* **27**, 286–295.
Bauer, E., Danjon, A. and Langevin, J.: 1924, *Comptes Rendus Acad. Sci.* **178**, 2115–2117.
Borovicka, J., Stork, R. and Bocek, J.: 1999, *Meteoritics Planet. Sci.* **34**, 987–994.
Bradin, O.A. and Orishich, T.I.: 1989, *Handbook for MAP* **32**, p.178.
Brownlee, D.E., Bates, B. and Schramm, L.S.: 1997, *Meteoritics* **32**, 157–175.
Brunner, W.: 1935, Publ. *Eidgenossischen Sternwarte Zürich* **6**, 1–120.
Divari, N.B.: 1967, (in Russian), *Kosmicheskiye issledovaniy*, **5**, 475–477.
Divari, N.B. and Mateshvili, Yu.D.: 1973, *Astron.Tsirk.***744**, 5–7 (in Russian).
Divari, N.B., Zaginailo, Yu.I. and Kovalchuk, L.V.: 1973, Astron.Vestnik.7, 223–230 (in Russian).
Fehrenbach, M. and Link, F.: 1974. *Comptes Rendus Acad. Sci.,* **1** (B279), 687–689.
Fesenkov, V.G.: 1924, *Astron. Nachr.* **220**, 33–42.
Gelinas, L.J., Lynch, K.A., Kelley, M.C., Collins, S., Baker, S., Zhou, Q. and Friedman, J.S.: 1998, *Geophys. Res. Lett.* **25**, 4047--4050.
Gray, C.R., Malshow, H.L. and Merrit, D.C.: 1971, AIAA Paper **N1111**, 1–17.

Gray, C.R., Malshow, H.L. and Merrit, D.C.: 1973, *NASA, Final Report CR-11231, 1973*, 211p.
Gumbel, J., Stegman, J., Murtagh, D.P., and Witt, G.: 2000, *COSPAR -2000 abstracts*.
Gusakovskaya, L.B.: 1973, *'Photoelectric photometry of twilight'*, Ph.D. Dissertation, University of Kazan.
Hofman, D.J. and Rosen, J.M.: 1982, *J. Geophys. Res.* **87**, 11039–11061.
Hunten, D.M., Turco, R.P., and Toon, O.B.: 1980, *J. Atmos. Sci.* **37**, 1342–1357.
Link, F.: 1971, *Planet. Space Sci.* **19**, 1585–1587.
Link, F.: 1975, *Planet. Space Sci.,* **23**, 1011–1012.
Link, F. and Gilbert, W.: 1970, *Comptes Rendus Acad.Sci.* **271**, B974–B976.
Love, S. G. and Brownlee, D.E.: 1991, *Icarus* **89**, 26–43.
Love, S. G., Joswiak, D.J., and Brownlee, D.E.: 1994, *Icarus* **111**, 227–236.
Mateshvili, G.G. and Mateshvili, Y.D.: 1986a, *Bull. Abastumani Astr. Obs* **61**, 188–198 (in Russian).
Mateshvili, G.G. and Mateshvili, Y.D.: 1986b, *Bull. Abastumani Astr. Obs* **61**, 167–188 (in Russian).
Mateshvili, G.G. and Mateshvili, Y.D.: 1988, in A.G.G.M.Tielens and L.J.Allamandola (eds.) *Interstellar Dust*, Proc. IAU Symposium, 135, NASA CP-3036, pp.463-468.
Mateshvili, G.G., Mateshvili, Y.D., and Megrelishvili, T.G.: 1986, *Optica atmospheri I aerosol. Moskva. Nauka.* 133-149 in Russian).
Mateshvili, G.G., Mateshvili, Y.D., and Mateshvili, N.Y.: 1997, *Sol. Syst. Res.* **31**, 483–488.
Mateshvili, I.D., Mateshvili, G.G., and Mateshvili, N.I.: 1998, *J. Aerosol Sci.* **29**, 1189–1198.
Mateshvili, N., Mateshvili, G., Mateshvili, I., Gheondjian, L., and Avsajanishvili, O.: 1999, *Meteoritics Planet. Sci.* **34**, 969–973.
McCartney, E.J.: 1977, *Optics of the atmosphere, scattering by molecules and particles*, John Wiley & Sons, New York, pp. 421.
Mikirtumova, G.G.: 1974, *Astron.Tsirk.* N.849, 5–7 (in Russian).
Murray, I. S., Hawkes, R., and Jenniskens, P.: 1999, *Meteoritics Plan. Sci.* **34**, 949–958.
Murrell, M.T., Davis, Jr., P.A., Nishiizumi, K., and Millard,Jr., H.T.: 1980, *Geochim. Cosmochim. Ac*ta **44**, 2067–2074.
Penndorf, R.: 1957, *J. Opt. Soc. Amer.* **47**, 76–182.
Reid, G.C.: 1997, *Geophys. Res. Lett.* **24**, 1095–1099.
Rietmeijer, F.J.M.: 1998, in J.J. Papike (ed.), *Planetary Materials, Reviews in Mineralogy*, **36**, The Mineralogical Society of America, Washington (DC), pp. 1–95.
Rietmeijer, F.J.M.: 1999, AIAA paper **94-0502**, 1–12.
Rietmeijer, F.J.M, Nuth, J.A., and Karner J.M.: 1999, *Astrophys. J.* **527**, 395–404.
Rosinski, J. and Snow, R.H.: 1961, *J. Meteorol.* **18**, 736–745.
Rozenberg, G.V.: 1966, *Twilight*. Plenum Press, New York, 1–380.
Spurny, P., Betlem, H., Van't Leven, J., and Jenniskens, P.: 2000, *Meteoritics Planet. Sci.* **35**, 243–249.
Steinhorst, G.: 1971, *Beit. zur Phys. Atmosph., Bd* **44**, 279–292.

RECOGNIZING LEONID METEOROIDS AMONG THE COLLECTED STRATOSPHERIC DUST

FRANS J.M. RIETMEIJER
Institute of Meteoritics, Department of Earth and Planetary Sciences,
University of New Mexico, Albuquerque, NM 87131, USA
E-mail: fransjmr@unm.edu

and

PETER JENNISKENS
SETI Institute, NASA ARC, MS 239-4, Moffett Field, CA, 94035, USA
E-mail: pjenniskens@mail.arc.nasa.gov

(Received 28 June 2000; Accepted 31 July 2000)

Abstract. Three chemical groups of primary "silicate" spheres <30 μm in diameter of cometary origin were collected in the lower stratosphere between 1981 May and 1994 July. The "silicate" sphere abundances represent an annual background from contributions by sporadic meteor and weak annual meteor shower activities. During two collection periods, from 06/22 until 08/18, 1983 (U2015), and from 09/15–12/15, 1981 (W7027/7029), a higher number of spheres was collected compared to other periods of the year represented by the other collectors studied here. This study links two different data sets, viz. the NASSA/JSC Cosmic Dust Catalogs and peak activities of annual meteor showers, and identified high-velocity cometary sources for collected stratospheric "silicate" spheres. The majority of spheres on flag U2015 may originate from comet P/Swift-Tuttle (Perseids), while the majority of spheres on flags W7027/7029 could be from comet P/Halley (Orionids) or comet P/Tempel-Tuttle (Leonids). Variations in relative proportions of the Mg,Si,Ca ± Al, Mg,Si ± Fe and Al, Si,Ca spheres may offer a hint of chemical differences among high-velocity comets. Proof for the findings reported here might be obtained by targeted cosmic dust collections in the lower stratosphere including periods of meteor shower and storm activity.

Key words: Ablation, chemical composition, comets, cosmic dust, Leonids, interplanetary dust particles (IDPs), meteor showers, silicates, spheres

1. Introduction

The recent period of Leonid storms that began in 1996 and is expected to continue until 2003, not only created intense public awareness but has also revived the moribund meteor science. Much of this revival is due to the intense 1998 and 1999 Leonid multi-instrument aircraft campaigns (Rietmeijer, 1999a, 2000a). These campaigns brought together researchers from a wide range of disciplines and built a bridge between the observational and analytical sciences. The latter are represented by cosmochemists who analyze collected extraterrestrial materials in the laboratory. So far the only attempt to collect surviving Leonid meteors was initiated through the NASA Cosmic Dust Program that, since May 1981, has made numerous successful samplings of stratospheric dust between 17–19 km altitudes (Zolensky *et al.*, 1994). The collected stratospheric dust particles are classified into four major groups, viz. (1) cosmic [C, or C?], (2) terrestrial contamination natural [TCN or TCN?; e.g. volcanic ash], (3) terrestrial contamination artificial [TCA or TCA?], such as Al-oxide spheres that are the solid fuel effluents of the Space Shuttle (Rietmeijer and Flynn, 1996), and (4) unknown (Mackinnon *et al.*, 1982; Fraundorf *et al.*, 1982). Most particles in the cosmic group, i.e. interplanetary dust particles (IDPs), have physical, mineralogical and chemical properties, such as D/H and $^3He/^4He$ ratios, proving an extraterrestrial origin (Bradley et al, 1988; Brownlee, 1985; Mackinnon and Rietmeijer, 1987; Pepin *et al.*, 2000; Sandford, 1987). Collected IDPs entered the Earth's atmosphere with velocities of tens of km s^{-1} and they decelerated by collisions with air molecules between 120 – 80 km to speeds of cm s^{-1}. The resulting concentration factor makes it possible to collect them efficiently after they have settled to the lower stratosphere. The IDPs are typically about 10–20 μm in size (Rietmeijer, 1998; Rietmeijer and Warren, 1994) and include three major subgroups:
(1) Chondritic aggregate IDPs,
(2) Non-aggregate, type CI and CM, chondritic IDPs, and
(3) Non-chondritic IDPs that occur as fragments of silicate-mineral and Fe,Ni-sulfide grains and aggregates thereof and as spheres.

Melting of fragments decelerating in the Earth's atmosphere will result in the formation of quenched-melt spheres (Brownlee, 1985; Love and Brownlee, 1991). Flash heating induces volatile element loss and chemical fractionation but the major elements in collected spheres should still have some 'memory' of the progenitor particles. Rietmeijer

(2000b) showed that within reason we could use the measured physical, chemical and mineralogical properties of the surviving IDPs and micrometeorites to constrain these properties in micrometeoroids that do not survive during atmospheric entry.

Spheres in the NASA Cosmic Dust Program have not received much attention although they may include dust that entered the atmosphere with cometary velocities (i.e. >20 km s^{-1}) from near-Earth asteroids and comets. The surviving dust will most likely be in the form of "silicate" spheres. We use the opportunity of the current Leonid storm activity to review "silicate" sphere data in the NASA Cosmic Dust Catalogs. We will attempt to identify meteor showers, and thus sources, that could have produced these spheres and their precursor fragments.

2. The NASA Cosmic Dust Collection

Since May 1982, the NASA Cosmic Dust Program has routinely collected stratospheric dust at altitudes between 17 – 19 km altitude using silicone oil-coated, inertial-impact, Lexan flat-plate collectors that are housed in pylons mounted underneath the wings of high-flying U2, ER-2 and W57B aircraft. Pressure sensors on the aircraft activate the collectors at the appropriate altitudes to avoid sampling the dirty troposphere. On average each collector is exposed to the stratosphere for an accumulated total exposure time of 30 – 40 hours but longer exposures have occurred (Rietmeijer and Warren, 1994). All pre- and post-flight handling of individual collectors (or flags), as well as sample curation, occurs in a dedicated Class 100 clean-room (Warren and Zolensky, 1994).

The sources used in this study are the Cosmic Dust Catalogs volumes 1 through 15 published by the NASA Johnson Space Center (Appendix). In these catalogs each particle is identified by a scanning electron microscope image (aggregate, sphere, fragment), its physical properties (e.g. size, color, shape), and an energy dispersive spectrum (EDS). The EDS spectrum shows the presence of major rock-forming elements from Na to Ni with abundances above the detection limit (typically a few percent) (Figure 1). These properties can be unambiguously determined, and allow classification of almost every particle in one of only four major groups (Mackinnon *et al.*, 1982). Stratospheric dust classification using these simple properties was proven to be highly accurate, consistent and effective (Rietmeijer, 1998 for a review).

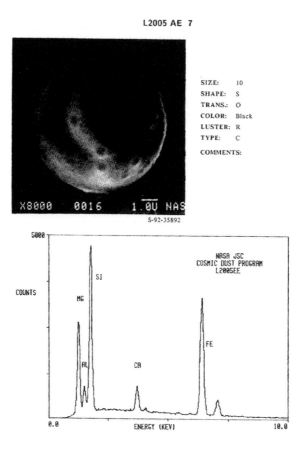

Figure 1. A typical page in the NASA/JSC Cosmic Dust Catalogs showing a scanning electron microscope image of a collected "silicate sphere, its elemental composition in an energy dispersive spectrum page and its physical properties. Mg,Si,Fe "silicate" sphere L2005AE7 with minor amounts of Al and Ca (Cosmic Dust Catalog, vol. 13; see appendix). Reproduced by courtesy of the National Aeronautics and Space Administration.

3. "Silicate" Spheres

The identification of a sphere is without prejudice and unambiguous. The peak heights in a EDS spectrum (Figure 1) are a measure of the elemental abundances in and can be reduced to a quantitative bulk composition of the analyzed material. This procedure is routinely used in analytical electron microscope studies but it is purposely not followed for the particles that are listed in the NASA Cosmic Dust Catalogs (see

Appendix). The relative peak heights offer a qualitative estimate of the elements present. In this manner, the major elements define three distinct chemical groups for the "silicate" spheres, viz. (1) Mg,Si ± Fe, (2) Mg,Si,Ca ± Al and (3) Al,Si,Ca. Elements with almost equal estimate abundance are listed. The '±' sign indicates elements that occur with variable abundances in a chemical group. Minor amounts of elements from one group may occur in another group, such as small amounts of Mg and Fe in Al,Si,Ca spheres. This chemical grouping is not an exact procedure but is does confirm the presence of only three major chemical groups of collected "silicate" spheres.

Sphere allocation to the TCN group is a default assignment. That is, its elemental composition does not resemble the elemental distribution of particles associated with collected chondritic aggregate IDPs and cluster IDPs that that are proven to be extraterrestrial (Brownlee, 1985; Rietmeijer, 1998; Zolensky *et al.*, 1994). We re-allocated the rare TCN "silicate" spheres in the NASA Cosmic Dust Collection to the C group because there are no known natural terrestrial processes that could produce "silicate" spheres that reach the lower stratospheres by a natural process. Silicate spheres are only known among the shards and spheres in the ejecta plume of the mount Etna volcano (Italy) with activity limited to the troposphere (Lefèvre *et al.*, 1985, 1986). These volcanic spheres contain abundant Al, Mg and Fe, which do not occur simultaneously in the stratospheric spheres, and they have high alkali contents (Lefèvre *et al.*, 1985, 1986).

The numbers of collected "silicate" C-spheres are listed in Table I, which shows the collection periods and number of cosmic "silicate" spheres with three distinct elemental compositions from the NASA Cosmic Dust Collection (Appendix). The actual collection dates are not known for most collectors (or flags) (Jack Warren, NASA-JSC Curatorial Facility, written comm.). All LAC abundances are normalized for comparison with the SAC abundances. This Table I has multiple entries when two or three individual collectors were flown simultaneously on the airplane. The number of particles on simultaneously flown collectors used by the NASA Cosmic Dust Program (see, appendix) are listed separately in Table II and show that variability of collected spheres among simultaneously flown collectors. This variability is an observed fact but it is a not understood phenomenon that might be related to the location of collectors on either the left or right wing of the aircraft (Rietmeijer and Warren, 1994). It is also possible that it reflects small scale, transient heterogeneity in particle number density in the lower stratosphere conceivably caused by the aircraft itself. In this paper we will adopt the average (rounded off to the closed integer) of the number of particles of the collectors as the abundance of spheres during the time of collection. The NASA Cosmic Dust Program initially deployed small-area collectors (SACs) of 30 cm^2 but later switched to large-area collectors (LACs, 'L' in Tables 1 and 2) with a 300 cm^2 surface. The LAC abundances are normalized to a 30-cm^2 area for comparison with particle abundances on the SACs (Rietmeijer and Warren, 1994). When this procedure yields a number < 1, one sphere is listed in Table I.

TABLE I

Collectors	Collection period	Number of spheres		
		Mg, Si ± Fe	Mg, Si, Ca ± Al	Al, Si, Ca
W7013	05/22-07/06, 1981*	1	3	1
W7017	07/07-09/15, 1981	1	2	1
W7027/7029	09/15-12/15, 1981	3	6	2
U2001	03/13-04/08, 1982	1	1	0
U2015	06/22-08/18, 1983	6	2	1
U2022	04/09-06/26, 1984	0	0	0
U2034	04/27-08/28, 1985	2	0	1
W7074	Aug, 1987 and early July, 1988	2	1	0
L2005/2006/2008	10/03-10/13, 1989**	3	1	1
L2009/2011	June and July, 1991	1	0	0
L2021#	01/07 and 02/01, 1994	1	1	0
L2036#	06/07 and 07/03, 1994	1	0	0

* Its total collection time of 65 hours is an anomaly of almost twice the average collection time (Rietmeijer and Warren, 1994); its re-normalized abundance is three spheres;

** The exact number of hours of collection is unknown (Rietmeijer and Warren, 1994);

Two collection days in January and February; *ibid* two days in June and July.

TABLE II

Collectors	Total number of spheres
W7027	3
W7029	18
L2005	15
L2006	5
L2008	11
L2009	4
L2011	1

TABLE III

Composition	Mean	Standard deviation	Range	Number of spheres
Mg,Si ± Fe	10.6	5.2	2 – 30	46
Mg,Si,Ca ± Al	10.5	4.2	3 – 17	26
Al,Si,Ca	11.2	5.3	3 – 20	14
Total	10.7	4.8	3 – 30	86

Table IV

Composition	Mean	Standard deviation	Range	Number of fragments
Mg,Si ± Fe	19.1	12.1	5.0 – 64.0	34
Mg,Si,Ca ± Al	34.1	7.0	28.3 – 44.0	4
Al,Si,Ca	22.8	12.8	8.4 – 52.5	13
Total	21.2	12.5	5.0 – 64.0	51

Table III shows the mean diameter (micrometers), the standard deviation and size range of the "silicate" spheres listed the NASA Cosmic Dust Catalogs that were collected between May 22, 1981 and July 3, 1994. The "silicate" sphere diameters form normal distributions (at a 95% confidence limit) with remarkably similar size ranges. There is only one Mg,Si ± Fe sphere with a 30 μm diameter while all other spheres in this chemical group are <20 μm in diameter. The mean diameters among the chemical groups are identical (Table III). We point out that our statistical treatment assumes that each collector is a complete representation of the stratospheric dust during the time of collection and the published catalogs are without any curatorial bias. This may not be true because (1) the largest spheres may be lost during SEM characterization and will not be listed in the catalogs, and (2) individual collectors could be incompletely sampled for characterization and publication in the NASA Cosmic Dust Catalogs. Nevertheless, recognizing this situation and other problems associated with this collection program (Rietmeijer and Warren, 1994), we proceed with our analyses.

4. Discussion

4.1. THE FORMATION OF PRIMARY "SILICATE" SPHERES

Three chemical groups of "silicate" spheres with Mg,Si ± Fe spheres being the most abundant ones occurred in the lower stratosphere between May 22, 1981 and July 3, 1994. They could be either quenched-melt residues (primary spheres) of progenitors melted by atmospheric entry flash heating or secondary spheres, that is, spall droplets from meteoroids >1 cm in size. In a study of thermal interactions of incoming micrometeoroids with the atmosphere as a function of particle size and mass, entry velocity and angle, Love and Brownlee (1991) predicted that the diameter of a progenitor particle is typically 1.5 to 2 times the diameter of the resulting primary sphere. This relationship predicts that the average size of the progenitors of the collected "silicate" spheres is 16 – 22 μm. Love and Brownlee (1991) also showed that the diameters of progenitor of a 10-μm sphere are a skewed distribution between ~15 and ~100 μm.

In order to verify these predicted relationships among the collected silicate dust, we reviewed all "silicate" fragments labeled "C" or "C?" in the Cosmic Dust Catalogs (Appendix) following the same procedure used for the "silicate" spheres. An extraterrestrial origin of many fragments is indicated by aggregate IDP-like material that is attached to their surface. The EDS spectra for the "silicate" fragments define the same chemical groups that were found for the "silicate" spheres. This similarity points to a genetic relationship between the fragments and spheres. Table IV shows the mean of the root-mean-square (rms) size (micrometers), the standard deviation and size range of "silicate" fragments collected in the lower stratosphere between May 22, 1981 and July 3, 1994. The rms size is calculated as $(a^2 + b^2)^{1/2}$, whereby a and b are the orthogonal longest and shortest particle dimensions. The size and size range of the fragments (Table IV) are in excellent agreement with the progenitor-sphere size relationships predicted by Love and Brownlee (1991) for primary spheres formed by melting and ablation of larger progenitor particles.

The modeling study by Love and Brownlee (1991) also showed that the ratio comet/(comet + asteroid) spheres < 30 μm in diameter is unity for primary spheres. Thus, the collected Mg,Si ± Fe, Mg,Si,Ca ± Al and Al,Si,Ca "silicate" spheres are the surviving remnants of micrometeoroids that entered the Earth's atmosphere at velocities >20 km s^{-1} (Brownlee et al., 1995), which would include a fraction of asteroidal asteroids (22–25 km s^{-1}), low-velocity comets (25-36.5 km s^{-1}) and all other comets (Rietmeijer, 2000b). The progenitors occur as individual IDPs (Table IV), as mineral grains (< 5 μm) embedded in aggregate IDPs and as large (>10 and up to ~50 μm) grains among the fragments of cluster IDPs (Rietmeijer, 1998). They are mostly Mg-rich olivines, Ca-poor and low-Ca Mg-rich pyroxenes and Ca,Mg-rich clinopyroxenes (Rietmeijer, 1998, 1999b; Thomas et al., 1995; Zolensky and Barrett, 1994). The major element abundances of these grains classify them among the Mg,Si ± Fe and Mg,Si,Ca ± Al chemical groups of fragments. Plagioclase fragments found in some aggregate IDPs but of much smaller size (Rietmeijer, 1998) could be progenitors of the Al,Si,Ca "silicate" spheres. The collected spheres might also be refractory melt residues of progenitors with a chondritic bulk composition such as meteorite matrix fragments or chondrules (Rietmeijer and Nuth, 2000). Chondrules are probably not present in comet nuclei but "silicate "fragments and fine-grained chondritic materials will be as predicted by a hierarchical dust accretion model of cycles of proto-planet modification and disruption (Rietmeijer, 1998; Rietmeijer and Nuth, 2000). In the earliest stages these processes will lead to the formation of porous cluster IDPs 60 – 100 μm in diameter, or even larger (as yet uncollected) "giant cluster IDPs". Cluster IDPs with "silicate" fragments will brake up during atmospheric entry, or will melt and quench as "silicate" spheres. The "silicate" spheres could represent (1) individual "silicate" meteoroids, (2) large "silicate" fragments of cluster IDPs, (3) quenched melt residues of aggregates that collapsed due to the surface tension of the earliest-formed melt, or (4) the residual cores that remain after ablation of a surface layer of a progenitor particle.

4.2. SETTLING RATES

Is it possible that individual showers can be responsible for the high incidence of spheres on some collectors? In order to address this question it is necessary to know the atmospheric residence time of the "silicate spheres". Their settling rates are obtained by the Stokes-Cunningham law (Kasten, 1968), which for an average "silicate" sphere 10.4 µm in diameter yields a settling time of ~12 days using the properties of the standard U.S. atmopshere-1976. In this calculation we assumed a density of 2.5 g cm^{-3} for quenched-melt spheres listed in Table III [Note: the fragments have a higher density, viz. ~3.2 g cm^{-3} for Mg-rich olivine and pyroxene and diopside, and ~2.8 g cm^{-3} for plagioclase]. The calculated residence time is consistent with the results of photometric measurements of aerosol loading in the atmosphere between 140 – 20 km altitudes during the 1998 November 14–20 Leonid meteor shower associated with comet P/Tempel-Tuttle with peak activity on November 17 (Mateshvili *et al.*, 1999). The measurements showed ambient aerosol loading of the atmosphere with a massive aerosol incursion on the morning of November 17 and had returned to almost ambient conditions on November 22, 1998 (Mateshvili *et al.*, 1999). The aerosol profiles are due to primary meteoroid injections and secondary particles from fragmentation and meteoric vapor condensation (Mateshvili *et al.*, 1999). The inferred aerosol residence time is > 5 days. Similar measurements for the annual η-Aquarid meteor shower linked to comet P/Halley showed residence times of 17 days between 130 – 50-km altitude (Mateshvili *et al.*, 1997) but the interpretations of these observations are complicated by multiple meteoroid injections on May 3, 6 and 7 (Mateshvili *et al.*, 1997). We adopt a settling time of 12 days for an average "silicate" sphere to the collection altitudes in the lower stratosphere.

4.3. SHOWER IDENTIFICATION

The incidence of "silicate" spheres listed in Table I is shown in Figure 2 as a function of time in the year when the collector plates were exposed.

Error bars show the accuracy of sampling (vertical, proportional to the square root of the number of collected spheres), and the period of time over which the collectors were exposed.

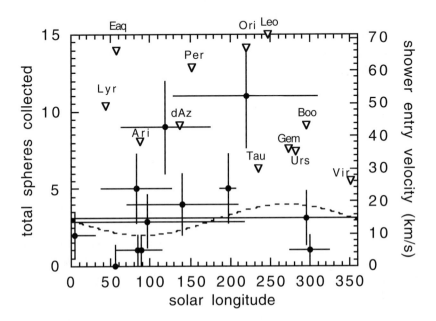

Figure 2. Number of spheres on individual collectors, (shown with error bars; see Table I) and the collection times as solar longitude. The sinuous curve in the lower portion of the diagram represents the possible sporadic background for Northern Hemisphere locations (Jenniskens, 1994). The possible contributing meteor showers (inverted triangles) are shown as a function of solar longitude and entry velocity (shower identifications as in Jenniskens, 1994).

The actual collection dates and times of the NASA Cosmic Dust Program during the period discussed here were somewhat haphazard. Individual collectors have variable lengths of their collection period and the actual times and dates of collection within these periods are randomly distributed. During the period May 1981 to July 1994, collectors flown in different years may have overlapping collection periods but not necessarily also overlapping collection dates (Rietmeijer and Warren, 1994). These, and other factors, make it difficult, but not impossible, to search for annual showers as the sources of the collected "silicates" spheres. The collection dates of flags U2001 and L2036 suggest that they were exposed to ambient dust conditions. Similarly, flag U2022 with zero "silicate" spheres was exposed to ambient dust conditions. Sphere abundances of two or less will be accepted as a background value of sporadic meteor activity. The background sporadic influx varies by less than 20 percent from year to year and is mostly continuous throughout the year, but with a sinusoidal variation due to changing positions of the apex for any given location on the Earth. A dashed line in Figure 2 shows the variation in the sporadic influx for a typical collection site on the Northern Hemisphere. The collection times of the flags W7017, U2034, W7074, L2009/2011 and L2021 (Table I) overlap with multiple annual meteor showers and their typically 3-4 "silicate" spheres is interpreted as a background of sporadic meteor activity plus many weak annual showers.

TABLE V

Cause	Reason	Examples of showers
High sphere formation efficiency	More fragile particles	Leonids '99, Draconids '98
	Large dust sizes	Leonids '98, major showers
	High velocity	Perseids, Orionids, Leonids
	Variable ablation intensity due to different entry angles within a shower	All showers
Higher mass influx	Overall	Leonids '66, Draconids '33/'46
	in specific mass ranges	Leonids '98

Flags W7027/7029 and U2015 stand out significantly from this sporadic background. If a meteor shower is responsible for such an increase of spheres on a dust collector, then it has to stand out from the sporadic meteor influx in one (or both) of two possible ways. Possible sources of the primary "silicate" sphere anomalies are shown in Table V. First, it could be a significant anomaly in mass influx. This is expected to occur during a rare meteor storm. The Leonid shower, for example, represents only about 5 hours of sporadic mass influx during a 1-hour storm even if the rate of observable meteors by visual observers (< +6 magnitude) is equivalent to 2000 hours of sporadic activity (Jenniskens et al. 1998). The second possibility is that the shower represents an anomaly in the efficiency of sphere production. We can envision three possible reasons. Firstly, large recently ejected Leonid meteoroids were found to fragment more easily than ordinary meteors (Borovicka et al. 1999, Murray et al. 1999). In that case, also less intense showers can dominate the production of spheres. Secondly, it is possible that sphere formation is efficient mainly for large > 1 cm sized meteoroids, for which fragmentation is often observed in the wake of the meteor. Meteor showers tend to dominate the mass influx of large meteoroids. An intense rain of fireballs was observed during the Leonid outburst of 1998, with much less an increase in the population of smaller meteors. Thirdly, it is possible that sphere formation is a signature of the ablation process itself, which may vary as a function of velocity. Quenching may be more rapid and efficient for the fastest meteors that ablate well above the mesopause. Meteor showers dominate the influx of fast meteors. That could even be annual showers, especially the eta-Aquarids, Perseids, Orionids and Leonids. Also, showers will always lead to relatively high incidences of grazing meteoroids at specific positions on the Earth, a process that may also enhance the efficiency of sphere formation. This process is especially relevant for the most active showers, because the meteor rate is much reduced at grazing incidence. The last mechanism could work for both slow and fast meteors. However, for meteor streams with velocities of $24 - 27$ km s^{-1} that could have entered the atmosphere at $60 - 80°$ (almost skipping) angles, the results obtained by Love and Brownlee (1991) then indicates they could survive without melting.

Figure 2 shows the time in the year (and the entry velocity) of a number of meteor showers that stand out above the sporadic background (Cook 1973; Jenniskens 1994). We adopted a 12-day settling time, and all times of the meteor showers were shifted accordingly. Annual showers typically vary in activity by less than 20 percent from year to year. Significant meteor outbursts are occasionally observed when the earth crosses the recent dust trail of a comet (Jenniskens 1995, Jenniskens et al. 1997a, -b), but no known meteor outbursts had recently occurred during the time of collection (Jenniskens 1995).

Atmospheric entry velocity provides another source-constraint because flash-heating (5–15 s) temperatures must be high enough to cause melting of Mg-rich olivine and pyroxene and plagioclase progenitor fragments. These melting temperatures are ~1,890°C for forsterite (pure Mg olivine and lower temperatures with iron in the structure), ~1,500–1,400°C for pyroxenes (iron similarly lowers the melting temperature), and ~1,100–1,550°C for plagioclase. Aggregates of these minerals will melt at lower temperatures while the small size (i.e. high surface free energy) of these fragments might facilitate flash melting. Adopting a conservative melting point of 1,100°C for the sphere progenitors (Rietmeijer, 1996), the work by Love and Brownlee (1994) then indicates that asteroidal entry velocities < 25 km s^{-1} (assuming a 45° entry angle) will be too low to cause melting of "silicate" fragments 20 – 34 µm in size.

Comparing the dust collection dates and dates of maximum activity of annual meteor streams, we found no significant increase of the detection of spheres as a result of the Daytime Arietid (Ari), Geminid (Gem) and Quadrantid (Boo) showers. These showers represent the highest mass influx for intermediate velocity showers (38, 36 and 43 km s^{-1} respectively). The Arietid and Geminid showers pass close to the Sun and Geminids are known to ablate more like a solid asteroidal particle than the typical cometary fluff ball. Flares are not observed. Sporadic meteors dominate the mass influx of very slow particles, except during the rare occasions of meteor storms caused by short period comets such as the Draconid (20.4 km s^{-1}) and Beilid showers. During the times of the sphere collections, no such storms were observed (Jenniskens 1995).

The two periods of high "silicate" sphere detection coincide with the occurrence of the summer Perseid (comet 109P/Swift-Tuttle) during the early summer of 1983 and the Orionid (comet 1P/Halley) and Leonid showers (comet 55P/Tempel-Tuttle) during the autumn of 1981, which are high entry velocity showers. Those are the likely source of the spheres if the retention time is correct. The eta-Aquarid shower (comet 1P/Halley) is also a candidate, but that shower has high incidence only on the Southern Hemisphere (Figure 2). Most of the dust in these meteor streams is > 150 μm in size, which is considerably larger than the fragments and spheres considered here, and larger than the typical sporadic meteoroid (150 μm).

Here we reached an interesting junction in the arguments. We used the predicted relationships between the diameters of the "silicate" progenitor fragments and the "silicate" spheres to claim that they are primary spheres of cometary origin. Using the information on the size distributions in cometary meteor streams and comet dust trials we find that the collected "silicate" progenitors are seemingly too small. In other words, the mechanism of sphere formation must not be the ablation of these individual progenitors, but result from the fragmentation of much larger meteoroids, or 'giant" cluster IDPs containing fragments > 50 μm in size. In that case, future-sampling efforts during meteor showers that are known to be rich in bright fireballs may reveal considerable numbers of spheres. This work predicts that some fraction of these spheres will be 10 μm-sized Mg,Si,Ca ± Al, Mg,Si ± Fe and Al, Si,Ca "silicate" spheres but their relative abundance in these showers is at this time unknown.

TABLE VI

Composition	All spheres	Flag U2015	Flags W7027/7029
Mg,Si ± Fe	54	67	27
Mg,Si,Ca ± Al	30	22	55
Al, Si,Ca	16	11	18

4.4. Were Different Comets Sampled?

If we accept that flags W7027/7029 may have sampled dust from comets 1P/Halley and 55P/Tempel-Tuttle, while collector U2015 may have sampled dust from comet P/Swift-Tuttle, we can examine possible differences between the dust from different high-velocities (>61 km/s) cometary sources. Table VI lists the relative proportions (%) of the three chemical groups of "silicate" spheres on flag U2015 and the simultaneously flown flags W7027 and W7029. These proportion on flag U2015 resemble these proportions found on all collected spheres (Table VI). The flags W7027/W7029 contain a higher than average proportion of Mg,Si,Ca ± Al spheres and a less than average amount of Mg,Si ± Fe spheres (Table VI). This difference could suggest that "silicate" spheres during the fall of 1981 came from sources rich in diopside (Mg,Si,Ca ± Al) fragments. All surviving fragments on these collectors belong to the Mg,Si ± Fe group which suggests that diopside fragments were preferentially melted compared to olivine and Ca-free and low Ca-pyroxene fragments, possibly because diopside fragments were structurally weakened by irradiation exposure in space prior to atmospheric entry (Rietmeijer, 1999b). Another explanation could be that fractional evaporation that concentrates refractory elements (e.g. Ca) in residual melt was more efficient for debris from sources active during the fall of 1991.

Primary "silicate" spheres collected in the lower stratosphere were linked to specific comets using an "approach of averages". This is possible despite uncertainties in catalog data reduction introduced by the irregular sampling strategy of the Cosmic Dust Program, when using average IDP properties such as diameter, an average 45° atmospheric entry angle, and an average dust settling rate. The presence of progenitor fragments and melt spheres highlights subtleties in the interactions of the parameters that determine the atmospheric peak-heating temperature. We note that the average sphere diameters on the flags U2015 and W7027/W7029 are smaller than the average diameter of the whole "silicate" sphere population. On U2015 the mean sphere diameter is 8.7 µm (stand. dev. = 3.3; range = 3–13; $N = 9$) and on W7027/W7029 it is 7.8 µm (stand. dev. = 4.4; range = 2–17; $N = 23$). A Student's t-test of these populations shows that there is no evidence to suggest that these two groups of spheres come from populations with different means. However, a similar test suggests that the mean diameter of "silicate" spheres on the collectors W7027/W7029 and U2015 is smaller than the mean diameter (Table III) of the entire "silicate" sphere population. These smaller diameters will increase the settling time from ~12 to ~20 days, which does not change our conclusions.

5. Conclusions

We used two different data sets, i.e. the NASA/JSC Cosmic Dust Catalogs and the times of peak activity of annual meteor showers, to explore any correlations between the occurrence of "silicate" spheres in the lower stratosphere between May 1981 and July 1994 and annual meteor showers. The collected primary "silicate" spheres < 30 μm in diameter and their progenitor fragments show the predicted size ratios for cometary debris entering the Earth's atmosphere. The compositions of the spheres and fragments each define three qualitatively similar chemical groups. The "silicate" sphere abundances on most collectors indicate an annual background of 2–4 "silicate" spheres < 30 μm in diameter per collector normalized to a 30 cm^2-area and an average 35-hours collection time. This background is composed of contributions from sporadic meteor activity and many weak annual meteor showers. During two collection periods, from 06/22 until 08/18, 1983 (U2015), and from 09/15-12/15, 1981 (W7027/7029), higher numbers of "silicate" spheres were collected with a smaller average diameter then during the times of the year represented by the other collectors listed in Table I. This study combined the times of "silicate" dust collection and peak shower activity, and constraints on dust properties of individual meteor showers. For the first time we identified specific showers, and thus cometary sources, for collected "silicate" IDPs. The majority of spheres on flag U2015 may originate from comet P/Swift-Tuttle, while the majority of spheres on flag W7027/7029 are from either comet 1P/Halley or comet 55P/Tempel-Tuttle. We have to admit that our statistical analyses alone does not, and can not, prove these associations of these stratospheric IDPs and high-velocity comets conclusively. This work stresses a need for targeted cosmic dust collections such as following the November 1999 peak in the Leonid storm activity. If we can be sure that this dust are quenched-melt Leonid meteors this would be the first experimental verification that it is possible to link collected dust in the lower stratospheres to periodic events. Improvement in the frequency of targeted dust collections will offer great opportunities to explore differences and similarities among IDP-producing sources.

Acknowledgments

We thank Mike Zolensky and an anonymous reviewer for their comments to improve our paper. We are grateful to Jack Warren (NASA/JSC Curatorial Facility) for providing the collection times. This work was supported by a grant from the National Aeronautics and Space Administration. *Editorial handling:* Noah Brosch.

References

Borovicka, J., Stork, R., and Bocek, J.: 1999, *Meteoritics Plan. Sci.* **34**, 987–994.
Bradley, J.P., Sandford, S.A., and Walker, R.M.: 1988, in J.F. Kerridge and M.S. Matthews (eds.), *Meteorites and the Early Solar System*, University Arizona Press, Tucson (AZ), pp. 861–898.
Brownlee, D.E.: 1985, *Annual Reviews in Earth Planetary Science* **13**, 147–173.
Brownlee, D.E., Joswiak, D.J., Schlutter, D.J., Pepin, R.O., Bradley, J.P., and Love, S.G.: 1995, *Lunar and Planetary Science* **XXVI**, 183–184.
Cook, A.F.: 1973, in C.L. Hemenway, P.M. Millman and A.F. Cook (eds.), *Evolutionary and Physical Properties of Meteoroids*, NASA SP-319, Washington (D.C.), pp. 183–191.
Fraundorf, P., McKeegan, K.D., Sandford, S.A., Swan, P., and Walker R.M.: 1982, *Proceedings 13th Lunar and Planet Science Conference, Journal of Geophysical Research*, **87**, *Supplement*, A403–A408.
Jenniskens, P.: 1994, *Astronomy and Astrophysics* **287**, 990–1013.
Jenniskens, P.: 1995, *Astronomy and Astrophysics* **295**, 206–235.
Jenniskens, P., Betlem, H., de Lignie, M., and Langbroek, M.: 1997a, *The Astrophysical Journal* **479**, 441–447.
Jenniskens, P., Betlem, H., de Lignie, M., Langbroek, M., and van Vliet, M: 1997b, *Astronomy and Astrophysics* **327**, 1242–1252.
Jenniskens, P., de Lignie, M., Betlem, H., Borovicka, J., Laux, C.O., Packan, D., and Kruger, C.H.: 1998, *Earth, Moon and Planets* **80**, 311–341.
Kasten, F.: 1968, *Journal of Applied Meteorology* **7**, 944–947.
Lefèvre, R., Gaudichet, A., and Billon-Galland, M.-A.: 1985, *Comptes Rendus Académie des Sciences. Paris, Tome 301, Série II*, **20**, 1433–1438.
Lefèvre, R., Gaudichet, A., and Billon-Galland, M.-A.: 1986, *Nature* **322**, 817–820.
Love, S.G. and Brownlee, D.E.: 1991, *Icarus* **89**, 26–43.
Love, S.G. and Brownlee, D.E.: 1994, *Meteoritics* **29**, 69–70.
Mackinnon, I.D.R. and Rietmeijer, F.J.M.: 1987, *Reviews of Geophysics* **25**, 1527–1553.
Mackinnon, I.D.R., McKay, D.S., Nace, G., and Isaacs, A.M.: 1982, *Proceedings 13th Lunar and Planet Science Conference, Journal of Geophysical Research*, **87**, *Supplement*, A413–A421.
Mateshvili, G.G., Mateshvili, Y.D., and Mateshvili, N.Y.: 1997, *Solar System Research* **31**, 483–488.
Mateshvili, N., Mateshvili, G., Mateshvili, I., Gheondjian, L., and Avsajanishvili, O.: 1999, *Meteoritics Plan. Sci.* **34**, 969–973.
Murray, I.S., Hawkes, R.L., and Jenniskens, P.: 1999, *Meteoritics Plan. Sci.* **34**, 949–958.
Pepin, R.O., Palma, R.L., and Schlutter, D.J.: 2000, *Meteoritics Plan. Sci.* **35**, 495–504.
Rietmeijer, F.J.M.: 1996, *Meteoritics and Planetary Science* **31**, 237–242.
Rietmeijer, F.J.M.: 1998, in J.J. Papike (ed.), *Planetary Materials*, Reviews in Mineralogy, **36**, The Mineralogical Society of America, Washington (DC), pp. 1–95.
Rietmeijer, F.J.M.: 1999a, *Meteoritics Planet. Sci.* **34**, 495.

Rietmeijer, F.J.M.: 1999b, *American Mineralogist* **84**, 1883–1894.
Rietmeijer, F.J.M.: 2000a, *Meteoritics Plan. Sci.* **35**, 647.
Rietmeijer, F.J.M.: 2000b, *Meteoritics Plan. Sci.* **35**, 1025–1042.
Rietmeijer, F.J.M. and Flynn, G.J.: 1996, *Meteoritics Plan. Sci.* **31**, A114–A115.
Rietmeijer, F.J.M. and Nuth III, J.A.: 2000, *Earth, Moon and Planets* **82–83**, 325–350.
Rietmeijer, F.J.M. and Warren, J.L.: 1994, in M.E. Zolensky, T.L. Wilson, F.J.M. Rietmeijer and G.J. Flynn (eds.), *Analysis of Interplanetary Dust, Amer. Inst. Physics. Conf. Proc.* **310**, Amer. Inst. Physics, New York (NY), pp. 255–275.
Sandford, S.A.: 1987, *Fundamentals of Cosmic Physics* **12**, 1–73.
Thomas, K.L., Blanford, G.E., Clemett, S.J., Flynn, G.J., Keller, L.P., Klöck, W., Maechling, C.R., McKay, D.S., Messenger, S., Nier, A.O., Schlutter, D.J., Sutton, S.R., Warren, J.L., and Zare, R.N.: 1995, *Geochimica et Cosmochimica Acta* **59**, 2797–2815.
Warren J.L., and Zolensky, M.E.: 1994, in M.E. Zolensky, T.L. Wilson, F.J.M. Rietmeijer and G.J. Flynn (eds.), *Analysis of Interplanetary Dust, Amer. Inst. Physics. Conf. Proc.* **310**, Amer. Inst. Physics, New York (NY), pp. 105–114.
Zolensky, M. and Barrett, R.: 1994, *Meteoritics* **29**, 616–620.
Zolensky, M.E., Wilson, T.L, Rietmeijer, F.J.M., and Flynn, G.J. (eds.): 1994, *Analysis of Interplanetary Dust., Amer. Inst. Physics. Conf. Proc.* **310**, Amer. Inst. Physics, New York, (NY), 357 + xiii pp.

Appendix

Clanton, U.S., Dardano, C.B., Gabel, E.M., Gooding, J.L., Isaacs, A.M., Mackinnon, I.D.R., McKay, D.S., Nace, G.A., and Warren, J.L.: 1982, Cosmic Dust Catalog 1(1) (particles from collection flag W7017), Curatorial Branch Publication 59 (JSC 17903), 111 p.
Clanton, U.S., Dardano, C.B., Gabel, E.M., Gooding, J.L., Isaacs, A.M., Mackinnon, I.D.R., McKay, D.S., Nace, G.A., and Warren, J.L.: 1982, Cosmic Dust Catalogs 2(1) and 2(2) (particles from collection flag W7029), Curatorial Branch Publication 62 (JSC 18221), 168 p.
Clanton, U.S., Gabel, E.M., Gooding, J.L., Isaacs, A.M., Mackinnon, I.D.R., McKay, D.S., Nace, G.A., and Warren, J.L.: 1982, Cosmic Dust Catalog 3(1) (particles from collection flag U2001), Planetary Materials Branch Publication 63 (JSC 18622), 95 p.
Clanton, U.S., Gabel, E.M., Gooding, J.L., Isaacs, A.M., Mackinnon, I.D.R., McKay, D.S., Nace, G.A., and Warren, J.L.: 1983, Cosmic Dust Catalog 4(1) (particles from collection flag W7027), Planetary Materials Branch Publication 65 (JSC 18928), 140 p.
Clanton, U.S., Gooding, J.L., McKay, D.S., Robinson, G.A., and Warren, J.L., Watts, L.A.: 1984, Cosmic Dust Catalog 5(1) (particles from collection flag U2015), Planetary Materials Branch Publication 70 (JSC 20111), 111 p.
McKay, D.S., Schramm, L.S., Ver Ploeg, K.L., Warren, J.L., Watts, L.A., and Zolensky, M.E.: 1985, Cosmic Dust Catalog 6(1) (particles from 13 collection flags), Planetary Materials Branch Publication 73 (JSC 20916), 75 p.

Zolensky, M.E., McKay, D.S., Schramm, L.S., Ver Ploeg, K.L., Warren, J.L., and Watts, L.A.: 1985, Cosmic Dust Catalog 7(1) (particles from collection flag U2022), Planetary Materials Branch Publication 74 (JSC 20917), 134 p.

Zolensky, M.E., McKay, D.S., Schramm, L.S., Ver Ploeg, K.L., Warren, J.L., and Watts, L.A.: 1986, Cosmic Dust Catalog 8(1) (particles from collection flag W7013), Planetary Materials Branch Publication 75 (JSC 22425), 129 p.

Zolensky, M.E., Barrett, R.A., McKay, D.S., Thomas, K.L., Warren, J.L., and Watts, L.A.: 1987, Cosmic Dust Catalog 9(1) (particles from collection flag U2034), Planetary Materials Branch Publication 77 (JSC 22744), 105 p.

Zolensky, M.E., Barrett, R.A., McKay, D.S., Thomas, K.L., Warren, J.L., and Watts, L.A.: 1988, Cosmic Dust Catalog 10(1) (particles from collection flag W7074), Planetary Materials Branch Publication 80 (JSC 23367), 153 p.

Zolensky, M.E., Barrett, R.A., Dodson, A.L., Thomas, K.L., Warren, J.L., and Watts, L.A.: 1990, Cosmic Dust Catalog 11(1) (particles from collection flag L2005), Planetary Materials Branch Publication 83 (JSC 24461), 170 p.

Zolensky, M.E., Barrett, R.A., Dodson, A.L., Thomas, K.L., Warren, J.L., and Watts, L.A.: 1991, Cosmic Dust Catalog 12 (particles from collectors L2005 and L2006), Planetary Materials Branch Publication 85 (JSC 25121), 351 p.

Barrett, R.A., Dodson, A.L., Thomas, K.L., Warren, J.L., Watts, L.A., and Zolensky, M.E.: 1992, Cosmic Dust Catalog 13 (particles from collectors L2005 and L2011), Office of the Curator #85 (JSC 25980), 339 p.

Warren, J.L., Barrett, R.A., Dodson, A.L., Watts, L.A., and Zolensky, M.E.: 1994, Cosmic Dust Catalog 14 (particles from collectors L2008 and L2009), Office of the Curator #91 (JSC 26678), 568 p.

Warren, J.L., Zolensky, M.E., Thomas, K., Dodson, A.L., Watts, L.A., and Wentworth, S.: 1997, Cosmic Dust Catalog 15 (parts 1-3) (particles from collectors L2036 and L2021), Office of the Curator #93 (JSC 27897), 477 p.

PRELIMINARY DATA ON VARIATIONS OF OH AIRGLOW DURING THE LEONID 1999 METEOR STORM

JOSEPH KRISTL, MARK ESPLIN, THOMAS HUDSON,
AND MICHAEL TAYLOR

Space Dynamics Laboratory, Utah State University Research Foundation,
Logan, UT 84301
E-mail: jkristl@sdl.usu.edu

and

CARL L. SIEFRING

Plasma Physics Division, Naval Research Laboratory, Washington, DC 20375

(Received 7 July 2000; Accepted 26 July 2000)

Abstract. As part of the 1999 Leonid MAC Campaign an extensive set of infrared (1.00 – 1.65 µm) airglow spectra and imaging data were collected from onboard the USAF FISTA aircraft. These data will permit a detailed study of the upper atmospheric conditions over a several day period centered on the Leonid meteor storm of 17/18 November, 1999 as well as during the meteor storm itself. We describe initial results of a spectral analysis that indicates a small but significant enhancement in the OH airglow emission during the peak of the storm but we cannot yet be certain of a cause and effect relationship. No similar systematic enhancement was observed in the O_2 (1.27 µm) airglow emission recorded with the same instrument.

Keywords: Ablation, airglow, Leonids 1999, lower thermosphere, O_2, OH, mesosphere, meteor

1. Introduction

During the 1999 Leonid meteor storm that occurred on 17-18 November 1999, two US Air Force aircraft conducted optical measurements from approximately 11.5 km altitude (Jenniskens *et al*, 2000a). The Leonid campaign consisted of five consecutive nighttime flights including stops in the United States, England, Israel, and the Azores. The Space Dynamics Laboratory of Utah State University operated several instruments in the visible and infrared spectral bands. One system on the FISTA (Flying Infrared Signatures Technology Aircraft) obtained high-resolution (4 cm^{-1}) measurements of the night sky emission spectra in the 1 to 1.65-micrometer band. Measurements were obtained above the clouds providing exceptional viewing conditions. The OH airglow emission layer originates at an altitude of ~87 km and has a half-width of typically 8–10 km. Its behavior during the storm night of 17/18 November 1999 was of particular interest because the OH airglow emission may be affected by the Leonid meteor ablation products that can penetrate to altitudes as low as 80 to 90 km altitudes. We note that typical Leonid meteor end-heights are much higher above ~100 km. We measured variability of the OH emission to investigate any changes that may result from meteor interactions with the atmosphere that could cause changes in the natural airglow emission via excitation caused by the meteor ablation products. It is also possible that organic materials in the meteors could be broken down into simpler products that include the OH hydroxyl radical.

To search for these effects an interferometer capable of continuous OH airglow measurements was operated from the FISTA aircraft both before and during the Leonid storm. Airglow data were collected to create a baseline measurement of the nightly variations in emission intensity that were due to natural variations and unrelated to the meteor shower. Past observations of OH airglow have shown considerable variability both at a fixed location that are primarily due to gravity waves and tides during the course of a single night. Significant variations in airglow emission intensity and rotational temperature have been observed with latitude and longitude primarily from spacecraft but also from airborne missions (Leinert et al., 1998). The data taken during the 1999 Leonid MAC Campaign provides an exceptional opportunity to investigate variations in the mid-latitude OH emission over a large longitudinal range (~160°) during a short period of time (<1 week). For this study we have not attempted to de-couple the temporal and spatial effects due to the aircraft motion, but simply report on some of the most interesting observations.

2. Instrumentation

The instrument used to collect this data is a Bomem Michelson M-150 interferometer, modified for installation and use on the FISTA aircraft.

This interferometer operates at 4 cm^{-1} resolution (apodized) with a scan rate of about 1 scan every 3 seconds. The interferometer field of view is 1.5° and it is sensitive from 1 to 1.65 micrometers. An intensified Xibion camera recorded the instrument field of view during the flight, providing information on the pointing elevation and azimuth. Figure 1 shows the instrument installed on the FISTA aircraft. It was mounted at 45° elevation relative to the aircraft's floor during most of the flights. On two occasions the camera was unmounted to track persistent meteor trains. This sensor operated almost continuously during the entire 1999 Leonid MAC campaign and collected an extensive set of night airglow spectra.

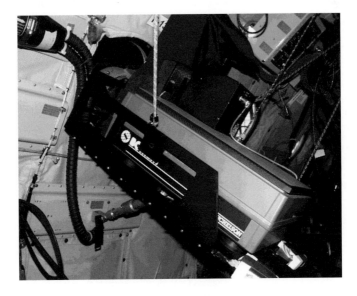

Figure 1. Bomem Interferometer Installed on the FISTA Aircraft.

3. OH Airglow Data

While the interferometer collected continuous data, the very low signal levels of the airglow background require the averaging of multiple samples of data to get an adequate signal to noise to distinguish the airglow emission from background noise. Because every unprocessed interferogram is recorded, the level of averaging can be adjusted during post-flight analysis.

528 KRISTL ET AL.

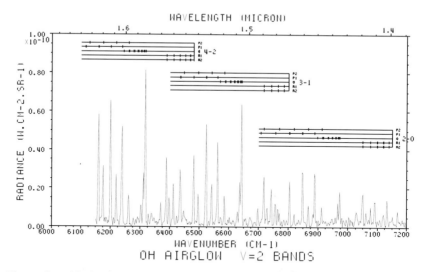

Figure 2a. Airglow spectrum measured from the FISTA aircraft with the $\nu = 2$ OH Bands Identified. Calculated OH line positions are shown above the observed emission lines.

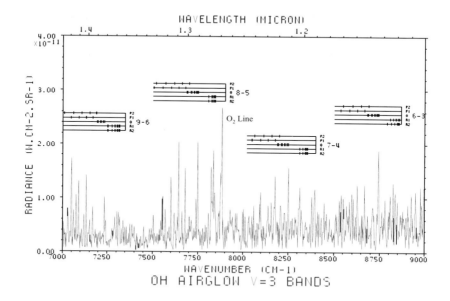

Figure 2b. Airglow spectrum measured from the FISTA aircraft with OH $\nu = 3$ Bands Identified. Also shown is the O_2 singlet-delta.

Figure 2 shows an example of an airglow spectrum measured during the 1999 Leonid MAC campaign from 17/18 November on the flight from Israel to the Azores (Jenniskens *et al.*, 2000a) with a high level of averaging (228 interferograms). The two plots show spectral detail of the $v = 2$ and $v = 3$ transitions that are within the range of the spectral response of the instrument. At 4 cm^{-1} resolution, the individual P and R branches are easily resolved, allowing studies of the OH rotational temperatures. The excellent atmospheric transmission from the aircraft altitude to the OH layer results in a very good measurement of the (2–0) $v = 2$ band and the (9–6) $v = 3$ bands that are difficult to observe from the ground due to atmospheric water absorption. The large peak at 7778 cm^{-1} (1.27 μm) in Figure 2B is the O_2 singlet-delta line, which is also difficult to observe from the ground due to self-absorption from the O_2 at lower altitudes.

The spectral data set can be used to search for any changes linked to the occurrence of the Leonid meteor storm. One method we used to look for systematic changes in the airglow intensity was to select a well-resolved transition with very good signal to noise ratio. We then integrate its intensity over the duration of the flight. Much smaller data averages, viz. approximately one-minute time intervals, were used in this analysis. As a result, only the strongest lines can be evaluated at this time spacing and the resulting integral over the Q branch transition of the (4–2) line is shown in Figure 3. The upper curve in Figure 3 is the Q branch integral; the lower curve is a baseline integral of a band with no OH emission features. The time gaps in Figure 3 are due to either data gaps or the removal of some time intervals when excessive, non-optical noise occurred in the data.

This increase in airglow intensity matches the time interval of the actual meteor influx over the region as measured from FISTA (Figure 4). Rates have not yet been corrected for radiant altitude or any other effects that might affect changes in the airglow layer and could mimic a correlation with the peak of the Leonid storm. The onset of the airglow increase at 00:50 – 01:00 UT corresponds with the onset of the meteor storm. Meteor activity had returned to pre-storm activity levels at 03:10 – 03:20 UT, which coincided with a return to the OH emission background level. This correlation does not prove a link between the meteor storm intensity and the OH airglow emission but is at least very suggestive. Still, we have not been able to identify other processes that could cause the observed transient enhanced OH emission at the time of the Leonid storm. An initial fit of rotational temperatures was made to the flight spectra including the Leonid meteor event but no significant temperature changes above the noise level were observed.

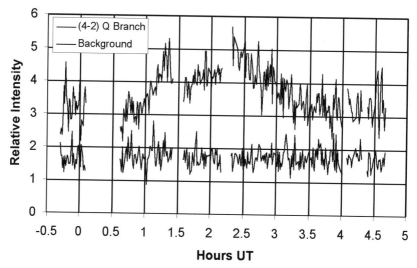

Figure 3. Integral of the $\nu = 2$ (4–2) Q branch of OH emission during the FISTA flight on 17/18 November 1999 as a function of Universal Time (UT). The upper curve is the OH emission integral; the lower curve is a non-OH band showing the data noise limit.

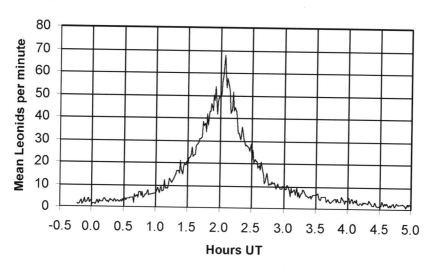

Figure 4. Observed meteor rate (not corrected for ZHR) (from Jenniskens *et al.*, 2000b)

4. Other Results

In the airglow spectrum in Figure 2, the 1.27 μm O_2 singlet-delta emission line is the only non-OH feature with strong signal. Our data in Figure 5 show that the O_2 emission does not track the average meteor flux, that is we find no increase at the time of the peak in the meteor storm. Several sharp emission peaks can be seen in Figure 5. Examination of the intensified video camera that was co-aligned with the interferometer reveals that meteors were passing through the field of view. We suggest that the O_2 emission peaks might be caused by meteor head or trail emissions at these wavelengths.

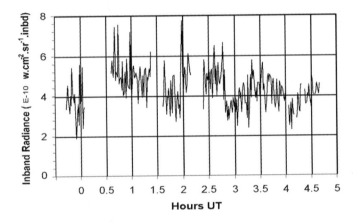

Figure 5. Time-history of spectral integral on 1.27 μm O_2 singlet-delta line

5. Discussion

The main uncertainty when drawing any definitive link between changes in the airglow intensity and the occurrence of the high Leonid meteor influx is caused by uncertainties in the natural nighttime fluctuations in OH emissions during the times and over the areas covered by the 1999 Leonid MAC campaign. For example, Figures 6 show the time history of the same integral before and after the night of the peak of the Leonid storm when the observed meteor fluxes were low. A steady drift in the increase in the emission is observed on both nights that were similar in magnitude to the drifts observed during the peak of the storm activity caution when postulating a link between the OH airglow emission and the occurrence of the storm. Possible other explanations for the increased OH airglow include increased gravity wave activity associated with storm complexes in the region (e.g., Swenson and Espy, 1995).

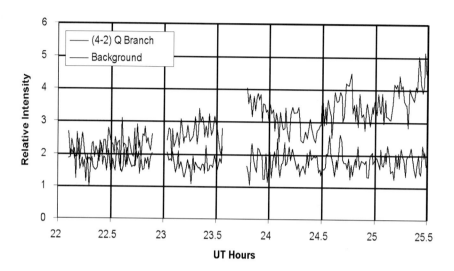

Figure 6a. OH airglow variability on the non-peak night of Nov 16/17 (RAF Mildenhall to Ben Gurion airport). The upper curve is the OH emission integral; the lower curve is a non-OH band showing the data noise limit.

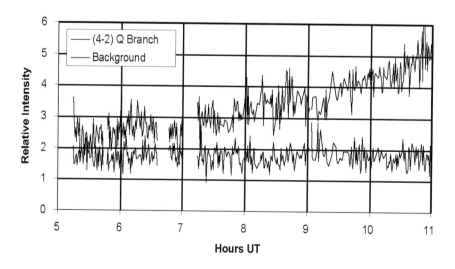

Figure 6b. OH airglow variability on Nov 18/19 (Lajes AFB to Patrick AFB).

If the increase in OH airglow is indeed due to a high influx of Leonid meteors, it must be associated with the input of kinetic energy of these meteors. We did not detect enhanced OH emission in the meteor spectra themselves. The total amount of mass represented by the meteor storm is relatively small compared to that of the daily sporadic meteor background. Most of the mass (and therefore the kinetic energy) of the Leonids is locked in the larger particles that contribute to the flux curve in Figure 4. We note that the number of large fireballs was much smaller during the 1999 storm than during the Leonid storm of 1998 (Jenniskens *et al.*, 2000a).

It is difficult to understand how the fast Leonid meteors could affect the atmospheric airglow levels at altitudes where OH is typically found. Optical emissions from the decay of excited OH molecules is at a maximum around a mean altitude of 87 ± 1 km with a layer thickness of 6–10 km (Baker and Stair, 1988). The visible Leonid meteors tend to ablate around 100-km altitude, but with peak brightness in a fairly wide range from 108 to 95 km (Jenniskens *et al.*, 1998). The rapid response of airglow to meteor rates does not suggest the operation of a vertical diffusion mechanism. Rather the increase of OH emission associated with the meteor activity could have originated from altitudes (slightly) above the normal airglow layer.

Excited OH in the nighttime mesosphere is thought to arise from the reaction of atomic hydrogen with ozone. This raises the possibility that an increase of the atomic hydrogen abundance could account for the observed effect. Atomic hydrogen diffuses rapidly and may explain the rapid decline of OH airglow with decreasing meteor rates. The atmospheric abundance of water is known to decay rapidly above 90-km altitude due to photolysis and freezing. It is possible that the hydrogen for this process originated from organic matter in the incoming meteoroids. This proposition finds some support in the observations of interplanetary dust particles with carbon contents 2 to 3 times the solar system value (Rietmeijer, 1998). This possibility requires that the very fast moving Leonid meteoroids are especially rich in organic materials. While the sporadic meteoroids might include debris with similar high carbon contents, their generally much lower entry velocity will prevent the efficient ablation of their organic materials. Alternatively, an increase of ozone from the dissociation of atmospheric O_2 in the incoming Leonid meteors may account for the observed increase in OH emission. Ozone plays an essential role in the chemistry that produces the typical OH airglow emission.

6. Conclusions

A considerable database of infrared airglow data was collected during the 1999 Leonid MAC campaign. This data allows investigations of the effects in the upper atmosphere that might be induced by this intense meteor storm. Enhanced OH airglow intensity was observed during the peak of the 1999 meteor storm occurrence but we can not yet prove this apparent, yet intriguing, correlation.

Acknowledgements

We thank Peter Jenniskens and two anonymous referees whose comments helped improve the presentation of this paper. The 1999 Leonid MAC mission was supported by NASA's Suborbital MITM, Exobiology, and Planetary Astronomy programs, by NASA's Advanced Missions and Technologies program for Astrobiology, NASA Ames Research Center, and the US Air Force/XOR. *Editorial handling*: Frans Rietmeijer.

References

Baker, D.J. and Stair Jr., A.T.: 1988, *Physica Scripta* **37**, 611–619.
Jenniskens, P., de Lignie, M., Betlem, H., Borovicka, J., Laux, C.O., Packan, D., and Kruger, C.H.: 1998, *Earth, Moon and Planets* **80**, 311–341.
Jenniskens, P., Butow, S.J., and Fonda, M.: 2000a, *Earth, Moon and Planets* **82-83**, 1--26.
Jenniskens, P., Crawford, C., Butow, S.J., Nugent, D., Koop, M., Holman, D., Houston, J., Jobse, K., Kronk, G., and Beatty, K.: 2000b, *Earth, Moon and Planets* **82-83**, 191–208.
Leinert, Ch., Bowyer, S., Haikala, L.K., Hanner, M.S., Hauser, M.G., Levasseur-Regourd, A.-Ch., Mann, I., Mattila, K., Reach, W.T., Schlosser, W., Staude, H.J., Toller, G.N., Weiland, J.L., Weinberg, J.L., and Witt, A.N.: 1998, *Astron. Astrophys. Suppl. Ser.* **127**, 20–30.
Rietmeijer, F.J.M.: 1998, in J.J. Papike (ed.) *Planetary Materials, Reviews in Mineralogy* **36**, 1–95.
Swenson, G.R. and Espy P.J.: 1995, *Geophys. Res. Lett.* **22**, 2845–2848.

AIRGLOW AND METEOR RATES OVER ISRAEL DURING THE 1999 LEONID SHOWER

NOAH BROSCH and OHAD SHEMMER

School of Physics and Astronomy and the Wise Observatory, Beverly & Raymond Sackler Faculty of Exact Sciences, Tel Aviv University, Tel Aviv 69978, Israel
E-mail: noah@wise.tau.ac.il; ohad@wise.tau.ac.il

(Received 01 June 2000; Accepted 29 August 2000)

Abstract. We present observations of the sky brightness obtained with a high-speed, wide field of view photometer and with a spectrometer at the Wise Observatory during the 1999 Leonid meteor campaign. The photometric measurements detected meteors and searched for overall increases in sky brightness. Spectroscopic measurements made at the zenith were used to measure the brightness of the Na I 5890Å doublet during the two nights before and during the Leonid peak. No significant enhancement took place in these transitions. The lack of increased Na emission is consistent with the estimates of mass influx during the storm.

Keywords: Airglow, Leonids 1999, meteor ablation, meteor storm

1. Introduction

The light of the night sky consists partly of strong emissions of sodium and oxygen atoms in a region in the atmosphere where meteors are ablated (Roach and Gordon, 1973). A layer of neutral sodium atoms peaks at about 95 km altitude (Cox *et al.*, 1993). The sodium emission is thought to be due to catalytic recombination of ozone and oxygen atoms, whereby sodium is released in an excited state. The resulting airglow has a typical intensity of 50 Rayleighs ($3.2\ 10^{-16}$ erg cm^{-2} sec^{-1} arcsec^{-2}), but varies by a factor of three with season.

The intensity of this airglow may change during a meteor storm, either because of an increase in the abundance of sodium or because molecular oxygen is dissociated by the kinetic energy of the meteors. Indeed, Batista *et al.* (1989) have claimed that sudden sodium layers are more abundant during meteor showers.

We report here measurements performed at the Wise Observatory (WO) in Israel at the time of the 1999 Leonid meteor storm. Both photometric and spectroscopic monitoring of the sky brightness was performed. The results indicate no significant effect in the observed pass bands.

2. Observations

The sky brightness above the site of the WO has been monitored from 13 to 20 November 1999. We used two independent monitoring systems, one being observations with a wide-field photometer of the sky brightness in the B and V standard bands and the other a spectroscopic system. The photometric observations provide a sensitive continuous measure of sky brightness, while the spectroscopic measurements are more intermittent but provide information on variations in individual emission lines. We concentrate here on observations obtained on the night preceding the Leonid 1999 storm and on the night of the storm itself.

2.1. Photometry

The photometric observations were performed with the WO two-star photometer (Brosch, 1995). Instead of attaching the photometer to the 1.0-meter telescope, we operated it outside the WO building, with no imaging optics. With this configuration, the sky patch at which the main channel of the photometer looks is defined by a round aperture in a folding mirror and corresponds to a circular area $\sim 8°$ in diameter.

The WO photometer has two channels, the second of which is limited by a mechanical aperture and by a second folding mirror, but has approximately the same field of view as the main channel. For monitoring the sky brightness enhancements during the Leonids, we selected a V-band filter in the first channel of the photometer and a B-band filter in the second channel. Note that the two channels do not have matched sensitivities. Our focus here will not be on changes in the absolute response of each channel or on the relative response between the channels, but rather on relative intensity changes within one channel that are correlated temporally with similar changes in the second channel.

At the beginning of each night the internal clock of the controlling computer was set manually to the GPS time, read off the WO master computer. The accuracy of this procedure is estimated at $\pm$ 1–2 seconds. The integration time was set to one second and the observations were started as soon as it became dark enough for observations. The photometer was first leveled with a bubble level, so that its optical axis would be approximately vertical. Unfortunately, there appear to have been some small undefined motions in the viewing direction, which may have been caused by accidental disturbances of the instrument by the many people present at the WO site during the night.

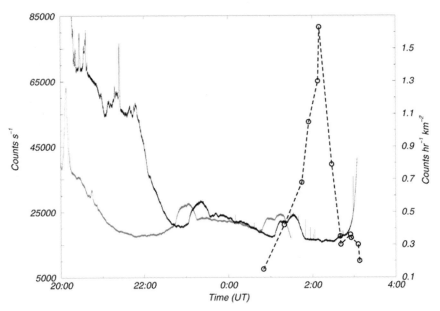

Figure 1. Photometric traces in the V-band on the two nights of monitoring. Note the similar levels of sky background flux on the two nights, after the Moon has set. The night of 16-17 November 1999 is plotted with a lighter trace than the plot for 17-18 November.

2.2. Spectroscopy

We used the FOSC (Faint Object Spectrometer Camera) of the WO described by Brosch and Goldberg (1994) for spectroscopic monitoring of the airglow during the same period. The FOSC was operated at the f/7 Ritchey - Chrétien focus of the 1.0-meter telescope for a different observing program, in which 10"-wide long-slit spectra were obtained of AGNs in the first half of each observing night. These spectra contain a significant amount of sky emission, which is recorded every time an exposure is obtained. Instead of discarding the sky contribution of each observation, as normally done for astronomical observations, we retained these spectra and reduced them to flux density (erg cm^{-2} Å^{-1} sec^{-1}) per square arcsecond along the slit.

The second halves of the nights November 16-17 and 17-18 were allocated to the Leonid sky monitoring project. During these periods the telescope was maintained in tracking mode close to the zenith for airglow monitoring, in a sky region free of noticeable stars. The observations were performed with a 2"-wide slit and were reduced in a uniform way.

3. Results.

3.1. Photometry

We collected photometric observations for each one-second integration time on November 16–17, 1999 from 15:53:10 ± 05 sec to 01:29:36 UT, and on November 17-18, 1999 from 20:14:01 ± 05 sec to 03:04:05 UT. The uncorrected count rates collected by the V channel of the photometer are plotted in Figure 1.

The two nights when the sky brightness was monitored started with the Moon in the sky at dusk. The Moon set at 20:28 UT on November 16 and at 21:25 UT the following night. The sky was partly cloudy on both nights, mainly in the first halves of the nights, causing random declines of sky brightness intensity. Because of the special occasion of the Leonid meteor shower, with a predicted storm component, the municipality of Mizpe Ramon switched off the city lights from about 21:00 UT (23:00 local time); the time when Leo and the shower radiant were above the horizon was, therefore, "dark". However, other installations within 20-km of the observatory kept their illumination on; some of this light may have contributed to the night sky brightness over the WO.

The figure shows a steady decrease in sky brightness from the beginning of each run. This is the signature of the setting Moon, approximately at first-quarter phase. Note that the sky brightness, particularly toward the end of both observing nights, remained approximately constant. Occasionally, bright stars would drift through the photometer field of view and cause a periodic increase in sky brightness. The larger peaks are due to ρ Gem and α Gem (Castor). At the end of the night, a rapid increase in sky brightness marks the onset of morning twilight.

There is no sign of a sky brightness increase at the time of the meteor storm, from 1 to 3h UT (13–15 in Figure 1). This indicates that, to a first approximation, the skies did not "lit up" because the Leonid strom was taking place. The sky brightness at the Leonid maximum between 2 and 2:30 UT, is constant to $\sim$3% in B and $\sim$5% in V.

We do, however, notice several sharp excursions at that time, about 1–2 seconds long, which are due to (Leonid) meteors passing the photometer field of view. The photometric response to the bright stars was used to calibrate the V-band response of the photometers to meteors. The intensity estimates are listed in Table I. We find a relative response of B/V $\approx$ 0.5, with no significant difference between meteors of magnitude +1.2 and +3.2.

These results can be used to provide intensity calibration for meteors observed with the Canadian - USAF intensified video cameras that

Table I.

Number	Time (UT)	V	B	Cam & no.
1	00:10:19	2.9		
2	00:36:44	3.2		
3	01:28:36	2.7	1.5	
4	01:40:26	3.4		O298
5	01:49:54	1.8		
6	01:57:11	2.4		
7	02:01:22	3.2	1.6	E233
8	02:03:10	2.6		E241
9	02:42:48	3.4		
10	02:44:30	3.6		
11	02:47:11	4.9		
12	02:59:17	1.2	0.6	

Note. Photometric meteors. The third and fourth columns list the approximate V-band and B-band magnitides of the meteors, estimated as explained in the text. The last column identifies meteors observed with intensified video cameras operated at the WO by Hawkes et al.(priv. corresp.), giving their internal numbering system.

were operated at WO at the same time. The simultaneously observed meteors are marked in Table I.

3.2. SPECTROSCOPY

A representative spectrum of the sky, obtained by the Faint Object Spectrometer Camera, is shown in Figure 2. It shows the usual strong airglow lines: [OI]λ5577Å, the Na λ5890Å blend, and [OI]λ6300Å, as well as many other fainter lines. It also shows a broad feature centered approximately at 5900Å, which we identify tentatively as sky glow from the high-pressure Sodium lamps operating within a few tens of km from the observatory. We caution that part of the narrow Na line may also originate from this source, *i.e.*, artificial light scattered or reflected into the telescope by clouds, as the spectrum of high-pressure Sodium lamps contains both broad and narrow components of the line.

In order to obtain the spectral sky brightness, the segments of the spectra with no object visible were averaged and the flux density was derived as the average flux density per pixel, across the dispersion. One should note that, in this case, the effective aperture in the second part of the night during the Leonid shower is narrower by half an order of magnitude than in the first half of the night, during the AGN observations. The measured flux densities of the spectra derived from

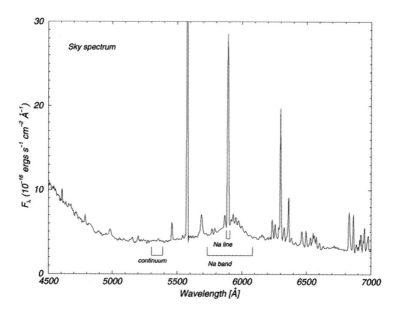

Figure 2. Airglow spectrum near zenith at the WO, obtained on November 16, 1999 with a two-arcsec wide slit. The spectral segments selected for further analysis are marked on the image.

the AGN monitoring program were therefore scaled down by a factor of five.

The spectra were flux calibrated using the WO standard sensitivity function and extinction curve, which generally do not change from night to night and are updated from time to time using spectrophotometric standard stars.

We binned the flux densities in spectral segments of interest, from the sparse spectral information. One synthesized band is a continuum stretch in the spectrum of the sky background free of obvious emission lines, from the neighborhood of ~ 5300Å, defined as $\lambda\lambda 5303$-5390Å, another contains the narrow component of the Na I doublet (blended into a single line at our resolution) and covering $\lambda\lambda 5879$-5903Å, and a third has the broad component centered at ~ 5900Å ($\lambda\lambda 5734$-6085Å).

The results are shown in Figures 3 and 4. The background continuum behaves much in the same way as the broad emission assigned to high pressure sodium lamps and is shown in Figure 3 in flux units per square arcsec. The sodium line emission shows a rapid decrease when city lights are turned off, then stays constant throughout the night. There is no significant increase during the Leonid shower (meteor count shown as a grey line in Figure 4).

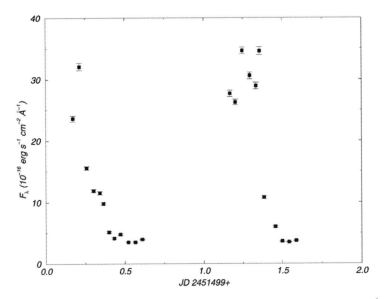

Figure 3. Sky surface brightness in a continuum band centered at λ5350Å.

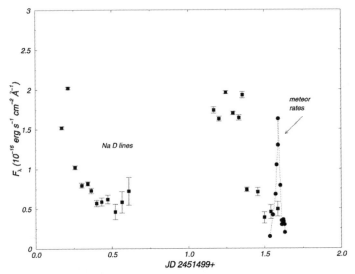

Figure 4. Sky surface brightness in a narrow band containing the Sodium Na I doublet. The shower activity, derived from intensified video observations at the WO, is indicated by a solid line.

Figure 4 shows that close to the peak of the Leonid storm there was no significant enhancement of the Na I λ5890Å emission. We also note that our observations show no significant change of the [OI] line emission dusing the Leonid peak.

4. Discussion

The activity of the Leonid shower is best described by the meteor flux for bodies brigther than +6.5-mag that would impact on a square km surface perpendicular to the Leonid radiant. We use here the meteor rates derived from the wide-field intensified video cameras that operated at the WO during the Leonid campaign (Brown *et al.*, 2000; figure 8). This activity profile is shown in Figure 4.

Let us consider the anticipated enhancement of the sodium abundance. Arlt *et al.*(1999) estimate that each square km of the atmosphere received 1.4 ± 0.3 Leonids of magnitude +6.5 and brighter at the peak of the storm, with a magnitude distribution index of r = 2.2. Similarly, Gural and Jenniskens (2000) have 0.82 ± 0.19 Leonids of +6.5 and brighter per hour at the peak of the storm. The peak value of activity derived by Brown *et al.*(2000) from the WO counts and to the same magnitude limit is 1.6 ± 0.2 km^{-2} hr^{-1}.

For a deposition altitude of 100 km, our 8-degree wide photometer aperture sampled an area of $\sim$150 km^2. Assuming a rate of 1 km^{-2} hr^{-1}, a typical particle mass of 0.07 g for a +0th magnitude meteor, a brightest meteor of +1.2 mag detected in the photometer field of view (Table I), and a typical storm duration of 0.7 hours, an additional mass of of 5 mg of sodium should have been added by the meteors in the segment of the atmosphere sampled by us. We adopt here the cosmic abundance of sodium: 2% (Hunten *et al.*, 1981). The corresponding additional mass seen by the 2-arcsec spectrometer aperture would be 10^{-6} times smaller, some 1×10^{14} atoms contributing to the line emission.

The steady-state atmosphere contains the Na atoms in a layer about 2-km thick, where the typical density is $\sim$7,000 atoms cm^{-3} (Cox *et al.*, 1993), corresponding to about 2×10^{15} atoms in the line of sight of the spectrometer. The expected sodium enhancement in the line of sight would, therefore, be only 0.5% of the steady-state column density. Our non-detection of a Sodium enhancement is therefore not surprising. Moreover, Borovicka *et al.*(1999) find some evidence that faint Leonids may be less abundant in Na than the cosmic abundance and that the sodium is deposited rather high in the atmosphere.

5. Conclusions

We were looking for an enhacement of the airglow emission, specifically meteoric sodium, in the higher atmosphere due to the influx of Na-bearing Leonid meteors. This was not detected, indicating that the Leonid storm did not cause a significant anomaly of the sodium abundance in the airglow layer. A more intense shower would have been needed to acheive a detectable Na enhancement.

Acknowledgements

We are grateful to the Israel Science Foundation, the Israel Space Agency, the Sackler Institute for Astronomy at Tel Aviv University, and Prof. Dan Maoz, for facilitating the WO observations of the Leonid shower in 1999. We thank Bob Hawkes for information about the timing of meteors observed with his cameras at the WO. We are grateful for the two nights of darkness provided by the Mizpe Ramon municipality.
Editorial handling: P. Jenniskens.

References

Arlt, R., Rubio, L.B., Brown, P., and Gyssens, M.: 1999, *WGN, Journal of the IMO* **27**, 286–295
Batista, P.P., Clemesha, B.R., Batista, I.S., and Simonich, D.M.: 1989, *JGR* **94**, 15349–15358.
Borovicka, J., Stork, R., and Bocek, J.: 1999, *Meteoritics Planet. Sci.* **34**, 987–994.
Brosch, N.: 1995, *MNRAS* **276**, 571–578.
Brosch, N. and Goldberg, Ye.: 1994, *MNRAS* **268**, L27–L28.
Brown, P., Campbell, M.D., Ellis, K.J., Hawkes, R.L. *et al.*: 2000, *Earth, Moon and Planets* **82–83**, 167–190.
Cox, R.M., Plane, J.M.C., and Green, J.S.A.: 1993, *GeoRL* **20**, 2841–2844.
Gural P. and Jenniskens P.: 2000, *Earth, Moon and Planets* **82-83**, 221–248.
Hunten, D.M.: 1981, *GeoRL* **8**, 369–372.
Rietmeijer, F.J.M.: 1999, *ApJL* **514**, L125–L127.
Roach, F.E. and Gordon, J.L.: 1973, "The Light of the Night Sky", Dordrecht: Reidel, p.62
Verani, S., Barbieri, C., Benn, C., and Cremonese, G.: 1998, *Plan. Space Sci.* **46**, 1003–1006.
Wilson, J.K., Smith, S.M., Baumgardner, J., and Mendillo, M.: 1999, *DPS* **31**, 3803.

ELF/VLF RADIATION PRODUCED BY THE 1999 LEONID METEORS

COLIN PRICE AND MOSHE BLUM

Department of Geophysics and Planetary Science, Tel Aviv University, Ramat Aviv 69978, ISRAEL
E-mail: cprice@flash.tau.ac.il

(Received 4 June 2000; Accepted 18 August 2000)

Abstract. For more than 200 years large meteors entering the atmosphere have been observed to produce audible sounds simultaneously with the optical flash. Since sound waves travel much slower that visible light, the only explanation was that electromagnetic waves produced by the meteors induce a vibration in a transducer close to the observer, producing an audible sound, known as electrophonics. To check this hypothesis, continuous measurements of low frequency electromagnetic waves were performed during the Leonids meteor storm on the night of 18 November, 1999. The analyses of the data indicate distinct electromagnetic pulses produced by the incoming meteors. Many of the weaker incoming meteors that could not be seen visibly were also detected electromagnetically, with a peak rate of approximately 15,000 meteors per hour occurring at the peak of the storm, nearly 50 times the visible rate.

Keywords: Electrophonics, ELF, Leonids 1999, meteors, radio waves, VLF

1. Introduction

For generations there have been claims that meteors entering the earth's atmosphere produce an audible sound simultaneously with the optical signature produced by the incoming meteor (Blagdon, 1784; Udden, 1917; Romig and Lamar, 1963; Andres *et al.*, 1969; Keay, 1980; Keay, 1993). It was difficult to explain these sounds being produced by shock waves or other acoustic signals produced by the meteorite itself, since given the distance of the meteor from the observer, there should always be a time delay between the optical and audible signals. However, if the meteor produced an electromagnetic wave in the audible frequencies, then this wave would reach the observer at the same instance as the visible light (Hawkins, 1958; Beech *et al.*, 1995). This low frequency wave could induce oscillations, vibrations, and sounds from many objects near the observer. Hence, any electrically conducting body (plants, hair, wires, metal sheets, speakers, fences, spectacles, etc.) could vibrate at audible frequencies, giving the observer the perception that the sound was produced by the meteor (Udden, 1917). This phenomenon is known as electrophonics. It is interesting to note that some observations mention sounds being heard before any optical flash in the sky (Nininger, 1939; Keay, 1980), allowing the observers to focus their attention in a particular direction before seeing the meteor burning up in the atmosphere.

We have tested this hypothesis by attempting to measure the ELF/VLF radiation from the meteors during the 1999 Leonid meteor storm. Here, we report a strong increase of VLF detections with unusual spectral signature that coincide with the peak of the storm.

2. Measurements

During the Leonids meteor storm on 18 November, 1999, electromagnetic measurements were continuously recorded to try and detect these radio waves produced by meteors. Since the best viewing location for the 1999 meteor shower was the Middle East, we were ideally located for this task. A permanent field site for observing ELF/VLF signals is located at the Desert Research Institute of Ben-Gurion University, at Sde Boker in the Negev Desert (30 N, 34 E). The antenna is designed to pick up very weak signals in the extremely low frequency (ELF: 100 Hz $<$ f $<$ 3000 Hz) and the very low frequency (VLF: 3 kHz $<$ f $<$ 50 kHz) range for use in lightning research. However, these frequencies are exactly those expected to be produced by meteors (Keay,1980) and, therefore, our setup was ideal for studying the meteor signals. The ELF/VLF antenna is 10 meters high, with two orthogonal triangular loops, each with a baseline of 18 meters, height 9 meters, giving an area of approximately 81 m^2 for each loop. One loop is aligned in the magnetic north-south direction, with the other along the magnetic east-west bearing. The sensitivity of the system in the broadband range (0.1–50 kHz) is 6 µV/meter. The dynamic range of the antenna/preamp set is approximately 100 dB, allowing us to detect lightning discharges from great distances. The data were collected on digital audio tapes (DAT) with GPS timing, to correlate with the optical measurements. The data were later digitized at 100 kHz.

3. Results

Since our antenna is sensitive to both lightning discharges and possible meteor pulses, we needed to differentiate between the lightning and meteor signals. In Figure 1 we see an example of the north-south magnetic field time series, in the 0.1--50 kHz range, produced by a lightning discharge (Figure 1a) and a meteor (Figure 1b) on the night of the meteor storm. There are two distinct differences between the lightning and the meteor signal. First, while the lightning pulse lasts no longer than 1 millisecond, the meteor pulse continues for up to 10 milliseconds. Although this is longer that the lightning pulse, this is much shorter than the duration of the optical meteor trail which can last for seconds. Second, the amplitude of the lightning pulse is much larger than the meteor pulse. These time series are from the same data file and, therefore, the relative amplitudes can be directly compared. It is important to note that at the time of sampling thunderstorms were observed over the Balkans by the Leonid MAC team, but there were no thunderstorms within a 2,000 km radius of the Negev field site.

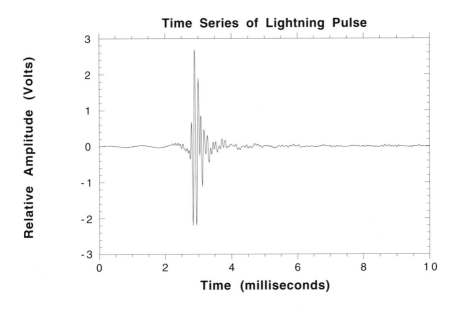

Figure 1a. Time series of lightning on the night of 18 November, 1999.

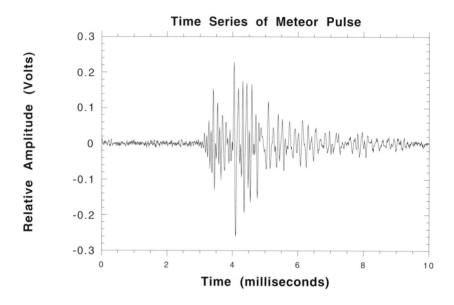

Figure 1b. Time series of a meteor electromagnetic pulses on the night of 18 November, 1999.

In addition to the obvious differences between the lightning and meteor time series, their respective spectra also show significant differences. It is well documented that the spectrum of distant lightning shows a maximum near 6 kHz (Volland, 1982). This is shown from our measurements in the Negev desert during August, 1999, when no precipitation or lightning activity occur in the Middle East (Fig 2a). This spectrum is an average of approximately 35 individual spectra. The maximum around 5 kHz agrees well with measurements of lightning from other parts of the globe. During the night of the Leonids meteor storm (18 November, 1999), very different spectra were obtained due to the incoming meteors (Fig 2b). Here too the spectrum represents an average of approximately 35 events.

Unlike the lightning spectra, the meteor spectra shows a minimum near 5kHz, with a large maximum in the ELF range (0.3-1.5 kHz) and an additional weaker maximum around 2 kHz. In the VLF range there appears a weaker, broader maximum between 6-15 kHz. No signal was observed above 20 kHz. The characteristic differences between the lightning and meteor spectra allow for the automatic determination of whether the electromagnetic signal is caused by lightning or by meteors. This enabled us to label the meteor pulses and, therefore, count the number of ELF/VLF meteor signals observed during the night of the 17-18 November, 1999. An example of the dynamic spectrum at the peak of the meteor shower is shown in Figure 3.

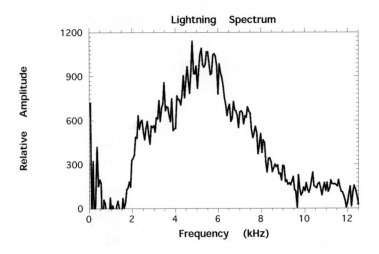

Figure 2a. The mean spectrum of lightning pulses.

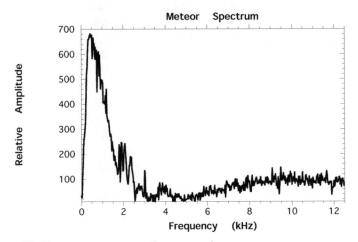

Figure 2b. The mean spectrum of meteor pulses.

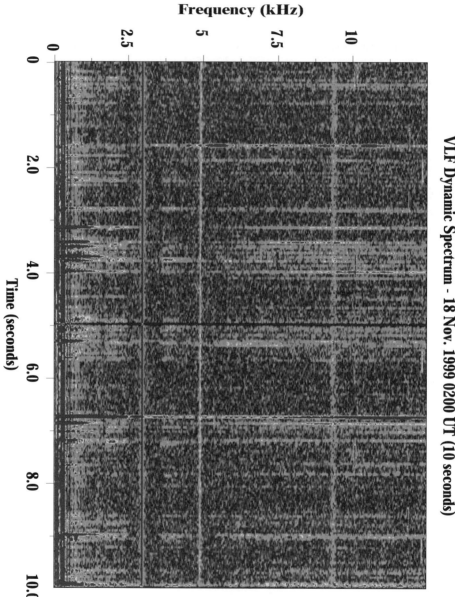

Figure 3. The dynamic spectrum during peak of Leonid meteor shower, around 02:00 UT, 18 August, 1999.

The dynamic spectrum represents only 10 seconds of data, where the spectrum is calculated every 10 milliseconds. The frequency of the 10 millisecond windows is shown on the vertical axis between 0-12.5 kHz, while the color code represents relative amplitude of the signals, red being the largest values. A few features are clearly seen in this 10-second snapshot. The horizontal red lines between 0-0.5 kHz represent the large noise produced by the electric power lines that operate at 50 Hz in Israel, together with all the higher harmonics. The horizontal lines shown at higher frequencies represent the anthropogenic signals from VLF transmitters around the globe, used for navigational purposes. The Russian VLF signals are transmitted in a pulsed format, as can be seen in Figure 3 above 10 kHz. The vertical lines represent the pulses for the individual meteors entering the earth's atmosphere. The mean spectrum of these events is shown in Figure 2b. Up to thirty VLF pulses are observed within this 10-second period. Whether all these VLF pulses are produced by individual meteors, or whether each meteor produces a series of pulses, is still unknown. Correlations with optical measurements will allow us to decipher this uncertainty in the future.

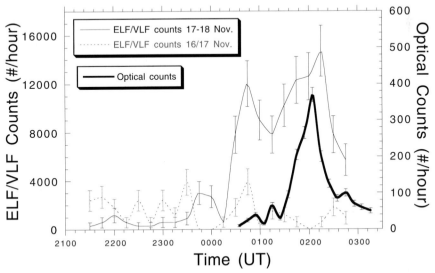

Figure 4. Hourly counts of optically observed meteors during the night of 17-18 November, 1999 (bold line), and the electromagnetically observed meteor counts during the night preceding the meteor shower (16-17 November: dotted line) and the night of the shower (17-18 November: solid thin line).

Based on the spectrum shown for the meteors in Figure 2b we used the 1.2 kHz frequency band to automatically identify the presence of a meteor in the dynamic spectrum. As shown above, this is exactly where the lightning signal is weakest. Although the meteor signal is stronger at lower frequencies (0.5 kHz), noise interference from the power line harmonics produces problems deciphering weak meteor signals at these frequencies. Using a specified threshold for the meteor signal at 1.2 kHz, we were able to count the number of electromagnetic pulses produced by the meteors. Since we have six hours of continuous recordings from 21:30 UT on the 17th November through 03:30 UT on the 18th, we were able to produce a time series of the hourly rate of electromagnetic meteor pulses, to compare with the local incident optical meteor observations (Brosch et al., this issue)(Figure4). The ELF/VLF hourly rate obviously depends on the threshold chosen, making our algorithm more or less sensitive to weak pulses.

As is clearly shown using the ELF/VLF method of counting the meteor flux, a peak flux of 15,000 per hour was detected, relative to 350 per hour using optical methods. Therefore, the ELF/VLF method detected nearly 50 times more meteors than the optical method. It should be pointed out that the radio pulse counts were obtained by sampling small segments of data (10 seconds) at 15 minute intervals. This was done to save time in data analysis, since each 10 seconds of ELF/VLF data represents 1 million data points. Analysis at finer temporal resolution will be done in the future. The ELF/VLF count maximum was observed in the sample taken at 02:15 UT, five to ten minutes after the optical peak of the meteor shower. This time correlation confirms that the electromagnetic pulses observed were produced by the incoming meteors. A similar analysis for the previous night (16-17 November, Figure 4) shows no such enhancement of the pulse counts. Although the ELF/VLF antenna observes signals from all directions, and from greater distances than the optical measurements, it is very likely that many weak meteors that cannot be seen optically still produce electromagnetic signals. However, with all the observers in the field during this night, no reports of audible sounds associated with the meteors could be found.

It is possible to estimate the effective area of detection at the peak of the shower, if we know the count rate, and the limiting magnitude of the meteors we detect. It is normally assumed that the limiting magnitude for observing optical meteors is +6.5. However, for the ELF/VLF meteors the limiting magnitude may be higher (smaller meteors). The effective area can be calculated as A = counts / 0.82 x $r^{\Delta m}$ x sin(hr) km^2, where 0.82 (km^{-2} hr^{-1}) is the peak influx of Leonids brighter than +6.5 magnitude (Gural and Jenniskens, this issue), r is the magnitude distribution index of approximately 2.1, hr is the height of the radiant position (70°), and Δm is the magnitude difference between our ELF/VLF limiting magnitude and the standard +6.5 limiting factor for the optical meteors. If we see only the meteors brighter than +6.5 then $\Delta m = 0$, and the effective area is 17,000 km^2. If we manage to detect meteors brighter than +7.5, then our effective area of detection is 36,000 km^2.

The electromagnetic flux rate shows an additional interesting feature not shown in the optical counts. A secondary peak of the shower is shown at 00:45:00 UT, an hour and a half before the optical peak. It is possible that the visible meteors represent only a small subset of the total meteors. From the ELF/VLF counts it appears that there existed a maxima of small sub-visible meteors 90 minutes before the optical peak.

4. Discussion

In addition to the advantage of being able to detect weak meteors, the electromagnetic method of determining the meteor fluxes can also be used during daylight hours, and in all weather conditions. Our measurements provide convincing proof that meteors do produce electromagnetic radiation as they enter the atmosphere, which can explain the sounds heard during observations of large fireballs (electrophonics).

The only theoretical explanation of how these radio waves are produced has been presented by Keay (1993, 1995). However, our measurements challenge the theory with new questions: How do sub-visible and small meteors produce radio signals? The theory applies only to large bolides (fireballs). Why do the radio signals never last more than 10 milliseconds? The theory explains radio signals lasting up to tens of seconds. Why do some people hear sounds before seeing the optical meteor? The theory describes the radio signals produced simultaneously with the bright optical signal. It is clear that more work is needed in this field.

Acknowledgments

We thank George J. Drobnock and an anonymous referee for comments that improved the manuscript. We thank Dr. David Faiman and the Solar Energy Research Center, Sde Boker, for allowing us to use their facility for our measurements; Boris Starobinets for assisting with the data collection; and Dr. Noah Brosch for inviting us to participate in the 1999 Leonid campaign. Support for construction of the VLF antenna was provided by the Israel Science Foundation. *Editorial handling*: Peter Jenniskens.

References

Andrews, A.D., Rackham, and T.W., Wayman, P.A.: 1969, *Nature* **222**, 727–730.
Beech, M., Brown, P., and Jones, J.: 1995, *Earth, Moon and Planets* **68**, 181–188.
Blagdon, C.: 1784, *Philos. Trans. R. Soc. London* **74**, 201–232.
Drobnock, G.J.: 1992, *Sky and Telescope* **77**, 329–331.
Hawkins, G.S.: 1958, *Nature* **181**, 1619–1611.
Keay, C.S.L.: 1980, *Science* **210**, 11–15.
Keay, C.S.L.: 1993, *J. Scientific Explor* **7**, 337–354.
Keay, C.S.L.: 1995, *Earth, Moon and Planets* **68**, 361–368.
Nininger, H.H.: 1939, *Popular Astronomy* **47**, 97–99.
Romig, M.F. and Lamar, D.L.: 1963, *RAND Memo. RM-3724-ARPA*.
Udden, J.A.: 1917, *Science* **46**, 616–617.
Volland, H.: 1982, in H. Volland (ed.), *Handbook of Atmospherics,* CRC Press, Boca Raton, Fla., pp. 179–250.

NOTE ON THE REACTION OF THE UPPER ATMOSPHERE POTASSIUM LAYER TO THE 1999 LEONID METEOR STORM

JOSEF HÖFFNER, CORD FRICKE-BEGEMANN, AND ULF VON ZAHN

Leibniz-Institute of Atmospheric Physics, Schloss-Str. 6, 18225 Kühlungsborn, Germany
E-mail: hoeffner@iap-kborn.de

(Received 15 June 2000; Accepted 15 August 2000)

Abstract. We have performed an experiment to study the impact of the 1999 Leonid meteor storm on the upper atmosphere potassium layer. For the experiment, we used a ground-based K-lidar located at the Observatorio del Teide on the island of Tenerife (28°N, 17°W). As is now known from external sources, the activity of the storm exhibited a sharp peak of activity on November 18, 1999 close to 02:05 UT with a full width at half maximum of only 45 min. Due to unfavorable weather conditions at our lidar site, we could not observe the behavior of the K layer immediately before or during the meteor storm. However, about 1 h past the peak of the Leonid storm, the clouds above the site started to show small holes. Hence, between 03:24 and 05:22 UT, we were able to perform lidar soundings of the K layer above our site. From those we can conclude that the 1999 Leonid meteor storm has not led to an outstanding enhancement of the upper atmosphere potassium layer.

Keywords: lower thermosphere, Leonids 1999, mesosphere, meteors, neutral atom debris layer, Potassium

1. Introduction

In 1996, the IAP initiated a systematic study by ground-based lidar of meteor trails and of the metal atom abundances within these trails. So far, these observations were performed at the site of the IAP (54°N, 12°E). The 1996 Leonids were observed with a single K lidar by Höffner *et al.* (1999). The 1997 Leonids could not be observed because our site was clouded over all night. The 1998 Leonid shower was studied, along with other meteor showers, with three metal resonance lidars (for Ca, Fe, and K) by von Zahn *et al.* (1999). One of their outstanding results was that the Ca/Fe abundance ratio in the Leonid meteor trails was found to be much lower than the CI meteoritic composition would indicate. We note, though, that in both 1996 and 1998, the Leonid showers produced only weak events at our site: The 1996 shower reached a peak zenithal hourly rate (ZHR) of only 86 ± 22 (Brown and Arlt, 1997) and the now famous Leonid fireball shower of 1998 peaked in fact over Europe having there a peak ZHR of 340 ± 20 (Arlt, 1998). However, only the tail of this peak could be observed because of bad weather that day.

For the 1999 Leonids, conditions for lidar-observations over middle Europe were expected to be much better than in previous years: McNaught and Asher (1999) predicted a genuine meteor storm (with ZHR > 1000), centered over the Near East. This put our Institute's site as well as our new field station at Tenerife (28°N, 17°W) reasonably close to the predicted Leonids storm center. Naturally, our lidar observations are always at the mercy of the local weather conditions, in particular of any cloud cover. In Europe, this poses a general problem for lidar observations of the Leonids as the latter occur in mid-November when there is a lot of wet and stormy weather. To eliminate this risk factor for some of our observations, the IAP acquired in 1999 a state-of-the-art meteor radar. It was installed on time for observations of the 1999 Leonids at our Juliusruh field station (55°N, 13°E). Singer *et al.* (2000) have reported interesting short-term temporal variations in ZHR observed with this radar during the 1999 Leonids storm. Unfortunately, our IAP site was totally clouded over during the night of November 17/18, 1999 and no lidar observations of the 1999 Leonids storm could be obtained there.

We had an additional K lidar located at Tenerife island. Yet, during the 1999 Leonids storm, the weather at this lidar site was marginal, too. During the period of maximum storm activity no lidar measurements could be performed. We were able to get some lidar observations of the K layer, starting 1.3 h past the storm peak. These observations shall be reported and discussed in the following chapters.

2. Instrumentation and Observation Site

We use a ground-based lidar working on the $K(D_1)$ resonance line of potassium (at 770 nm) in order to determine (a) altitude profiles of the absolute number densities of K atoms in the altitude region 80 to 110 km (Eska *et al.*, 1998; 1999), (b) temperature profiles throughout the same altitude region (von Zahn and Höffner, 1996; She and von Zahn, 1998), and (c) the K atom densities deposited in meteor trails (Höffner *et al.*, 1999; von Zahn *et al.*, 1999).

In our lidar, altitude profiles of the number of backscattered photons are obtained simultaneously in two ways: (1) after integration over 4000 laser pulses with 200 m altitude binning and (2) from each individual laser pulse with 7.5 m altitude binning. Profiles of type 1 are used to study the regular metal layers and the air temperature as well as to provide real-time displays at the lidar controls. Profiles of type 2 are used for analysis of meteor trails (their analysis being based typically on running means of either 0.5 or 4 s length). Table I summarizes the technical specifications of the lidar.

The lidar-observed meteor trails occur at altitudes between 75 and 120 km and they are observed to contain atom number densities of K up to 3000 cm^{-3} (Gerding *et al.*, 1999). During their flight through the Earth's upper atmosphere, very few meteoroids will cross the laser beam directly. However, the trails of neutral atoms left behind by the ablating meteoroids have considerable lifetimes (typically tens of minutes). Therefore, some of these trails are blown by the upper atmosphere winds through our laser beam, which is directed vertically. At, say, 90 km altitude, the wind speeds are of the order 15 m/s and our laser beam has a diameter of about 50 m. Hence, during its passage through the laser beam, the trails become observable by the lidar for a few seconds or more, depending on the trail ages and hence their diameters.

Our K lidar is mounted in a 20 ft standard container. Since spring 1999, we operate this lidar at the Observatorio del Teide at the Tenerife island (28°18'N, 16°31'W), where it is located 2390 m above sea level.

TABLE I

Parameter	Type / quantity
type of laser	alexandrite ring laser
wavelength	769.898 nm (in air); $K(D_1)$ line
tuning range	±1.6 pm
spectral width	± 20 MHz
pulse energy	175 mJ
pulse length	150 ns
pulse rate	30 pulses per second
laser beam divergence	0.5 mrad
telescope primary mirror	80 cm diameter
photon detection	single photon counting

3. Observations

3.1. THE FIELD CAMPAIGN

The primary purpose of our November 1999 field campaign at Tenerife was to study the thermal structure of the mesopause region under winter conditions at low latitudes. Naturally, the chance to observe the reaction of the K layer to the predicted Leonid storm was much welcomed as well. Our K lidar was available for nighttime observations from the evening of November 5 until the morning of November 29, 1999; that is for 24 nights. During this period, we could obtain atmospheric data during 15 nights out of which 8 nights yielded continuous observations of more than 10 hours.

3.2. LIDAR-OBSERVED METEORS

Throughout our lidar operations, we collected the high-temporal resolution data for searches of meteor trails. Although, unfortunately, weather conditions did not allow us to observe genuine Leonids trails during the night of the Leonids storm, we list in Table II those nights in which we observed substantial potassium meteor trails.

TABLE II

Date of night	Number of meteors	Length of trails [s] [km]	Altitude of trails
Nov 6/7	4	all < 10	80 ... 85
Nov 8/9	1	51	87
Nov 10/11	1	8	83
Nov 11/12	4	all < 10	78 ... 83
Nov 20/21	1	5	85
Nov 25/26	5	8 ... 480	88 ... 100
Nov 26/27	2	5 , 1600	82 , 85
Nov 27/28	3	10 ... 68	79 ... 86
Nov 28/29	2	3 , 5	80 , 81

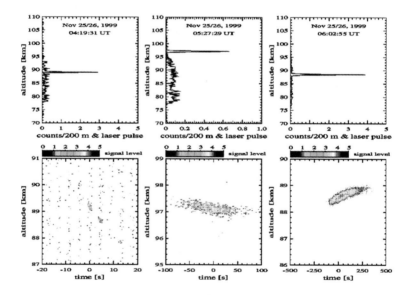

Figure 1. Examples of three meteor trails, as observed with our K lidar during the night Nov 25/26, 1999. The upper panels give count rate profiles at the time of maximum K densities in the trails. Integration times are 0.5 s, 4 s, and 4 s for the left, middle, and right panels, respectively. The lower panels show the passage of the wind-driven meteor trail through the laser beam on a much longer time scale, but more focussed altitude scale. Here, times "zero" are identical to the times given in the upper panels. The "signal" levels are proportional to the ratio of the trail density over the K layer density.

From their altitudes, one can already distinguish different types of meteors. Trails at and below 85 km are produced by sporadics (order of 15 km/s), trails in the range 85 to 95 km by medium fast meteors (order of 30 km/s), and those above 95 km by fast meteors like (non-fireball) Leonids (von Zahn et al., 1999).

In Figure 1 we show three lidar-observed meteor trails from the night Nov 25/26, 1999. The upper panels show the profiles of photon counts during the time of maximum K density in the trail. At the rather short integration times, the normal K layer deteriorates to the shallow feature with about 0.1 to 0.2 photon counts/200m and laser pulse between 80 and 95 km altitude. The lower panels show the passage of the wind-driven meteor trail through the laser beam on a much longer time scale, but more focussed altitude scale. Altitude resolution in these panels is 35 m. In the left lower panel, the apparent noise is actually caused by photons back scattered from single K layer atoms. The latter become noticeable due to the fact that the only moderately strong meteor trail is embedded well within the K layer. The back scattered photons are densest along vertical lines (= fixed times) due to a periodic scan of the laser wavelength across the $K(D_1)$ fine structure line for Doppler temperature measurements. This feature of our lidar can be recognized only in presentations with a time resolution of a second and better. Figure 1 demonstraste the excellent capabilities of our K lidar for meteor trail observations.

3.3. The night of the Leonid storm

Today we know that the 1999 Leonid storm peaked November 18, 1999, about 02:05 UT ± 5 min with a peak equivalent zenithal hourly rate (ZHR) of 3,700 ± 100 (Arlt et al., 1999; Singer et al., 2000). From photographic records, the mean radiant position of the Leonids was determined to be R.A. = 153.67 ± 0.05° and DEC = +21.70 ± 0.05° (Betlem et al., 2000). The 1999 Leonid storm showed an obvious lack of bright meteors, much different from the 1998 Leonid (Arlt et al., 1999).

At the location of our lidar at Tenerife, the Leonid radiant rose at 00h 44m UT and set 14h 27m UT. At the time of the peak of storm activity, the radiant altitude was 17° above the horizon. Even though this is a comparatively low radiant altitude for meteor observations, it should have produced a local maximum ZHR of larger than 1000. Most unfortunately, however, the Observatorio del Teide was in a strong northwesterly flow of moist air throughout the days near the Leonid storm. At the site of our lidar, this flow produced a topographic, near-continuous cloud deck, combined with frequent drizzle. Yet, this condition was co-existing with that of rapid evaporation of the clouds on the leeward site of the Tenerife island which allowed for excellent visual and camera observations of the 1999 Leonids from the Southern part of Tenerife.

It was about 1 h past the peak of the Leonid storm, that the clouds above the lidar started to show a few small holes. Hence, between 03:24 and 05:22 UT, we were able to perform a few lidar soundings of the K layer above our lidar. These will be presented in the next section.

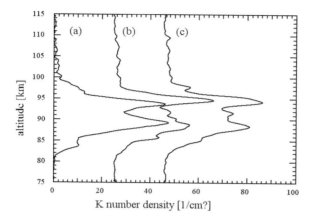

Figure 2. Three K density profiles measured during the morning of Nov 18, 1999. The integration periods were, from left to right, 03:24 to 03:39 UT, 04:12 to 04:44 UT, and 04:52 to 05:21 UT. Absolute number densities are calibrated against the Rayleigh-backscattered photons from 30 km altitude. The zero lines for profiles (b) and (c) are offset to the right in proportion to their temporal separation from profile (a).

3.4. K DENSITIES

In our standard mode of lidar observations, we integrate for each individual altitude channel the back scattered photons from 4,000 laser pulses before we store a profile of integrated photon counts. Under clear skies and with our standard pulse rate of 30 pulses/s, such a single data file is collected and stored in 2.2 min. There is, however, one step of sophistication in this process: Before we store the photon counts from any individual laser pulse into the growing data file, the lidar data collecting programme checks internally whether or not the laser pulse produced a signal from Rayleigh scatter in the lower stratosphere above a pre-selected signal level. If that signal level is not reached, e.g. by the presence of clouds, all photons back scattered from this laser pulse are discarded. This avoids adding just background counts to the altitude channels of the K layer in the case that the clouds become too thick or the laser energy unexpectedly low for useful data collection. This mode of data collection becomes absolutely essential under weather conditions as we encountered during the night of the 1999 Leonid storm. In fact, we could assemble just 3 data files (each from 4,000 good laser pulses through holes in the clouds) which lasted from 03:24 to 03:39 UT, 04:12 to 04:44 UT, and 04:52 to 05:21 UT. Nevertheless, these three soundings of the K layer afford us an interesting look at the K layer within a few hours after passage of the 1999 Leonid storm.

Figure 2 shows the three K density profiles obtained during the morning hours of Nov 18, 1999. The absolute number densities are shown vs. altitude, after a binomial filter over ± 5 altitude channels of

200 m each had been applied. The 2nd and 3rd profiles are offset from the 1st one in proportion to their difference in sampling time. The statistical error of the number densities due to the limited photon counts is approximately ± 1 atom cm^{-3}. No significant deviations of these profiles from common K density profiles over Tenerife are recognizable. That statement applies to the densities above 100 km where most of the Leonid meteoroids should have ablated as well as to the double-humped layer shape. The upper layer, maximizing near 95 km altitude, is just one of the quite common sporadic K layers, while the maximum near 89 km represents the peak of the normal K layer. These statements could be supported by numerous examples from our earlier K lidar observations at Tenerife and elsewhere (see e.g. Eska, 1998). In Figure 3, we compare the mean density profile of the morning of Nov 18 with that of the monthly mean November profile, the latter being based on well over 100 h of our lidar observations. We point out that above 98 km altitude, the two profiles fall rather close to each other. The small differences above 98 km of the two number density profiles will not contribute significantly to any difference between the K column densities of the two scenarios.

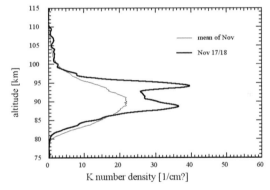

Figure 3. Mean K number density profiles for the morning of Nov 18 (bold line) and for the month of November (thin line).

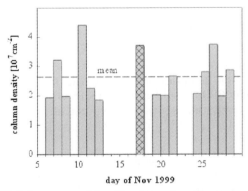

Figure 4. Nightly means of the K column densities measured during 15 nights in November 1999. The Leonid storm occurred in the morning of Nov 18.

We can also test the K column densities for any peculiarity of the Leonid storm night. To this end, we provide Fig. 4, in which we have assembled the mean nightly K column densities as measured throughout our November 1999 campaign at Tenerife. The mean from the 15 nights of observations is 2.7 x 10^7 cm^{-2}, the column density right after the Nov 18 Leonid storm is 3.7 x 10^7 cm^{-2}. The latter value is obviously somewhat larger than the November mean, but well within the natural variability of this parameter. Therefore our data hardly support the assumption that the slight Nov 18 surplus in column density was produced by the 1999 Leonid storm.

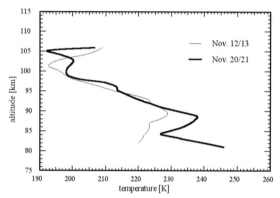

Figure 5. Nightly mean temperature profiles before (Nov 12/13, thin line) and after (Nov. 20/21, bold line) the Leonid storm.

3.5. AIR TEMPERATURES

Speculations have been aired that the material ablating from the Leonid meteoroids may re-condense to so many "smoke particles" (so designated by Hunten *et al.*, 1980) that their absorption of solar energy may lead to a measurable rise in upper air temperatures. If such an effect indeed exists, it could be observed only during the daytime following the Leonid storm or even later. Just in order to compare the air temperatures in the metal layers (80 to 105 km altitude) during nights before and after the Leonid storm, we present in Figure 5 our mean nightly temperature profiles for the two nights closest to that of the Leonid storm. Data of Nov 12/13 are based on more than 10 h, data of Nov 20/21 are based on more than 8 h of lidar observations. The error bars on the temperatures are typically ± 3 K near 90 km altitude and ± 6 K where the profiles end. The differences between the two profiles fall well within the normal variability of our nightly mean temperature profiles obtained at Tenerife. Therefore, no direct influence of the Leonid storm on these profiles can be identified.

4. Discussion

As shown in Section 3, neither the K number density profiles nor the K column densities measured by our K lidar at Tenerife can give definite answers to the question whether or not the 1999 Leonid storm has left a significant, though certainly transient mark on the K layer. This situation is in large measure due to the weather-caused lack of information on the K layer shortly before and during the actual storm period. A conclusion, that such an impact of meteor showers on the metal layers does not exist, would be contrary to lidar observations of the K layer during the 1996 Leonid shower by Höffner et al. (1999) and of the Ca layer during a 1997 March shower by Gerding et al. (1999). In either case, the authors observed considerable increases of metal atom column densities within about an hour of the onset of the shower. We note, though, that Chu, et al. (2000) conclude from their lidar observations that the 1998 Leonid shower did not have a significant impact on the abundance of the background Fe layer above Okinawa.

One unusual feature of the observation conditions at Tenerife has been the following: Due to the low altitude of the Leonid radiant (about 17°) in combination with the high entry velocity of the Leonid meteoroids and the comparatively low boiling temperature of K-containing minerals, the potassium contained in the Leonid meteoroids was very likely totally ablated and/or evaporated at altitudes above 100 km (McNeil et al., 1998; von Zahn et al., 1999). Above 100 km, the time constant for conversion of K atoms into K ions is of the order of 1 day (at night). Time constants for chemical conversion and for vertical transport are even much longer. Thus, we would expect an accumulation of meteoric K atoms at these altitudes at least over a few hours after the Leonid meteor storm. Why we fail to observe such an increase of K remains to be studied.

This brings us to a comparison of our measured K column densities with literature values. In June 1996, Eska et al. (1999) measured K column densities of 5.5×10^7 cm^{-2} at 28°N, whereas in June 1997 and at 54°N, Eska et al. (1998) measured 4.5×10^7 cm^{-2}. These results have been interpreted by the authors as indicating a considerable latitude dependence of the K column density. In addition, for 54°N Eska et al. (1998) reported a strong semiannual variation of the K column density, with the mean November column density being about 67% of that in June. In case that the percentage semiannual variation at 28°N is the same as in 54°N (which we do not know), the Eska, et al. results would imply a mean November column density of 3.7×10^7 cm^{-2} at Tenerife. Our own measured mean November K column density is 2.7×10^7 cm^{-2} (see Figure 4). Considering in particular the small data base of Eska et al. (1999) for 28°N and the uncertainties in extrapolating the seasonal variation from 54°N to 28°N, we consider our Tenerife data in satisfactory agreement with the extrapolation of the Eska et al. results.

5. Conclusions

The 1999 Leonid meteor storm has not led to an outstanding enhancement of the upper atmosphere potassium layer. This statement reflects the state of the K layer about 2 to 3 hours after the peak of the storm activity. Due to unfavorable weather conditions, the behavior of the K layer could not be observed immediately before and during the meteor storm. Therefore, an actual quantification of the changes in the K layer due to the meteor storm could not be obtained.

Acknowledgments.

We gratefully acknowledge the friendly and efficient support of the Instituto de Astrofisica de Canarias (La Laguna, Tenerife, Spain) and in particular of its J.J. Fuensalida for our observations at the Observatorio del Teide. Setting up of our lidar at the Observatorio and its daily operations were also greatly helped by the staff of the Vacuum Tower Telescope of the Kiepenheuer Institut für Sonnenphysik (Freiburg, Germany). *Editorial handling:* Mark Fonda.

References

Arlt, R.: 1998, *WGN, Journal of the IMO* **26**, 239–248.
Arlt, R., Bellot Rubio, L., Brown, P., and Gijssens. M.: 1999, *WGN, Journal of the IMO*, **27** 286–295.
Betlem, H., Jenniskens, P., Spurny, P., Docters van Leeuwen, G., Miskotte, K., ter Kuile, C., Zerubin, P., and Angelos, C.: 2000, *Earth, Moon and Planets* **82–83**, 277–284.
Brown, P. and Arlt, R.: 1997, *WGN, Journal of the IMO* **25**, 210–214.
Chu, X., Pan, W., Papen, G., Gardner, C.S., Swenson, G., and Jenniskens,P.: 2000, *Geophys. Res. Lett.* **27**, 1807–1810.
Eska, V.: 1998, *Die Kaliumschicht in der oberen Atmosphäre (75-110 km): Beobachtungen, Analysen und Modellierung*, Ph.D. Dissertation, Rostock University, Rostock, Germany.
Eska, V., Höffner, J., and von Zahn, U.: 1998, *J. Geophys. Res.* **103**, 29207–29214.
Eska, V., von Zahn, U., and Plane, J.M.C.: 1999, *J. Geophys. Res.* **104**, 17173–17186.
Gerding, M., Alpers, M., Höffner, J., and von Zahn, U.: 1999, *J. Geophys. Res.* **104**, 24689–24698.
Höffner, J., von Zahn, U., McNeil, W.J., and Murad, E.: 1999, *J. Geophys. Res.* **104**, 2633–2643.
Hunten, D.M., Turco, R.P., and Toon, O.B.: 1980, *J. Atmos. Sci.* **37**, 1342–1357.
McNaught, R.H. and Asher, D.J.: 1999, *WGN, Journal of the IMO* **27**, 85–102.
McNeil, W.J., Lai, S.T., and Murad, E.: 1998, *J. Geophys. Res.* **103**, 10899–10911.
She, C.Y. and von Zahn, U.: 1998, *J. Geophys. Res.* **103**, 5855–5863.
Singer, W., Molau, S., Rendtel, J., Asher, D., Mitchell, N.J., and von Zahn, U.: 2000, *MNRAS,* in press.
von Zahn, U., Gerding, M., Höffner, J., McNeil, W.J., and Murad, E.: 1999, *Meteorit. Planet. Sci.* **34**, 1017–1027.
von Zahn, U. and Höffner, J.: 1996, *Geophys. Res. Lett.* **23**, 141–144.

MESOSPHERIC AND LOWER THERMOSPHERIC WINDS AT MIDDLE EUROPE AND NORTHERN SCANDINAVIA DURING THE LEONID 1999 METEOR STORM

WERNER SINGER AND PETER HOFFMANN
*Leibniz-Institut für Atmosphärenphysik, Schloss-Str. 6,
18225 Kühlungsborn, Germany
E-mail: singer@iap-kborn.de*

NICHOLAS J. MITCHELL
*Dep. of Physics, University of Wales, Aberystwyth,
Ceredigion SY23 3BZ, Wales, UK*

and

CHRISTOPH JACOBI
Institut für Meteorologie, Universität Leipzig, 04103 Leipzig, Germany

(Received 5 July 2000; Accepted 30 August 2000)

Abstract. We report observations of winds in the mesosphere and lower thermosphere during the Leonid meteor storm of November 17/18, 1999. The observations were obtained at five radar sites in Middle Europe and Northern Scandinavia using meteor radars in Germany and Northern Sweden, Medium Frequency (MF) radars in Germany and Northern Norway and Low Frequency (LF) wind measurements in Germany. We present hourly means of zonal and meridional winds covering the altitude range 82 km to 106 km. At mid-latitudes (52°- 54°N) we observe strong eastward and southward directed winds during the storm phase of the Leonid shower in the early morning hours of November 18 whereas eastward and northward directed winds are dominating at high latitudes (67°- 69°N). Strong semidiurnal and weaker diurnal tidal oscillations are observed in the wind field at both latitudes at altitudes above 90 km.

Key Words: Leonids 1999, Lower thermosphere, mesosphere, meteors, winds

1. Introduction

The Leonid meteor storm is caused by meteoroid streams associated with the comet Temple-Tuttle. Meteoroid streams with high dust flux rates generate meteor storms with spectacular displays of numerous persistent trails. The knowledge of the background wind field at mesospheric/lower thermospheric altitudes can contribute to improve the understanding of trail motions and to separate between intrinsic motion of the trail and motion of the heated gas. Ground-based wind measurements by MF and meteor radars allow a continuous and reliable monitoring of the upper atmospheric wind field in the altitude range from 70 to about 100 km and can provide appropriate data. After presenting the radar systems, we discuss our observational results obtained during the main phase of the 1999 Leonid meteor storm.

2. Observation sites and radar experiments

Radar observations of the mesospheric and lower thermospheric wind field were carried out during the 1999 Leonid meteor storm in Germany at Juliusruh (meteor and MF radar) and Collm (LF wind profiler) as well as in Northern Sweden at Kiruna (meteor radar) and in Northern Norway at Andenes (MF radar). The observations at Middle Europe are about 1600 km apart from those at Northern Scandinavia.

We will discuss in more detail the meteor radar experiments as these systems provide besides the winds also height-dependent meteor flux rates. The meteor radar of the Leibniz-Institute of Atmospheric Physics is located at Juliusruh on the island Rügen, Germany (54°38'N, 13°24'E). It is a commercially produced all-sky interferometric meteor radar ("SkiYMet") operating at a frequency of 32.55 MHz and a peak envelope power of 12 kW. The radar uses a single crossed-element antenna on transmission and an interferometer on reception consisting of five crossed-element antennas (Hocking *et al.*, 2000). The antenna system provides a near uniform angular distribution of detections. A range accuracy of 2 km and angular accuracy of 1 to 2 deg in meteor location were possible. The accuracy of the estimated mean winds is generally in the order of 5 to 7 m/s (Hocking and Thayaparan, 1997). In the Leonid experiment the objective was to locate as many meteors as possible as well as to determine entrance velocities. A two-point coherent integration was used in order to optimise the time resolution for entrance speed estimation, resulting in a sample resolution of 0.94 ms at a pulse repetition frequency of 2144 Hz. The radar was configured for a cut-off altitude of 120 km during the Leonid shower and was operated in this mode from November 15, 21:00 UT until November 20, 14:00 UT. Before and after that time the radar was operated in the optimised mode for wind measurements up to 110 km altitude. The meteor data have been analysed for the mean winds using 1-hour bins of data at each height of observation. The data are also binned according to height with an approximate vertical resolution of 3 km (Hocking and Thayaparan, 1997).

The University of Wales Aberystwyth meteor radar is located at Esrange (67°53'N, 21°06'E) near Kiruna in Northern Sweden. This radar is a SkiYMet system too. It operates, however, with a peak power of 6 kW. The radar uses crossed-element antennas to ensure a near uniform azimuthal sensitivity to meteor echoes. During the time of the Leonid shower the radar was operating in a mode optimised for meteor drift measurements of mesospheric winds and so it did not reliably record meteors at heights greater than 110 km. The same data analysis has been applied as for the Juliusruh radar.

The MF radars are operated by the Leibniz-Institute of Atmospheric Physics at Juliusruh on 3.18 MHz and at Andenes, Norway (69°18'N, 16°01'E) on 1.98 MHz (Kremp *et al.*, 1999; Singer *et al.*, 1997). Both systems apply the spaced antenna method and horizontal winds are obtained on the basis of the full correlation analysis in an altitude range between 70 km and 92-94 km. Hourly mean height profiles of horizontal wind vectors are estimated from samples of 2-5 minutes time resolution and a vertical resolution of 2 km at Juliusruh and 4 km at Andenes with an accuracy comparable with the meteor winds. The observations at Juliusruh are restricted to daytime and twilight conditions due to the high night-time noise level at Central Europe.

The LF wind profiler at Collm (52°N, 15°E) performs total reflection radio wind measurements at oblique incidence using the ionospherically reflected sky wave on 177 kHz, 225 kHz and 270 kHz (Schminder *et al.*, 1997). The height estimation is done on the 177 kHz measuring path. The system applies also the spaced antenna method. The high radio wave absorption on daytime restricts the measurements to night-time and twilight conditions. As result of the diurnal course of the reflection height the winds at different altitudes (85-105 km) are measured at different times.

3. Upper mesospheric winds during Leonid 1999 meteor storm

The meteor observations of the Leonid 1999 storm by the meteor radars indicate that the storm maximum occurs in the period 01:00 and 04:00 UT of November 18, 1999 as shown in Figure 1 from the observations of Juliusruh at altitudes between 75 and 120 km. The presented hourly meteor rates contain sporadic, shower, and storm meteors, not only Leonids. The meteor storm activity peaks at

Juliusruh more precise on 02:08 UT as shown in a detailed study of the fine structure of the Leonid 1999 storm at three sites (Singer *et al.*, 2000).

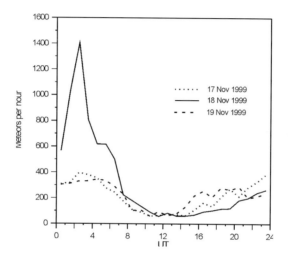

Figure 1. Meteor rates [hour^{-1}] during the Leonid meteor storm 1999 observed by meteor radar at Juliusruh, Germany.

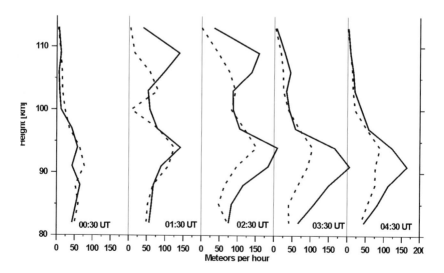

Figure 2. Hourly meteor rates during the maximum of the Leonid meteor storm on November 18, 1999 obtained at Juliusruh (full lines) and Kiruna (dashed lines).

by the different modes of radar operation and the differences in transmitted power.

In the following we concentrate our study of upper atmospheric winds to the period between 17 and 18 November as well as to the peak of the Leonid storm in the morning hours of November 18, 1999. The term zonal wind stands for winds directed towards east (positive sign) or towards west (negative sign), the term meridional wind stands for winds directed towards north (positive sign) or towards south (negative sign).

TABLE I

	SKiYMET Juliusruh 54.3°N	MF radar Juliusruh 54.3°N	LF profiler Collm 52°N	SKiYMET Kiruna 67.9°N	MF radar Andenes 69.3°N
85 km					
U_0	7.5	2.5		-10.4	-2.3
V_0	-3.5	-0.1		-6.6	-2.1
U_{12}/P_{12}	2.9 / (6.3)	1.9 / (8.7)		5.3 / 5.7	9.3 / 8.2
U_{24}/P_{24}	1.8 / (17.7)			2.8 / (8.6)	3.5 / 7.3
91 km					
U_0	12.8	6.3		2.8	-5.5
V_0	-4.9	1.1		-1.5	-0.3
U_{12}/P_{12}	12.2 / 11.7	1.4 / (11.2)		24.3 / 10.4	13.0 / 9.4
U_{24}/P_{24}	16.0 / 8.0			12.3 / 6.7	9.4 / 11.0
94 km					
U_0	15.5		17.5	5.9	
V_0	-4.7		-1.6	-4.4	
U_{12}/P_{12}	30.5 / 10.4		4.4 / 8.8	21.0 / 9.7	
U_{24}/P_{24}	14.2 / 6.6			8.6 / 7.8	

Explanations (mean winds on Nov. 17-19, 1999):
U_0[m/s] = prevailing zonal wind, positive towards east;
V_0[m/s] = prevailing meridional wind, positive towards north;
U_{12} /U_{24}[m/s] = amplitude of the semidiurnal/diurnal tidal wind (zonal component);
P_{12} / P_{24}[UT] = phase of the semidiurnal/diurnal tidal wind (zonal component), defined as time of the occurrence of the eastward wind maximum, values in brackets are uncertain due to low tidal amplitudes.

3.1. UPPER MESOSPHERIC WINDS

The variability of the upper mesospheric wind field in height and time has been studied for a 48 hour period on 17/18 November centred to the storm maximum. Hourly mean winds are presented for the altitudes 85 km, 91 km, and 94 km where both meteor radars provide a good data coverage. In addition, the prevailing winds and the mean diurnal and semidiurnal tidal wind components are estimated for a 3-day period from 17 to 19 November and summarized in Table I. For clearness, the tidal winds are restricted to the zonal components, the meridional components not shown here are of comparable magnitude. The tidal variability is noticeable above 90 km altitude.

Hourly means of zonal and meridional winds obtained at mid-latitudes by the meteor radar and MF radar at Juliusruh as well as by the LF profiler at Collm are presented in Figure 3. Wind speeds up to 45 m/s are observed during the storm phase. MF (*) and LF (•) winds were included if their altitudes were within ±1 km of meteor wind altitude. The dashed lines represents the tidal fits to the hourly mean values. The tidal fits in general reproduce well the measured diurnal variability except for the meridional component at 91 km during the storm phase.

Figure 4 depicts the hourly mean winds obtained with the Kiruna meteor radar and the Andenes MF radar for the same period with zonal winds in the left panel and meridional winds in the right panel. The MF radar data (+) were selected using the same criteria as at mid-latitudes. Also at high latitudes, the observed diurnal variability is generally well reproduced by the tidal fit. The zonal winds vary in the same way at middle and high latitudes during the storm phase whereas the meridional winds show an opposite variation.

The hourly mean winds from the different observations are in general agreement. Some of the details show differences, but the zonal winds agreement between meteor and MF radar observations at 85 km is good.

3.2. MESOSPHERIC/LOWER THERMOSPHERIC WINDS DURING THE STORM MAXIMUM

Both meteor radars at Juliusruh and Kiruna have detected meteor echoes above 100 km during the activity maximum of the Leonid 1999 storm as shown Figure 2. The obtained meteor rates allow the estimation of height profiles up to 106 km altitude. Hourly mean profiles of the zonal and meridional winds from both sites are presented in Figures 5 and 6. Zonal winds agree in shape and magnitude between 90 and 100 km.

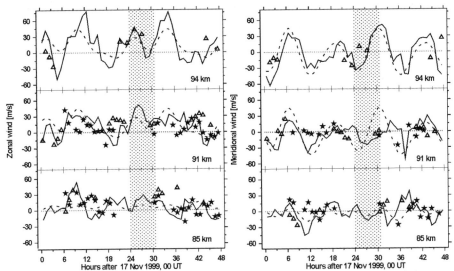

Figure 3. Hourly mean zonal winds (left panel) and meridional winds (right panel) after meteor radar (full lines), MF radar (*) and LF drift observations (•) at Middle Europe on 17/18 November 1999. The hatched bar indicates the Leonid storm phase. The dashed lines represent tidal fits to the meteor winds (for details see text).

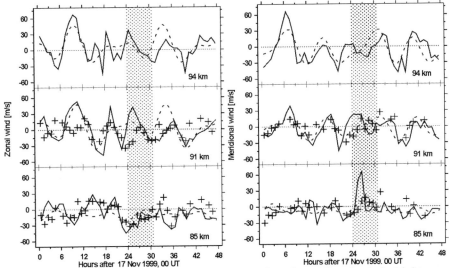

Figure 4. Hourly mean zonal winds (left panel) and meridional winds (right panel) after meteor radar (full lines) and MF radar observations (+) at Northern Scandinavia on 17/18 November 1999. The hatched bar indicates the Leonid storm phase. The dashed lines represent tidal fits to the meteor winds (for details see text).

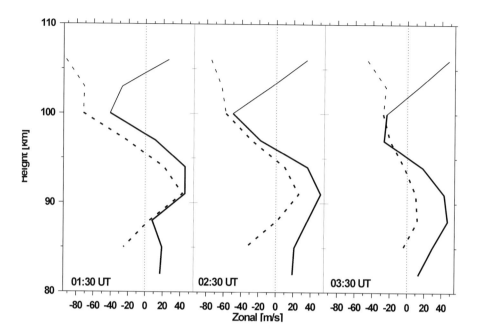

Figure 5. Hourly mean height profiles of zonal wind during the maximum of the Leonid meteor storm on November 18, 1999 obtained by meteor radars at Juliusruh (full lines) and Kiruna (dashed lines). Winds above 100 km (thin lines) should be treated cautiously due to possible electric field influences.

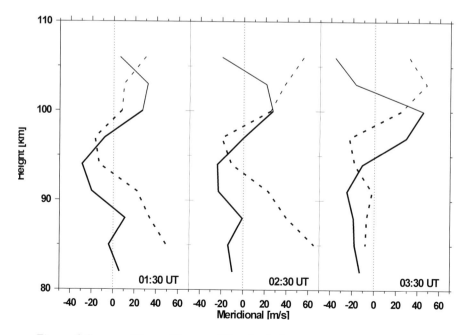

Figure 6. Same as Figure 5 but meridional wind.

The meridional winds are different below about 95 km with strong northward directed winds up to 70 m/s at 85 km altitude at high latitudes. The winds above 100 km are given for information only and should be treated cautiously. At these altitudes the trail motion may be strongly influenced by electric fields and we can not say how real the data are.

4. Summary

Observations of winds in the upper mesosphere/lower thermosphere have been carried out at five radar sites in Middle Europe and Northern Scandinavia during the Leonid meteor storm of November 17/18, 1999 using meteor radars, MF radars, and a LF wind profiler. 48-hour records of hourly mean winds collected by meteor radars at 54.3°N and 67.9°N are presented for 85, 91, and 94 km altitude and are supplemented by MF radar observations as well as LF wind profiler data from the same geographical region.

The observations from the different experiments are in general agreement. Some of the details show differences, but the agreement between the zonal winds observed by the meteor and MF radars at 85 km is good. Similar results of meteor and MF wind comparisons based on a 3-years data set were reported by Hocking and Thayaparan (1997).

We observe eastward directed zonal winds of about 40 m/s above 90 km in mid- and high latitudes at the beginning of the activity maximum on November 18, 00 UT. The winds are decreasing and changing to westerlies between 03 and 04 UT. The meridional winds are of opposite direction at mid- and high latitudes with southward directed winds of 40 m/s at mid-latitudes and a reversal of the wind direction is again observed between 03 and 06 UT.

At both latitudes tidal wind oscillations are evident above 90 km with a strong semidiurnal tidal wind amplitude between 10 and 30 m/s and a diurnal tidal wind amplitude between 10 and 15 m/s.

Acknowledgements

The authors thank J. Weiss, D. Keuer and R. Latteck from the Institut für Atmosphärenphysik for operating the Skiymet radar at Juliusruh and the MF radars at Juliusruh and Andenes as well as D. Kürschner from the Institut für Geologie und Geophysik der

Universität Leipzig for performing LF drift observations. W.S. and P.H. express their gratitude to W.K. Hocking for most helpful support in software development for the SKiYMET radar. This study was partially supported by the Deutsche Forschungsgemeinschaft by contract BR 2023/1-1. *Editorial handling:* Mark Fonda.

References

Hocking, W.K. and Thayaparan, T.: 1997, *Radio Sci.* **32**, 833–865.
Hocking, W.K., Fuller, B., and Vandepeer, B.: 2000, *J. Atmos. Solar-Terr. Phys.*, in press.
Kremp, Ch., Berger, U., Hoffmann, P., Keuer, D., and Sonnemann, G.R.: 1999, *Geophys. Res. Lett.* **26**, 1279–1282.
Schminder, R., Kürschner, D., Singer, W., Hoffmann, P., Keuer, D., and Bremer, J.: 1997, *J. Atmos. Terr. Ph.* **59**, 2177–2184.
Singer, W., Keuer, D., and Eriksen, W.: 1997, In *Proc. 13th ESA Symp. on European Rocket and Balloon Programmes and related Research*, **ESA SP-397**, 101–103.
Singer, W., Molau, S., Rendtel, J., Asher, D.J., Mitchell, N.J., and von Zahn, U.: 2000, *MNRAS.*, in press.

OBSERVATION AND INTERPRETATION OF METEOROID IMPACT FLASHES ON THE MOON

LUIS R. BELLOT RUBIO
Instituto de Astrofísica de Canarias, La Laguna, Tenerife, Spain
E-mail: lbellot@ll.iac.es

JOSE L. ORTIZ
Instituto de Astrofísica de Andalucía, CSIC, Granada, Spain
E-mail: ortiz@iaa.es

PEDRO V. SADA
Universidad de Monterrey, Monterrey, México

(Received 21 June 2000; Accepted 8 August 2000)

Abstract. The first unambiguous detection of meteoroids impacting the night side of the Moon was obtained during the 1999 Leonid storm. Up to eight optical flashes were recorded with CCD video cameras attached to small telescopes on November 18, 1999. Six impacts were videotaped by at least two independent observers at the same times and lunar locations, which is perhaps the strongest evidence for their collisional nature. The flashes were clearly above the noise and lasted for less than 0.02 s. Although previous observational efforts did not succeed in detecting impact flashes, additional candidates have been reported in the literature. The evidence accumulated so far implies that small telescopes equipped with high speed cameras can be used as a new tool for studying meteoroid streams, sporadic meteoroids, and hypervelocity collisions. In this review we discuss the various intervening parameters for detectability of flashes on the night side of the Moon (geometrical effects, contamination by scattered light from the day side, and properties of the meteoroids such as speed and flux of particles). Particular emphasis is placed on the analysis of the observations in order to derive relevant physical parameters such as luminous efficiencies, impactor masses, and crater sizes. Some of these parameters are of interest for constraining theoretical impact models. From a simple analysis, it is possible to derive the mass distribution of the impactors in the kg range. A more elaborate analysis of the data permits an estimate of the fraction of kinetic energy converted to radiation (luminous efficiency) if the meteoroid flux on the Moon is known. Applied to the 1999 lunar Leonids, these methods yield a mass index of 1.6 ± 0.1 and luminous efficiencies of 2×10^{-3} with an uncertainty of about one order of magnitude. Predictions of visibility of the major annual meteor showers are given for the next few years. These include the forthcoming 2001 Leonid return, for which we estimate detection rates in the visible.

Keywords: Hypervelocity impacts, Leonids 1999, luminous efficiencies, lunar craters, meteoroids, meteors, Moon

1. Introduction

The search for meteoroid impacts on atmosphereless bodies has attracted some interest because of its potential for deriving information on the physical properties of the impactors (chemical composition, density, structure, etc), their mass distribution and fluxes, and, even, the properties of the target surface material. Planets and satellites may be regarded as huge detectors whose collecting areas permit the observation of very large meteoroids in much less time than that required by ground-based monitoring of the Earth's atmosphere. In this regard, the Moon is the first natural body to scrutinize, not only because it is the object closest to Earth, but also because the meteoroid population in the neighborhood of the Moon is reasonably well known from the observation of terrestrial meteors. This implies that results from lunar impacts can be directly compared with results from more conventional techniques.

Meteoroids impacting the Moon give rise to a variety of essentially different phenomena that allow their detection. These include seismic waves, enhancements of the tenuous lunar atmosphere, and light flashes. Analysis of data from the Apollo lunar seismic network (Oberst and Nakamura, 1991) established that the 1974 Leonid shower produced signals consistent with impacts of meteoroids in the mass range from 0.1 to 1 kg. Other meteor showers could also have been detected by this network. Transient enhancements in the constituents of the tenuous lunar atmosphere may reveal the occurrence of meteoroid impacts. The Moon cannot retain any gaseous species for a long time, so continuous resupply is necessary. Impact-driven vaporization has been proposed as the most likely source of sodium and potassium in the lunar atmosphere, and indeed significant increases in lunar sodium were detected at the time of the 1997 and 1998 Leonids (Hunten *et al.*, 1998; Verani *et al.*, 1998; Wilson *et al.*, 1999). However, monitoring of the Moon during other meteor showers such as the Quadrantids did not reveal any variation of sodium in the atmosphere (Verani *et al.*, 1999).

It has been known for some time that impacts of meter-sized bodies on the Moon should cause optical flashes detectable with photometer technology (Melosh *et al.*, 1993), but the population of objects that are big enough is low and no unambiguous impact flashes have been recorded. Inspired by the successful detection of comet Shoemaker-Levy 9 collision with Jupiter in 1994, a systematic search for fainter events on the Moon was started in 1997 using more sensitive techniques based on CCD technology (Ortiz *et al.*, 1999). Although no impacts were unambiguously detected, it was stressed that a small 0.25 m telescope could easily observe flashes from meteoroids releasing energies well below 5×10^6 J in the visible range. It was also pointed out that the 1999 Leonid storm would provide a unique opportunity to record the optical flashes associated with meteoroids impacting the Moon due to the greatly enhanced flux of particles expected and the favorable geometrical conditions of the encounter. Other attempts to detect lunar impact flashes include the ALPO's Lunar Meteor Search program from 1955 to 1965 (see Westfall 1997) and the optical transient survey of the Moon conducted by Beech and Nikolova (1999). Unfortunately, none of these efforts led to unequivocal observations of light flashes on the Moon.

The first unambiguous detection of lunar impact flashes was indeed obtained during the 1999 Leonid shower. These observations open the door to the remote sensing of objects and physical processes that would be difficult to observe otherwise. In particular, lunar impact flashes may provide important information on the physics of hypervelocity impacts. Experimental work has been carried out for low velocity collisions (e.g., Schultz, 1996, Kadono and Fujiwara, 1996), but high velocity impacts such as those involving meteoroids are much more difficult to reproduce in the laboratory. For this reason, knowledge of the characteristics of such events is primarily based on numerical simulations. Most of these studies consider impactors of asteroidal composition and, therefore, their results are not directly applicable to collisions involving cometary material. The analysis of real lunar impacts makes it possible to estimate key parameters that help constrain numerical simulations.

In this review we address the observational aspects of meteoroid impact flashes on the Moon. Section 2 deals with the detection of lunar flashes during the 1999 Leonids. The interpretation of the observations is the subject of Section 3. We describe how to estimate luminous efficiencies, mass distribution indices, impactor masses, and crater sizes. In Section 4, some results from numerical simulations of hypervelocity impacts are presented. Section 5 deals with the various conditions for the visibility of meteor showers on the Moon. Finally, Section 6 is devoted to calculating impact detection rates for the major annual meteor showers, with emphasis on the forthcoming Leonid showers. Hints for successful observations are given throughout the paper.

2. Observation of the 1999 lunar Leonids

Numerical simulations of the Leonid stream evolution (McNaught and Asher, 1999, Brown, 1999) suggested the possibility of storm level activity from the shower on November 18, 1999 at the time when the Earth was to cross the dust trail generated by 55P/Tempel–Tuttle in 1899. Ortiz et al. (1999) had previously pointed out that geometric conditions would be favorable during the night of maximum activity. These expectations prompted several groups to monitor the night side of the Moon in search for Leonid impacts. Soon after the 1999 Leonid

Table I.

Impact number	Time (UTC)	Magnitude	Lunar position Latitude	Longitude	Observers
1	03:05:44.89	+5	40 ± 1 N	65 ± 1 W	DP, DD
2	03:49:40.38	+3	3 ± 1 N	48 ± 1 W	DP, DD, PS, RF
3	04:08:04.10	+5	15 ± 1 S	78 ± 1 W	DP, DD
4	04:32:50.79	+4	21 ± 3 N	51 ± 3 W	PS
5	04:34:49.52	+7	21 ± 3 N	38 ± 3 W	PS
6	04:46:15.52	+3	14 ± 1 N	71 ± 1 W	BC, DD
7	05:14:12.92	+6	15 ± 1 N	58 ± 1 W	PS, DD
8	05:15:20.92	+5	21 ± 1 N	59 ± 1 W	PS, DD

storm materialized on Earth, reports on the detection of lunar flashes were issued (Dunham, 1999). Although several 1 m telescopes were scheduled for observing the Moon at Calar Alto Observatory, Sierra Nevada Observatory, and Teide Observatory (all three in Spain), bad weather or technical problems prevented their use. Fortunately, positive observations came from smaller telescopes operated by B. Cudnik (0.36 m aperture, Columbus, TX, USA), D. Dunham (0.13 m, Mount Airy, MD, USA), R. Frankenberger (0.2 m, San Antonio, TX, USA), D. Palmer (0.13 m, Greenbelt, MD, USA), and P.V. Sada (0.2 m, Monterrey, México). At least eight impact flashes were videotaped by the last four observers, all using CCD cameras attached to their telescopes. For a complete description of the observations, we refer the reader to Dunham et al. (2000) and Ortiz et al. (2000). Observers used different, sometimes overlapping, fields of view. This turned out to be useful for confirming impact flashes, but implies that the individual observations cannot be combined into a single analysis due to the different lunar areas monitored.

Table I summarizes the observational data for the eight light flashes found by visual inspection of the tapes. The last column gives the initials of the observers who recorded the flashes. Their maximum magnitudes (in the wavelength range 0.4 to 0.9 μm) were obtained from comparison with the signals of reference stars and should be accurate to within $\pm 1^{m}$. The selenographic locations of the flashes in Dunham's records were determined by fitting the limb, and are uncertain by $\pm 1°$. The locations of the impacts detected by Sada were calculated by means of interpolation in time between two recognizable lunar features that drifted within the field of view due to inaccurate tracking of the telescope. The events summarized in Table I were very brief. They are

Figure 1. Half-frame images of the five light flashes recorded by P.V. Sada on November 18, 1999 (flashes 2,4,5,7, and 8 in Table I, respectively). These are 53 × 53 arcsec enlargements of the original 8 × 6 arcmin field of view.

mainly seen in half-frames (0.0167 s), the brightest flashes showing a much fainter (typically 3–4 mag) afterglow in the following half-frame.

All the impact flashes, except 4 and 5, were videotaped by at least two independent observers. Both times and selenographic positions are coincident, making a strong case that the flashes are indeed the result of Leonids colliding with the Moon. Alternative explanations not related to meteoroid impacts include cosmic rays and specular reflection of sunlight from artificial satellites or space debris. Cosmic rays can be ruled out because they usually affect a few pixels of the detector, whereas the observed flashes span a larger detector area. Another proof that cosmic rays were not responsible for the flashes comes from the fact that Sada's telescope was somewhat defocused at the time of the two last impacts, with the result that the central obstruction of the secondary mirror is clearly seen in the images (Fig. 1). This feature is very difficult to explain in terms of cosmic rays hitting the detector. On the other hand, the flashes occurred close to local midnight with the Moon at high altitude above the local horizon, which strongly suggest that the events are not due to objects in low orbit around the Earth. Moreover, Dunham *et al.* (2000) note that none of the known geosynchronous satellites were near the Moon at the time of the observations. These considerations, together with the fact that the lunar Leonids peaked at about 04:02 UT according to numerical calculations (Asher, 1999), provide compelling evidence for the impact origin of the flashes.

3. Analysis of the observations

High speed collisions such as those of meteoroids striking the Moon are difficult to reproduce in the laboratory because we still do not have means of accelerating the required masses to velocities typical of meteoroids. As a result, the physics of hypervelocity impacts is studied through both numerical simulations and scaling of low-velocity experiments. The analysis of lunar flashes makes it possible to improve this situation by providing empirical values of key parameters describing these events. In addition, light flashes allow us to monitor the meteoroid population in a mass range hitherto unreachable from Earth by conventional techniques. All these advances hinge on a relatively accurate knowledge of the properties of the impactors, most notably their velocities and fluxes on Earth. This section is devoted to providing examples of the capabilities of the analysis of lunar flashes by focusing on the inference of luminous efficiencies, the mass distribution index of the particles, impactor masses, and crater sizes.

3.1. DETERMINATION OF LUMINOUS EFFICIENCIES

For several reasons, a key parameter in hypervelocity impacts is the luminous efficiency η —the fraction of the initial kinetic energy converted into radiation. Knowledge of this parameter allows us, for example, to estimate impactor masses. It also permits the inference of meteoroid bulk densities by constraining theoretical impact models. Prior to 1999, the emphasis of numerical simulations was on particles of asteroidal composition moving at several km s^{-1}. Depending on the properties of the projectile and target material (chemical composition, impact velocity, etc), the resulting luminous efficiencies varied from 10^{-5} to 10^{-3} (Nemtchinov et al., 1998b). Very few simulations had been carried out for particles of cometary composition, and these were invariably restricted to small velocities. Under such conditions, no reliable estimate of η in impacts involving Leonid meteoroids was available prior to the 1999 return of the shower. In addition, theoretical models had suggested that, for fixed velocities and meteoroid bulk densities, luminous efficiencies might depend on impactor mass, incidence angle, and even the lunar relief (Nemtchinov et al., 1998a).

Clearly, theoretical models may benefit from luminous efficiencies derived empirically. Under certain conditions, it is possible to infer reliable values of η from the analysis of optical flashes on the Moon. This was done for the first time by Ortiz et al. (2000) and Bellot Rubio, Ortiz, and Sada (2000). The basic idea is that the observed cumulative number of impacts within the field of view will match the expected number only when the true luminous efficiency is used to calculate the latter. For this method to work, it is necessary to know the meteoroid flux on the Moon at the time of the observations.

The cumulative flux distribution of meteoroids of mass m is given by

$$F(m) = F(m_0) \left[\frac{m}{m_0}\right]^{1-s}, \qquad (1)$$

where $F(m)$ represents the flux of particles whose mass is higher than m, m_0 is the mass of a shower meteoroid producing a (terrestrial) meteor of magnitude +6.5, and s is the so-called mass index. For most meteor showers, both $F(m_0)$ and s are well known from visual observations on Earth.

Substituting m in Eq. 1 by $2E/V^2$, with V the meteoroid velocity, the cumulative flux of particles as a function of their kinetic energy E can be written as

$$F(E) = F(m_0) \left[\frac{m_0 V^2}{2}\right]^{s-1} E^{1-s}. \qquad (2)$$

On the other hand, the energy per unit area reaching the Earth can be approximated by

$$E_\mathrm{d} = \frac{E}{f \pi R^2} \eta, \qquad (3)$$

where η is the luminous efficiency, and R is the Moon–Earth distance. The coefficient f describes the degree of anisotropy of light emission. It should be 2 if light is emitted isotropically from the surface, or 4 if light is emitted from very high altitude above the Moon's surface.

The number of events above an energy per unit area E_d reaching the telescope in a time interval Δt can therefore be expressed as

$$N(E_\mathrm{d}) = \int_{t_0}^{t_0 + \Delta t} F(m_0, t) \left[\frac{2 f \pi R^2}{\eta m_0 V^2}\right]^{1-s} E_\mathrm{d}^{1-s} A \, dt, \qquad (4)$$

where A is the lunar area perpendicular to the radiant direction within the field of view. $N(E_\mathrm{d})$, which depends on η, is the quantity to be compared with the observations.

Figure 2 presents the results of this method applied to the 1999 lunar Leonids (Bellot Rubio, Ortiz, and Sada, 2000). The flux profile entering the calculations is taken to be gaussian in shape with a peak of 10 km^{-2} hour^{-1} and a FWHM of 45 min. The mass of a Leonid meteoroid producing a meteor of magnitude +6.5 is $m_0 = 2 \times 10^{-8}$ kg according to Hughes (1987). For the mass index we use $s = 1.83$ in the magnitude range -1 to $+6$ and $s = 1.87$ for brighter meteoroids. Inspection of Fig. 2 reveals that the observational data at the high-energy end are best matched by a luminous efficiency $\eta = 2 \times 10^{-3}$. We estimate this value to be uncertain by an order of magnitude or

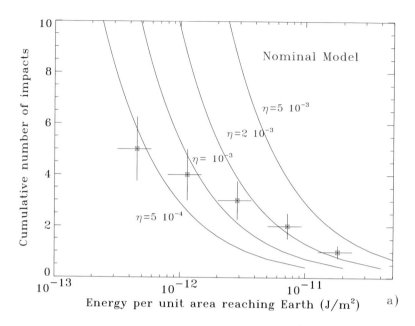

Figure 2. Comparison between observed (open squares) and expected (solid lines) cumulative number of 1999 Leonid impacts as a function of the energy received on Earth for several luminous efficiencies. From Bellot Rubio *et al.* (2000). Note that some of the faintest flashes may have been missed (see text for details).

less (see the discussion in Bellot Rubio *et al.*, 2000). In interpreting Fig. 2 it is necessary to bear in mind that some faint flashes may have been missed as they occurred near the sensitivity limit of the camera. This probably accounts for the deviation of experimental points with respect to the $\eta = 2 \times 10^{-3}$ curve at the lower-energy end. However, such deviation might also reflect a possible variation of η with mass, the smaller impactors converting less kinetic energy into light and vice versa. Unfortunately, the small number of flashes available for analysis implies that the issue of a possible mass dependence of η cannot be settled at the moment.

The 1999 lunar Leonids demonstrate the feasibility of estimating luminous efficiencies from real impacts. The value obtained so far, $\eta \sim 2 \times 10^{-3}$, applies only to Leonid meteoroids because η may be highly dependent on velocity. It would be desirable to carry out the same analysis for other meteor showers and impact geometries in order to investigate the dependence of η on parameters such as velocity and incidence angle. Other investigations appear to require similar or slightly larger luminous efficiencies for explaining an additional impact flash that might have occurred on the Moon in July 1999 (Ortiz *et* al., 2000), but lack of knowledge of the meteoroid velocity complicates the interpretation to a large extent. The advantage of monitoring the major annual meteor showers on the Moon is that the meteoroid velocity is known from Earth-based photographic observations. For reliable inferences of the luminous efficiency, a statistically significant number of events need to be accumulated. Systematic campaigns involving telescopes of various sizes can certainly provide the necessary observations.

3.2. Determination of the meteoroid mass distribution

The Moon, as a huge collecting area, permits the detection of very large particles in reasonable time intervals. This makes the characterization of such particles possible, thereby extending our knowledge of the properties of meteoroid streams to the high-mass end. From sufficient observational data it would be possible, for instance, to determine the mass of the largest particles present in the dust trails that give rise to meteor showers on Earth. Whipple's (1951) comet model provides an estimate of this limit as a function of certain comet and meteoroid parameters (see also Jones, 1995), but testing this formulation empirically has proved difficult. Not only can the upper mass limit be obtained from the analysis of lunar impacts, but also the mass index s describing the meteoroid population according to Eq. 1. Very remarkably, the inference of s is independent of η provided the luminous efficiency does not vary with m. This makes it possible to estimate the mass index directly from the observations *without any explicit knowledge of* η. Such a mass index may be necessary to evaluate Eq. 1 if the indices derived from visual observations do not apply to the larger lunar impactors.

Equation 1, with the help of Eq. 3, can be rewritten as

$$F(m) = F(m_0) \left[\frac{\eta m}{\eta m_0}\right]^{1-s} = F(m_0) \left[\frac{E_d(m)}{E_d(m_0)}\right]^{1-s}. \qquad (5)$$

By taking the logarithm of the above expression we arrive at

$$\log F(E_d) = C + (1-s) \log E_d, \qquad (6)$$

where C embodies constants that are not relevant for the analysis. From this equation it is clear that the logarithm of the observed cumulative number of impacts vs the logarithm of the energy received on Earth can be fitted by a straight line whose slope yields the mass index s. Caution must be taken in the analysis because the probability of detection of flashes decreases with E_d. In particular, the true number of faint events will be larger than the observed one. At present it is difficult to estimate

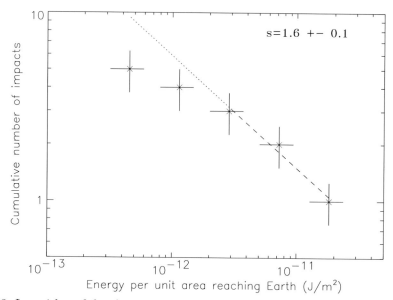

Figure 3. Logarithm of the observed cumulative number of impacts vs. the logarithm of E_d for the 1999 lunar Leonids. The straight line is the best fit to the points with $E_d > 2 \times 10^{-12}$ J m^{-2}. The statistics are poor, but the cumulative number of impacts is reasonably well described by a mass index $s = 1.6 \pm 0.1$.

such detection probabilities, but a rigorous treatment must include a proper correction, much as it is done in the analysis of visual meteor observations (e.g., Koschack and Rendtel, 1990).

Figure 3 shows the logarithm of the cumulative number of impacts as a function of $\log E_d$ for the 1999 lunar Leonids. In order to minimize the effect of our lack of knowledge of detection probabilities, the fit is performed only for the brightest events ($E_d > 2 \times 10^{-12}$ J m^{-2}), that is, those that cannot be missed because of their high signal-to-noise ratios. Despite the small number of events available, the distribution is well described by a mass index $s = 1.6 \pm 0.1$. This value is somewhat smaller than the index $s = 1.83$ derived from the 1999 Leonid fireballs in the magnitude range -1 to -6 (Arlt, *et al.*, 1999), but the agreement is remarkable in view of the limited data set on which our analysis is based.

3.3. Impactor masses

Once luminous efficiencies are known, impactor masses can readily be obtained from Eq. 3 by inserting the measured value of E_d and noting that $E = mV^2/2$. The mass of the Leonid meteoroids that produced the brightest lunar flash in 1999 turns out to be 4.9 kg if $\eta = 2 \times 10^{-3}$ is assumed (Bellot Rubio, Ortiz, and Sada, 2000). Arguments supporting the view that impactor masses are uncertain by less than a factor of 10 have been given by these authors. Such particles correspond roughly to terrestrial Leonids of magnitude -10 according to Hughes (1987).

An additional, alternative method can be envisaged for determining the total mass of the meteoroids striking the Moon. This technique consists in the monitoring of the lunar sodium during moments of high meteoroid activity to search for changes in the Moon's tenuous sodium atmosphere, which is believed to be partially maintained by impact-driven vaporization of surface material (see, for example, Morgan *et al.*, 1989, and references therein). The usefulness of this method is somewhat dependent on reliable models of the impact process as well as on a detailed treatment of the dynamical evolution of sodium in the lunar atmosphere. In spite of these difficulties, however, the method shows great promise, as transient enhancements of atmospheric sodium have already been detected at the time of the 1997 and 1998 Leonid showers (Hunten *et al.*, 1998; Verani *et al.*, 1998; Wilson *et al.*, 1999).

3.4. CRATER SIZES

Knowledge of impactor masses makes it possible to estimate crater sizes. Although no experiments on hypervelocity impacts involving meteoroids such as those striking the Moon have been conducted, the results of more conventional experiments can be scaled for a prediction of crater diameters resulting from lunar impacts. Gault's (1974) scaling law for craters up to 100 m in diameter in regolith reads

$$D = 0.25 \, \rho_p^{1/6} \, \rho_t^{-1/2} \, g^{-0.165} \, W^{0.29} \, \sin^{1/3}\theta, \tag{7}$$

where D is the (transient) crater diameter, ρ_p and ρ_t are the projectile and target bulk densities, respectively, g is the gravitational acceleration, W is the impactor's kinetic energy, and θ the incidence angle with respect to the vertical (all in mks units). For the Moon, appropriate values are $g = 1.67$ m s^{-2} and $\rho_t = 3000$ kg m^{-3}. Another estimate has been provided by Schmidt and Housen (1987):

$$D = \gamma^{-0.26} \, m^{0.26} \, V^{0.44}, \tag{8}$$

where m and V are the mass and speed of the impactor, respectively, and

$$\gamma = 0.31 \, g^{0.84} \, \rho_p^{-0.26} \, \rho_t^{1.26} \, (\sin 45/\sin \theta)^{1.67}. \tag{9}$$

As pointed out by Melosh (1989), these formulae result in similar crater diameters in spite of their different origins. This is especially true for small energy events, i.e., the case of meteoroids impacting the Moon. However, they must be applied with care because they were obtained for intermediate impact velocities of 10–20 km s^{-1}.

Assuming a bulk density $\rho_p = 1000$ kg m^{-3}, the above expressions lead to crater diameters of 11 and 32 m, respectively, for the biggest 1999 Leonid impactor ($m = 4.9$ kg). With $\rho_p = 100$ kg m^{-3}, one finds 7 and 27 m, respectively. Craters of this size are well below the resolution capabilities of telescopes on Earth, but may be detected on high resolution images by spacecraft orbiting the Moon. Three such missions are scheduled for 2002 and 2003: ESA Smart 1 (about 50 m/pixel resolution), ISAS Lunar A, and ISAS Selene. Only in exceptional cases should we expect larger craters, as the diameter is mainly determined by the velocity of the impactor and the Leonids possess the highest speed among the various meteoroid streams.

4. Numerical simulations

The detection of Leonid flashes on the Moon has triggered some very recent impact modeling efforts. Contrary to previous simulations, the new ones include the basic properties of meteoroid particles (comet-like composition, low density, high velocity, etc). The work of Artemieva, Shuvalov, and Trubeskaya (2000) deserves special mention. These authors simulated the vertical collision of Leonid particles on the Moon by means of a 2D hydrodynamical code. Vertical instead of oblique incidence was assumed on the basis of previous simulations where the luminous efficiency did not vary much with the entry angle.

According to the results of Artemieva et al. (2000), the flashes are mainly the result of thermal emission from hot plasma plumes created by vaporized meteoroid and target material. The whole process takes place on a very short time interval, of the order of 10^{-3} s. The first stage is characterized by the plasma being optically thick. The temperature drops rapidly and the gas becomes optically thin, leading to increased radiation fluxes. Artemieva et al. (2000) find some evidence that meteoroid bulk densities of 100 kg m^{-3} are to be preferred with respect to 1000 kg m^{-3} in order to explain the observed duration of the flashes. By integrating the radiative flux over time, they obtain theoretical luminous efficiencies of 10^{-3} for 1000 kg m^{-3} particles and 2×10^{-3} for 100 kg m^{-3} meteoroids. Moreover, the luminous efficiency is found to vary little (to within 10–20%) with impactor mass.

The uncertainties in this kind of simulation may be reduced to some degree by observational input. For example, the short duration of the flashes suggests that Leonids are very low density meteoroids. Larger densities would lead to longer durations that are not consistent with the observations. Another example comes from the remarkable similarity between the luminous efficiencies resulting from the Artemieva et al. simulations and the analysis of the 1999 lunar flashes by Bellot Rubio et al. (2000). This agreement suggests that the experimental value of η is essentially correct (at least for the Leonid lunar impacts), which in turn validates the results of the numerical calculations.

Impact flashes contain much more information than can be extracted at the present time. The shape of the light curve is determined by, among other factors, the chemical composition of the lunar soil and the meteoroid. Obtaining such light curves with sufficient temporal resolution would allow us to infer these compositions, but this is a difficult observational endeavor due to the extremely short duration of the flashes. Spectroscopic analyses of the radiation generated during impacts are also of great interest for determining chemical compositions. Advances in these directions can be expected in the future as new instrumentation becomes available.

5. Detectability of impacts on the Moon

The probability of detecting optical flashes on the Moon at the time of a meteoroid shower is determined by several factors, among them geometrical conditions (position of the subradiant point on the Moon and lunar phase), the specifics of the observational technique (telescope optics and background illumination from the day side of the Moon), and properties of the meteoroids themselves (such as particle speed and spatial density). In this section we describe these contributing factors in some depth.

5.1. Geometric considerations

Not all meteoroids striking the Moon can be observed from Earth. First, it is necessary that the subradiant point position on the Moon allow impacts to occur on the lunar hemisphere facing the Earth. This condition is always fulfilled unless the radiant lies at selenographic longitude 180° and latitude 0°. However, the lunar area perpendicular to the meteoroid direction (the quantity determining the efficiency of the Moon as a particle collector) will decrease with increasing angular distance of the subradiant point to the Earth-facing hemisphere. Second, it is necessary that impacts take place on the night side of the Moon as seen from Earth. This constraint stems from the need of a high signal-to-noise ratio for unambiguous detection of the light flashes.

5.1.1. Subradiant point on the Moon

For a given shower, the selenographic coordinates of the subradiant point (φ, λ) are calculated from the equatorial coordinates of the radiant (α, δ) as determined from Earth, the location of the Earth in its orbit with respect to the Sun (via the solar longitude $l_\odot$), and the location of the Moon with respect to our planet at the time of the shower's maximum activity. In order to simplify the calculations, we make the approximations that the inclination of the lunar orbit to the ecliptic is zero and that the rotation axis of the Moon points exactly toward the ecliptic north pole. These approximations are reasonable because the inclination of the mean lunar equator to the ecliptic is of the order of 1.5° and will allow us to compute the position of the Moon by using the lunar phase only. The first step is to transform the equatorial coordinates of the radiant to ecliptic coordinates (l, b) by means of standard formulae. Both l and b define the direction of the meteoroid trajectories in the vicinity of the Earth. The ecliptic coordinates of the Earth are given by $l_E = l_\odot + 180°$ and $b_E = 0°$. Finally, the position of the Moon with respect to the Earth is given by the lunar phase angle χ, defined to be zero at new Moon, 90° at first quarter, and so on. Our simplifying assumptions imply that the ecliptic latitude of the Moon is zero.

Figure 4 displays the encounter geometry. It is clear that the selenographic longitude of the subradiant point is $\varphi = l - l_E - \chi$, whereas the selenographic latitude must coincide with the ecliptic latitude of the radiant (i.e., $\lambda = b$) because of our approximation that the lunar equator has a zero inclination to the ecliptic. At this point we note that selenographic longitudes are measured from the central meridian counterclockwise as seen from the lunar north pole.

5.1.2. Lunar area subject to impacts

The total area of the night side of the Moon perpendicular to the meteoroid direction and visible from Earth, $A_\perp$, is a measure of the efficiency in detecting impacts. Other parameters have been proposed to indicate how favorable the encounter geometry is (e.g., Beech and Nikolova, 1998), but $A_\perp$ is more intuitive and useful for further calculations.

$A_\perp$ depends on the selenographic latitude λ and longitude φ of the subradiant point, as well as on the lunar phase. As before, we make the simplifying assumption that the inclination of the Moon's orbit to the ecliptic is zero. Given the complex geometry, a fast and efficient way to compute $A_\perp$ is Monte Carlo simulations. Let xyz be a reference frame with origin at the center of the Moon, the x-axis pointing towards the Earth, and the z-axis pointing towards the ecliptic north pole.

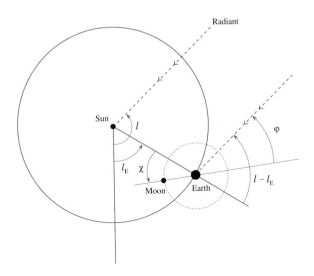

Figure 4. Geometry for the calculation of the selenographic coordinates of the subradiant point on the Moon. The Earth–Moon distance has been exaggerated for clarity (indeed, the ecliptic longitudes of the Moon and the Earth are the same for practical purposes). The selenographic coordinates refer to a cartesian coordinate system centered on the Moon with the x-axis pointing to the Earth and the z-axis to the ecliptic north pole. The meteoroid's direction is indicated by the dashed lines.

This system defines the selenographic coordinates of any point on the Moon's surface. Meteoroids strike the Moon homogeneously distributed in planes perpendicular to the radiant direction. Hence, we define the auxiliar coordinate system XYZ by rotating the xyz system until the z-axis points toward the radiant. This is equivalent to a rotation of angle φ around the z-axis and a rotation of angle $\theta = \pi/2 - \lambda$ around the y-axis. In this coordinate system, we generate a sufficiently large number of particles uniformly distributed in the XY plane and such that their coordinates verify the condition $X^2 + Y^2 \leq 1$. Next, the corresponding (positive) Z coordinates are obtained by means of the equation of a sphere. The set of XYZ coordinates represents the locations of particles impacting the Moon. In order to compute $A_\perp$, it is necessary to know how many such particles are visible from Earth. To that aim, the XYZ coordinates are transformed back to the xyz system,

$$\begin{pmatrix} x \\ y \\ z \end{pmatrix} = \begin{pmatrix} \cos\theta\cos\varphi & -\sin\varphi & \sin\theta\cos\varphi \\ \cos\theta\sin\varphi & \cos\varphi & \sin\theta\sin\varphi \\ -\sin\theta & 0 & \cos\theta \end{pmatrix} \begin{pmatrix} X \\ Y \\ Z \end{pmatrix} \qquad (10)$$

and their selenographic longitudes calculated. Only those particles lying on the night side of the Moon (i.e., whose longitudes are between that of the terminator and the non-illuminated limb) are counted. The number of such particles over the total number of particles, multiplied by πR^2 (with R the Moon's radius), gives the lunar area of the night side of the Moon perpendicular to the radiant direction. Note that optimum geometric conditions ($\lambda = 0$, $\varphi = 0$, and new Moon) imply $A_\perp = \pi R^2$. This is the maximum cross-section of the Moon as a particle detector. Obviously, the larger the value of $A_\perp$, the better the observability of the shower from Earth. Throughout we have assumed the meteoroid velocity is big enough (> 40 km s^{-1}) so that no gravitational deflection and focusing by the Moon or the Earth occurs. This is actually the case for all meteoroid streams whose observation is of interest.

The method used to compute $A_\perp$ can be extended to derive the lunar area perpendicular to the meteoroid direction within the field of view, A, which is necessary for the calculation of the expected cumulative number of impacts (Eq. 2) for a given telescope setup. Both the position and size of the camera field of view need to be considered for calculating A, but this is the only modification required to apply the above procedure.

Increasing the detection rate may be achieved by centering the field of view as close as possible to the subradiant point on the Moon while keeping the terminator at the greatest distance possible, since this will increase A. If the subradiant point lies on the hidden lunar hemisphere (a situation often met), then the angular distance of the center of the field of view to the subradiant point should be minimized. In that case, observations near the limb are recommended.

5.2. Signal-to-noise considerations

The signal-to-noise ratio (SNR) for the detection of an impact flash in a single detector element is, approximately,

$$\frac{S}{N} = \frac{0.5 m V^2 \eta \pi r^2 Q}{2\pi R^2 h\nu \left[\frac{0.5 m V^2 \eta \pi r^2 Q}{2\pi R^2 h\nu} + d_c(t) + M_b(t) + R_n^2\right]^{1/2}} \quad (11)$$

with η the luminous efficiency, R the Earth–Moon distance, m the meteoroid mass, V the meteoroid speed, h Planck's constant, ν the frequency of the radiation, Q the quantum efficiency of the detector, M_b the Moon brightness (in electrons), R_n the read-out noise (in electrons), d_c the dark current (in electrons), r the telescope aperture radius, and t the integration time.

According to Melosh et al. (1993), a single element detector such as a photomultiplier would be able to record impacts of meter-sized meteoroids. They quoted a threshold sensitivity of 10^{-6} W m^{-2} for a photomultiplier coupled to a 1 meter telescope. CCD arrays are far more sensitive because M_b is considerably smaller in each pixel (due to the small angular size of a single pixel compared to the photometer's aperture) and also because quantum efficiency is usually higher in CCDs. For a typical image scale of 1 arcsec/pixel, M_b is several orders of magnitude smaller using CCDs than using a photometer with a 10 arcmin aperture.

The background brightness is mainly due to scattered light from the Moon's day side, but it also has a contribution from the Earth-lit surface of the lunar night side, which is not completely dark. Scattered light from the day side depends on the atmospheric conditions at the observing site as well as on the optics of the telescope. For typical observing conditions, with a lunar illumination of 20–30 per cent, the brightness of the night side of the Moon is of order $m_v = 12$ mag arcsec^{-2}, but reaches brighter magnitudes as the lunar phase increases. For the 1999 Leonid campaign, it was close to 8 mag arcsec^{-2}.

The easiest way to increase the SNR is to decrease M_b. This can be achieved by using short integration times. Since the light flashes are very brief (of the order of 0.02 s), reducing the exposure time results in better SNRs because the signal is not modified. Obviously, the larger the distance between the subterrestrial and subsolar points, the smaller the background brightness. This means that the best viewing conditions occur at new Moon, but the angular separation between the Moon and the Sun is too small to allow observations. The optimum viewing conditions are therefore those with the Moon being a few days before or after new Moon, with phase angles between 70° and 90° or between 270° and 290°.

In addition, the SNR might be higher in the near infrared than in the visible because more energy should be radiated as a result of the larger plume size when the plasma is cold enough to emit in the near IR; that is, the luminous efficiency should be higher. Another advantage of the infrared is that the scattered radiation from the day side of the Moon is lower than in the visible. The main drawback of infrared observations is the fast variation of the sky brightness, which often shows significant differences on time scales of the order of minutes.

5.3. OTHER CONSIDERATIONS

A number of additional factors influence the visibility of lunar impact flashes. First, it is clear that the likelihood of detecting impacts depends on the amount of energy released in the process. Impacts caused by meteoroids moving at high speeds will be much easier to detect simply

Table II.

Shower	Max	$l_\odot$ (deg)	Radiant coordinates				V	s	$F_{6.5}$
			α	δ	l	b			
QUA	Jan 4	283.16	230	49	201	63	41	1.91	0.03
ETA	May 5	45.5	338	−01	339	8	66	1.99	0.04
PER	Aug 12	139.8	46	58	62	39	59	1.95	0.07
ORI	Oct 21	208.0	95	16	95	−07	66	2.06	0.01
LEO	Nov 17	235.27	153	22	147	10	71	1.92	0.03
GEM	Dec 13	262.0	112	33	109	11	35	1.95	0.06

because the energy reaching Earth goes as the velocity squared. Second, it is necessary that the spatial density of meteoroids is sufficiently large to ensure that particles will strike the Moon during the observations. Not all meteor showers produce high fluxes of meteoroids, so monitoring of the Moon is preferable when the major annual showers peak on Earth.

Another consideration is integration time. Although the events are bright (the magnitude of the brightest flash during the 1999 Leonids was +3), the use of magnitudes can be misleading. Indeed, the optical flashes are very intense, but only *during extremely short time intervals*. It is therefore convenient to reduce the integration time as much as possible in order for the background not to hide the signal coming from the impact. In this regard, CCD video cameras or very fast readout CCDs are necessary for increasing the detection probability.

6. Estimating impact detection rates

In this section we use the previous results for examining the observability of a number of annual showers on the Moon during the next five years. The Quadrantids, η Aquarids, Perseids, Orionids, Leonids, and Geminids have been selected because of their high velocity and/or high flux of meteoroids. As mentioned before, high velocities imply that more energy is radiated, making the light flashes easier to detect from Earth. High fluxes mean more particles colliding with the Moon and higher likelihood of observing impacts. Special attention is paid to the forthcoming Leonid showers in view of the greatly enhanced fluxes expected in 2001 and 2002.

Table II summarizes basic observational parameters for the Quadrantids (QUA), eta Aquarids (ETA), Perseids (PER), Orionids (ORI),

Leonids (LEO), and Geminids (GEM) according to the IMO meteor shower working list (Rendtel et al., 1995). The second column gives the date of maximum activity. Solar longitudes ($l_\odot$, J2000.0) and radiant coordinates refer to this date. V is the meteoroid's velocity, s the mass index in the visual range, and $F_{6.5}$ the maximum flux of meteoroids brighter than magnitude +6.5 in km^{-2} h^{-1}.

From the values of s and $F_{6.5}$, it is possible to estimate the number of events detectable from Earth during one our of observing with different instruments. Table III shows the results for 0.2 m f/10 and 1 m f/2 telescopes (columns N_1 and N_2, respectively) assuming that the lunar area perpendicular to the meteoroid's direction within the field of view is $A \sim 10^6$ km^2. These figures have been computed according to Equation 2 with $\eta = 2 \times 10^{-3}$ and threshold energies taken from Fig. 5. It is very important to stress here that the number of events is strongly dependent on the value adopted for s. The mass indices quoted in Table II refer to the average value of s during the period of shower activity, but very often s decreases at the time of maximum activity. Smaller mass indices mean that large particles are more abundant, leading to increased detection rates. For this reason, values in Table III must be taken as rough lower limits. According to our estimates, one may expect of the order of 4–11 and 1–2 impact flashes per hour in the field of view during the Quadrantid, Perseid, and Geminid maxima with 1 m f/2 and 0.2 m f/10 telescopes, respectively.

In order to quantify the visibility of the showers in terms of $A_\perp$, the time of maximum activity is calculated from the solar longitude $l_\odot$ for each year from 2001 to 2005. The phase of the Moon at that time is also computed to derive the selenographic coordinates of the subradiant point. The results are presented in Table IV. The fifth and sixth columns give the selenographic latitude and longitude of the subradiant point at the time of maximum activity. χ is the lunar phase (0° for new Moon, 90° for first quarter Moon, 180° for full Moon, and 270° for last quarter Moon). $A_\perp$ (normalized to πR^2) varies from 0 (impacts not visible) to 1 (best geometrical conditions). For comparison, $A_\perp$ was $0.44 \pi R^2$ at the time of the 1999 Leonid shower maximum (with $l = 147.7°$, $b = 10.2°$, $l_\odot = 235.367°$, and $\chi = 111.8°$). The data for the 2001 and 2002 Leonids refer to the peak times on Earth predicted by McNaught and Asher (1999) plus the time needed by the Moon to reach the same ecliptic longitudes (2.5 h in 2001 and 0.5 h in 2002).

Favorable conditions will occur for the showers whose $A_\perp$ values are in bold type. Normally, showers with very high $A_\perp$ values occur near new Moon. They are of no interest because the Moon cannot be observed for a long time under dark skies. The best geometrical conditions occur for the Quadrantids in 2001, for the Perseids in 2002

Table III.

Shower	N_1	N_2	Shower	N_1	N_2
QUA	7.7	1.8	ORI	0.06	0.01
ETA	0.9	0.2	LEO	2.4	0.6
PER	4.1	0.9	GEM	10.7	2.3

and 2005, for the Orionids in 2001 and 2004, for the Leonids in 2001 and 2004, and for the Geminids in 2002. The visibility of η Aquarid impact flashes on the Moon is not good in any year except perhaps 2001.

6.1. EXPECTED DETECTION RATES FOR THE 2001 LEONIDS

According to McNaught and Asher (1999), the Earth–Moon system will cross in 2001 the dust trails generated by comet Tempel–Tuttle nine and four revolutions ago. Maximum activities are somewhat uncertain at the moment, but ZHR values of about 15000 for each trail have been suggested. The closest approach of the Earth to the two trails is predicted for November 18, 2001 at 17:31 and 18:19 UT, respectively. The Moon will reach the same ecliptic longitudes about 2.5 hours later. As can be seen in Table IV, the lunar phase will be 42°, while geometric conditions are rather favorable with $A_\perp = 0.50\,\pi R^2$. This makes the 2001 Leonid shower an excellent candidate for producing lunar impact flashes. Europe is badly placed because the maximum will occur at moonset or later, whereas Brazil is probably the best location for recording the peak with the Moon at a sufficient altitude above the horizon. Even if the Moon cannot be observed from a given location at the time of maximum activity, observations before or after the peak will be very valuable, as some impacts might still be detected (see Table III for detection rates of Leonid flashes under non-storm conditions). In any case, the Moon will be in a dark sky for a short time period, so infrared observations may be advantageous.

Figure 5 shows the cumulative number of impacts detectable from Earth per hour of observation of a lunar area of 10^6 km^2, as a function of the energy reaching Earth. The lunar area of 10^6 km^2 is in the direction perpendicular to the radiant; this area is close to that observed using a conventional CCD video camera attached to a 0.2 m, f/10 telescope aimed at the subradiant point on the Moon. The three curves are predictions for several luminous efficiencies (5×10^{-3}, 2×10^{-3}, and 10^{-3}). The vertical lines represent the sensitivity thresholds of

Table IV.

Shower	Year	Date	Hour (UT)	Subradiant λ	Subradiant φ	χ (deg)	$A_\perp$ $(\times \pi R^2)$
QUA	2001	Jan 3	12	63	1	97.3	**0.33**
	2002	Jan 3	18	63	−144	241.9	0.08
	2003	Jan 4	00	63	85	13.4	0.45
	2004	Jan 4	06	63	−49	146.8	0.16
	2005	Jan 3	12	63	−169	266.8	0.12
ETA	2001	May 5	23	08	−48	272.3	0.13
	2002	May 6	06	08	−178	53.7	0.00
	2003	May 6	12	08	57	188.7	0.30
	2004	May 5	18	08	−7	317.9	0.00
	2005	May 6	00	08	144	83.1	0.10
PER	2001	Aug 12	12	39	−170	272.3	0.04
	2002	Aug 12	18	39	48	53.7	**0.40**
	2003	Aug 12	23	39	−87	188.7	0.00
	2004	Aug 12	06	39	144	317.9	0.17
	2005	Aug 12	12	39	19	83.1	**0.38**
ORI	2001	Oct 21	09	−7	11	56.2	**0.69**
	2002	Oct 21	15	−7	−114	180.8	0.00
	2003	Oct 21	21	−7	114	313.0	0.30
	2004	Oct 21	03	−7	−26	92.6	**0.64**
	2005	Oct 21	09	−7	−160	226.1	0.00
LEO	2001	Nov 18	20	10	49	41.6	**0.50**
	2002	Nov 19	11	10	−83	173.1	0.06
	2003	Nov 18	02	10	170	282.3	0.01
	2004	Nov 17	08	10	28	64.4	**0.48**
	2005	Nov 17	14	10	−105	197.9	0.00
GEM	2001	Dec 13	23	11	38	348.4	0.89
	2002	Dec 14	06	11	−92	118.7	**0.42**
	2003	Dec 14	12	11	142	244.4	0.11
	2004	Dec 13	18	11	4	22.9	0.94
	2005	Dec 14	00	11	−133	159.3	0.11

different telescopes. Asterisks mark the predicted cumulative number of detections using the different instruments. Note that 1 m f/10 telescopes would record a smaller number of flashes because the total area comprised in the field of view is considerably smaller than with the other instruments. For these calculations a background brightness of

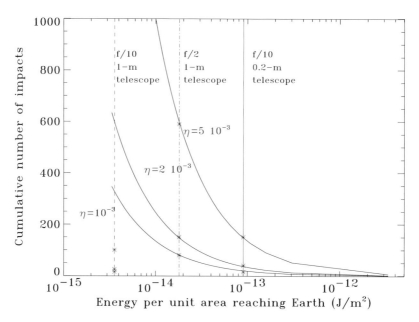

Figure 5. Expected number of detections of 2001 Leonids during one hour of observing for several telescopes and luminous efficiencies (solid lines). The sensitivity thresholds of the different telescopes are indicated by the vertical lines. They have been obtained assuming typical values of quantum efficiency, readout noise, etc. The actual number of detections depends on the field of view of the telescope, and is represented by asterisks. A Leonid flux of 10 km^{-2} h^{-1}, equivalent to ZHRs of about 30000 on Earth, and $s = 2.0$ have been used for the calculations.

12 mag arcsec^{-2} and a cumulative flux of 10 meteoroids km^{-2} h^{-1} with masses higher than 0.02 mg have been assumed. This flux would be equivalent to a zenithal hourly rate of roughly 30,000 on Earth. The calculations can be scaled to different areas and different fluxes by simply multiplying the curves by the appropriate factor. With small 0.2 m telescopes, up to 40 impact flashes can be expected within the field of view during one hour of observing. The number of flashes increases for 1 m telescopes at f/2. These may record up to 150 flashes in one hour if the flux of Leonid meteoroids reaches the predicted value.

7. Concluding remarks

The 1999 lunar Leonids have demonstrated that CCD cameras attached to telescopes of even 0.2 m in diameter can successfully detect light flashes of meteoroids impacting the Moon. Careful analyses of the observations provide a great deal of information on the physics of hypervelocity impacts and the properties of meteoroids and meteor streams. The new technique, however, still awaits full exploitation. Observations with different telescope setups and in different wavelength ranges will certainly increase the number of events available for analysis. Only when a sufficiently large database has been accumulated will the investigation of topics such as the dependence of the luminous efficiency on velocity and mass be possible.

Almost all meteor showers visible from Earth can be observed on the Moon. However, it is necessary that the flux of particles and the meteoroid velocity be large enough to ensure high detection rates. As a consequence, only the major annual showers deserve close scrutiny. We have described in detail the various conditions for the visibility of impact flashes on the Moon with a view to provide predictions for the next few years. The most promising showers are the Quadrantids in 2001, the Perseids in 2002 and 2005, and the Geminids in 2002. No doubt, the 2001 Leonid return will be the best candidate if the meteoroid flux turns out to be as high as expected. Concerning the 2002 Leonid shower, a nearly full Moon along with very bad geometric conditions will render any observational effort almost worthless.

Acknowledgements

We would like to thank David Martínez-Delgado and Terry Mahoney for their careful reading of the paper and Pere Lluis Pallé for his continuous support. This work has been partially funded by project ESP97-1773-C03-01 and by the Instituto de Astrofísica de Canarias under project 310400. *Editorial handling:* Peter Jenniskens.

References

Arlt, R., Bellot Rubio, L.R., Brown, P. and Gyssens, M.: 1999, *WGN, Journal of the IMO* **27**, 286–295.
Artemieva, N.A., Shuvalov, V.V., and Trubeskaya, I.A.: 2000, *Lunar Planet. Sci. Conf.* **31**, Abstract 1402.
Asher, D.J.: 1999, *IAUC*, 7320.
Beech, M. and Nikolova, S.: 1998, *Nuovo Cimento* **21C**, 577–581.
Beech, M. and Nikolova, S.: 1999, *Meteoritics & Plan. Sci.* **34**, 849–852.
Bellot Rubio, L.R., Ortiz, J. and Sada, P.V.: 2000, *Astrophys. Journal Letters*, submitted.
Brown, P.: 1999, *PhD Thesis*. University of Western Ontario, Canada.
Dunham, D.W.: 1999, *IAUC*, 7320.
Dunham, D.W., Cudnik, B., Palmer, D.M., Sada, P.V., Melosh, H.J., Beech, M., Frankenberger, R., Pellerin, L., Venable, R., Asher, D., Sterner, R., Gotwols, B.,

Wun, B., and Stockbauer, D.: 2000, *Lunar Planet. Sci. Conf.* **31**, Abstract 1547, 2000.

Gault, D.E.: 1974, in R. Greeley, P.H. Schultz (eds.), *A primer in lunar geology*, NASA Ames, Moffet Field, p. 137–175.

Hughes, D.W.: 1987, *Astron. Astroph.* **187**, 879–888.

Hunten, D.M., Sprague, A.L., and Cremonese, G.: 1998, *Bull. Am. Astron. Soc.* **30**, 1115–1115.

Jones, J.: 1995, *MNRAS* **275**, 773–780.

Kadono, T. and Fujiwara, A.: 1996, *J. Geophys. Res.* **101**, 26097–26109.

Koschack, R. and Rendtel, J.: 1990, *WGN, Journal of the IMO* **18**, 44–58 and 119-140.

McNaught, R.H. and Asher, D.J.: 1999, *WGN, Journal of the IMO* **27**, 85–102.

Melosh, H.J.: 1989, *Impact Cratering: A Geologic Process.*, Oxford University Press, New York, 120–121.

Melosh, H.J., Artemieva, N.A., Golub, A.P., Nemtchinov, I.V., Shuvalov, V.V., and Trubetskaya, I.A.: 1993, *Lunar Planet. Sci. Conf.* **24**, 975–976.

Morgan, T.H., Zook, H.A., and Potter, A.E.: 1989, *Lunar Planet. Sci. Conf.* **19**, 297–304.

Nemtchinov, I.V., Shuvalov, V.V., Artemieva, N.A., Ivanov, B.A., Kosarev, I.B., and Trubetskaya, I.A.: 1998a, *Astr. Vest.* **32**, 116–132.

Nemtchinov, I.V., Shuvalov, V.V., Artemieva, N.A., Ivanov, B.A., Kosarev, I.B., and Trubetskaya, I.A.: 1998b, *Lunar Planet. Sci. Conf.* **29**, Abstract 1032

Oberst, J. and Nakamura, Y.: 1991, *Icarus* **91**, 315–325.

Ortiz, J., Aceituno, F.J. and Aceituno, J.: 1999, *Astron. & Astrophys.* **343**, L57–L60.

Ortiz, J., Sada, P.V., Bellot Rubio, L.R., Aceituno, F.J., Aceituno, J., Gutiérrez, P.J., and Thiele, U.: 2000, Optical detection of meteoroidal impacts on the Moon, *Nature*, in press.

Rendtel, J., Arlt, R., and McBeath, A.: 1995, *Handbook for Visual Meteor Observers.*, International Meteor Organization, Potsdam (Germany), 280–281.

Schmidt, R.M. and Housen, K.R.: 1987, *Int. J. Impact Eng.* **5**, 543–560.

Schultz, P.H.: 1996, *J. Geophys. Res.* **101**, 21117–21136.

Verani, S., Barbieri, C., Benn, C., and Cremonese, G.: 1998, *Planet. Space Sci.* **46**, 1003–1006.

Verani, S., Barbieri, C., Benn, C., Cremonese, G., and Mendillo, M.: 1999, Observations of the Lunar Sodium Atmosphere during the 1999 Quadrantid Meteor Shower., *DPS Meeting 31*, Abstract 38.01

Westfall, J.E.: 1997, *Worthy of Resurrection: Two past ALPO Lunar Projects.*, ALPO Monograph No 7

Whipple, F.L.: 1951, *Astrophysical Journal* **113**, 464–474.

Wilson, J.K., Smith, S.M., Baumgardner, J., and Mendillo, M.: 1999, *Geophys. Res. Let.* **26**, 1645–1648.

SUBJECT INDEX

A

Δa - range in semi-major axis............159
A2 effect......................................153,161
ablation...110,481
    metallic species concentration.....482
    secondary 399,427
    vapor density..........................106,111
accretion of dust.................................336
aerosol loading................... 489,500,514
aerothermochemistry 65,79,425,454
afterglow
    leading edge descent..... 407,409,420
    optical spectrum.....................407,411
    spatial structure............................407
    temporal evolution................416,423
airglow
    definition 535
    link with shower activity 525,530
    OH 529,430
    optical intensity............................536
    sodium emission...................430,540
    spectrum, near-IR 528
    spectrum, optical...................432,540
air-meteoroid interaction 99,116
air-vapor cloud (see vapor cloud)
angular velocity..................................233
aqueous alteration 343
ARIA...3,7
astrobiology (overview)......................15
atmospheric
    ambient temperatures...................562
    chemistry, induced..... 65,79,425,454
    conditions at 95 km........................94
    extinction model 213,233
    heating.................................482,451
    pressure increase..........................482
    sciences (overview) 21
    transmission in near-UV..............393
atomic line emission
    intensities 420
    Leonid meteor..............................404
automated detection...... 44,173,223,251

B

B parameter ZHR curve..............195,301
beading...383
beginning height 183,280,299
    distribution 184
    Draconids 146
    increase with mass 185
blob...395
Boltzmann constant 420
brightness
    mid-IR..86
    optical................................ 61,88,178
buoyant air 468

C

C - cosmic 506
C I 65,73
    chondrite.......................................327
C_2 66,73,145
C-H stretch vibration 452
Ca I.. 412-416
Ca II 76,405,416
CAI inclusions 339
camera pointing..................................239
carbonaceous meteorites
    fraction of fall 60,327
carbonization......................................72
CH_4 444,452
chalcophile 340
Chapman airglow mechanism 480
chondrite...339
    melting point 331
chondrules..338
    composition..................................339
    diameter..341
cluster analysis...................................237
clustering in meteor incidence rate
    temporal (< 1 second)..................237
CN 16,66,72,131,376
    abundance in meteor plasma 77
    spectrum at 387 nm.......................72

SUBJECT INDEX

CO 65,451
    mid-IR emission 451
    sub-mm emission 133
CO_2 448,452,487
    mid-IR emission 450
collision model for vapor cloud
    Variable Hard Sphere 97
comet
    color 144
    composition 18,311,317,337
    influx 59
    orbit 55P/Tempel-Tuttle 151
    organic matter 72,333,144
    satellite model 149
comet dust
    boulders 343,583
    density 319,328
    mid-IR emission 440
    outflow speed 150,320
    polarization (optical) 143
    porosity 319,328
    size distribution 145
comet dust trail (see dust trail)
comet individual
    1P/Haley 318,518
    19P/Borrelly 319
    21P/Giacobini-Zinner 141,321
    55P/Tempel-Tuttle 321,440
    109P/Swift-Tuttle 518
    C/1995 O1 Hale-Bopp 320
cometesimals 342
communications 8,41
composition principal components .. 332
    carbon bearing ferromagn. 333
    carbonaceous 333
    ferromagnesiosilica 333
condensation experiments 497
continuous flow 111
    onset at altitude 111
continuum radiation
    mid-IR 453
    optical 409
control center Leonid MAC 9
counting tool 192
Cr I 405
crater sizes 585
cumulative flux (see flux)
CVF 441

D

delivery of organics 65
density
    comet dust 319
    meteoroid, IDP 327
diffusion coefficient 484
dispersion of meteoroids 153
distance to meteor 216,233,310
diurnal variation
    of atmospheric HCN 134
drag coefficient 481
DSMC technique 96,114
dustball model 110,188,362,386
    size distribution 363
dust trail
    number density 205,270
    cross section 191
    mass of largest particle present ... 585
    position shift 205
    width 202, 205,270

E

Earth limb
    distance to 310
echo duration 267
Einstein coefficients 420
ejection speed (see comet outflow)
electrophonic noise 545
ELF - extremely low frequency 546
ELF/VLF detections 549
elves
    definition 15
    seen from ARIA 13
    seen from FISTA 14
ending height 183,280,299
    change with mass 185
energy
    emitted by +0 magn. meteor 178
    excitation 405
    per unit area reaching Earth 581
    transfer during impact 112
evaporation rate 111
excitation energy 405
exobiology (overview) 15
exposure
    to interplanetary space 366,520
extinction, atmospheric 232,213

SUBJECT INDEX

F

F - skew parameter..........................356
f_m - mean anomaly factor.................158
false detection rate...........................175
Fe I..405,434,480
FeO (persistent train).........425,434,480
Fe(OH)$_2$..482
Filament (see Leonid Filament)
fireball
    all-sky image 1998.......................287
    classification.........................327,342
FISTA..3,7
flight path......................................4,5,49
flow field..63,96
    mean free path............................111
    near meteoroid 111,121
    number density.............100,102,114
    wake temperatures............62,99,108
flux (see also rate, mass influx):
    1997..311
    1998 fireballs........................288,293
    1999 storm........19,195,237,262,181
    1-minute counts...........................195
    comparison ground-airborne.......216
    function of azimuth.....................241
    function of elevation.....219,236,243
    function of magnitude.....20,258,272
    function of mass....................186,254
    height dependent (radar)..............568
    modeling......................................230
    near real time reporting....10,39,169
    peak time.........160,181,194,262,568
    profile shape.........................163,200
    second peak (1866 trailet)...........197
    temporal variations.......226,237,568
forward meteor scatter......................257
fragmentation
    grain sizes....................................362
    meteor............................362,382,345
    meteorite......................................345
free molecular flow...........................111

G

glue..360,387
grain model.......................................315
green line.......408,432,483,373,410,426
GSM..45

H

H I.............................16,72,125,406,533
H$_2$O..452,487
    atmosphere...................137,452,487
    comet.....................................318,320
    mid-IR emission..........................452
    submillimeter observations..........137
HALOE...137
HCN..131
    atmospheric abundance...............132
    influx..136
    spectroscopic parameters.............132
heat of sublimation...........................482
heat transfer coefficient....................482
heating.................................97,110,451
height
    mass dependence meteors...........186
    distribution..................................183
heliocentric distance of ecliptic
    crossing.......................................150
High-Definition TV.......................8,371
HPS - high pressure sodium.............435

I

ICCD...210
    photon counting mode (UV).......393
    quantum efficiency................372,393
IDP...331
images (see meteor images)
impacts on the Moon (see Moon)
influx (see flux)
initial radius..................63,102,107,113
INMARSAT..9
intense evaporation...........................111
intensity of a line..............................420
interplanetary dust particle (IDP)
    (see also meteoroid)
    aggregate IDP.............................337
    anual influx............................59,67
    bulk composition.........................337
    cluster IDP..................................337
    definition.....................................331
    density...328
    porosity.......................................328
interstellar dust
    model..315
    organic mantle material...............315

J

jet-like features 380,383

K

K I.........................411,414,418,555,560
Knudsen number.................................94

L

LEOC 36,169
Leonid Filament................. 199,302,306
Leonid MAC
    coordination ground observations.48
    definition...2
    flight trajectory 4,5,49
    layout of instruments 6
    participants.......................................6
Leonid mass 179,253
Leonid meteoroids
    in cosmic dust catalogs.................515
Leonid shower history
    1974 (impacts on Moon) 576
Leonids - individual
    00:37:15 UT beading....................383
    01:31:16 UT, Nov. 17 1998 473
    02:09:26 UT...................................75
    02:36:40 UT...................................75
    03:08:48 UT jets 384
    03:24:40 UT.................................372
    03:30:33 UT train 431
    04:00:29 UT.................. 401,443,457
        trajectory 401,462
        lightcurve 402,459
    10:05 UT train IR..........................446
    12:26 UT train IR..........................446
    17:47:06 UT, Nov. 17 1998 61
    17:54:08 UT, Nov. 17 1998 85
    18:08:47 UT, Nov. 17 1998 61
    20:24:40 UT, Nov. 17 1998 86
Leonid spectrum 76
    near-UV..76
    mid-IR 90
    optical................................ 61,374,404
    Very Low Frequency....................549
lidar
    altitude profiles 558
    K-trail density 559
    meteor trail altitude.....................558

lidar
    neutral Fe-layer 1998....................17
    neutral K-layer density profiles...560
light curve 87,259,351,402
    classical 352
    function of height..........................87
    sodium emission 360
    symmetry.....................................354
lightning discharge
    VLF spectrum 549
limiting meteoroid mass 179
line population 420
    non-equilibrium 63,422
Liquid Mirror Telescope..................250
lithification processes 330
lithophile 340
Lorentz profile 162,200
Low-Light Level TV camera............210
luminous efficiency
    optical emission Moon impacts...581
    optical emission of meteors.........178
luminous region (meteor)
    size................................. 63,79,102
lunar (see also Moon)
    area subject to impacts.................588
    impact light flashes..............575,579
Lunar A mission 586
Lunar Meteor Search program 577

M

Mach number 94
magnitude
    determination 177
magnitude distribution index
    (see also mass distribution)
    1999.......................... 20,258,252,235
    1998 fireballs 293
    1997..310
    1995-1999 (forward radio MS)...274
magnitude loss
    due to angular velocity 233
    due to distance 233
    due to extinction 232,213
MALC 40
mass of a Leonid meteoroid
    +6.5............................ 178,253,581
    as function of magnitude 179

SUBJECT INDEX

mass distribution index
   (see magnitude distribution index)
   definition 582
   1999 250,255,260,272,582
   1998 260
   1997 311
   1995-1999 274
   from echo duration........................272
mass influx 67
   current annual 59
MDS (Meteor Detection Site) 42
mean anomaly factor 158
mean free path....................................111
meridional wind 569
mesosphere...567
Meteor Alert Center.............................40
Meteor Detection Sites 42
meteor (also see Leonid)
   definition 58
meteor count (see flux)
   end points only..............................225
meteor detection software
  44,173,251,223
meteor flux (see flux)
meteor height (see beginning height;
   and see ending height)
meteor image
   in mid-IR 85,86
   in near-UV 395
   in visible................................371,383
meteor outburst
   duration 162,200
meteor rate (see flux)
meteor shower - individual
   Draconids 141,353,491,518
   eta Aquarids 490,518
   Leonids (see Leonids)
   Orionids................................518,490
   Perseids 61,352,490,518
meteor storm 10
meteor trail
   neutral atom debris 555
   visible on sky, record...................227
meteor train (see persistent train)
meteor wake (see wake)
meteoric debris................... 489,499,505
   mean radius 500
   settling rate..................................503

meteorite
   chemical subtypes........................329
   composition..................................337
   fragmentation 345
   influx 59
   micro 327
meteoroid (see interplanetary dust
   particle)
meteoroid stream
   model..158
MeteorScan 173
meter-sized meteoroids
   annual influx 60
MetRec 41
Mg I ....
  61,125,337,375,380,404,412,436
Mg(^{3}S-^{3}P) line..................................432
micrometeorites
   definition 326
microwave measurements 129
Midcourse Space Experiment...........306
MILSTAR 9
minerals...332
MIRIS...82, 441
MM...326
momentum transfer............................118
MONACO..97
Moon 574
   brightness of night side................591
   sodium atmosphere......................576
Moon impacts
   detectability............................587,594
   detection rate annual showers 592
   telescope sensitivity threshold.....595
MS-Track II 9

N

N I.. 61,375,404
N II 405,406
Na I.........360,375,405,415,416,432,535
   early release 360
   neutral atom debris layer 542
   persistent trains 433,480,426
NaHCO$_3$ 482
nanometeoroids..................................331
narrow band imaging 381,353,392
NASA Cosmic Dust Program 507

SUBJECT INDEX

nebulosity ... 383
neutral atom debris layer ... 560
NH ... 144
NH_2 ... 144
NO_2 ... 425,436
number density (see flux)

O

O I ... 62,125,405,410,431,483
    777.4 nm line emission ... 63
    forbidden line emission
      (see green line)
O_2 ... 483
    singlet-delta line ... 531
O_3
    persistent train depletion ... 480
    submillimeter observations ... 137
observability function ... 267
observation sites individual
    Canada ... 167
    Georgia ... 489
    Germany ... 565
    Hawaii ... 129,167
    Israel ... 47,172,357,391,545,535
    Italy ... 265
    New Mexico ... 249
    Scandinavia ... 565
    Slovak Republic 1998 ... 265,285
    Spain ... 39,209,277
    Tenerife ... 167,555
    UK (1998) ... 471
observations
    hybrid visual/video ... 191
    optimal pointing ... 239
    space based ... 305
OH hydroxyl radical ... 433,525
    airglow layer altitude ... 533
    variation during shower ... 530
Olympus satellite ... 30
Operational Monitoring Program ... 33
OPlan ... 32
optical thickness vapor cloud ... 115
orange arc FeO emission ... 434
orbits
    1998 Leonids video ... 295
    1999 storm photographic ... 277

organic carbon
    delivery to early Earth ... 65,78,129
    mass fraction in cometary dust
      ... 72,78,317,318,337
    mass influx ... 67
origin of life ... 15,58
outflow velocity (see comet dust)

P

p-parameter ... 159
P55/Tempel-Tuttle (see comets)
    recovery ... 47
particle (see interplanetary dust)
    beam model ... 112
peak time (see flux)
penetrating fluence ... 31
persistent train
    3-D structure ... 462
    99 Leonid MAC, list of trains ... 12
    billowing ... 462,479
    chemistry (optical luminosity) ... 480
    Chippenham train ... 473
    cross-section ... 485
    dark center ... 481,484
    drift by horizontal winds ... 464
    expansion velocity ... 476
    initial decay of emission ... 478
    initial radial expansion ... 477
    mid-IR emissions ... 439
    model ... 480
    morphology ... 474
    of 04:00:29 UT fireball ... 460
    optical spectrum ... 410,424,433
    separation of walls ... 477,482
    temperature ... 482
    tubular - thickness walls ... 477
    vertical motion ... 468
    Y2K train ... 11,12,399,443,457
photodissociation ... 485
photometric mass ... 178
    as function of magnitude ... 179
photometry ... 489,536,590
planetary sciences (overview) ... 18
plate constants approach ... 176
pointing direction ... 239
    time variation on aircraft ... 224
    recommendations ... 239